P. Shlyakhin

STEAM TURBINES

Theory and Design

University Press of the Pacific
Honolulu, Hawaii

Steam Turbines:
Theory and Design

by
P. Shlyakhin

ISBN: 1-4102-2348-5

Reprinted from the original edition

University Press of the Pacific
Honolulu, Hawaii
http://www.universitypressofthepacific.com

П. Н. ШЛЯХИН

ПАРОВЫЕ ТУРБИНЫ

Translated from the Russian by A. Jaganmohan, B. E. (Hons)

This volume is intended as a text-book for the students of technical colleges as well as engineers and designers specialising in turbine building.

Basic theoretical concepts of the thermodynamic processes of stationary steam turbines have been dealt with in detail. Variable load operation of these turbines has also been considered.

The reader will find here enough material concerning the basic concepts of gas dynamics as applied to steam turbines as well as design and construction of steam turbines and their details with regard to mechanical strength. Considerable space has been devoted to the description of turbines of various manufacture.

The book contains a profusion of tables, diagrams and illustrations which, it is hoped, would enable the reader to acquire a better understanding of the theory and design of steam turbines.

CONTENTS

PART ONE

THEORY AND DESIGN OF STEAM TURBINES

PART TWO

DESIGN AND CONSTRUCTION OF STEAM TURBINE COMPONENTS

PART ONE

THEORY AND DESIGN OF STEAM TURBINES

Chapter One

INTRODUCTION

1-1. FUNDAMENTAL PRINCIPLES OF TURBINE DESIGN AND A SHORT INTRODUCTION OF THEIR DEVELOPMENT

The steam turbine is a prime-mover in which the potential energy of the steam is transformed into kinetic energy and the latter in its turn is transformed into the mechanical energy of rotation of the turbine shaft. The turbine shaft, directly, or with the help of a reduction gearing, is connected with the driven mechanism. Depending on the type of the driven mechanism a steam turbine may be utilised in the most diverse fields of industry, for power generation and for transport.

Transformation of the potential energy of steam into the mechanical energy of rotation of the shaft is brought about by different means.

Steam turbines are broadly classified into three types: impulse, reaction, and combined (impulse-reaction), depending upon the way in which the transformation of potential energy into the kinetic energy of a steam jet is achieved.

The idea of steam turbine is of a remote origin. It is a well-known fact that in about 120 B.C.

Fig. 1-1. Hero's engine

1— source of heat; *2*— water vessel; *3*— spherical receiver; *4*— supporting tubes; *5*— nozzles; *6*— hollow axles

Fig. 1-2. Branca's machine

1— boiler; *2*— nozzle; *3*— wheel; *4*— shaft; *5*— toothed transmission gears; *6*— pounding mill

Hero of Alexandria made the first prototype of a turbine working on the reaction principle. This device embodying a primitive steam powered installation (Fig. 1-1) consisted of a source of heat *1*, a vessel *2*, filled with water, a spherical receiver *3* with special tubes (nozzles) *5*. Under the action of the source of heat the water in the vessel heated up and evaporated producing saturated steam, which flowing through the vertical tubes (uprights) *4* and the horizontal tubes (axles) *6* entered into the spherical receiver. With the increase of pressure, the steam in the spherical receiver passed out into the atmosphere through the nozzles *5*.

The jets of steam issuing out from these nozzles exerted on them a reactive force and compelled the sphere to rotate around its horizontal axis.

Many centuries later, in the year 1629, Giovanni Branca gave the description of a machine, built by him, shown in Fig. 1-2. It consisted of a steam boiler *1*, the lid of which was made in the shape of a man, a long tube (nozzle) *2*, a hori-

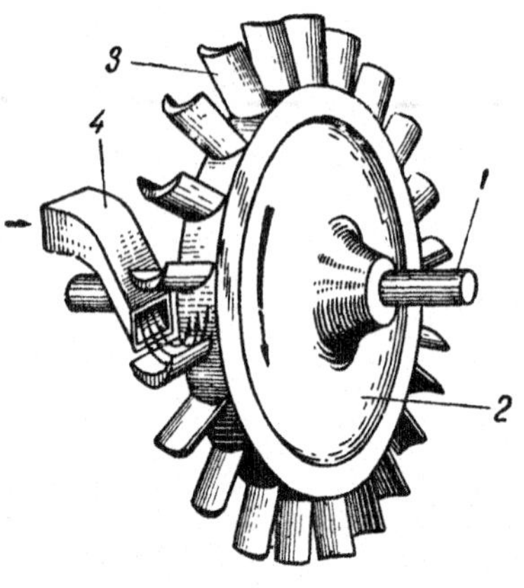

Fig. 1-3. Simple impulse turbine
1— shaft: *2*— disc; *3*— blade; *4*— nozzle

zontal wheel *3* with blades, a shaft *4* and a toothed gear transmission *5* for driving a pounding mill *6*. The steam generated in the boiler after having expanded in the nozzle *2* attained a high velocity. This high-velocity steam jet striking on the blades of the wheel *3* rotated the latter. The speed of rotation of the wheel *3* and its turning moment on the shaft *4* depended on the velocity and the quantity of steam flow per unit of time. Branca's machine, from its principle of action, is a prototype of the impulse turbine.

In the years 1806-13, at the Suzunsky factory in Altai, Russian inventor Polikarp Zalesov built a number of models of steam turbines.

In the thirties of the 19th century the workers of Nizhny Tagil built steam turbines which, however, did not find much industrial usage[1].

Substantial progress in the development and construction of steam turbines was noticed at the end of the 19th century. In the year 1890, Swedish engineer Gustav de-Laval built a single-disc steam turbine of 5 h. p. capacity, with a flexible shaft and a disc of equivalent strength.

The simplest single-disc steam turbine consists of the following principal parts (Fig. 1-3): expansion nozzle *4*, shaft *1* and disc *2*, with blades *3* fixed on its periphery. The shaft *1* along with the disc *2* mounted upon it comprises the most important part of the turbine and is known as the rotor, which is housed in the turbine casing. The journals of the shaft are placed in bearings which are located in the base of the turbine casing. (The turbine casing and its base are not shown in the figure.)

In turbines of these types the expansion of the steam is achieved from its initial pressure to its final one in a single nozzle or a group of nozzles situated in the turbine stator and placed in front of the blades of the rotating disc. The decrease of steam pressure in the nozzles is accompanied by a decrease of its heat content; this decrease of

[1] For detailed data about the development of steam turbines please see A. A. Radtsig, *The Development of Steam Turbines*, 1934; Lev Gumilevsky, *The Creators of Steam Turbines*, ONTI, 1936.

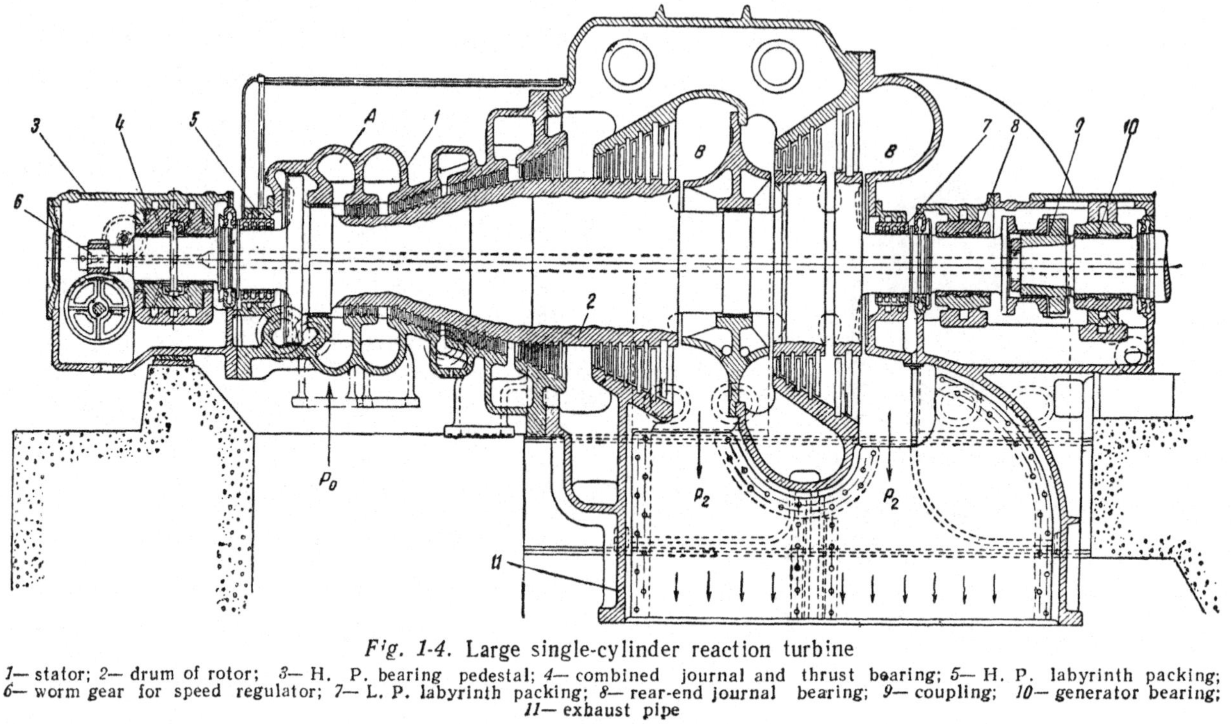

Fig. 1-4. Large single-cylinder reaction turbine
1— stator; *2*— drum of rotor; *3*— H. P. bearing pedestal; *4*— combined journal and thrust bearing; *5*— H. P. labyrinth packing; *6*— worm gear for speed regulator; *7*— L. P. labyrinth packing; *8*— rear-end journal bearing; *9*— coupling; *10*— generator bearing; *11*— exhaust pipe

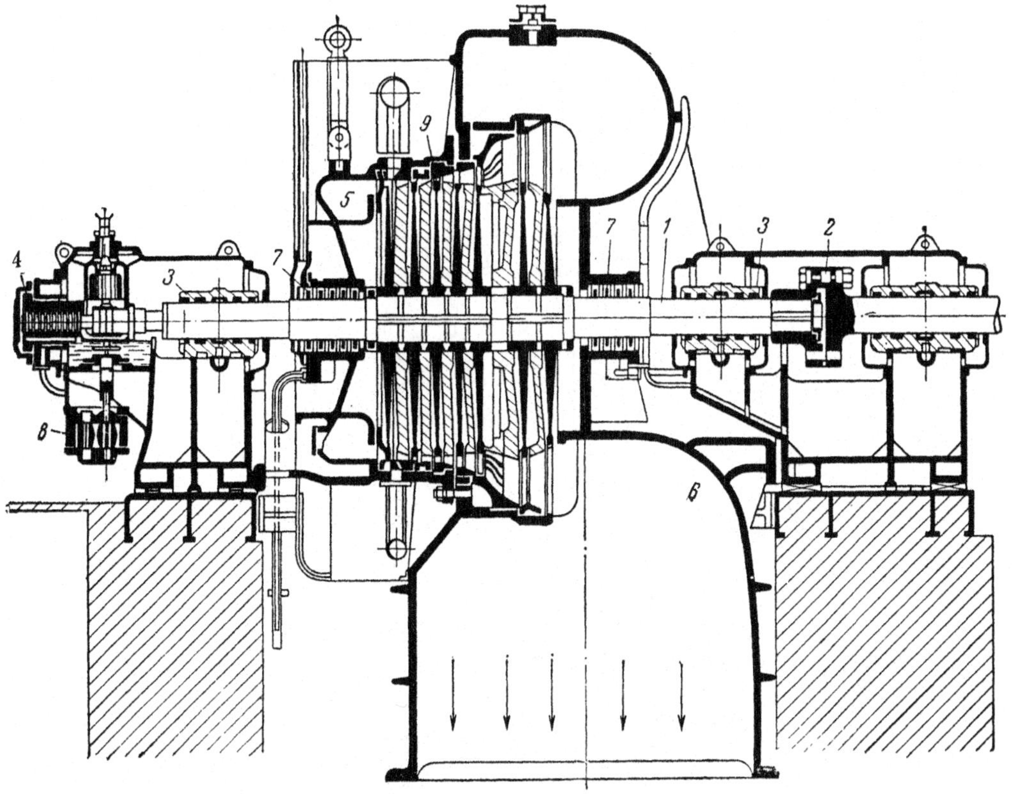

Fig. 1-5. Multistage impulse turbine

1— shaft with seven discs; *2*— coupling; *3*— front and rear journal bearings; *4*— collar-type thrust bearing; *5*— annular chamber supplying steam to the first stage, *6*— exhaust pipe; *7*— shaft gland packing; *8*— gear oil pump; *9*— turbine casing

heat content achieved in the nozzles subsequently accounts for the increase in the velocity of the steam issuing from the nozzles. The velocity energy of the steam jets exerts an impulse force on the blades and performs mechanical work on the shaft of the turbine rotor.

The turbines in which the complete process of expansion of steam takes place only in stationary canals (nozzles), and the velocity energy is transformed into mechanical work on the turbine blades (without any further expansion taking place in them) are known as impulse turbines. Steam velocity at the exit of the nozzles in such turbines reaches a value of about 1,200 m/sec and over.

Small single-stage impulse turbines were and are being built with high speed. The first turbine of this type built by de-Laval operated at a speed of 30,000 r.p.m., and the turbine was provided with a reduction gear for transmitting the turning moment to the driven mechanism (e. g., an electrical generator, etc.).

The power of a single-stage impulse turbine even at circumferential velocities reaching a value of 350 m/sec does not exceed 500 to 800 kW. Small capacity of a single unit, low efficiency, and in the majority of cases the necessity to utilise a reduction gear limit the usage of single-stage steam turbines.

The steam turbine shown in Fig. 1-4 works on a different principle. Fresh steam from the chamber *A* enters the guide blade passages (nozzles) which are located in stationary turbine casing *1*. Moving blades are fixed on the periphery of rotor *2* forming passages for the flow of steam through them. Steam from chamber *A*, after passing through the guide and moving blades, enters chambers *B*.

In its passage from chamber *A* to chambers *B* steam continuously undergoes expansion from the initial pressure p_0 to its final state p_2. Expansion and the decrease in heat content of steam per unit occurs continuously in all the blade passages, both guide and moving. To begin with fresh steam from chamber *A* enters the interblade passage of the first row of guide blades, fixed to the turbine stator *1*. From here steam enters the first row of moving blades fixed on the periphery of the rotor *2*. From the interblade passages of the first row of moving blades steam enters the second row of guide blades and so on, passing successively through the blade passages of all

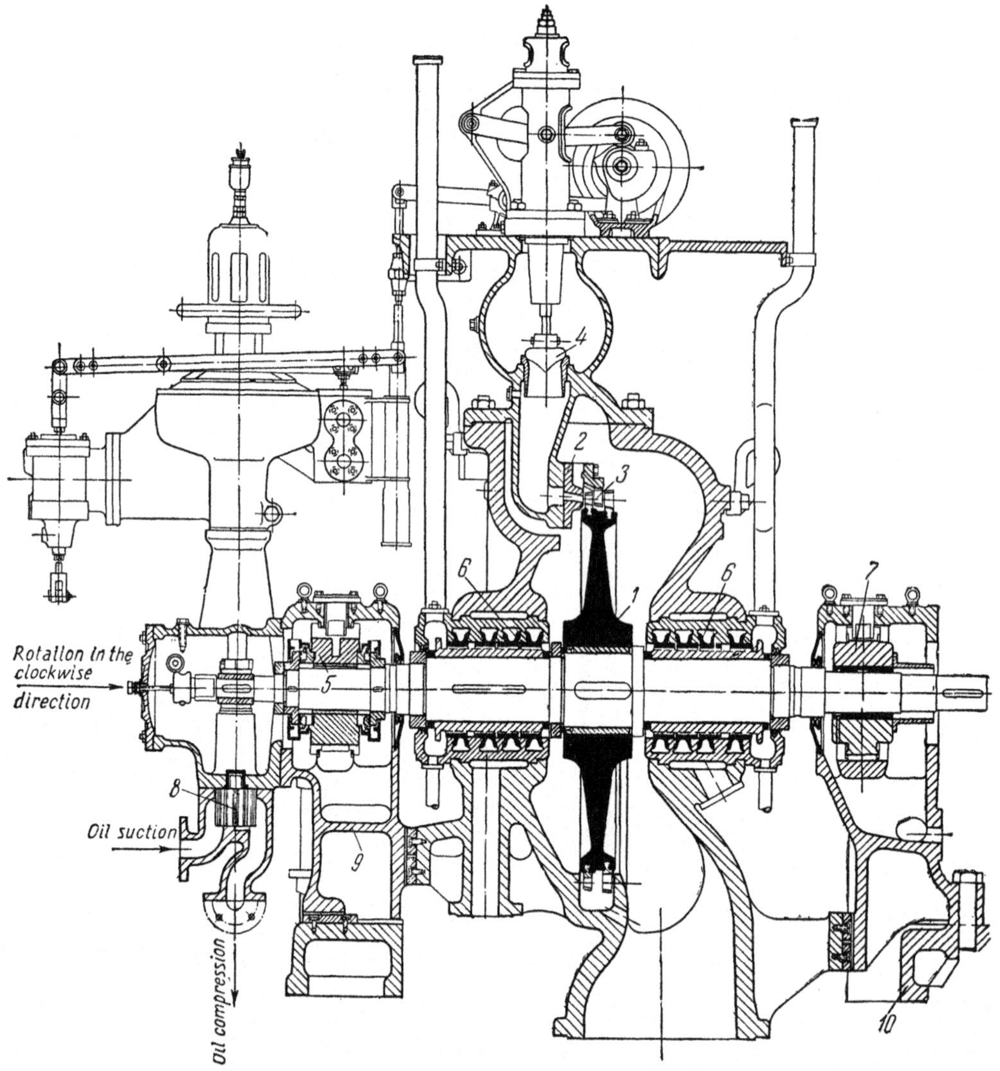

Fig. 1-6. Steam turbine built by the Lenin Nevsky Works

1— two-row velocity stage disc; *2*— nozzles; *3*— guide blades; *4*— regulating valves; *5*— combined journal and thrust bearing; *6*— front- and rear-shaft labyrinth packings; *7*— rear-end journal bearing; *8*— gear oil pump; *9-10*— front and rear supporting stools

the rows of moving and guide blades. Steam leaving the last row of moving blades, known as exhaust steam, from chambers *B* enters the condenser, through exhaust pipe *11*, where it condenses.

Two adjacent rows of blades, fixed in the casing and on the drum (rotor) correspondingly, form a stage of turbine. Accordingly a turbine which has several such successive rows of guide and moving blades, is known as a multistage turbine.

In the turbine under consideration, in contrast with the previous one, the expansion of steam takes place both in the stationary or guide blades and in the rotating or moving blades.

Those turbines, in which the expansion of steam occurs not only in the passages of the guide blades, but also in the passages of the moving blades, so that the overall decrease in heat content in all the stages is, more or less, uniformly distributed between them, are known as reaction turbines.

Steam turbines working on the above principle for industrial purposes were first suggested in 1884 by a British engineer C. A. Parsons.

The velocity of steam flow in multistage reaction turbines (excepting the last few stages of a condensing turbine) is relatively small (about 100 to 200 m/sec).

Further development of the impulse principle of the de-Laval turbine led to the emergence of multistage impulse turbines in the year 1900.

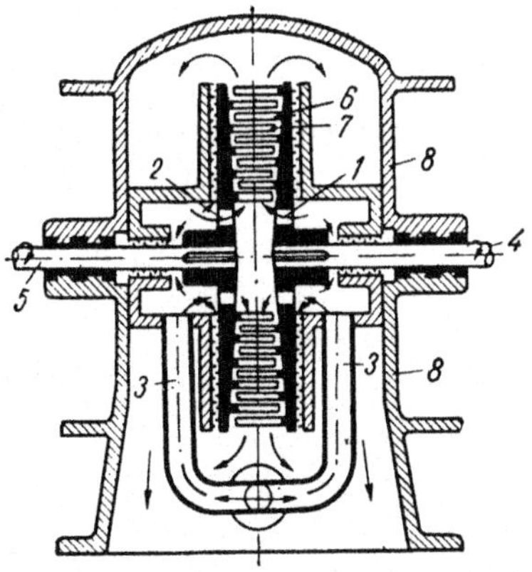

Fig. 1-7. General arrangement of a radial flow turbine
1 and *2*—discs; *3*—fresh steam supply; *4* and *5*—shafts; *6* and *7*—moving blades; *8*—casing

The main idea in the constructon of such turbines was to utilise, instead of one group of nozzles, several such consecutive groups, which along with the discs of moving blades formed several stages. Expansion of steam occurred in each of the group of nozzles of a stage, whereas in the moving blades only the direction of the steam flow was altered, and consequently the velocity energy of the steam was transformed into mechanical work.

A turbine of this type is shown in Fig. 1-5. The turbine shaft *1* carries seven discs, which have moving blades fixed on their periphery. Partitions, known as diaphragms, are placed between each pair of adjacent discs carrying moving blades. These diaphragms are fixed to the stator *9* of the turbine; nozzles for the expansion of steam are placed in these diaphragms. Fresh steam from the annular chamber *5* enters the nozzles of the first stage and thereafter consecutively passes through the passages between the moving blades and the nozzles of the following stages of the turbine.

Exhaust steam from the moving blades of the last stage enters pipe *6* and is led away to the condenser.

In the same year, 1900, Curtis expounded a turbine with a single pressure stage, but with two or three velocity stages.

A single-disc turbine with two velocity stages constructed by the Lenin Nevsky Works is shown in Fig. 1-6.

All the turbines so far considered are known as axial turbines, since the flow of steam is in a direction along (parallel) the axis of the turbine. Along with axial turbines, in the first quarter of the 20th century, also were built radial turbines, in which the flow of steam is achieved in a direction perpendicular to its axis, i. e., radially. The first radial turbine was proposed and made by the Ljungström brothers in the year 1910. The schematic diagram of such a turbine is shown in Fig. 1-7.

The turbine consists of two discs *1* and *2*, fixed at the ends of the shafts *4* and *5*. The lateral surfaces of these discs, facing each other, carry the moving blades fixed on them in circular arcs of different radii. Steam from the steam main *3* enters the central chamber through the aperture in the discs and thereafter passing through all the rows of moving blades is led away to the exhaust pipe *8* of the turbine.

The expansion of steam from its initial pressure p_0 to the final pressure p_2 occurs in the passages of all the rows of the revolving blades. In this turbine there are no guide blades, all the blades being of the moving type. The shafts *4* and *5* together with the discs mounted on them rotate in opposite directions. Such turbines are known as pure reaction turbines.

Radial turbines are also built with fixed guide blades. In this case they operate on a principle similar to that of the axial reaction turbines considered previously.

1-2. DEVELOPMENT OF THE TURBINE BUILDING INDUSTRY IN THE U.S.S.R.

Prior to the Great October Socialist Revolution turbine building industry in Russia developed at a very slow pace. The first turbine with a capacity of 200 kW was built in the year 1907 by the Petersburg Metal Factory (now Leningrad Metal Works), the only factory working on the construction of steam turbines at that time. From the year 1907 to 1913 this factory built a total of 26 turbines with a maximum capacity of 1,250 kW in a single unit. With the beginning of the First World War turbo-construction in Russia was practically at a standstill.

This industry was restarted only in the year 1923; in 1924 the Leningrad Metal Works (L.M.W.) built a turbine with a capacity of 20,000 kW. Turbine building in the U.S.S.R. began developing at a fast pace during the first and the subsequent five-year plans, especially in the latter period. During the First Five-Year Plan period this works built turbines of medium and large power capacities (up to 50,000 kW).

Beginning with the Second Five-Year Plan L.M.W. switched over to the construction of condensing turbines with capacities of 24,000, 50,000, and 100,000 kW, rated at steam pressures of 29 ata and a temperature of 400°C. During the same period L.M.W. worked out an original method of construction of bleeder (pass-out) turbines with capacities of 25,000 kW at the same initial steam conditions, and made their first prototypes as well. Besides, L.M.W. built a topping turbine (a turbine operating with high initial steam pressure and temperature and delivering steam to a condensing turbine from its exhaust or from one of its intermediate stages) with steam pressures of 125 ata and a temperature of 450°C. Construction of a steam turbine with a capacity of 100,000 kW at 3,000 r. p. m., a unique feat for that period of turbo-construction in the world, and the construction of bleeder turbines of large capacities were some of the attainments of the L.M.W., which placed it in the ranks of the foremost turbine building factories.

The Kharkov Turbine Works, one of the largest of its kind, was put into commission in the year 1934 which began the construction of single-cylinder steam turbines with capacities of 50,000 kW, and in the year 1938 it brought out a turbine with a capacity of 100,000 kW at 1,500 r. p. m. with initial steam conditions of 29 ata and a temperature of 400°C.

In recent years L.M.W. built a series of condensing steam turbines with capacities of 25,000 to 100,000 kW at steam conditions of 90 ata and a temperature of 500°C running at speeds of 3,000 r.p.m. Fig. 1-8 (see appendix at the end of the book) shows a high-pressure steam turbine of 50,000 kW, built by the L.M.W., the description of which is given in Chapter 10. Further, in the year 1952 L.M.W. built a steam turbine of 150,000 kW at 3,000 r.p.m. at supercritical steam conditions: initial steam pressure of 170 ata and a temperature of 550°C. At the present time the L.M.W. has produced a series of bleeder turbines with one or two specified steam regulations having capacities of 50,000 kW at initial steam conditions of 130 ata and 565°C. In 1957 L.M.W. built a turbine of 200,000 kW capacity rated at initial steam conditions of 130 ata and 565°C, with intermediate reheating of steam to a temperature of up to 535°C. In the year 1958 L.M.W. completed the project work of a unique condensing turbine with a capacity of 300,000 kW at steam pressure and temperature of 240 ata and 580°C respectively, with an intermediate reheating of steam up to a temperature of 565°C. At present, the first prototype of such a unique turbine has been built and a turbine of similar capacity has also been manufactured at the Kharkov Turbine Works (Kh. T.W.).

At this works a turbine, type SKR-100, with a capacity of 100,000 kW designed for initial steam conditions of 300 ata and 650°C is now under construction. This turbine shall deliver steam to a usual condensing turbine.

The construction of large-capacity gas turbines has been started at the L.M.W. A gas turbine of 12,000 kW capacity was built here in 1957 earmarked for the system of underground production of gas from coal. In the same year the project work on a gas turbine of 25,000 kW capacity was carried out, and its construction was completed in 1959.

Similarly, the Kharkov Turbine Works has designed and is constructing turbines operating at high steam pressures with capacities of 25,000 to 100,000 kW both of the condensing and the topping types, rated at initial steam conditions of 90 ata and 500-535°C as also 130 ata and 565°C. In the year 1956 Kh.T.W. built a turbine, type K-100-90 (VKT-100) with a total capacity of 100,000 kW at steam conditions of 90 ata and 535°C. A turbine, type K-150-130 of 150,000 kW capacity with initial steam pressure and temperature of 130 ata and 565°C respectively, built by the Kh.T.W., is at present operating at one of the large-power stations. In addition this works is building large-capacity condensing turbines for atomic power stations. A unique steam turbine, type SKK-300, designed for supercritical steam conditions, and a topping turbine, type SKR-100 with capacity of 100,000 kW operating at steam pressure and temperature conditions of 300 ata and 650°C are now on the production lines. A gas turbine of 50,000 kW capacity is also under construction.

High-pressure steam turbines constructed by the L.M.W. and the Kh.T.W. have many of their details in common, which simplifies production techniques, lowers production costs and facilitates the replacement of worn-out parts at the time of major repairs.

The Ural Turbomotor Works (U.T.W.) was put into operation just before the beginning of the Great Patriotic War. This works is building turbines with steam regulation with capacities of 12,000; 25,000; and 50,000 kW as also gas turbines.

In 1950 the Kaluga Turbine Works was commissioned, which has since then designed and built a series of condensing and bleeder steam turbines of small capacity rated at 35 ata and 435°C.

Turbine building was started at the Bryansk Locomotive Works in the year 1952, where turbines of medium capacity (25,000 kW) are under production.

The turbine building industry of the U.S.S.R. is now developing at a fast pace, and this tempo will be accelerated all the time. It has been resolved that the installed capacity of thermal power stations in the U.S.S.R. should be more than doubled by the end of the Seven-Year Plan period.

The commissioning, in 1954, of the first nuclear power station supplying electric energy on an industrial scale reflects the meritorious work done by Soviet scientists and engineers in the field of power generation. Here too the steam turbine is the basic prime-mover.

A new Soviet nuclear power station with an installed capacity of 100,000 kW was commissioned in the year 1958. On completion the total installed capacity of this station will be as high as 600,000 kW.

1-3. CLASSIFICATION OF STEAM TURBINES

Steam turbines may be classified into different categories depending on their construction, the process by which heat drop is achieved, the initial and final conditions of steam used and their industrial usage as follows:

1. According to the number of pressure stages:

a) single-stage turbines with one or more velocity stages usually of small-power capacities; these turbines are mostly used for driving centrifugal compressors, blowers and other similar machinery;

b) multistage impulse and reacton turbines; they are made in a wide range of power capacities varying from small to large.

2. According to the direction of steam flow:

a) axial turbines in which the steam flows in a direction parallel to the axis of the turbine;

b) radial turbines in which the steam flows in a direction perpendicular to the axis of the turbine; one or more low-pressure stages in such turbines are made axial.

3. According to the number of cylinders:

a) single-cylinder turbines,

b) double-cylinder turbines,

c) three-cylinder turbines, and

d) four-cylinder turbines.

Multicylinder turbines which have their rotors mounted on one and the same shaft and coupled to a single generator are known as single shaft turbines; turbines with separate rotor shafts for each cylinder placed parallel to each other are known as multiaxial turbines.

4. According to the method of governing:

a) turbines with throttle governing in which fresh steam enters through one or more (depending on the power developed) simultaneously operated throttle valves;

b) turbines with nozzle governing in which fresh steam enters through two or more consecutively opening regulators;

c) turbines with bypass governing in which steam besides being fed to the first stage is also directly led to one, two or even three intermediate stages of the turbine.

5. According to the principle of action of steam:

a) impulse turbines in which the potential energy of the steam is converted into kinetic energy in nozzles or passages formed by adjoining stationary blades, and in the moving blades the kinetic energy of the steam is converted into mechanical energy; according to the present day practice of impulse turbines this classification is relative, since these turbines operate with quite a significant degree of reaction in moving blades increasing in succeeding stages (in condensing turbines);

b) axial reaction turbines in which the expansion of steam between blade passages both of the guide and moving blades of each stage takes place nearly to the same extent;

c) radial reaction turbines without any stationary guide blades;

d) radial reaction turbines having stationary guide blades.

6. According to the heat drop process:

a) condensing turbines with regenerators; in these turbines steam at a pressure less than atmospheric is directed to a condenser; besides, steam is also extracted from intermediate stages for feed water heating, the number of such extractions usually being from 2-3 to as much as 8-9. The latent heat of exhaust steam during the process of condensation is completely lost in these turbines. Small-capacity turbines of earlier designs often do not have regenerative feed heating;

b) condensing turbines with one or two intermediate stage extractions at specific pressures for industrial and heating purposes;

c) back pressure turbines, the exhaust steam from which is utilised for undustrial or heating purposes; to this type of turbines can also be added (in a relative sense) turbines with deteriorated vacuum, the exhaust steam of which may be used for heating and process purposes;

d) topping turbines; these turbines are also of the back pressure type with the difference that the exhaust steam from these turbines is further utilised in medium- and low-pressure condensing turbines. These turbines, in general, operate at high initial conditions of steam pressure and temperature, and are mostly used during extension of power station capacities, [1] with a view to obtain better efficiencies;

e) back-pressure turbines with steam extraction from intermediate stages at specific pressures; turbines of this type are meant for supplying

[1] By "extension of power stations capacities" here is meant additional installation of high-pressure boilers (critical and supercritical pressures) and topping turbines as additional units, delivering steam to the already existing medium-pressure turbines from the exhaust of the topping turbines.

Table 1-1

Stationary Steam Turbines Operating at Steam Pressures of 35 to 130 ata. Types and Basic Parameters. GOST 3618-58

Type of turbine	Type designation: New design	Type designation: Former design	Nominal values of parameters: Capacity (while under continuous operation), kW	Initial conditions: Pressure, ata	Initial conditions: Temperature °C	Temperature, °C. Intermediate reheating	Temperature, °C. Regenerative feed heating	Temperature, °C. Cooling water
Condensing, without intermediate regulation	K-6-35 K-12-35	AK-6 AK-12	6,000 12,000	35	435	—	145	25 20
	K-25-90 K-50-90 K-100-90	VK-25 VK-50 VK-100	25,000 50,000 100,000	90	535	—	215	15 10; 15 10; 15
	K-100-130 K-150-130* K-200-130	PVK-100 PVK-150* PVK-200	100,000 150,000 200,000	130	565	565	230	10; 15

* Turbine K-150-130 (PVK-150) may be built for a capacity of 160,000 kW according to the requirements of the customer.

Table 1-2

Stationary Steam Turbines Operating at Steam Pressures of 35 to 130 ata. Types and Basic Parameters. GOST 3618-58

Type of turbine	Type designation: New design	Type designation: Former design	Nominal values of parameters: Capacity (while under continuous operation), kW	Initial conditions: Pressure, ata	Initial conditions: Temperature, °C	Bleeding steam pressure, ata: 1.2	5	7	10	13	Temperature, °C. Regenerative feed heating	Temperature, °C. Cooling water
						Value of bleeding, tons/hour						
Condensing, with bleeding for reheating	T-2.5-35 T-4-35 T-6-35 T-12-35	AT-2.5 AT-4 AT-6 AT-12	2,500 4,000 6,000 12,000	35	435	14 22 30 65	— — — —	— — — —	— — — —	— — — —	145	25 25 25 20
	T-25-90	VT-25	25,000	90	535	90	—	—	—	—	215	20
Condensing, with bleeding for industrial purposes (process steam)	P-0.75-35/5 P-1.5-35/5 P-2.5-35/5 P-4-35/5 P-6-35/5	AP-0.75 AP-1.5 AP-2.5 AP-4 AP-6	750 1,500 2,500 4,000 6,000	35	435	— — — — —	5 10 18 25 40	— — — — —	— — — — —	— — — — —	145	25
Condensing, with bleeding, both for reheating and industrial purposes (process steam)	PT-12-35/10	APT-12	12,000	35	435	40	—	—	50	—	145	20
	PT-12-90/7 PT-12-90/10 PT-25-90/10 (PT-50-90/13)**	— VPT-12 VPT-25 (VPT-50)**	12,000 12,000 25,000 50,000	90	535	25* 25 50 100	— — — —	30 — — —	— 35 70 —	— — — 140	215	20
	PT-50-130/7 PT-50-130/13	— VPT-50	50,000 50,000	130	565	80* 90	— —	120* —	— —	— 115	230	20

* The value of bleedings given in the table is not exact. This value would be corrected while designing.
** The manufacture of these types of turbines shown in brackets will be continued up to the year 1965 (inclusive).

the consumer with steam of various pressure and temperature conditions;

f) low-pressure (exhaust-pressure) turbines in which the exhaust steam from reciprocating steam engines, power hammers, presses, etc., is utilised for power generation purposes;

g) mixed-pressure turbines with two or three pressure stages, with supply of exhaust steam to its intermediate stages.

The turbines enumerated under 'b' to 'e' usually have extractions for regenerative feed-heating, in addition to the extraction of steam at specific pressures for other purposes.

7. According to the steam conditions at inlet to turbine:

a) low-pressure turbines, using steam at pressures of 1.2 to 2 ata;

b) medium-pressure turbines, using steam at pressures of up to 40 ata;

c) high-pressure turbines, utilising steam at pressures above 40 ata;

d) turbines of very high pressures, utilising steam at pressures of 170 ata and higher and temperatures of 550°C and higher;

e) turbines of supercritical pressures, using steam at pressures of 225 ata and above.

8. According to their usage in industry:

a) stationary turbines with constant speed of rotation primarily used for driving alternators;

b) stationary steam turbines with variable speed meant for driving turbo-blowers, air circulators, pumps, etc.;

c) nonstationary turbines with variable speed; turbines of this type are usually employed in steamers, ships and railway locomotives (turbo-locomotives).

All these different types of turbines described above depending on their speed of rotation are either coupled directly or through a reduction gearing to the driven machine.

In the present course of *Steam Turbines* only the steam turbines of the stationary type operating at constant speeds of rotation will be considered. Tables 1-1, 1-2 and 1-3 give the specifications of the various types of steam turbines manufactured in the Soviet Union.

Examples of turbines of revised design with new designations.

Condensing turbine without extraction with a capacity of 6,000 kW at an initial steam pressure of 35 ata:

Steam turbine K-6-35 GOST[1] 3618-58.

Similar to the one given above but with extraction:

[1] GOST: The U.S.S.R. State Standards.

Steam turbine T-6-35 GOST 3618-58.

Same as above having a capacity of 6,000 kW with extraction for the supply of process steam at a pressure of 5 ata.

Steam turbine P-6-35/5 GOST 3618-58.

Same as above but with a capacity of 50,000 kW at an initial steam pressure of 130 ata and extractions for process steam at a pressure of 7 ata as well as feed heating.

Steam turbine PT-50-130/7 GOST 3618-58.

Back-pressure turbine of 12,000 kW capacity operating at an initial steam pressure of 90 ata and an exhaust pressure of 13 ata:

Steam turbine P-12-90/13 GOST 3618-58.

Table 1-3

Stationary Steam Turbines Operating at Pressures of 35 to 130 ata. Types and Basic Parameters. GOST 3618-58

Type of turbine	Type designation: New design	Type designation: Former design	Nominal values of parameters: Capacity (while under continuous operation), kW	Initial conditions: Pressure, ata	Initial conditions: Temperature, °C	Back pressure ata
Back-pressure turbines	R-1.5-35/3	AR-1.5-3	1,500			3
	R-1.5-35/5	AR-1.5-5	1,500			5
	R-1.5-35/10	AR-1.5-10	1,500			10
	R-1.5-35/15	AR-1.5-15	1,500			15
	R-2.5-35/3	AR-2.5-3	2,500			3
	R-2.5-35/5	AR-2.5-5	2,500			5
	R-2.5-35/10	AR-2.5-10	2,500	35	435	10
	R-2.5-35/15	AR-2.5-15	2,500			15
	R-4-35/3	AR-4-3	4,000			3
	R-4-35/5	AR-4-5	4,000			5
	R-4-35/10	AR-4-10	4 000			10
	R-4-35/15	AR-4-15	4,000			15
	R-6-35/3	AR-6-3	6,000			3
	R-6-35/5	AR-6-5	6,000			5
	R-6-35/10	AR-6-10	6,000			10
	R-6-90/31	VR-6-31	6,000			31
	R-12-90/7	—	12,000			7
	R-12-90/13	—	12,000			13
	R-12-90/18	VR-12-18	12,000	90	535	18
	R-12-90/31	VR-12-31	12,000			31
	(R-25-90/18)*	(VR-25-18)*	25,000			18
	(R-25-90/31)*	(VR-25-31)	25,000			31
	R-25-130/7	—	25,000			7
	R-25-130/13	—	25,000			13
	R-25-130/18	PVR-25-18	25,000			18
	R-25-130/31	—	25,000	130	565	31
	R-50-130/7**	—	50,000			7
	R-50-130/13**	—	50,000			13
	R-50-130/18	—	50,000			18

* The manufacture of these types of turbines shown in brackets will be continued up to the year 1965 (inclusive).

** Turbines R-50-130/7 and R-50-130/13 can be constructed with lower capacities if desired by the customer, in which case they will be made on the basis of the parameters of turbine type PT-50-130/7.

1-4. PRINCIPLE OF ACTION OF A STEAM TURBINE

The steam jet, issuing from a stationary nozzle or group of nozzles exerts on the blades of the turbine a force P_u (kg) in the direction of their rotation. The force P_u developed by the steam while in its passage through the turbine blades is transformed into mechanical work at the blade rim. The work done by 1 kg of steam on the blades in one second is

$$L = P_u u \text{ [kg m/sec]}, \qquad (1\text{-}1)$$

where u—circumferential velocity of the blades in m/sec.

The force exerted by the steam on the blades may be found as follows. We know from mechanics that the change in momentum during a definite period of time is equal to the force exerted, and therefore

$$P_u \tau = \frac{G}{g}(c_{1t} - c_2), \qquad (1\text{-}2)$$

where τ — period for which the force P_u of the steam acts on the blades, sec;

G—quantity of steam flow from the nozzle or group of nozzles for the same interval of time, kg;

g—9.81 m/sec^2 —acceleration due to gravity for a freely falling body;

c_{1t}—theoretical velocity of steam at the nozzle exit (without taking into consideration the energy losses existing in practice which will be considered later), m/sec;

c_2—velocity of steam after undergoing a change in its direction of flow, m/sec.

Substituting in equation (1-2) $\tau = 1$ sec and $G = 1$ kg we have

$$P_u = \frac{1}{g}(c_{1t} - c_2). \qquad (1\text{-}2a)$$

Fig. 1-9 shows the principle of action of steam on bodies of various shapes. It may be shown that the force P_u exerted by the steam jets on bodies of various shapes, under similar conditions of flow, would not be the same. For the different types of flow shown in Fig. 1-9 these forces can be very easily evaluated.

Let the initial velocity of the steam at the nozzle exit for all the three cases be the same, equal to c_{1t}, but in different directions in conformity with the receiving surfaces. For this specific case let the velocity c_{1t} of the steam be 196.2 m/sec. Further we shall consider the three bodies A, B and C to be fixed.

Case (a)

Steam with an initial velocity of c_{1t} strikes body A in a direction normal to its receiving surface and undergoes a change in its direction of flow through an angle of 90° while spreading in all directions along the surface of the body, so that the projection of velocity c_2 in the direction of action of force P_1 of the steam jets equals zero. Substituting in equation (1-2a) the initial and final velocities of steam c_{1t} and c_2 we find the force exerted in the direction of velocity c_{1t}

$$P_1 = \frac{1}{9.81}(196.2 - 0) = 20 \text{ kg}.$$

Case (b)

Neglecting losses due to friction on the curved surface of the blade we have $c_2 = -c_{1t}$. Hence the force P_2 exerted in the direction of velocity c_{1t}, from equation (1-2a), will be equal to

$$P_2 = \frac{1}{9.81}(196.2 + 196.2) = 40 \text{ kg}.$$

Case (c)

Neglecting losses due to friction on the blade surface as in case (b) we have once again $c_2 = -c_{1t}$.

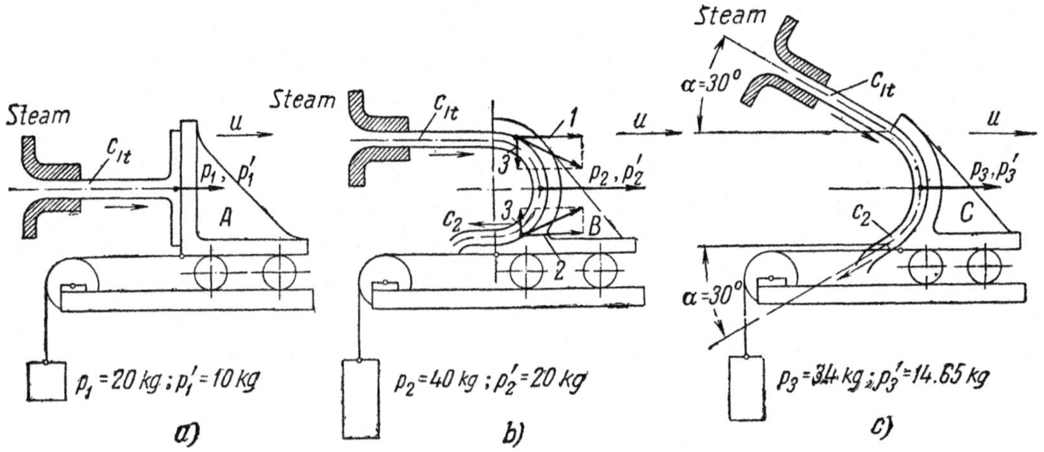

Fig. 1-9. Principle of action of steam jet on bodies of various shapes
1, 2 and *3*— forces acting on the body

In this case the steam jet at entry to the blade surface as well as at the exit from it is not flowing in a direction parallel to the direction of the force P_3 exerted on the body, and therefore it immediately follows that in equation (1-2a) in place of velocities c_{1t} and c_2 their projections in the direction of the force P_3 must be substituted.

Components of the velocities c_{1t} and c_2 in the direction of the force P_3 will accordingly be equal to

$$c'_{1t} = c_{1t} \cos 30^\circ = 196.2 \times 0.866 = 170 \text{ m/sec};$$

$$c'_2 = c_2 \cos 30^\circ = -196.2 \times 0.866 = -170 \text{ m/sec}.$$

Thus the force P_3 according to the equation (1-2a) will be equal to

$$P_3 = \frac{1}{9.81}(170 + 170) = 34.7 \text{ kg}.$$

From the above illustrations it follows that the maximum force is obtained for case (b) where the steam jet flowing along the blade surface undergoes a complete reversal in its direction, i. e., through 180°. However, in the construction of steam turbines such a flow of steam is not possible, and therefore, as shown in case (c), steam jets are directed at an angle both from the exit of the stationary nozzles and the moving blades. However, this angle of inclination α to the plane of rotation of the blades is made as small as possible (usually from 11 to 16°).

To be able to obtain useful work from the action of the steam it is necessary that the body be free to move. If we suppose that the bodies A, B and C, under the action of the steam, are displaced in the direction of the arrows, then knowing the velocity of displacement u, we can easily calculate the force P' and the work L. Supposing that, under the action of the steam jets, the bodies A, B and C are displaced in the direction of the force P' with similar velocities u. Then the force P' in all the three cases for $\tau = 1$ sec and $G = 1$ kg will be determined from the following considerations:

Case (a)

The steam velocity relative to moving body A will be

$$w_1 = c_{1t} - u.$$

Velocity w_1 is known as the relative velocity.

The relative velocity of steam after change of direction of flow on body A will be

$$w_2 = c_2 = 0.$$

Whence the force exerted by the steam is determined from the equation

$$P'_1 = \frac{1}{g}(w_1 - w_2) = \frac{1}{g}(c_{1t} - u). \qquad (1\text{-}3)$$

Case (b)

The relative velocity of steam jet striking the concave surface of body B will be equal to

$$w_1 = c_{1t} - u.$$

The relative velocity of steam leaving the blade surface of body B will be equal to

$$w_2 = -w_1 = -c_{1t} + u.$$

Consequently the force exerted by the steam jet on body B will be

$$P'_2 = \frac{1}{g}(w_1 - w_2) = \frac{2}{g}(c_{1t} - u). \qquad (1\text{-}3a)$$

Case (c)

The projection of the relative velocity of steam jet striking body C on the direction of velocity u will be equal to

$$w_{1u} = c_{1t} \cos 30^\circ - u.$$

The component of the relative velocity of steam leaving the blade surface C will be determined by the equation

$$w_{2u} = w_{1u} = -c_{1t} \cos 30^\circ + u$$

and the force exerted

$$P'_3 = \frac{1}{g}(w_{1u} - w_{2u}) = \frac{2}{g}(c_{1t} \cos 30^\circ - u). \qquad (1\text{-}3b)$$

If now we suppose that the initial velocity of steam c_{1t} and the velocity u of displacement of the three bodies A, B and C in all the three cases are equal, i. e., $c_{1t} = 196.2$ m/sec and $u = 98.1$ m/sec, then substituting these velocities in equations (1-3), (1-3a) and (1-3b) we obtain the numerical values of force P'

for Case (a)

$$P'_1 = \frac{1}{9.81}(196.2 - 98.1) = 10 \text{ kg};$$

for Case (b)

$$P'_2 = \frac{2}{9.81}(196.2 - 98.1) = 20 \text{ kg};$$

for Case (c)

$$P'_3 = \frac{2}{9.81}(196.2 \times 0.866 - 98.1) = 14.65 \text{ kg}.$$

The work done by 1 kg of steam during 1 sec in all the three cases of displacing the three bodies with a velocity u is determined with the help of equation (1-1).

Case (a)

$$L_1 = P'_1 u = 10 \times 98.1 = 981 \text{ kg m/sec};$$

Case (b)

$$L_2 = P'_2 u = 20 \times 98.1 = 1{,}962 \text{ kg m/sec};$$

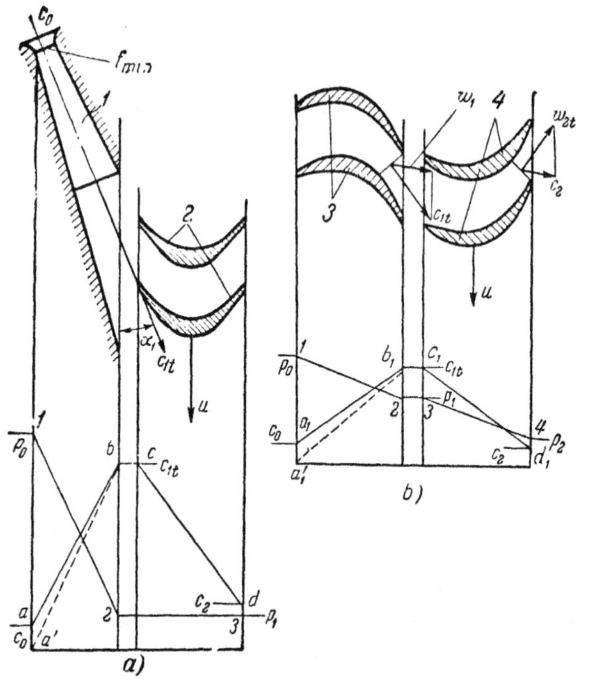

Fig. 1-10. Impulse (*a*) and reaction (*b*) stages of a turbine
1— nozzle; *2* and *4*— moving blades; *3*— guide blades

Case (c)

$L_3 = P'_3 u = 14.65 \times 98.1 = 1{,}445$ kg m/sec.

It follows, therefore, from equation (1-3b) that the force P'_3 of the steam jet depends on the value of cosine of the angle α_1. With the decrease of angle α_1 to its minimum value—zero, force P'_3 will reach its limiting value P'_2. In the case of increase of the angle α_1 the force exerted in the direction of rotation will go on diminishing until at $\alpha_1 = 90°$ this force will entirely vanish. Hence great attention is paid to the proper selection of angle α_1 for steam turbine nozzles and blades. In steam turbines the usual practice is to use an angle α_1 from 11 to 14° for the initial stages, and 20 to 30° for the low-pressure stages of large-capacity condensing turbines.

Let us now consider the operating processes in a steam turbine stage. Fig. 1-10 shows the diagrammatic arrangement of an impulse (*a*) and a reaction (*b*) stage (degree of reaction $\varrho = 0.5$) of a turbine.

The expansion of steam from its initial pressure p_0 to the final pressure p_1, in the case of an impulse stage, takes place completely in the nozzle *1*. The kinetic energy of the steam, obtained during its expansion in the nozzle, is converted into mechanical work in the moving blades *2*.

In Fig. 1-10, *a* the line *1-2-3* shows the variation of steam pressure in the moving blades of the turbine in an impulse stage. During expansion in the nozzle the velocity of steam increases from its initial value of c_0 to its maximum value c_{1t} at the nozzle exit. The line *a-b-c-d* shows the variation of absolute velocity in an impulse stage. If the velocity of steam at entrance to the nozzle $c_0 = 0$ then the variation of absolute velocity will be represented by the line *a'-b-c-d*. The velocity of steam at exit from the blades will be equal to c_2. The ideal and actual work done by 1 kg of steam in a turbine stage may be determined from the kinetic energy equations.

In a characteristic turbine stage the kinetic energy of 1 kg of steam will be equal to:

at entry to the nozzle $\frac{c_0^2}{2g}$ [kg m/kg];

at exit from the nozzle $\frac{c_{1t}^2}{2g}$ [kg m/kg];

at exit from the moving blades $\frac{c_2^2}{2g}$ [kg m/kg].

The increase in the available energy $\frac{c_{1t}^2}{2g} - \frac{c_0^2}{2g}$ [kg m/kg] obtained in the nozzle accounts for the decrease in heat content of steam from i_0 to i_{1t}. Consequently we may write an equation for the process in the nozzle

$$\frac{c_{1t}^2}{2g} - \frac{c_0^2}{2g} = \frac{1}{A}(i_0 - i_{1t}), \qquad (1\text{-}4)$$

where i_0—heat content in kcal/kg of steam before entering the nozzle;

i_{1t}—heat content in kcal/kg of steam at the nozzle exit after expansion without losses (adiabatic expansion);

$A = \frac{1}{427}$—thermal equivalent of work in kcal/kg m.

If at entry to the nozzle the steam has a velocity $c_0 = 0$ then the energy available for doing work is obtained due to the reduction of the heat content while expanding in the nozzle. In this case:

$$\frac{c_{1t}^2}{2g} = \frac{1}{A}(i_0 - i_{1t}) \text{ [kg m/kg]}. \qquad (1\text{-}4a)$$

While the steam flows through the blades of an impulse stage there is a decrease in its kinetic energy which is

$$\frac{c_{1t}^2}{2g} - \frac{c_2^2}{2g} \text{ [kg m/kg]}.$$

Mechanical work is obtained at the cost of this decrement in the energy of steam. Consequently the following relation may be formulated:

$$L_a = \frac{c_{1t}^2}{2g} - \frac{c_2^2}{2g} \text{ [kg m/kg]}, \quad (1\text{-}5)$$

where L_a—the work done by 1 kg of steam on the blades of an ideal impulse turbine or on the shaft of a single-stage single-row impulse turbine.

In reaction turbines the expansion of steam from its initial pressure p_0 to the final value p_2 is achieved both in the guide blades *3* and the moving blades *4*. Consequently in the moving blades of a reaction turbine there is a simultaneous conversion of heat energy into kinetic energy due to expansion of steam and kinetic energy into mechanical work at the turbine shaft.

The variation of steam pressure in a reaction turbine stage will be as shown by the line *1-2-3-4* (Fig. 1-10, *b*). The line a_1-b_1-c_1-d_1 shows the change in absolute velocities of steam in a reaction stage. If the velocity of steam at entry to the stationary (guide) blades is equal to $c_0=0$ then the change in velocity will be represented by the line a_1'-b_1-c_1-d_1.

If steam issuing from the guide blades enters the moving blades with a relative velocity of w_1 and after suffering a change in its direction of flow leaves the moving blades with a relative velocity of w_{2t}, then the kinetic energy of 1 kg of steam at entry to the blades will be $\frac{w_1^2}{2g}$ [kg m/kg], and at exit will be equal to $\frac{w_{2t}^2}{2g}$ [kg m/kg]. Thus the increase in the kinetic energy of 1 kg of steam directly in the moving blades will be equal to

$$\frac{w_{2t}^2}{2g} - \frac{w_1^2}{2g} \text{ [kg m/kg]}.$$

Besides, the steam enters the moving blades with a kinetic energy equal to $\frac{c_{1t}^2}{2g}$ [kg m/kg]. The loss of kinetic energy as a result of the leaving velocity c_2 for each kg of steam can be determined as equal to $\frac{c_2^2}{2g}$ [kg m/kg].

Consequently the work done by 1 kg of steam in an ideal stage of a reaction turbine would be obtained from the equation

$$L_p = \frac{c_{1t}^2}{2g} + \frac{w_{2t}^2 - w_1^2}{2g} - \frac{c_2^2}{2g} \text{ [kg m/kg]}. \quad (1\text{-}6)$$

Chapter Two

FLOW OF STEAM THROUGH A TURBINE STAGE

2-1. EXPANSION OF STEAM IN NOZZLES NEGLECTING LOSSES

A nozzle is a passage of varying cross-sectional areas in which the potential energy of the steam is converted into kinetic energy. The increase in velocity of the steam jet at the exit of the nozzle is obtained due to the decrease in heat content of the steam. The total energy of flowing steam, from thermodynamics, consists of the internal energy u, the kinetic energy $c^2/2g$ and the work pv of the steam in consequence of its pressure. Designating these quantities by suffixes 0 and 1 at entry to, and exit from the nozzle, we may represent the total energy of steam at entry to the nozzle equal to

$$u_0 + \frac{A}{2g} c_0^2 + Ap_0v_0$$

and at exit from the nozzle

$$u_1 + \frac{A}{2g} c_{1t}^2 + Ap_1v_1.$$

(Here the kinetic energy and the work of steam pressure are represented in heat units by multiplying these quantities with the thermal equivalent of work A).

If we consider the ideal case of steam flow where there is no exchange of heat with the surroundings, then according to the law of conservation of energy, the total energy before and after the nozzle must be equal, i. e.,

$$u_0 + \frac{A}{2g} c_0^2 + Ap_0v_0 = u_1 + \frac{A}{2g} c_{1t}^2 + Ap_1v_1.$$

But $(u+Apv)$ is the heat content i, and consequently

$$i_0 + \frac{A}{2g} c_0^2 = i_{1t} + \frac{A}{2g} c_{1t}^2.$$

From here we obtain the general equation for steam flow through nozzles

$$\frac{A}{2g} (c_{1t}^2 - c_0^2) = i_0 - i_{1t}, \quad (2\text{-}1)$$

i. e., the increase in the kinetic energy of the steam jet during its expansion in the nozzle is equal to the decrease in heat content so that, in the ideal case of steam flow (neglecting losses) under consideraton, the decrease in heat content is equal to the adiabatic heat drop

$$h_0 = i_0 - i_{1t}.$$

The velocity and the heat content of steam in an ideal flow process are designated here with the subscript t.

The theoretical velocity at the nozzle exit is easily determined from the equation (2-1)

$$c_{1t} = \sqrt{\frac{2g}{A}(i_0 - i_{1t}) + c_0^2} = 91.5\sqrt{(i_0 - i_{1t}) + \frac{c_0^2}{8,378}} = 91.5\sqrt{h_0 + \frac{c_0^2}{8,378}}. \qquad (2\text{-}2)$$

(The coefficients 91.5 and 8,378 are obtained as a result of the substitution of $g = 9.81$ m/sec^2 and

$A = \frac{1}{427}$ kcal/kg m, so that $91.5 = \sqrt{8,378}$).

If the initial velocity of the steam at entry to the nozzle is neglected (as is the usual case in turbine calculations) the above equation simplifies to

$$c_{1t} = \sqrt{\frac{2g}{A}(i_0 - i_{1t})} = 91.5\sqrt{i_0 - i_{1t}} = 91.5\sqrt{h_0}. \qquad (2\text{-}2a)$$

From theoretical and experimental investigations it has been found that the steam flowing through a nozzle of constant or convergent section while expanding in it reaches only up to a certain minimum value p_{cr}, equal to 0.577 of the initial pressure p_0 before the nozzle for dry saturated steam and $0.546p_0$ for superheated steam. The final pressure of steam after expansion in such nozzles does not depend on the pressure of the surroundings into which the steam is discharged if the surroundings are at a pressure equal to or less than p_{cr}. The relation $p_{cr}/p_0 = v_{cr}$ is known as the critical pressure and the velocity attained by steam at this particular pressure is known as the critical velocity c_{cr}.

If the pressure after the nozzle $p_1 > p_{cr} = v_{cr}p_0$, then the expansion of steam takes place only up to the pressure p_1; the corresponding velocity of steam at the exit from the nozzle at this pressure is less than the critical velocity c_{cr}. Special types of convergent-divergent nozzles are made use of to obtain an exit pressure $p_1 < p_{cr}$ and consequent supercritical velocity $c_1 > c_{cr}$. Thus depending on the values of the initial and final pressures of steam before and after the nozzle the steam flow may be divided into two types:

with a velocity of flow less or equal to the critical;

with a velocity higher than the critical.

Convergent nozzles [1] are used for the first type of flow (Fig. 2-1, *a*), and convergent-divergent nozzles for the second (Fig. 2-1,*b*). The size of the nozzle can be estimated if the following conditions of flow are known:

the quantity of steam flow through the nozzle,

the initial and final pressures of steam p_0 and p_1,

the initial temperature t_0 and

the heat drop in the nozzle h_0.

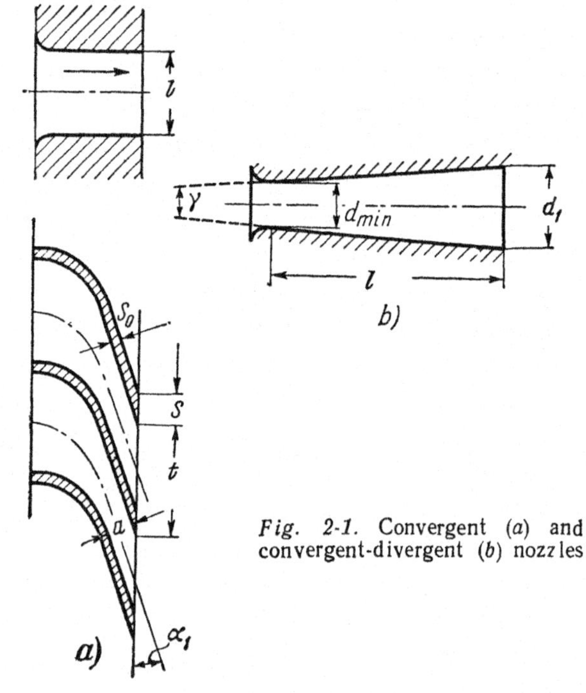

Fig. 2-1. Convergent (*a*) and convergent-divergent (*b*) nozzles

(a) Convergent Nozzles and Their Calculations

The convergent nozzle has a uniformly decreasing cross-sectional passage area, the exit end having the minimum area of section, the value of which depends on the conditions of flow. The expansion of steam from pressure p_0 to p_1 takes place along the full length of the nozzle. Nozzle calculations are made mostly with a view to obtain the cross-sectional areas of the steam passage through the nozzle. This is simply determined by making use of the equation of continuity.

$$Gv = fc, \qquad (2\text{-}3)$$

where G—quantity of steam flow, kg/sec;

[1] The possibility of obtaining supercritical velocities of steam using only convergent nozzles will be dealt with later in this book.

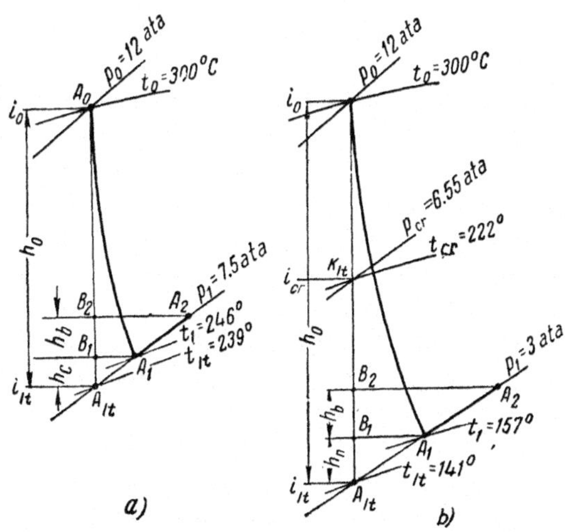

Fig. 2-2. I-s diagram for an impulse stage

f—cross-sectional area of the nozzle, m^2;
v—specific volume of steam at the given nozzle section, m^3/kg;
c—velocity of steam at the given nozzle section, m/sec.

In equation (2-3) the value of *G* is usually given and the values of *c* and *v* are found out by calculations. The velocity of steam c_{1t} is obtained from the equations (2-2) or (2-2a). The specific volume of steam *v* is determined either with the help of an *i-s* diagram or steam tables.

Example 2-1. Find the cross-sectional area of a convergent nozzle from the following data: the pressure and temperature before the nozzle $p_0 = 12$ ata and $t_0 = 300°C$; pressure after the nozzle 7.5 ata; quantity of steam flow through the nozzle 1.2 kg/sec. Neglect losses.

On the *i-s* diagram we find point A_0 conforming to the initial conditions of steam. Through this point we shall drop an adiabatic line up to a pressure of 7.5 ata—point A_{1t} (Fig. 2-2, *a*). The length A_0A_{1t}, according to a predetermined scale gives the heat drop in the nozzle. Measuring A_0A_{1t}, we find it to be equal to $h_0 = 28$ kcal/kg.

We shall now obtain the steam velocity from the relation (2-2a) $c_{1t} = 91.5\sqrt{28} = 485$ m/sec.

The conditions of steam at the nozzle exit are given by pressure 7.5 ata and temperature 239°C. The specific volume of steam conforming to this condition of steam $v_{1t} = 0.315$ m^3/kg, which may be obtained from steam tables if the lines of constant volume are not shown on the *i-s* diagram.

The cross-sectional area for the nozzle exit is determined from equation (2-3):

$$f = \frac{Gv_{1t}}{c_{1t}} = \frac{1.2 \times 0.315}{485} = 0.00078\ m^2 = 7.8\ cm^2.$$

(b) Convergent-Divergent Nozzles and Their Calculations

The expansion of steam in such nozzles takes place in the following manner.

In the convergent part of the nozzle steam expands from its initial pressure p_0 up to the critical pressure p_{cr}. At the smallest cross-section of the nozzle, known as the t h r o a t, steam attains the critical velocity c_{cr}. Expansion of steam from the pressure p_{cr} to the exit pressure p_1 takes place in the divergent part of the nozzle with a uniformly increasing steam velocity ultimately attaining the final value of $c_{1t} > c_{cr}$.

Steam pressure in the nozzle throat is determined from the equation

$$p_{cr} = \nu_{cr} p_0.$$

The process of expansion of steam from its initial pressure to the final one is shown on the *i-s* diagram (Fig. 2-2,*b*). On the adiabatic line A_0A_{1t} point k_{1t}, conforming to the critical pressure p_{cr}, is marked. The critical velocity of steam attained at the nozzle throat is determined from the equation

$$c_{cr} = 91.5\sqrt{i_0 - i_{cr}}, \qquad (2\text{-}4)$$

where i_{cr}—the heat content of steam at the throat, kcal/kg.

The specific volume of steam v_{cr} for the steam conditions at the point k_{1t} is found directly from the *i-s* diagram or from steam tables (from the known value of temperature t_{cr} or the dryness fraction x_{cr} read off the *i-s* diagram).

The cross-section of the nozzle at the throat is determined according to the equation (2-3) where c_{cr} and v_{cr} are substituted in place of *c* and *v*. The exit section of the nozzle is also found in like manner from the equation (2-3) where the specific volume of steam at the point A_{1t} (Fig. 2-2, *b*) will be substituted so that the steam velocity obtained will be

$$c_{1t} = 91.5\sqrt{i_0 - i_{1t}}.$$

The minimum cross-section of a convergent-divergent nozzle can also be determined by a different method as given below. In place of the velocity *c* in equation (2-3) the well-known expression

$$c_{cr} = \sqrt{2g\frac{k}{k+1}p_0v_0} \qquad (2\text{-}4a)$$

for the critical velocity is substituted.

The numerical values of *k* which are used in general are:

$k = 1.035 + 0.1x$—for saturated steam with a dryness fraction of *x*;

$k = 1.135$—for dry saturated steam;

$k = 1.3$—for superheated steam.

If we substitute the value c_{cr} from the equation (2-4a) in the equation (2-3) and rearrange we have

$$f_{min} = \frac{G}{199\sqrt{p_0/v_0}} \qquad (2\text{-}5)$$

for dry saturated steam and

$$f_{\min}=\frac{G}{209\sqrt{p_0/v_0}} \qquad (2\text{-}5a)$$

for superheated steam.

However, it is found from experience that the following equation gives values nearer to those achieved in reality both for saturated and superheated steam:

$$f_{\min}=\frac{G}{203\sqrt{p_0/v_0}} \qquad (2\text{-}5b)$$

(here p_0 is in kg/cm^2 and v_0—in m^3/kg).

The length of the divergent portion of the nozzle is determined by the equation

$$l=\frac{d_1-d_{\min}}{2\tan\gamma/2}, \qquad (2\text{-}6)$$

where d_1 and $d_{\min}$—diameters of the nozzle at the exit and throat sections (or the value of the sides, if the nozzle is square or rectangular);

γ—the angle of divergence of the nozzle; it is recommended that this value may be used within the limits of 6 to 12°.

With very small values of angle γ the length of the nozzle would be large whereas with larger values of γ there is a possibility of the steam jet detaching itself from the nozzle wall and consequently turbulence losses would make their appearance.

Example 2-2. Design a convergent-divergent nozzle from the following data: initial pressure of steam 12 ata; final pressure of steam 3 ata; steam temperature 300°C; quantity of steam flow 1.2 kg/sec. Losses may be neglected.

On the *i-s* diagram (Fig. 2-2, *b*) we shall first find the point (point A_0) conforming to the initial conditions of steam. Through this point we shall drop a vertical line (adiabatic) up to the back pressure of 3 ata (point A_{1t}). The length A_0A_{1t} gives the heat drop in the nozzle which is equal to

$$A_0A_{1t}=h_0=73 \text{ kcal/kg}.$$

The velocity of steam issuing from the nozzle

$$c_{1t}=91.5\sqrt{73}=782 \text{ m/sec}.$$

The condition of the steam at the point A_{1t} is determined by the pressure 3 ata and the temperature $t_{1t}=141$°C.

The specific volume of steam at the nozzle exit

$$v_{1t}=0.636 \text{ m}^3\text{/kg}.$$

The cross-section of the nozzle at the exit

$$f_1=\frac{Gv_{1t}}{c_{1t}}=\frac{1.2\times 0.636}{782}=0.000977 \text{ m}^2=9.77 \text{ cm}^2.$$

The minimum cross-section of the nozzle is determined from formula (2-5a):

$$f_{\min}=\frac{1.2}{209\sqrt{\frac{12}{0.223}}}=0.000777 \text{ m}^2=7.77 \text{ cm}^2,$$

where $v_0=0.223$ m^3/kg (according to V.T.I.[1] steam tables).

[1] Power Institute of the U.S.S.R.

2-2. EXPANSION OF STEAM IN NOZZLES CONSIDERING LOSSES

Because of losses due to friction and turbulence steam while flowing through a nozzle gains heat. Thus the actual heat content of steam at the nozzle exit will be slightly higher than the theoretical value, i. e., $i_1>i_{1t}$.

Consequently the actual process of expansion of steam in the nozzle will not be represented by the adiabatic A_0A_{1t}; but by a different curve (Fig. 2-2) from which it follows that the heat drop in the nozzle during the process of expansion, considering the losses, will be lower than in the case of an adiabatic expansion, i. e.,

$$i_0-i_1<i_0-i_{1t}.$$

The actual velocity of steam c_1 is therefore determined:

at $c_0=0$

$$c_1=91.5\sqrt{i_0-i_1}; \qquad (2\text{-}7)$$

at $c_0\neq 0$

$$c_1=91.5\sqrt{(i_0-i_1)+\frac{c_0^2}{8{,}378}}. \qquad (2\text{-}7a)$$

Comparing equations (2-2) and (2-2a) with equations (2-7) and (2-7a) and bearing in mind the condition $i_0-i_1<i_0-i_{1t}$, we conclude that $c_1<c_{1t}$. The actual velocity of steam at the nozzle exit is related with the theoretical one by the so-called velocity coefficient φ:

at $c_0=0$

$$c_1=\varphi c_{1t}=91.5\varphi\sqrt{i_0-i_{1t}}; \qquad (2\text{-}7b)$$

at $c_0\neq 0$

$$c_1=\varphi c_{1t}=91.5\varphi\sqrt{(i_0-i_{1t})+\frac{c_0^2}{8{,}378}}. \qquad (2\text{-}7c)$$

Usually the velocity coefficient of a nozzle is from 0.91 to 0.98; often a mean value of 0.95 for φ is used. The actual value of φ for different nozzles is different depending on its size and shape and can be determined only experimentally.

The presence of energy losses in the nozzle decreases the velocity of steam issuing from the nozzle exit. The kinetic energy losses in a nozzle may be found from the equation

$$\Delta L_n=\frac{c_{1t}^2}{2g}-\frac{c_1^2}{1g}=\frac{1}{A}(i_1-i_{1t})=\frac{1}{A}h_n \qquad (2\text{-}8)$$

in mechanical work units, or

$$h_n=\frac{A}{2g}(c_{1t}^2-c_1^2)=\frac{c_{1t}^2-c_1^2}{8{,}378}=(1-\varphi^2)\frac{c_{1t}^2}{8{,}378}=$$

$$=\left(\frac{1}{\varphi^2}-1\right)\frac{c_1^2}{8{,}378} \qquad (2\text{-}8a)$$

in heat units (here $c_1^2/2g$—kinetic energy of 1 kg of steam at the exit of the nozzle for the actual process of expansion of steam, kg m/kg.) Determination of the nozzle size taking into consideration the accompanying losses is carried out in the following manner.

Knowing the steam conditions before the nozzle and the exit pressure the adiabatic A_0A_{1t} can be found (Fig. 2-2). The adiabatic heat drop in the nozzle is determined as a difference of the heat contents

$$h_0 = i_0 - i_{1t}.$$

The theoretical exit velocity of the steam conforming to this heat drop is determined by equation (2-2) or (2-2a). The actual velocity of steam c_1 at the nozzle exit is thereafter found from the equations (2-7b) and (2-7c) for the particular assumed value of φ. The heat losses h_n in the nozzle are calculated from equation (2-8a). Setting off h_n upwards from the point A_{1t} up to B_1 and drawing a constant heat line therefrom (horizontally) until it intersects the constant pressure line or isobar p_1, the actual steam conditions at the nozzle exit are found (point A_1). The specific volume of steam v_1 at the nozzle exit is determined from the steam conditions at the point A_1 (thus accounting for the losses). The exit cross-section of the nozzle is calculated using equation (2-3) substituting the values c_1 and v_1 obtained as described above.

Example 2-3. Design a convergent nozzle using the data given in example 2-1, taking into consideration the losses. Assume $\varphi = 0.95$.

From example 2-1 we have $h_0 = 28$ kcal/kg; $c_{1t} = 485$ m/sec; actual steam velocity will be $c_1 = \varphi c_{1t} = 0.95 \times 485 = 461$ m/sec.

The losses in the nozzle according to formula (2-8a)

$$h_n = \frac{c_{1t}^2 - c_1^2}{8,378} = \frac{485^2 - 461^2}{8,378} = 2.75 \text{ kcal/kg.}$$

Setting off this quantity upwards from the point A_{1t} (Fig. 2-2,*a*) and drawing a horizontal line through the point B_1 up to the isobar $p_1 = 7.5$ ata we obtain point A_1 which determines the steam conditions after the nozzle which are: pressure 7.5 ata and temperature 246°C. From steam tables we now find the specific volume of steam v_1 at the nozzle exit; $v_1 = 0.319$ m³/kg. The exit cross-section of the nozzle

$$f_1 = \frac{1.2 \times 0.319}{461} = 0.000829 \text{ m}^2 = 8.29 \text{ cm}^2.$$

Example 2-4. Design a convergent-divergent nozzle using the data given in example 2-2 taking into account the losses, assuming $\varphi = 0.95$.

From the previous example we know the heat drop and the theoretical velocity of steam at the nozzle exit;

heat drop $i_0 - i_{1t} = 73$ kcal/kg;

theoretical velocity $c_{1t} = 782$ m/sec.

The actual velocity of flow

$$c_1 = \varphi c_{1t} = 0.95 \times 782 = 743 \text{ m/sec.}$$

The losses in the nozzle from equation (2-8a) are

$$h_n = \frac{c_{1t}^2 - c_1^2}{8,378} = \frac{782^2 - 743^2}{8,378} = 7.4 \text{ kcal/kg.}$$

Setting off the losses upwards from the point A_{1t} (Fig. 2-2,*b*) and drawing a horizontal line through the point B_1 up to the isobar $p_1 = 3$ ata we obtain point A_1 which gives the steam conditions at the nozzle exit: pressure 3 ata and temperature $t_1 = 157°$ C.

From steam tables the specific volume of steam at the nozzle exit is found to be

$$v_1 = 0.664 \text{ m}^3/\text{kg}.$$

The exit cross-section of the nozzle

$$f_1 = \frac{1.2 \times 0.664}{743} = 0.001072 \text{ m}^2 = 10.72 \text{ cm}^2.$$

The critical pressure attained at the throat of the nozzle will be

$$p_{cr} = 0.546 \times 12 = 6.55 \text{ ata}.$$

From the *i-s* chart we find

$$i_0 - i_{cr} = 35 \text{ kcal/kg}.$$

Assuming $\varphi = 1$ we determine the critical velocity of steam at the throat section

$$c_{cr} = 91.5\sqrt{35} = 541 \text{ m/sec.}$$

The specific volume of steam at the point k_{1t} for a pressure of 6.55 ata and temperature $t_{cr} = 222°$C

$$v_{cr} = 0.35 \text{ m}^3/\text{kg}.$$

Thus the minimum cross-section of the nozzle

$$f_{min} = \frac{1.2 \times 0.35}{541} = 0.000776 \text{ m}^2 = 7.76 \text{ cm}^2.$$

The calculation of f_{min} in the manner shown above, assuming $\varphi = 1$, agrees well with the result obtained by using equation (2-5a).

2-3. EXPANSION OF STEAM IN NOZZLES AT CONDITIONS OTHER THAN DESIGNED

The process of expansion of steam in a nozzle will differ from the design conditions conforming to a particular back pressure, if the back pressure varies, with the steam conditions before the nozzle remaining constant.

Fig. 2-3 shows two different types of nozzles: *a*—convergent nozzle; *b*—a convergent-divergent nozzle. The exit cross-section of the convergent nozzle is equal to the minimum or throat section of the convergent-divergent nozzle. The steam conditions before the nozzles in both cases are the same. The design back pressures for the two nozzles are $p_1 > p_{cr}$ for the convergent nozzle and $p_{1d} < p_{cr}$ for the convergent-divergent nozzle. The convergent nozzle has been designed for a steam flow of G_1[kg/sec], and the divergent one for a flow of G_0[kg/sec].

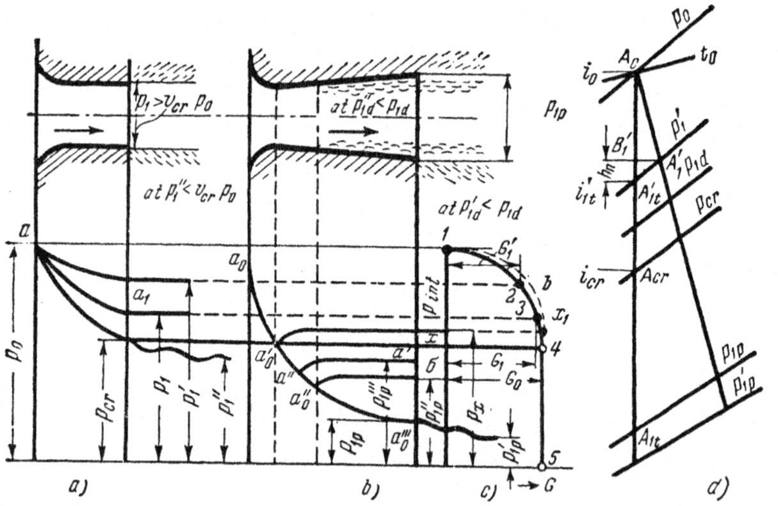

Fig. 2-3. Expansion of steam in convergent and convergent-divergent nozzles

The variations of pressure along the longitudinal axis of the nozzles for the calculated (design) back pressures (exit pressures) are shown by curves aa_1 for the convergent nozzle and by the curve $a_0a_0^{I}\,a_0^{II}\,a_0^{III}$ for the convergent-divergent nozzle.

Let us consider the process of expansion of steam at back pressures other than the design pressures.

a) The Convergent Nozzle

The variations of the quantity of steam flow through the nozzle at various back pressures (curve *1-2-3-4-5*), with the steam conditions before the nozzle remaining constant, are shown in Fig. 2-3,*c*. On this diagram pressures are measured along the ordinates and the quantity of steam flow along the abscissa. These curves have been obtained in the following manner.

The exit cross-section of the nozzle f_1, for the given steam flow and back pressure G_1 and p_1 respectively, is determined as in Example 2-3. Thereafter for different arbitrary back pressures the mass flow of steam is determined. For example, if we assume the pressure after the nozzle equal to p_1' and p_{cr} and set these off across the line A_0A_{1t} on the *i-s* diagram (Fig. 2-3,*d*) we have the heat drop conforming to these pressures equal to $i_0 - i_{1t}'$ and $i_0 - i_{cr}$.

The actual value of steam velocity issuing from the nozzle is determined in accordance with equations (2-7,b) and (2-4)

$$c_1 = 91.5\varphi\sqrt{(i_0 - i_{1t}')} \text{ and } c_{cr} = 91.5\sqrt{(i_0 - i_{cr})}.$$

Losses in the nozzles are determined according to formula (2-8a)

$$h_n = \left(\frac{1}{\varphi^2} - 1\right)\frac{c_1^2}{8,378}.$$

Setting off the losses h_n from the point A_{1t}' and drawing a horizontal through the point B_1' until it intersects the isobar p_1' we find the point A_1' which gives the steam conditions after the nozzle for the pressure p_1'. From the *i-s* diagram the specific volume of steam v_1 and v_{cr} for the steam conditions at the points A_1' and A_{cr} (point of intersection on the critical pressure line) are determined.

The quantity of steam flow through the nozzle is determined according to equation (2-3)

$$G = \frac{fc}{v}.$$

Substituting in the above equation the values c_1, v_1 and c_{cr}, v_{cr} we obtain the quantity of steam flow G_1' and G_0 respectively. At the point *1* in the diagram (2-3,*c*) the steam flow through the nozzle is equal to zero since the steam conditions before and after the nozzle are the same. The points *2* and *4* in the diagram are obtained by calculating for back pressures of p_1' and p_{cr} whereas point *3* conforms to the given conditions of flow.

Thus the curve *1-2-3-4* indicates the variations of mass flow of steam through the nozzle for various back pressures varying within the limits of p_0 to p_{cr}, with the steam conditions before the nozzle remaining constant. Consequently the variations of back pressure, within the limits p_0 and p_{cr}, are accompanied by variations of exit velocity of steam as also the mass flow.

If the back pressure $p_1'' < p_{cr}$, then the pressure of steam in the exit section of the nozzle does not alter remaining equal to the critical pressure. With further decrease in the back pressure after the nozzle the pressure in the exit section of the nozzle continues to remain unaltered and equal to p_{cr} (the critical pressure), as also the critical velocity of steam flow. In the above cases the mass flow also remains unaltered and equal to the maximum mass flow G_0 (line *4-5*). Consequently with back pressures after the nozzle exit less than the critical p_{cr} the expansion of steam in the nozzle takes place only up to the critical pressure. The expansion of steam from the critical pressure p_{cr} to the back pressure of the surroundings beyond the nozzle takes place without any further increase in the velocity of flow. Thus in a simple convergent nozzle[1] the exit velocity of steam from the nozzle cannot be greater than the critical velocity.

b) Convergent-Divergent Nozzle

At the design conditions of steam p_0, t_0, and p_{1d} the variations of pressures along the axis of the nozzle will be as shown by the curve $a_0a_0^{I}a_0^{II}a_0^{III}$ and the mass flow of steam through the nozzle is equal to G_0. With decrease in back pressure, i. e., when $p_{1d}^{I}<p_{1d}$ the process of expansion of steam within the confines of the nozzle remains unaltered. Expansion of steam from the pressure p_{1d} to p_{1d}^{I} takes place beyond the limits of the nozzle accompanied by a dispersal of the jet. The mass flow of steam with decrease in back pressure, however, remains the same and equal to G_0.

If steam issues from the nozzle with a pressure higher than the rated value, i. e., if $p_{1d}^{II}>p_{1d}$ then the expansion of steam takes place only in the nozzle; further the steam jet detaches itself from the nozzle wall giving rise to considerable turbulence in the flow. At the place where the steam jet detaches from the wall the steam pressure falls below the pressure beyond the nozzle. Starting from the point where this dissociation occurs the pressure builds up quickly again and there appears a so-called shock wave. Thus the process of expansion of steam in a convergent-divergent nozzle in the case of flow of steam from the nozzle into the surrounding medium at a back pressure higher than the design value p_{1d}^{II} will be as shown by the line $a_0a_0^{I}a_0^{II}b$. The process of expansion of steam in the nozzle for any arbitrary pressure $p_{1d}^{III}>p_{1d}$ will be as shown by the line $a_0a_0^{I}a^{II}a^{I}$, etc. For every case of expansion of steam to a pressure $p_{mean}>p_{1d}$ a shock wave will appear at the place where the flow detaches itself from the nozzle wall accompanied by a considerable turbulence and growth of losses.

Thus with the formation of shock waves there is an increase in losses in the nozzle, the velocity coefficient φ decreases and the actual velocity of steam c_1 consequently is also lower. The mass flow, however, remains the same for these higher back pressures so long as the pressure in the nozzle throat does not increase beyond p_{cr}.

[1] Convergent nozzle without oblique exit.

With an increase in pressure beyond the exit of a convergent-divergent nozzle the maximum mass flow G_0 remains unaltered until a back pressure p_x is reached at which the shock wave appears in the minimum or throat section of the nozzle (line $a_0a_0^{I}x$). Increase in the back pressure beyond p_x will cause a gradual decrease in the mass flow G (broken line x_1b1).

The back pressure of the surroundings may be determined from the equation

$$P_x=\left(0.546+0.454\sqrt{\frac{f_1-f_{min}}{f_1}}\right)p_0. \qquad (2\text{-}9)$$

The variations of mass flow of steam through a convergent-divergent nozzle for back pressures from p_0 to p_x is shown by the broken line $1bx_1$, and for pressures p_x and lower, by the line x_15.

Thus the deviation of back pressure from the design values in a convergent-divergent nozzle brings forth a considerable increase in kinetic energy losses in it. Convergent-divergent nozzles are unsuitable for varying conditions and therefore (as also from constructional point of view) these nozzles find a limited usage in the present-day steam turbines. They are used mostly for the first few regulating stages of a multistage turbine.

2-4. EXPANSION OF STEAM IN THE OBLIQUE EXIT REGION OF NOZZLES

In steam turbines nozzles are placed at some angle with respect to the plane of rotation of the turbine discs, and therefore in the region of the nozzle exit an oblique section is obtained. Shapes of such nozzles are shown in Figs. 1-10, 2-1,*a* and 2-4. The process of expansion of steam in nozzles with oblique exit differs from that in nozzles with straight exit section.

a) Convergent Nozzle

The process of expansion of steam in a convergent nozzle with an oblique exit does not in any way differ from that in a nozzle with straight exit section, i. e., equation (2-3) holds good for nozzles with oblique exit section also, only if $p_1/p_0\geqslant\nu_{cr}$. However, with $p_1/p_0<\nu_{cr}$ the process of steam expansion in a nozzle with oblique exit will have the following special features. The expansion of steam from its initial pressure p_0 to the pressure p_{cr} takes place along the nozzle length exactly as in the case of a convergent nozzle without oblique exit. Consequently in the minimum exit section of the nozzle f_{min} (section *1-2* on Fig. 2-4,*a*) critical pressure p_{cr} is attained and also the critical velocity of flow c_{cr}. The expansion of steam from p_{cr} to p_1 with the accompanying increase in velocity from c_{cr} to c_1 takes place in the oblique exit region of the nozzle. During these processes, as is found from experimental investigations, the velocity c_1 obtained at the nozzle exit may well high exceed the critical value.

At the point *1* in the nozzle section *1-2* the jet of steam leaving the edge enters the surroundings which are at a pressure p_1. Consequently at the point *1* the pressure of steam suddenly falls from p_{cr} to p_1.

On portion *2-3* of the oblique nozzle exit expansion of steam takes place gradually from p_{cr} to p_1. Thus from the point *1* we may project a pencil of isobars within the limits of pressures p_{cr} to p_1. The isobars may be schematically represented by *1-2*, *1-2'*, *1-2''* and *1-3* on the basis of experimental results (Fig. 2-4,*a*). Expansion of steam in the oblique exit region is accompanied by a deflection of the steam jet from the axis of the nozzle which starts from that section where the critical velocity c_{cr} is attained. Here the direction of flow of steam at any section of the nozzle's oblique exit region depends on the direction of the isobars *1-2*, *1-2'*, *1-2''*, etc., and the so-called Mach angle θ which may be obtained from the equation

$$\sin\theta = \frac{c_s}{c_1},$$

where c_s— velocity of sound for the particular steam pressure.

Line *1-3* approximately shows the limiting pressure p_{1a}, up to which expansion of steam may be carried out in the oblique exit region.

If the pressure beyond the nozzle is less than p_{1a}, then the further expansion of steam will take place beyond the confines of the nozzle and will be accompanied by energy diffusion without any further increase in the velocity. If the pressure after the nozzle $p_1 > p_{1a}$, then the final pressure p_1 will be found in an intermediate section of the oblique exit (e.g., isobar *1-2''*). Such a case is shown in Fig. 2-4.

The displacement of isobars in the oblique exit region of the nozzle from section *1-2* is brought about by the increase in specific volume of steam during its expansion from p_{cr} to p_1 or p_{1a}, if $p_{1a} > p_1$. The width of the steam jet increases in conformity with the direction (displacement) of the isobars. Thus the expansion of steam in an oblique nozzle exit is subject to the very same laws of convergent-divergent nozzles.

It is very simple to show that the degree of expansion of steam in the oblique exit region will be greater, the greater the value of angle $90°-\alpha_1$, i. e., the smaller the angle of inclination of the nozzle α_1. At the same time when $90°-\alpha_1=0°$ angle $\alpha_1=90°$, i. e., there is now no oblique exit for the nozzle and thus there is no possibility of obtaining any further expansion of steam along with an increase in velocity of flow above critical. With an increase in angle $90°-\alpha_1$, angle α_1 diminishes resulting in an oblique exit section for the nozzle in which steam may be expanded further.

Thus the maximum possible expansion of steam in the oblique exit region of a nozzle depends on the angle of inclination α_1.

For the heat calculations of working (moving) blades in steam turbines it is indispensable that the direction of flow of steam from the nozzle exit be known. Hence, besides angle α_1, it is necessary to know ω, the angle of deflection of the steam jet from the nozzle axis while expanding in the oblique exit region.

The angle ω is determined in the following manner.

In Fig. 2-4,*a* we have

a—width of the nozzle at section *1-2*;

a_1—width of the steam jet at the nozzle exit;

l—nozzle height at section *1-2* (measured in a direction perpendicular to the plane of the paper).

l_1—height of the steam jet beyond the nozzle exit (section *3'-4*);

c_{cr} and v_{cr}—the critical velocity of flow and the specific volume of steam at the smallest nozzle section f_{min} (section *1-2*);

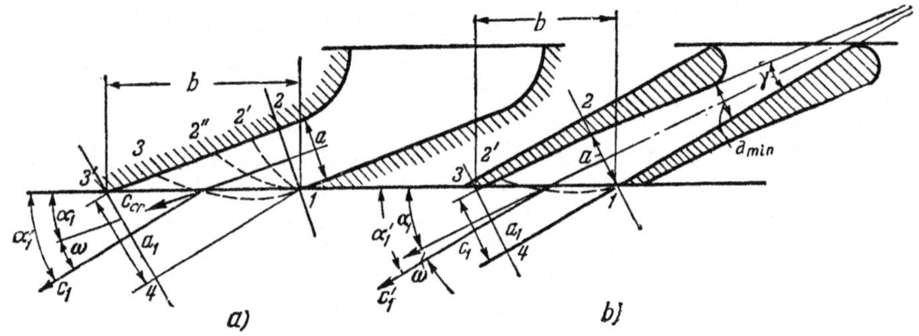

Fig. 2-4. Deflection of steam jet in the oblique exit region
a) — convergent nozzle; *b)* — convergent-divergent nozzle

c_1 and v_1—velocity of flow and specific volume of steam at the exit.

Since the quantity of steam flow through sections *1-2* and *3'-4* is the same we may write the equation

$$G = \frac{f_{min} c_{cr}}{v_{cr}} = \frac{f_1 c_1}{v_1}, \qquad (2\text{-}10)$$

where $f_{min} = al$ and $f_1 = a_1 l_1$ are the sectional areas at *1-2* and *3'-4* respectively.

Substituting in equation (2-10) the above in place of f_{min} and f_1 and eliminating l from both the left- and the right-hand sides of the equation (assuming that $l = l_1$) we have

$$\frac{a c_{cr}}{v_{cr}} = \frac{a_1 c_1}{v_1}. \qquad (2\text{-}11)$$

From the figure it is clear that $a = b \sin \alpha_1$ and $a_1 = b \sin(\alpha_1 + \omega)$.

Substituting the above values of a and a_1 in equation (2-11) we have

$$\frac{\sin \alpha_1 c_{cr}}{v_{cr}} = \frac{\sin(\alpha_1 + \omega) c_1}{v_1}. \qquad (2\text{-}11a)$$

From expression (2-11a) substituting α_1' for $(\alpha_1 + \omega)$ we finally obtain

$$\sin \alpha_1' = \sin(\alpha_1 + \omega) = \frac{c_{cr}}{c_1} \times \frac{v_1}{v_{cr}} \sin \alpha_1. \qquad (2\text{-}12)$$

All the quantities on the right-hand side of equation (2-12) are easily obtained and consequently it is easy to determine the angle of deflection α_1' and the exit angle ω.

Besides, it is obvious that,

$$\sin \alpha_1' = \frac{f_1}{f_{min}} \sin \alpha_1. \qquad (2\text{-}13)$$

When the expansion of steam in the oblique exit region of a nozzle is a maximum, i. e., when the pressure along the line *1-3* becomes equal to p_{1a}, the exit angle ω_{max} will be a maximum (limiting value). The limiting deflection of a steam jet in a nozzle is obtained as follows.

When the limiting pressure p_{1a} is reached on line *1-3* in the oblique exit region the angle of deflection of the steam jet $\alpha_1 + \omega_{max}$ may be approximately equated to the Mach angle θ (in all other sections of the nozzle in the oblique exit region the Mach angle is not equal to the sum $(\alpha_1 + \omega_{max})$. This angle is determined from the equation

$$\sin \theta = \frac{c_s}{c_{1a}}, \qquad (2\text{-}14)$$

where C_s—velocity of sound at pressure p_{1a};
c_{1a}—velocity of flow of steam issuing from the nozzle while expanding from the initial pressure to the pressure p_{1a}.

Hence for the limiting expansion of steam in the oblique exit region of a nozzle equation (2-12) will be as follows

$$\sin(\alpha_1 + \omega_{max}) = \sin \theta = \frac{c_{cr}}{c_{1a}} \times \frac{v_{1a}}{v_{cr}} \sin \alpha_1, \qquad (2\text{-}14a)$$

where v_{1a}—specific volume of steam at pressure p_{1a}.

It may be assumed with a good degree of approximation that $\frac{c_{cr}}{c_{1a}} \approx \frac{c_s}{c_{1a}}$; then on the basis of equations (2-14) and (2-14a) we may write

$$1 = \frac{v_{1a}}{v_{cr}} \sin \alpha_1, \qquad (2\text{-}15)$$

from where we have

$$v_{1a} = \frac{v_{cr}}{\sin \alpha_1}. \qquad (2\text{-}15a)$$

Thus equation (2-15a) may be used as the basic one for the calculations of limiting expansion of steam in the oblique exit region of a convergent nozzle. The expansion of steam up to the given back pressure p_1 will be possible only if v_{1a} is greater than or equal to v_1.

Example 2-5. Steam is expanded in a convergent nozzle with an oblique exit from pressure $p_0 = 16.5$ ata, $t_0 = 295°C$ to the final condition $p_1 = 5$ ata. The nozzle angle $\alpha_1 = 18°$, velocity coefficient $\varphi = 0.95$. Find ω, the angle of deflection of the steam jet in the oblique exit region of the nozzle.

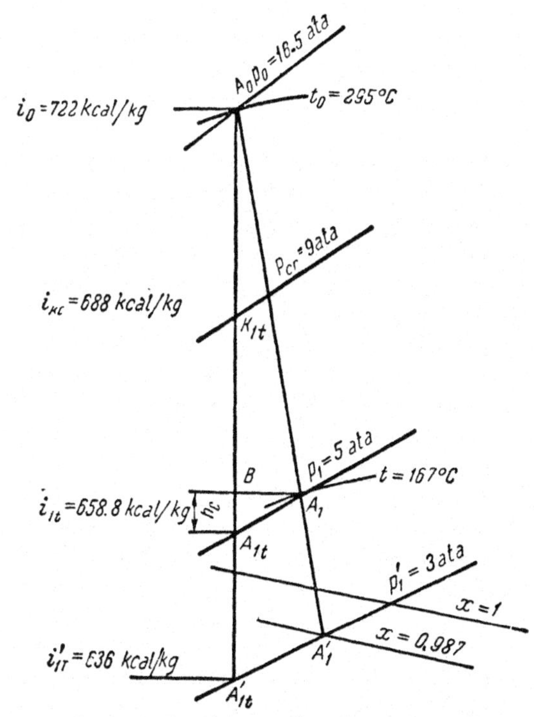

Fig. 2-5. I-s diagram for steam expansion

Let us first ascertain whether in the given case a convergent nozzle is suitable or not.

From the *i-s* diagram (Fig. 2-5) we find $i_0 = 722$ kcal/kg, the heat content of steam before the nozzle, point A_0; $i_{1t} = 658.8$ kcal/kg, heat content of steam at the point A_{1t}.

Theoretical velocity of steam

$$c_{1t} = 91.5\sqrt{i_0 - i_{1t}} = 91.5\sqrt{722 - 658.8} = 727 \text{ m/sec.}$$

The actual velocity of steam

$$c_1 = \varphi c_{1t} = 0.95 \times 727 = 690 \text{ m/sec.}$$

The heat losses in the nozzle according to equation (2-8a) are

$$h_n = \frac{727^2 - 690^2}{8{,}378} = 6.2 \text{ kcal/kg.}$$

Setting off the heat losses h_n upwards from the point A_{1t} and drawing a horizontal through the point B until it intersects with the isobar p_1, we obtain the point A_1 which gives the actual steam conditions after the nozzle. Joining the points A_0 and A_1 by a straight line we obtain an approximate picture of the process of expansion of steam in the nozzle.

The critical pressure of steam in the smallest nozzle cross-section is determined from the relation $p_{cr} = 0.546 \times 16.5 = 9$ ata, whence $i_{cr} = 689$ kcal/kg.

The point of intersection of the line A_0A_1 and the isobar p_{cr} gives the steam conditions at the exit from the minimum nozzle section (line *1-2* on Fig. 2-4, *a*).

The specific volume of steam at the exit from the nozzle

$$v_1 = 0.3996 \text{ m}^3\text{/kg.}$$

The specific volume of steam at the smallest cross-section is

$$v_{cr} = 0{,}249 \text{ m}^3\text{/kg.}$$

Let us now determine the specific volume up to which the steam may expand in the oblique exit region of the nozzle. From equation (2-15a)

$$v_{1a} = \frac{v_{cr}}{\sin\alpha_1} = \frac{0.249}{\sin 18^\circ} = 0.805 \text{ m}^3\text{/kg.}$$

Comparing the two specific volumes v_1 and v_{1a} we find that expansion of steam from 9 to 5 ata can be achieved in the oblique exit region of the nozzle.

The critical velocity of steam in the smallest nozzle section is determined from the equation

$$c_{cr} = 91.5\sqrt{i_0 - i_{cr}} = 91.5\sqrt{722 - 689} = 525 \text{ m/sec.}$$

The angle subtended by the steam jet is determined from the equation (2-12)

$$\sin(\alpha_1 + \omega) = \frac{c_{cr}}{c_1} \times \frac{v_1}{v_{cr}} \sin 18^\circ = \frac{525 \times 0.3996}{690 \times 0.249} 0.309 = 0.378.$$

Consequently

$$\alpha_1 + \omega = 22^\circ 12'.$$

The angle of deflection of the jet is therefore

$$\omega = 22^\circ 12' - 18^\circ = 4^\circ 12'.$$

b) Convergent-Divergent Nozzle

Additional expansion of steam to a pressure less than the design value can be achieved in the oblique exit region of a convergent-divergent nozzle as well, e.g., from p_1 to p'_1. Deflection of the steam jet in the oblique exit region of a convergent-divergent nozzle takes place beyond the exit section *1-2* of the nozzle (Fig. 2-4, *b*); in the case of a convergent nozzle the exit section coincides with the minimum nozzle section or the throat.

The deflection of jet in a convergent-divergent nozzle may be obtained from equation (2-12), substituting in it v_1 and c_1 in place of v_{cr} and c_{cr} for the section *1-2*, and v_1' and c_1' in place of v_1 and c_1 for section *3-4*

$$\sin\alpha_1' = \sin(\alpha_1 + \omega) = \frac{c_1}{c_1'} \times \frac{v_1'}{v_1} \sin\alpha_1. \quad (2\text{-}16)$$

As in the previous case of the convergent nozzle we can write the following relation for the limiting expansion of steam in the oblique exit region of a convergent-divergent nozzle

$$v'_{1a} = \frac{c_s'}{c_1} \times \frac{v_1}{\sin\alpha_1}, \quad (2\text{-}17)$$

where c_s' —velocity of sound for the steam conditions at the exit section *1-2*.

Velocity c_s' may be obtained from the well-known thermo-dynamic relation

$$c_s' = \sqrt{gkpv}, \quad (2\text{-}18)$$

where p and v must be replaced by the final pressure beyond the nozzle p'_1 and the final specific volume v_1'. Approximately we may take c_s' equal to the critical velocity in the throat section c_{cr}. Values c_1 and v_1 pertain to the design section *1-2* of the nozzle.

If from equation (2-17) we obtain a value $v_{1a}' \geqslant v_1'$ then the expansion of steam in the oblique exit region of a convergent-divergent nozzle may be achieved up to the final pressure p_1'. If $v_{1a}' < v_1'$ then expansion of steam up to this final pressure of p_1' will not be possible.

Example 2-6. A convergent-divergent nozzle is designed for the following steam conditions: $p_0 = 16.5$ ata, $t_0 = 295$°C, $p_1 = 5$ ata, nozzle angle $\alpha_1 = 22^\circ$. For the same conditions of steam as given above the pressure after the nozzle is reduced to 3 ata, i. e., $p_1' = 3$ ata. Find the angle of deflection of the steam jet in the oblique exit region of the nozzle.

From example 2-4 we have the data: $c_1 = 690$ m/sec; $\varphi = 0.95$; $v_1 = 0.3996$ m³/kg; $i_0 = 722$ kcal/kg. Specific volume of steam for $p_1' = 3$ ata at the point A_1' (Fig. 2-5) is $v_1' = v_s x = 0.6177 \times 0.987 = 0.61$ m³/kg.

Velocity c_s' is determined from equation

$$c_s' = 100\sqrt{gkp_1'v_1'} = 100\sqrt{9.81 \times 1.134 \times 3 \times 0.61} = 453 \text{ m/sec},$$ where $p_1' = 3$ kg/cm²; $k = 1.134$ (for wet steam).

From equation (2-17) we determine the specific volume of steam at the pressure up to which expansion of steam may be achieved in the oblique exit region of the nozzle, i. e., at section *1-2*

$$v'_{1\varphi}=\frac{453\times0.3996}{690\sin 22^\circ}=0.694 \text{ m}^3/\text{kg}.$$

Comparing the specific volumes v'_{1a} and v'_1 we conclude that it is possible to expand the steam further from 5 ata to 3 ata in the oblique exit region of the nozzle.

Velocity of steam at the nozzle exit with a back pressure of 3 ata

$$c'_1=91.5\varphi\sqrt{i_0-i'_{1t}}=91.5\times0.95x\times\sqrt{722-636}=$$
$$=807 \text{ m/sec}.$$

The angle subtended by the steam jet with respect to the plane of rotation of the turbine disc is determined from equation (2-16):

$$\sin\alpha'_1=\frac{690\times0.61}{0.3996\times807}\times\sin 22^\circ=0.488;\ \alpha'_1=29^\circ10'.$$

The angle of deflection of the steam jet in the oblique exit region of the nozzle will therefore be

$$\omega=29^\circ10'-22^\circ=7^\circ10'.$$

2-5. TRANSFORMATION OF ENERGY IN THE MOVING BLADES OF AN IMPULSE STAGE

Steam issuing from the nozzle with an absolute velocity of c_1 enters the blade passages at an angle α_1. Because of the rotation of the turbine disc velocity of steam at the entry to the blade passages will have, with respect to the passage walls, a different value and a different direction as well. This velocity is known as the relative velocity of entry of steam and is designated by the letter w_1. The magnitude and direction of w_1 can be very easily determined by drawing the parallelogram of velocities (Fig. 2-6). Graphically subtracting the velocity u (mean blade circumferential velocity) from the absolute steam velocity c_1 we obtain relative velocity w_1. In the above graphical solution c_1 will appear as the diagonal of the parallelogram, and u as one of its sides. Consequently the magnitude and direction of velocity w_1 will be represented by the second side of the parallelogram of velocities. Angle β_1 showing the direction of flow of steam at the blade passage entrance is known as the a n g l e o f e n t r y. To obtain a shock free entrance of steam into the blade passages the blades must be fixed on the disc[1] with an angle of entry equal to β_1.

The velocity may also be obtained analytically making use of the trigonometric equation for a triangle:

$$w_1=\sqrt{c_1^2+u^2-2uc_1\cos\alpha_1};$$

angle β_1 is found from the relation

$$\sin\beta_1=\frac{c_1}{w_1}\sin\alpha_1.$$

Due to the curvature of the blade passage steam suffers a change in the direction of flow and leaves the blades with a relative velocity of w_2 at an angle β_2 with respect to the plane of rotation of the disc. β_2 is known as the a n g l e o f e x i t. Angle β_2 is usually less than angle β_1. In practice $\beta_2=\beta_1-(2^\circ \text{ to } 10^\circ)$; sometimes $\beta_2=\beta_1$.

Due to the presence of losses in blade passages (see Chapters Three and Five) the relative velocity w_2 will be less than the relative velocity w_1, i. e.,

$$w_2=\psi w_1, \tag{2-19}$$

where ψ—velocity coefficient, taking into consideration the effect of undesirable resistances to the passage of steam through the moving blades. The absolute exit velocity of steam c_2 leaving the blades is also similarly determined by a graphical addition of the relative velocity w_2 and u, the circumferential velocity. In the parallelogram of velocities c_2 will appear as the diagonal, the other two sides being w_2 and u.

Velocity c_2 may also be determined analytically as

$$c_2=\sqrt{w_2+u^2-2uw_2\cos\beta_2}.$$

However, when calculating the various velocities for a turbine stage it is not necessary to draw the velocity parallelogram since a velocity triangle, as shown in Fig. 2-6, is sufficient for our purpose.

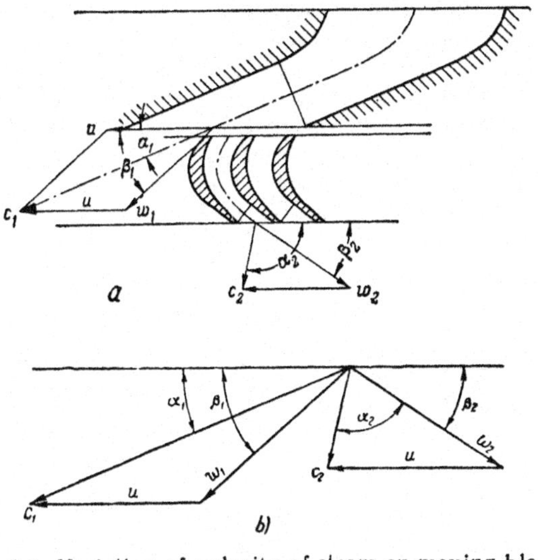

Fig. 2-6. Variation of velocity of steam on moving blades of an impulse stage

[1] In practice to ensure a shockfree entry of steam into the blade passages these blades are bevelled to an angle slightly greater than β_1 on their entry sides (leading edges).

Knowing the change in the velocity of flow of steam in the blades, we can easily determine the change in kinetic energy. A portion of the kinetic energy of steam, during its passage through the blades, is lost in overcoming resistances to its flow.

The loss in kinetic energy per kg of steam is determined as follows:

$$\Delta L_b = \frac{w_1^2 - w_2^2}{2g} \qquad (2\text{-}20)$$

(L_b—kinetic energy loss in moving blades) in mechanical or work units or

$$h_b = A\left(\frac{w_1^2 - w_2^2}{2g}\right) = \frac{w_1^2 - w_2^2}{8{,}378} = (1 - \psi^2)\frac{w_1^2}{8{,}378} =$$

$$= \left(\frac{1}{\psi_2} - 1\right)\frac{w_2^2}{8{,}378} \qquad (2\text{-}20a)$$

in heat units,

where $\frac{w_1^2}{2g}$ — kinetic energy of 1 kg of steam at entry to the blades, in kg m/kg;

$\frac{w_2^2}{2g}$ — kinetic energy of 1 kg of steam at the exit from the blades, in kg m/kg.

The actual amount of kinetic energy of steam per kg which can be used to obtain mechanical work on the blades of an impulse turbine taking into consideration the above losses will be:

$$L_{is} = L_{th} - \Delta L_n - \Delta L_b = \frac{c_{1t}^2 - c_2^2}{2g} - \frac{c_{1t}^2 - c_1^2}{2g} -$$

$$- \frac{w_1^2 - w_2^2}{2g} = \frac{c_1^2 - w_1^2 + w_2^2 - c_2^2}{2g} \text{ [kg m/kg]}. \qquad (2\text{-}21)$$

Energy loss h_b in blades which is caused as a result of the friction and sundry resistances to flow is transformed into heat. Thus while flowing through the blade passages the heat content of steam is increased by an amount equal to h_b. Consequently the actual condition of steam after its passage through the blades is determined by the point A_2 (Fig. 2-2) which is found by setting off the quantity h_b upwards from the point B_1 up to the point B_2 and drawing a horizontal through B_2 until it intersects with the isobar p_1.

2-6. TRANSFORMATION OF ENERGY IN THE MOVING BLADES OF A REACTION STAGE

The available heat drop $h_0 = i_0 - i_{1t}$ is nearly equally distributed between the fixed (guide) and the moving blades of an axial reaction turbine stage, i. e.,

$$h_0 = h_{01} + h'_{02}; \quad h_{02} \approx h'_{02}, \qquad (2\text{-}22)$$

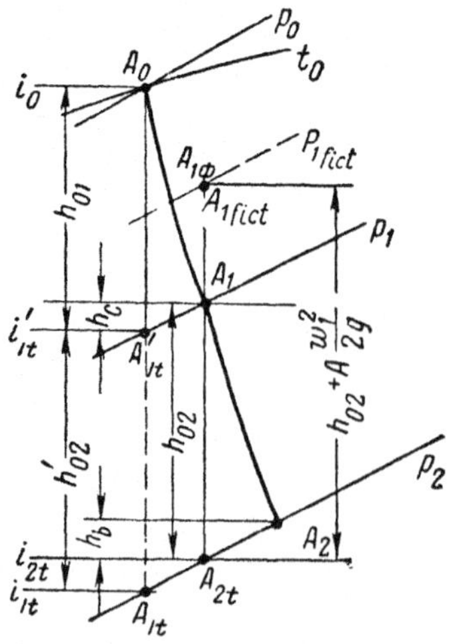

Fig. 2-7. I-s diagram for a reaction stage

where h_{01} and h'_{02}—the available heat drops for the guide and the moving blades on the basis of the adiabatic A_0A_{1t}.

Since there is a loss of heat h_n in the guide blades, the actual available heat drop in the moving blades will be h_{02} instead of h'_{02} (Fig. 2-7). The relation between the heat drop actually utilised for doing work and the available heat drop for a turbine stage is known as the **degree of reaction** and is designated by the letter ϱ,

$$\text{i. e.,} \quad \frac{h_{02}}{h_0} = \varrho, \qquad (2\text{-}23)$$

$$\text{or} \quad h_{02} = \varrho h_0. \qquad (2\text{-}23a)$$

The heat drop utilised in the guide blades, where the pressure drops from p to p_1, is $h_{01} = i_0 - i'_{1t}$. During the expansion of steam in the moving blades its pressure decreases from p_1 to p_2.

The velocity of steam at exit from the guide blades is calculated when $c_0 = 0$ in accordance with the equation (2-7b), and when $c_0 \neq 0$ according to equation (2-7c). Velocities w_1 and c_2, and the angles β_1 and α_2 are determined, as in the case of an impulse stage, either graphically with the help of velocity triangles or analytically (Fig. 2-8).

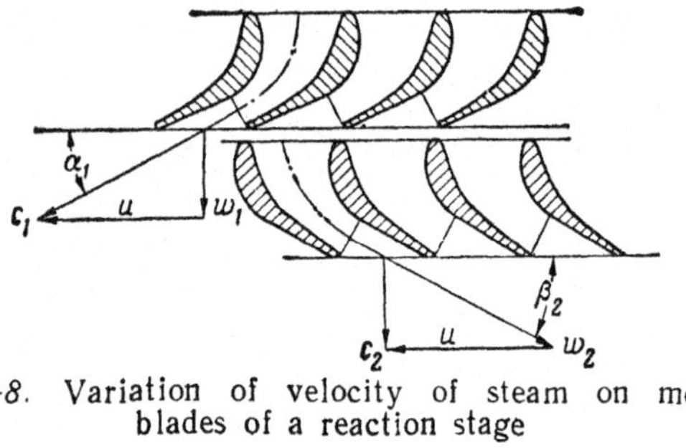

Fig. 2-8. Variation of velocity of steam on moving blades of a reaction stage

The available energy of 1 kg of steam in moving blades, which is equal to $\frac{w_{2t}^2}{2g}$ is determined from the equation

$$\frac{w_{2t}^2}{2g}=\frac{w_1^2}{2g}+\frac{h_{02}}{A}, \qquad (2\text{-}24)$$

where w_{2t}—theoretical relative velocity of steam at the exit from the moving blades without taking into consideration losses on blades. Velocity of steam is found from equations (2-23) and (2-24)

$$w_{2t}=\sqrt{w_1^2+\frac{2g}{A}h_{02}}=\sqrt{w_1^2+8{,}378\,h_{02}}=$$
$$=\sqrt{w_1^2+8{,}378\,\varrho h_0}. \qquad (2\text{-}25)$$

Due to the presence of losses on the moving blades (see Chapters Three and Five) the actual velocity with which steam issues from the moving blade passages will be

$$w_2=\psi w_{2t}=\psi\sqrt{w_1^2+8{,}378\varrho h_0}. \qquad (2\text{-}26)$$

Velocity c_2 and angle α_2 are obtained from the exit velocity triangle where w_2 and u are the relative exit velocity of steam and circumferential velocity of the blades respectively, and β_2 is the exit angle. In general for reaction turbine stages

$$\alpha_1=\beta_2 \text{ and } \alpha_2=\beta_1.$$

Kinetic energy losses per kg of steam in a reaction stage are determined from the equation

$$\Delta L_b=\frac{w_{2t}^2-w_2^2}{2g} \qquad (2\text{-}27)$$

in mechanical units or

$$h_b=A\left(\frac{w_{2t}^2-w_2^2}{2g}\right)=\frac{w_{2t}^2-w_2^2}{8{,}378}=(1-\psi^2)\frac{w_{2t}^2}{8{,}378}=$$
$$=\left(\frac{1}{\psi^2}-1\right)\frac{w_2^2}{8{,}378} \qquad (2\text{-}27a)$$

in heat units.

The actual amount of kinetic energy per kg of steam which may be utilised for doing mechanical work on the moving blades of a reaction stage with losses considered will be

$$L_{rs}=L_{th}-\Delta L_n-\Delta L_b=\left[\frac{c_{1t}^2}{2g}+\frac{w_{2t}^2-w_1^2}{2g}-\frac{c_2^2}{2g}\right]-\frac{c_{1t}^2-c_1^2}{2g}-\frac{w_{2t}^2-w_2^2}{2g}=\frac{c_1^2-w_1^2+w_2^2-c_2^2}{2g}$$
$$[\text{kg m/kg}] \qquad (2\text{-}28)$$

The expression obtained above coincides with the one obtained for an impulse turbine stage taking the losses into consideration(2-21) and consequently becomes a general expression for any stage with any degree of reaction.

The process of heat drop on an i-s diagram for a reaction stage is obtained in the following way (Fig. 2-7). From the point A_0, where steam is at the initial conditions of p_0 and t_0 an adiabatic is dropped until it intersects the final pressure line p_2 (after the moving blades). Heat drop $A_0A'_{1t}$ is used up in the guide blades. Point A'_{1t} indicates the steam conditions after the guide blades (without considering losses). If losses are taken into consideration the steam conditions after the guide blades will be determined by the point A_1; losses in the guide blades will be found using equation (2-8a).

At the point A_1 steam flowing with a relative velocity of w_1 has a kinetic energy equal to $A\frac{w_1^2}{2g}$ [kcal/kg].Marking off this quantity from the point A_1 we obtain the point A_{1f}, which gives the fictitious steam conditions before the moving blades, known as the stagnation conditions. If steam flowing with a relative velocity of w_1, at the steam conditions determined by the point A_1, is adiabatically brought to a standstill then the various steam parameters p, v and t will increase to those values which are stipulated by the point A_{1f}. The steam properties at the point A_{1f}, conforming to the pressure p_{1f}, specific volume v_{1f}, and temperature of steam t_{1f} are commonly known as the stagnation parameters. The available heat drop for conversion into work is given by the quantity $A_{1f}A_{2t}$. However, it may be noticed that the adiabatic A_1A_{2t} is slightly greater than $A_{1t}A_{1t}$ (shown by dotted line), the difference being very little it may be neglected.

In the present-day practice even for impulse turbines it is usual to give some degree of reaction (usually from $\varrho=0.05$ and higher). These turbines occupy an intermediate place with regard to pure impulse and pure reaction turbines. The conversion of energy in the moving blades of a stage with any degree of reaction takes place according to the same principle as in the case of the so-called reaction stages (with $\varrho=0.5$), with the difference that the major portion of energy conversion, in the case of an impulse stage with a small degree of reaction, takes place in the nozzle or fixed blades.

All the above derived relations for reaction stages are also applicable in full for a stage with any degree of reaction.

Representation of the heat drop process for a stage, with a small degree of reaction on the i-s diagram, will be dealt with in Chapter Seven.

Chapter Three

CONCEPTS FROM GAS DYNAMICS

3-1. GENERAL CONCEPTS OF FLUID[1] FLOW IN BLADE PASSAGES

Euler's theory of streamlined fluid flow forms the basis for the calculation of all turbine parts where the flow of steam comes into the picture. This theory is based on the assumption that the fluid flow is streamlined and that all the flow parameters vary only in one direction. Fluid flow in which the fluid properties intensively vary only in one direction with hardly any noticeable variations in the other two directions may be considered as one dimensional flow. The flow of fluid along a streamline conforms to the above conditions. A streamline is a line, chosen from the general flow, at every point of which the velocity vector is in a direction tangential to it. In gas dynamics, besides streamlines, flow of elementary stream tubes is also considered. The stream tube is that part of the fluid which is bounded by streamlines. Thus from the point of view of Euler's theory fluid flow must consist of a finite number of stream tubes. One of the biggest advantages of Euler's stream flow theory is its simplicity.

The application of the stream theory makes it possible to determine with ease the important flow parameters in a turbine stage, e. g., the mean direction of flow after the blades, its velocity, the turning moment exerted on the blades of a turbine stage, as also the work done by steam (gas) in each stage and the turbine as a whole. Because of its simplicity this theory has found wide usage with almost all turbine designers from the very beginning of turbine building and is used to this day for the theoretical calculations both for steam turbines and for compressors. However, the actual conditions of steam flow in a turbine do not conform to the stream theory in its entirity. The flow of steam in a turbine stage is in reality quite complex and has a three dimensional character.

The nature of steam flow in a turbine stage depends upon the shape of the blade passages, height of blades, conditions of flow at entry to the blade passages, etc. The interblade passages in a turbine stage are formed by the surfaces of adjacent blades mounted on the stator and the rotor. The flow of steam through the blades is accompanied by some loss in its velocity. The greater this loss, the lower is the efficiency of each stage and turbine as a whole. The magnitude of energy loss in a turbine stage basically depends on the geometric properties of the fixed and the moving blades, stage characteristic u/c_0 and the non-dimensional factors Re and M. Experimental investigations show that the nature of flow at entry to the blade passages has a good deal of influence on the losses of energy occurring in a turbine stage, e. g., nonuniformity of flow, dispersion of flow along the height and the pitch, interference of shrouding, etc.

3-2. GEOMETRICAL PROPERTIES OF BLADE PROFILES

In a steam turbine the fixed and the moving blades are arranged in such a way as to form endless cascades. Blade cascades are made up of an infinite number of similar blade profiles situated equidistantly from each other. The distance between two points situated at exactly the same place on each of two adjoining profiles is known as the pitch of the cascade and is designated by the letter t (Fig. 3-1). For two-dimensional cascades the pitch of the cascade remains constant along the height of the blades. In the case of circular cascades the blade pitch increases with the blade height. The dimension a_1 shows the width of the interblade passage at exit. Width of the blade b between the points k and n on the centreline of the profile (dotted line in Fig. 3-2) is known as the chord of the profile; it appears as one of the most important characteristic dimensions both for fixed and rotating blade profiles. In the present-day design practice it is usual to take the distance between the tangents drawn at the leading and trailing edges as the chord b (Fig. 3-1). Since, as a rule, the present-day blade profiles have a curvilinear shape, it is customary to take the angle formed by the tangential line *1* at the point k to the blade profile, and the line *2* drawn

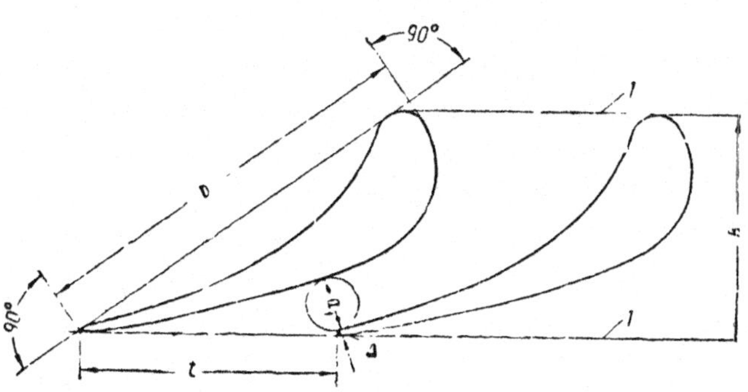

Fig. 3-1. Guide blade profile with nozzle shaped passage
1, 1— planes perpendicular to the turbine axis

[1] Fluid is considered here in its widest meaning; it may be a liquid, air, steam or any gas.

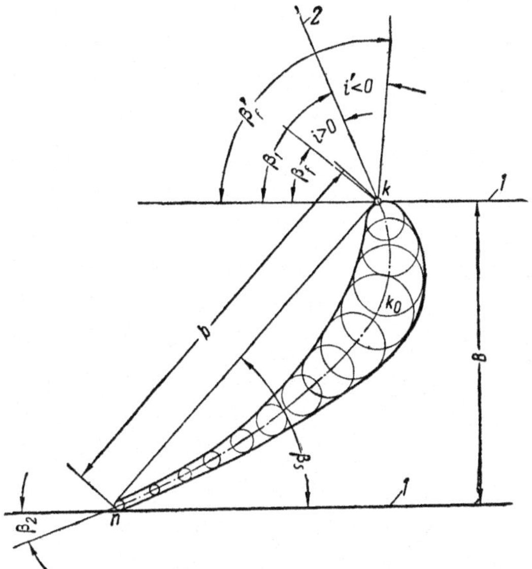

Fig. 3-2. Guide blade profile with angle of setting
1— plane perpendicular to the turbine axis; *2*— tangent to curve nk_0k at point k

tangential to the chord kk_0n at the point k as β_1, the theoretical angle of flow (Fig. 3-2). If the flow is directed parallel to this tangential line at entry to the blade passages then it is assumed to be shockfree. Any deviation of flow from this tangential line will not conform to the design conditions and is consequently accompanied by additional losses. The angle between the direction of flow and the tangential line *2* is known as the attack angle i. When the angle β_f of the flow is less than the theoretical angle β_1 then the angle of attack i is positive $i=\beta_1-\beta_f>0$ and conversely if $\beta_f'>\beta_1$ the angle of attack i' will have a negative value, i. e.,

$$i'=\beta_1-\beta_f'<0.$$

Experimental investigations show that a decrease in the angle of flow leads to a more intensive growth of losses than when it is increased. The angle of attack influences the losses to a greater extent in impulse blade cascades than in reaction blade cascades. Flow of steam or gas which is uniform at entry to impulse or reaction blade cascades meets with the leading blade tips in any turbine stage, the presence of which establishes further loss of energy. The magnitude of these losses depends upon the angle of attack at entry to the blade passages, their shape, etc. Blades of constant section along the height and constant angle of blade setting β_s are known as constant profile blades (Fig. 3-2). Blades of varying sections along their height but with a constant angle of blade setting are known as blades of varying profiles. Blades of varying sections as well as varying angles of blade setting along their height are known as twisted blades.

For the generalisation of experimental results obtained from the aerodynamic investigations of blade profiles two non-dimensional geometric factors are employed: relative blade pitch $\bar{t}=t/b$ and relative blade height $\bar{l}=l/b$. It is found from experimental investigations that, for a given blade profile, the efficiency and energy losses depend on $\bar{t}$, β_s, $\bar{l}$ and many other factors, where β_s—angle of blade setting.

3-3. FORCE EXERTED BY FLOW OF STEAM OR GAS (FLUID) ON BLADES

Let us consider a two-dimensional laminar flow of fluid through a blade cascade consisting of an infinite number of blade profiles, and try to examine the force interaction of the flow and the blade profiles (Fig. 3-3). We shall assume that at some distance, section *1-2*, from the blade cascade the leading blade tips do not influence the flow, so that we can assume the velocity of flow at this section to be one and the same for all profiles, equal to c_1 for guide blade profiles (Fig. 3-3,*a*) and w_1 for moving blades (Fig. 3-3, *b*). And further, it will be assumed that the flow is directed at an angle equal to α_1 and β_1 for the two cascades of guide and moving blades. After flowing through the blade cascades, at some distance from the blades, section *3-4*, the flow again reverts to plane laminar flow with velocity and direction given by c_2 and α_2 for guide blades (nozzles) and w_2 and β_2 for moving blades. Let the static pressure of flow at section *1-2* be equal to p_1, and at section *3-4* be equal to p_2. We shall separate out an elementary portion of the flow of width t_n, and t_b, the blade pitch, for the guide and moving blades and height equal to unity, to consider the forces acting on each of the blade profiles.

Thus, this isolated elementary flow stream flows through a single-blade passage of height equal to unity. To simplify the solution of the problem it is assumed that the flow is bounded by impervious streamlines along *1-1'-3* and *2-2'-4*. With the above assumptions and considering a steady flow it can be assumed that the mass flow through any cross-section of this elementary flow stream remains constant. The force exerted by the elementary stream on the blade will equal the force received by a blade of unit height.

In conformity with the assumed system of coordinates u, z for cascades of moving blades let us consider the interaction of flow with the blades in the two directions u and z. We shall assume that u, conforming to axial turbines, is the linear velocity of rotation of the blades. Force exerted

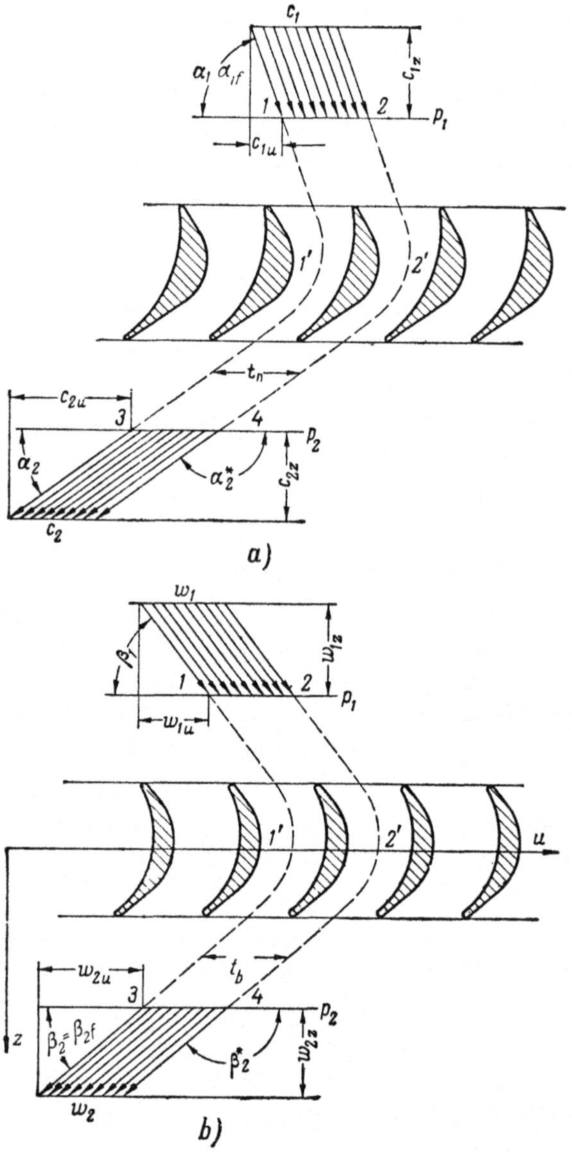

Fig. 3-3. Flow pattern through cascades
a — nozzle blade cascade; *b* — moving blade cascade

in the direction of the u axis will create useful turning moment on the turbine shaft. At the same time the force exerted on the blades in the direction of the z axis is transmitted as thrust on to the bearings through the turbine disc and the shaft. The nozzles and guide blades are stationary and therefore the forces exerted on them lead to stresses at the places where they are fastened to the turbine body.

The vector of the total force acting in the direction of flow will be equal to the vector sum of

a) for guide blades

$$\vec{P}_{on}=\frac{G}{g}(\vec{c}_1-\vec{c}_2)+t_n(\vec{p}_1-\vec{p}_2); \qquad (3\text{-}1)$$

b) for moving blades

$$\vec{P}_{ob}=\frac{G}{g}(\vec{w}_1-\vec{w}_2)+t_b(\vec{p}_1-\vec{p}_2), \qquad (3\text{-}2)$$

where $\frac{G}{g}(\vec{c}_1-\vec{c}_2)$ — the change in momentum while flowing through a guide blade passage of unit height;

G — the mass flow of steam (gas) through a guide blade passage of unit height;

$t_n(\vec{p}_1-\vec{p}_2)$ — vector of the force caused as a result of the fall of static pressure of steam or gas while flowing through the blades of unit height;

$\frac{G}{g}(\vec{w}_1-\vec{w}_2)$ — change in momentum while flowing through a moving blade passage of unit height;

$t_b(\vec{p}_1-\vec{p}_2)$ — vector of the force caused as a result of the fall in static pressure of steam or gas while flowing through a moving blade passage of unit height.

To determine the force exerted on the blades in the directions u and z we shall make use of the momentum equation for the flow of steam through the blade passages.

a) Projecting all the forces acting in a guide blade system on to the principal axes u and z we have

1. On the u axis

$$\frac{G}{g}(c_{1u}-c_{2u})-P_{un}=0, \qquad (3\text{-}3)$$

where $\frac{G}{g}(c_{1u}-c_{2u})$ — the change in the momentum of flow in the direction of the u axis while flowing through a blade passage of unit height;

P_{un} — force exerted by the blades on the flow in the direction of the u axis.

From equation (3-3) we obtain

$$P_{un}=\frac{G}{g}(c_{1u}-c_{2u}). \qquad (3\text{-}4)$$

2. On the z axis

$$\frac{G}{g}(c_{1z}-c_{2z})+t_n(p_1-p_2)-P_z=0, \qquad (3\text{-}5)$$

where $\frac{G}{g}(c_{1z}-c_{2z})$ — change in the flow momentum in the direction of the $\pm z$ axis while passing through a blade passage of unit height,

P_z — force exerted by the blade on the flow in the direction of the $\mp z$ axis.

From equation (3-5) this force will be equal to

$$P_z=\frac{G}{g}(c_{1z}-c_{2z})+t_n(p_1-p_2). \qquad (3\text{-}6)$$

b) Projecting all the forces acting in a moving blade system on to the axes u and z we have

1. On the u axis

$$\frac{G}{g}(w_{1u}-w_{2u})-P_u=0, \qquad (3\text{-}7)$$

where $\frac{G}{g}(w_{1u}-w_{2u})$ — change in the momentum of flow of the fluid while passing through an inter-blade passage of unit height for a moving blade system in the direction of the u axis,

P_u — force exerted by the blade in a direction opposite to the u axis.

From equation (3-7) we have

$$P_u=\frac{G}{g}(w_{1u}-w_{2u}). \qquad (3\text{-}8)$$

2. On the z axis

$$\frac{G}{g}(w_{1z}-w_{2z})+t_b(p_1-p_2)-P_z=0, \qquad (3\text{-}9)$$

where $\frac{G}{g}(w_{1z}-w_{2z})$ — change in the momentum of flow while passing through a moving blade of unit height in the direction of the $\pm z$ axis,

P_z — force exerted by the blade on the flow in the direction of the $\mp z$ axis.

Force P_z is determined from the equation (3-9):

$$P_z=\frac{G}{g}(w_{1z}-w_{2z})+t_b(p_1-p_2). \qquad (3\text{-}10)$$

The work done by steam on the blades of a turbine stage may be determined (Fig. 3-4, *a*) as follows.

Let the circumferential velocity of rotation of the blades be equal to u at the mean diameter of the turbine disc. We shall assume that the velocities c_1, c_2, w_1 and w_2 of steam are known. From Fig. 3-4, *b* it is clear that

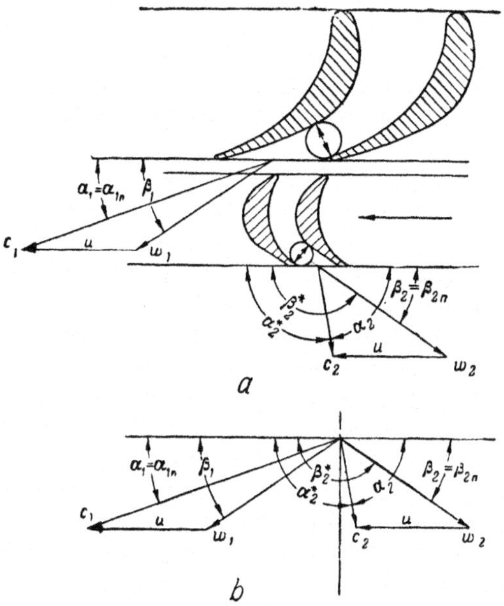

Fig. 3-4. Section through a turbine stage
a – turbine stage, *b* – velocity triangles

$$\bar{c}_1=\bar{w}_1+u \text{ and } \bar{c}_2=\bar{w}_2+u.$$

Projecting these vectors on to the u axis we find that

$$c_{1u}=w_{1u}+u \text{ and } c_{2u}=w_{2u}+u.$$

Whence

$$c_{1u}-c_{2u}=w_{1u}-w_{2u}. \qquad (3\text{-}11)$$

From equations (3-7) and (3-11) we obtain

$$P_u=\frac{G}{g}(c_{1u}-c_{2u})=\frac{G}{g}(w_{1u}-w_{2u}). \qquad (3\text{-}12)$$

The turning moment exerted by steam on the moving blades of a turbine stage is determined from the equation

$$M_u=P_u\nu z=\frac{G_0}{g}\nu(c_{1u}-c_{2u})=\frac{G_0}{g}\nu(w_{1u}-w_{2u}), \qquad (3\text{-}13)$$

where ν — the mean radius of the rotor disc carrying the moving blades,

G_0 — mass flow of steam per sec through a turbine stage equal to $G_0=Gl_bz$ (l_b — height of the blades),

z — number of moving blades per stage.

The work done by steam at the rim of the disc will be

$$L_u=M_u\omega \text{ [kg m/sec]}, \qquad (3\text{-}14)$$

where ω — angular velocity of the blades.

The power developed is

$$N_u = \frac{M_u \omega}{102} \text{ [kW]}. \quad (3\text{-}15)$$

Substituting in equation (3-15) the value of M_u obtained from equation (3-13) we have, since $r\omega = u$:

$$N_u = \frac{G_0 u}{102 g}(c_{1u} - c_{2u}) = \frac{G_0 u}{102 g}(w_{1u} - w_{2u}) \text{ [kW]}. \quad (3\text{-}16)$$

If the inflow and outflow of steam take place at different mean diameters for the moving blades then the power developed at the rim of the turbine disc will be

$$N_u = \frac{G_0}{102 g}(u_1 c_{1u} - u_2 c_{2u}) = \frac{G_0}{102 g}(u_1 w_{1u} - u_2 w_{2u}) \text{ [kW]}, \quad (3\text{-}17)$$

where u_1 and u_2 are the circumferential velocities of the blades at entry and exit.

The mass flow of steam through a turbine stage is usually a known quantity (see Chapter Five). Velocities u_1 and u_2 are obtained from the constructional details (See Chapter Twelve). Velocities of steam w_1 and c_2 as well as the angles β_1 and α_2 are determined from the velocity triangles or by analytical methods (See Chapter Two). Thus all the quantities contained in equation (3-17), excepting c_{1u} and w_{2u}, may be determined by simple calculations. To find the values of c_{1u} and w_{2u} it is necessary to know velocities c_1 and w_2 and the angles α_1 and β_2 at the exit of the fixed and the moving blades. To determine the values of c_1 and c_{1t} according to equation (2-7b) or (2-7c) it is necessary to know the value of φ and similarly the value of ψ must be known for the determination of w_2 from equation (2-19) or (2-26).

It is found from experience that the angles α_1 and β_2, the exit angles for the guide and moving blades respectively, in actual practice do not conform to the calculated values. So as to obtain reliable data for design purposes these four quantities—c_1, w_2, α_1 and β_2—have to be found experimentally.

3-4. EXPERIMENTS ON TURBINE BLADES

The flow of steam through a blade cascade, as mentioned above, is exceedingly complicated. It depends upon the flow conditions at entry to the blades, the shape of the blades, the non-dimensional geometric factors $\bar{l}$ and $\bar{t}$, the inlet angle of steam, etc. The main aim of experimental investigation of turbine blades is to study the various physical properties of flow of steam while passing through blades, the determination of coefficient of mass flow[1] μ, velocity coefficients φ and ψ as well as the angles α_1 and β_2.

Aerodynamic investigations of blade profiles not only enable us to determine the quantities μ, φ, ψ, α_1 and β_2 for a given blade arrangement, but also facilitate the construction of efficient blade profiles both for fixed and moving blades.

[1] Coefficient of mass flow is the name given to the value obtained from the relation between the actual mass flow through a blade passage or cascade and the theoretically calculated mass flow.

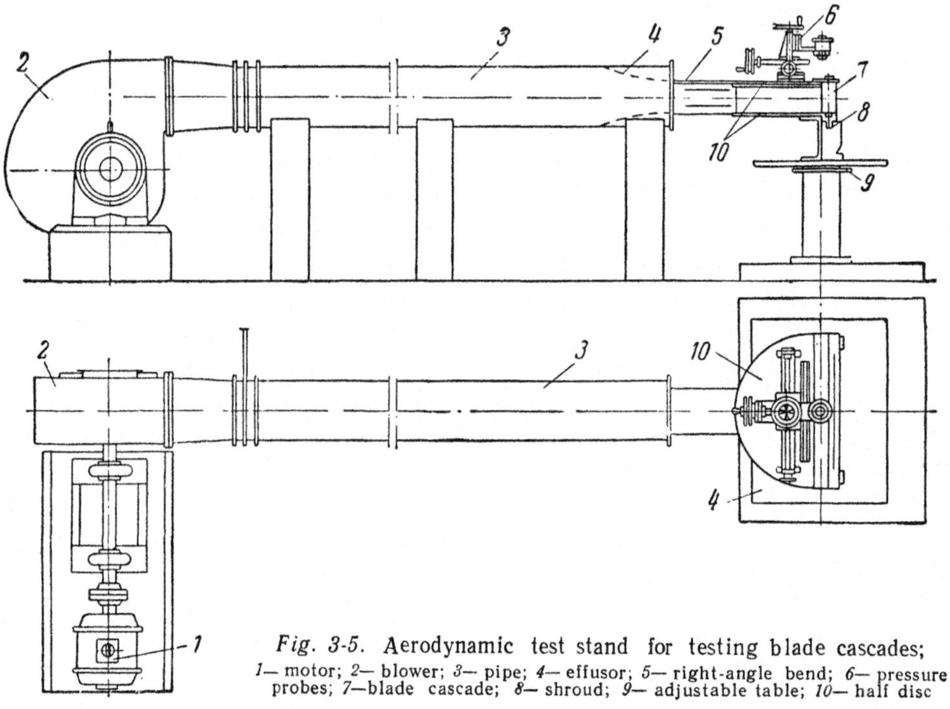

Fig. 3-5. Aerodynamic test stand for testing blade cascades;
1— motor; *2*— blower; *3*— pipe; *4*— effusor; *5*— right-angle bend; *6*— pressure probes; *7*—blade cascade; *8*— shroud; *9*— adjustable table; *10*— half disc

The efficiency of a blade profile depends on the energy loss of the steam flowing through it. The smaller the energy loss, the higher is the efficiency of the blade profile. If flow takes place without separation from the blade walls then the value of μ and the velocity coefficients φ and ψ will be higher with a consequent decrease in loss of energy resulting in higher efficiencies.

Experimental investigations on turbine blade profiles are carried out on static or dynamic models or on actual turbines. Static investigations of turbine blades, as a rule, are carried out with the help of compressed air. Investigations on dynamic models or turbines under normal conditions may be carried out either with air or superheated steam.

The general arrangement of the experimental set-up consists of a compressed air source (usually a blower or turbo-compressor), a receiver for damping fluctuations in air pressure, connecting pipes delivering compressed air to the blade cascades, the blade cascades and the various measuring instruments for observing the flow properties. Fig. 3-5 shows one such arrangement in use at the Bezhitsa Institute of Heavy Machine Building for the investigation of blade profiles. Air from blower *2* is led through receiver *3* and effuser *4*, made up of special aerodynamic profiles, from where it enters the rectangular tube *5* and ultimately blade cascade *7*. To obtain a uniform flow of air for the investigation of blade profiles tube *3* is provided with special types of filters and aerodynamic profiles forming the effusor part of the experimental set-up. The effusor not only helps in damping the fluctuations of pressure, but also reduces the boundary layer thickness[1]. The air flow before the blades must be uniform so as to enable the proper and accurate determination of profile losses. In addition to the above given arrangement for the damping of pressure variations it is usual practice to arrange for the reduction of boundary layer thickness by suction at the walls of the tube carrying air supply to the blade cascades (after the effusor). When investigating the effect of non-uniformity of flow before the blades on the magnitude of losses, it is customary to artificially create the required degree of turbulence of air by special methods.

Fig. 3-6 shows the experimental stand for the investigation of blade cascades under static conditions. Air from the compressor is led to the receiver *1* where it is damped. Screen *2* effects a further uniformity of flow. Movable fixtures *4* and *5* permit the variation of direction of flow of air so as to enable the determination of the effect of different attack angles i on the magnitude of profile losses. Measurements of the various flow parameters are carried out after ensuring a uniform air flow to the blade cascade for a particular given condition. Static and dynamic pressure heads in the air pipe are measured at a distance of about 200 mm from the blade cascade. If the flow before the blades is uniform then it is sufficient to measure the pressure and velocity at a single point in the flow. The pressure distribution after the blade cascade depends upon the distance of the section, where the pressures are measured, from the exit of the blade cascade. Near the exit of the blades the pressure field is very uneven, however, as the distance from the exit increases the pressure gradients even out. The trailing edge of each blade creates a wake in which the velocities are considerably lower than in the flow stream. The flow in the trailing edge wake is accompanied by a considerable amount of turbulence and vortices which leads to trailing edge losses or 'wake losses'. The last mentioned losses depend on the thickness of the trailing edge. With the increase in trailing edge thickness the wake losses increase. Experimental results show

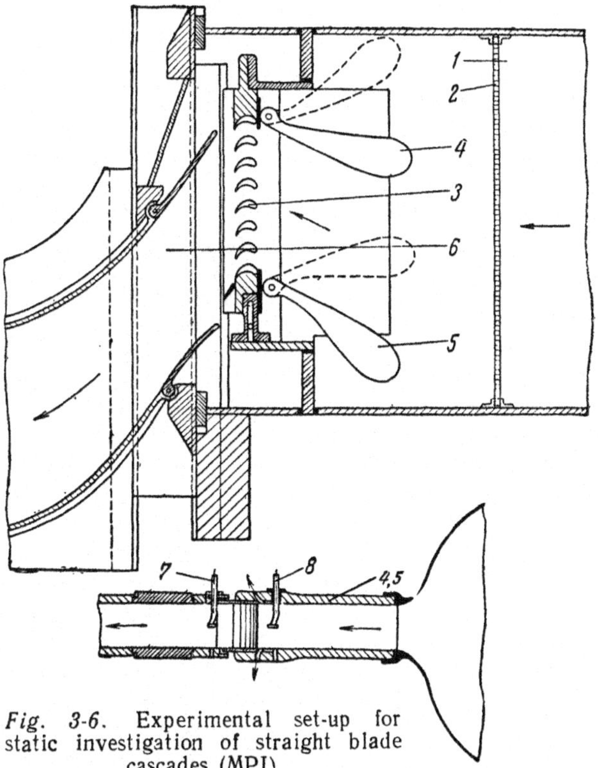

Fig. 3-6. Experimental set-up for static investigation of straight blade cascades (MPI)

1— receiver; *2*— damping net; *3*— blade cascade; *4* and *5*— adjustable fixtures; *6*— diffusor; *7* and *8*— direction and total pressure probes

[1] The layer of air or gas next to the tube wall where the velocity varies from zero at the wall to the full value of free stream velocity is known as the "boundary layer".

that the wake losses are proportional to the relation between the trailing edge thickness Δ and the width of the blade passage a_1 (Fig. 3-1) and further depend on the shape of blade profile, the attack angle, relative blade pitch, etc.

The velocity of flow in a blade pitch between two adjoining blade profiles varies from its minimum value in the trailing edge wake to its maximum at the centre. The direction of flow changes as well. Depending upon the relative blade height l the non-uniformity of flow after the blade cascade varies not only along the pitch, but also along the height of the blades. With the increase in distance from the blade exit the air stream leaving the cascade becomes more uniform in character which is accompanied by further loss of energy.

To exclude or at least to minimise the effect of trailing edge losses, while carrying out experimental investigations for the determination of profile losses, the relative blade height is made sufficiently large ($\bar{l} \geqslant 3\text{-}4$). Further it is necessary to choose correctly the section at which the pressure heads and the direction of flow are to be observed. Observations of the direction of flow and the pressure heads are carried out at the mean height of the blade passages, since measurements of these quantities at the boundaries may be affected by the turbulence of the trailing edge wake where secondary flows are established. It is found from experience that for obtaining reliable values of c_1, w_2, α_1 and β_2 the number of blades used in a cascade should not be less than seven.

When utilising a medium other than the one in actual practice for aerodynamic investigations of blade cascades use is made of the theory of similitude. Geometric similarity of profiles has to be maintained as also similarities of non-dimensional factors like M (Maiyevsky number), Re (Reynolds number) and adiabatic index.

For any fluid the criterion

$$M = \frac{c}{c_s}, \tag{a}$$

where c—velocity of flow on the blade profiles;
c_s— velocity of sound in the medium under consideration.

Velocity c_s depends on the properties of the medium and is determined from the equation

$$c_s = \sqrt{kgpv} \approx \sqrt{kgRT}, \tag{b}$$

where $g = 9.81$ m/sec—the acceleration due to gravity;
k—the adiabatic index for the given fluid (steam, gas, air, etc.);
p and v—the pressure and specific volume of the medium respectively;
R—gas constant;
T—absolute temperature of the flow, °K.

While designing the stream flow in an experimental blade cascade the dimensional factors must be maintained in accordance with the law of similarity, i. e., $M_m = M_p$ (M_m—Maiyevsky number for the blade model and M_p—Maiyevsky number for the prototype).

Equation (b) enables the determination of c_{sm} the velocity of sound for the medium in which the blades are being tested and c_{sp}, the velocity of sound for the actual fluid in use in the prototype.

On the basis of equality of the Maiyevsky numbers we may write

$$c_m = c_p \frac{c_{sm}}{c_{sp}}, \tag{c}$$

where c_m—velocity of flow in the model;
c_p—velocity of flow of gas or steam in the prototype;
c_{sm}—velocity of sound for the medium in which the blades are being tested in the model;
c_{sp}—velocity of sound for the medium in actual use in the prototype.

To satisfy the condition of similarity of flow of the fluid medium on the blades of the model as well as those of the prototype the non-dimensional factor Re must be maintained constant, i. e.,

$$\mathrm{Re}_m = \mathrm{Re}_p, \tag{d}$$

where Re_m—Reynolds number for the blades of the model;
Re_p—Reynolds number for the blades of the prototype.

For steam turbines the equation (d) may be written as

$$\frac{\varrho_m c_m b_m}{\mu_m} = \frac{\varrho_p c_p b_p}{\mu_p}, \tag{e}$$

where ϱ_m and ϱ_p—densities of the mediums in which the blades are placed for the model and the prototype respectively, kg sec²/m⁴;
b_m and b_p—characteristic linear dimensions conforming to the blades of the model and the prototype;
μ_m and μ_p—coefficients of dynamic viscosity for the medium in which the blades are placed in the model (usually air or steam) and the prototype (steam or gas), kg sec/m².

The characteristic dimension for turbine blades may be taken as equal to b (Fig. 3-1).

From equation (e)

$$b_m = b_p \frac{\varrho_p c_p \mu_m}{\varrho_m c_m \mu_p}, \tag{f}$$

and the scale for the model

$$\lambda = \frac{b_m}{b_p}. \qquad (j)$$

The criterion Re for models and prototypes working in the region where C_d, the drag coefficient, is independent of the value of Re is usually not maintained the same, as this does not affect the results obtained.

As has been observed before, for the determination of profile losses blades of reasonably large relative heights are made use of, both for guide and moving blade profiles ($\bar{l} \geqslant$ 3-4). The total pressure head and the direction of flow are measured both along the pitch and the blade height, as a rule, at the midsection and the mean blade height. For blades of large relative heights wake losses have hardly any effect on the total losses in a cascade, thus permitting the measurements of total pressure head and the direction of flow at the midsection of the canal.

In cascades of smaller relative blade heights ($\bar{l} \leqslant$ 1 to 1.5) the effect of wake losses is keenly felt. The total pressure head and the direction of flow after the cascade vary to a great extent not only along the pitch but also along the height of the blade passages. Hence in the case of cascades with smaller relative blade heights measurements of pressures and directions of flow are carried out at several sections along the height of the blade passage (Fig. 3-7). This figure shows the arrangement of the blade models along with the connecting piping. The pipe *1* serves both as an air lead

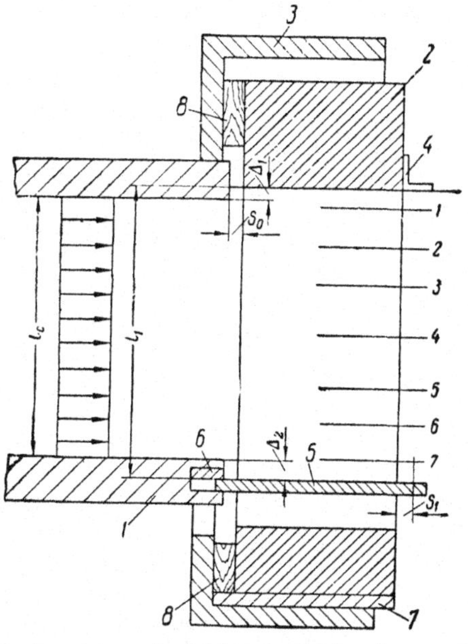

Fig. 3-7. Arrangement of blade models along with connecting piping

1— air-lead; *2*— blade cascade; *3*— box holding blades; *4*— cap; *5*— plate with slots for blades; *6*— liner for fixing plate *5*; *7*— liner for varying Δ_1; *8*— wooden packing

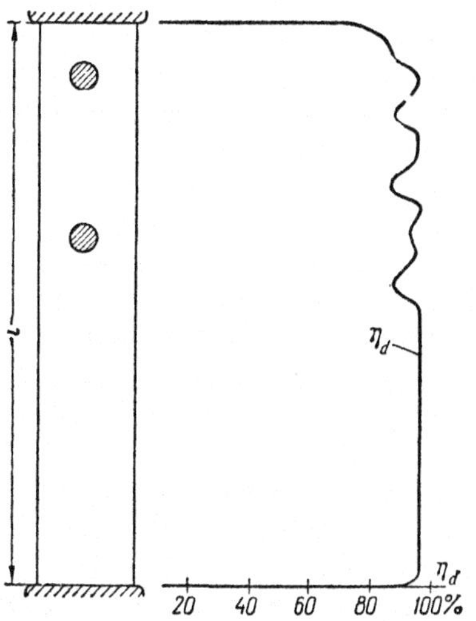

Fig. 3-8. Variation of efficiency along the blade height with fasteners

and as a damper where the pressure fluctuations are smoothed out before the air enters the blade cascade *2*. There are seven cross-sections shown in this figure for the given blade height. The larger the number of sections at which the fluid properties are measured, the greater will be the accuracy with which losses can be determined for blades of small relative heights.

Experimental results show that the nature of flow before the blades does influence the losses occurring in a blade cascade. Both non-uniformity of flow and the presence of shroudings Δ_1 and Δ_2 influence the growth of losses in the blade passages. For the arrangement shown in Fig. 3-7 it is possible to vary the thickness of the top and bottom shrouding by varying the thickness of the plates *5* and *6*, thus enabling the study of the influence of shroudings Δ_1 and Δ_2 on the losses in a blade passage.

Experimental investigations show that along with the wake losses also exist losses of energy due to the disturbance of flow caused by the presence of fastenings which join the blades into one cascade. Depending on the height of blades and the number of such fastenings losses due to disturbance of flow may lead to an appreciable reduction of efficiency. The effect of the presence of such fastenings on the efficiency of a moving blade cascade is shown in Fig. 3-8 according to the data given by V.T.I.[1] Similar investigations for determining the effect of fastenings on the efficiency of a blade passage were carried out at the Ts.K.T.I. (The Central Boiler and Turbine Research Institute). Use of such fastenings for

[1] Lagun, V. P., "Experimental Investigation of Reaction Blades", Tech. Memorandum, 1957. V.T.I. (Power Institute of the U.S.S.R.)

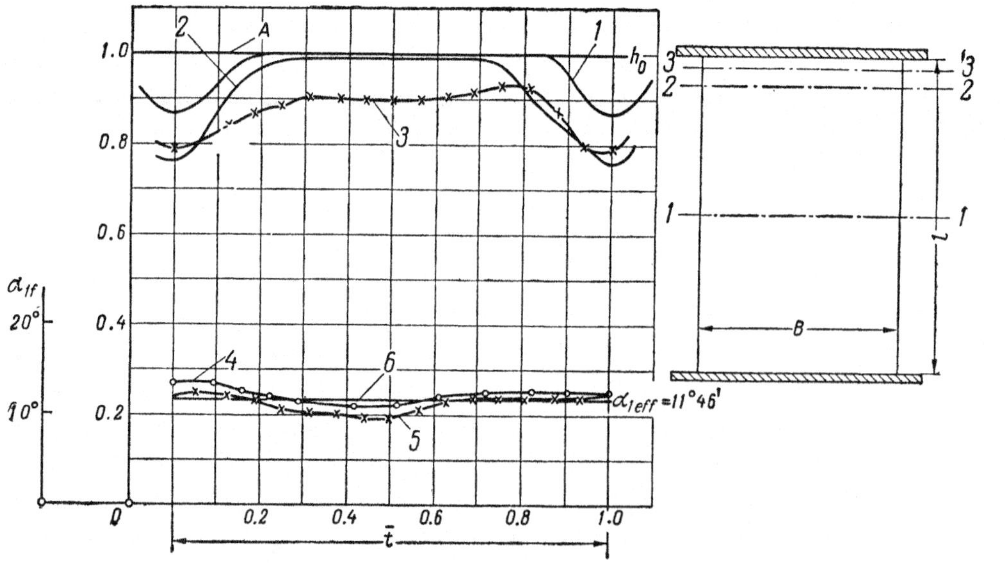

Fig. 3-9. Variation of total pressure and flow direction along the pitch and height of guide blade cascade

rigidity always leads to additional losses, lowering of efficiency and to the increase in non-uniformity of flow at entry to the next row of blades. As an illustration the results of experimental investigation of blade cascades are shown in Fig. 3-9 for guide blades and Fig. 3-10 for moving blades. Fig. 3-9 shows the results of total pressure heads along the pitch for a guide blade cascade of small relative blade height $\bar{l}=0.82$ and angle[1] $\alpha_{1eff}=10°46'$. Line *A* shows the total pressure before the blades equal to h_0. Curve *1* gives the change in total pressure along the pitch after the blades for section *1-1*. The mean direction of flow after the blades is given by an angle 11° 58′ at section *1-1*, i.e., 12′ more than the value of α_{1eff}. Lines *2* and *3* show the variation of total pressures after the blade passages for sections *2-2* and *3-3* along the pitch. Comparing the curves *1*, *2* and *3* it follows that the variation of total pressure head diminishes for sections nearer to the roof and floor of the blade passages. Decrease in total pressure in the blades is explained by the increase in losses at the blade passage boundaries (roof and floor of the blade passage). These losses are known as end losses. Lines *4* and *5* show the variation of the direction of flow along the pitch for various sections along the blade height. Thus, for example, the mean value of the angle of flow α_{1f} for section *2-2* is equal to 12°17′ and for section *3-3* 11°14′. Comparing the mean experimental values of the angle at which the flow takes place and taking their mean we find that this angle for all practical purposes does not differ much from the angle α_{1eff} (line *6*). Hence for all practical calculations of a turbine stage the direction of steam leaving the blade passage may be taken equal to the angle α_{1eff}. As a general rule we may assume that

$$\alpha_{1f} \approx \alpha_1 = \alpha_{1\,eff} = \arc\sin\left(\frac{a_1}{t}\right). \qquad (3\text{-}18)$$

Equation 3-18 enables the determination of $a_1 = t \sin\alpha_1$ knowing the values of α_1 and t. Fig. 3-10 shows similar results for a blade cascade made up of moving blade profiles with a relative blade height of $\bar{l}=1.68$. Line *A* conforms to the total pressure heads before the blades, h_0. Line *1* shows the total pressure variations after the blades at section *1-1*, and line *2* stands for the pressures at section *2-2*, all being measured along the pitch of the cascade. Losses of total pressure heads for the curve *2* are greater than those for curve *1*. The increase in losses at the roof and floor of the cascade, as stated before, is due to the influence of tip losses (wake losses). Lines *3* and *4* show the variation in the direction of flow along the pitch for sections *1-1* and *2-2*. The direction of flow leaving the blades given by β_{2f} for relatively long blades does not appreciably differ from the effective angle of exit β_{2eff}. However, for blades of small relative blade height the difference between β_{2f} and β_{2eff} could reach values of

[1] The sine of the effective angle of outflow is determined from the relation

$$\sin\alpha_{1\,eff} = \frac{a_1}{t}$$

and

$$\alpha_{1\,eff} = \arc\sin\left(\frac{a_1}{t}\right).$$

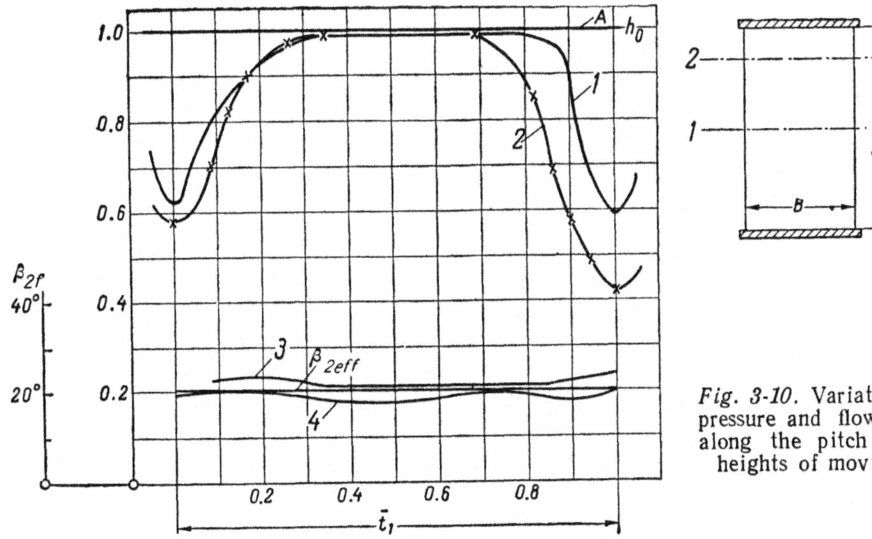

Fig. 3-10. Variation of total pressure and flow directions along the pitch for various heights of moving blades

appreciable importance. For the calculations of a turbine stage it is usual to take

$$\beta_{2f} \approx \beta_2 = \beta_{2eff} = \arc\sin\left(\frac{a_2}{t_1}\right), \qquad (3\text{-}19)$$

where a_2—width of the exit blade passage;
t_1—pitch of the moving blades;
β_2—direction of flow of steam after the blades.

The width of the exit blade passage is determined according to equation (3-19)

$$a_2 = t_1 \sin\beta_2.$$

3-5. CALCULATIONS OF TEST RESULTS

During the static investigation of turbine blade cascades the following measurements are made, on the basis of which all calculations of stage parameters are carried out.

The total pressure head before the blades is measured in millimetres of water, mercury or any other liquid. In the case of uniform flow field before the blades the total pressure head may be measured at a single point, however, if the flow is non-uniform measurements of total pressure heads have to be carried out at several points.

Beyond the blade cascade local dynamic pressure heads are measured along the pitch of the blades at the midsection of the blades, whereas for blades of small relative blade heights dynamic pressure heads are measured both along the pitch and the blade height as well, which facilitates the investigation of the influence of tip losses on the efficiency of the cascade. We shall designate the local dynamic pressures after the blades as h_{di}. Beyond the cascade the static pressure heads are also observed which we shall name as h_{sti}. Since the fluid flow beyond the blades is not only non-uniform but also divergent in character, i. e., the streamlines have different flow directions, it is necessary to measure the direction of flow along the pitch of the blades at various points; for blades of small relative blade heights this measurement is carried out along the blade height as well. Let us denote the local directions of flow after the blades as α_{1fi} for guide blades and β_{2fi} for cascades of moving blades.

These measurements permit the determination of the basic parameters indispensable for the calculations and design of both guide and moving blades.

The experimental data as obtained above may be treated either with or without taking into consideration the effect of compressibility of the fluid. At velocities of flow of fluid from the blade cascade $c > 0.5\, c_s$ (where c_s—velocity of propagation of sound in the given medium where the flow takes place) the compressibility of the fluid cannot be dispensed with since it leads to a considerable error in the ultimate results. However, if the velocity of outflow of the fluid from the blade cascade is less, i. e., $c \leq 0.5c_s$, it is quite possible to neglect the effect of compressibility, which in fact leads to a considerable simplification of the calculations. It may be mentioned that most of the static investigations of blade profiles are carried out at $c \leq 0.5c_s$. The design formulas are derived in conformity with the above given conditions.

The theoretical velocity of flow for nozzles and guide blades is obtained from the equation

$$c_{1t} = 100\sqrt{\frac{2g}{\gamma}(h_0 - h_{st})_{\text{mean}} C_c}, \qquad (3\text{-}20)$$

where g—acceleration due to gravity;
γ—specific weight of the fluid before the blades;

$(h_0-h_{st})_{mean}$—mean value of the total pressure head;

C_c—conversion coefficient for the conversion of a column of mercury, water or any other liquid into kg/cm².

The theoretical velocity of flow after the blades is similarly determined from equation (3-20), excepting that in place of c_{1t} w_{2t} has to be placed.

The actual local velocities of flow after the blades for guide blade cascades are calculated from the equation

$$c_{1i}=100\sqrt{\frac{2g}{\gamma_i}h_{di}C_c}. \quad (3\text{-}21)$$

The actual local velocities of flow after the blade cascades for moving blades as well are calculated according to equation (3-21) replacing in the above equation c_{1i} by w_{2i}.

The actual velocity of out flow after the blade cascade for guide blades averaged according to the equation of mass flow is

$$c_1^*=\sqrt{(c_{1a}^*)^2+(c_{1u}^*)^2}, \quad (3\text{-}22)$$

where c_{1a}^* and c_{1u}^* — the axial and circumferential components of velocity c_1^*.

For blade cascades of moving blades the very same formula is made use of with the proper velocities replacing c_{1a}^* and c_{1u}^*, i. e.,

$$w_2^*=\sqrt{(w_{2a}^*)^2+(w_{2u}^*)^2}. \quad (3\text{-}22a)$$

To determine the values of c_1^* and w_2^*, first the component velocities c_a and w_a (axial) and c_u and w_u (circumferential) are found as average values along the pitch of the blades

$$\left.\begin{aligned}
&c_a=\frac{1}{t}\sum_{i=0}^{i=n}c_{1i}\sin\alpha_{1fi}\Delta t_i\\
&\text{and}\quad c_u=\frac{1}{t}\sum_{i=0}^{i=n}c_{1i}\cos\alpha_{1fi}\Delta t_i\text{ — for guide blades,}\\
&w_a=\frac{1}{t}\sum_{i=0}^{i=n}w_{2i}\sin\beta_{2fi}\Delta t_i\\
&\text{and}\quad w_u=\frac{1}{t}\sum_{i=0}^{i=n}w_{2i}\cos\beta_{2fi}\Delta t_i\text{ — for moving blades,}
\end{aligned}\right\} \quad (3\text{-}23)$$

where c_{1i} and w_{1i} — the actual local velocities of flow after the blades taken as an average value within the interval Δt_i along the pitch; c_{1i} is determined according to equation (3-21) for mean values of h_{di} within the interval Δt_1;

α_{1fi} and β_{2fi} — the actual local directions of flow after the blade cascade taken as an average value within the interval Δt_i along the pitch,

Δt_i — elementary interval along the pitch where the local values of c_{1i}, w_{2i}, α_{1fi} and β_{2fi} are averaged;

t and t_1 — pitch of the blade cascade for guide and moving blades.

The axial and circumferential components c_{1a}^*, w_{1a}^* and c_{1u}^*, w_{1u}^* of the velocities c_1^* and w_2^* will be

$$\left.\begin{aligned}
&c_{1a}^*=\frac{1}{l}\sum_{i=0}^{i=k}c_{ak}\Delta l_k\ \text{and}\ c_{1u}^*=\\
&=\frac{1}{l}\sum_{i=0}^{i=k}c_{uk}\Delta l_k\text{ — for guide blades, and}\\
&w_{2a}^*=\frac{1}{l''}\sum_{i=0}^{i=k}w_{ak}\Delta l_k\ \text{and}\ w_{2u}^*=\\
&=\frac{1}{l}\sum_{i=0}^{i=k}w_{uk}\Delta l_k\text{ — for moving blades,}
\end{aligned}\right\} \quad (3\text{-}24)$$

where c_{ak}, w_{ak}, c_{uk}, w_{uk} — the axial and circumferential components of velocities after the blade cascades taken as average values along the blade pitch at different heights of the blades;

Δl_k — elementary interval of height of the blade cascade where the mean values of c_a, c_u, w_a and w_u are assumed constant along the pitch.

Substituting the values c_{1a}^*, c_{1u}^*, w_{2a}^* and w_{2u}^* obtained from equation (3-24) in equation (3-22) and (3-22a) the values of c_1^* and w_2^* are determined.

The above calculations enable us to determine:

1) the direction of flow after the blades

$$\left.\begin{aligned}
&\tan\alpha_{1f}=\frac{c_{1a}^*}{c_{1u}^*}\text{ — for guide blades,}\\
&\tan\beta_{2f}=\frac{w_{2a}^*}{w_{2u}^*}\text{ — for moving blades;}
\end{aligned}\right\} \quad (3\text{-}25)$$

2) coefficient of mass flow through the blade passages

$$\mu_n = \frac{G_a}{G_t} = \frac{\gamma_1 c_1^* \sin \alpha_{1f}}{\gamma_{1t} c_{1t} \sin \alpha_1} \text{ — for guide blades,}$$

$$\mu_b = \frac{G_a}{G_t} = \frac{\gamma_2 w_2^* \sin \beta_{2f}}{\gamma_{2t} w_{2t} \sin \beta_2} \text{ — for moving blades,} \quad (3\text{-}26)$$

where G_a—the actual mass flow through the blade passages;

G_t—theoretical mass flow through the blade passages;

γ and γ_t—specific weight of the fluid at the exit of the blades with and without losses;

α_{1f} and β_{2f}—the mean directions of flow after the blades for guide and moving blades;

α_1 and β_2—the directions of flow of the fluid at the axis of the blade passage at the exit.

The velocity coefficients φ and ψ as well as the efficiency of the blades are calculated by determining the mean velocities of flow c_1 and w_2 for guide and moving blades after the blades. As an example, these velocities are calculated as shown below for a guide blade cascade. The mean value of the mass flow after the cascade of guide blades, taking into consideration the variations of mass flow along the passage, may be determined with sufficient accuracy according to the following equation

$$\bar{m} c_1 = \frac{1}{F} \sum_{i=0}^{i=F} m c_{1i} \Delta F_i, \quad (3\text{-}27)$$

where $\bar{m}$—mass flow of the fluid through a cascade;

c_1—average value of velocity of flow after the blades;

m—mass flow of the fluid through an elementary area of the section after the blades;

c_{1_i}—the local velocity of flow of the fluid after the blades taken as an average value within the limits of the elementary area.

The elementary area ΔF_i is formed by an elementary length Δl_i along the blade height and elementary portion of the blade pitch Δt_i (Fig. 3-11). Since the area ΔF_i is at right angle to the velocity of flow after the blades c_{1_i} the numerical value of ΔF_i will be equal to $\Delta t_i \sin\alpha_{1fi} \Delta l_i$. Since ΔF_i is expressed in terms of the elementary lengths of blade pitch and blade height Δt_i and Δl_i, it is usual to measure the quantities h_{di}, h_{sti}, α_{1fi} and β_{2fi} separately; first along the pitch and then along the height. The calculations are carried out in like order.

Mass flow of fluid $\bar{m}$ and m are determined from the equations

$$\left.\begin{aligned} \bar{m} &= \frac{1}{g} \sum_{i=0}^{i=F} \gamma_i c_{1i} \Delta F_i; \\ m &= \frac{\gamma_i}{g} c_{1i} \Delta F_i. \end{aligned}\right\} \quad (3\text{-}28)$$

Substituting the values of $\bar{m}$ and m obtained from equation (3-28) in the equation (3-27) and rearranging we have

$$c_1 = \frac{\sum_{i=0}^{i=F} \gamma_i c_{1i}^2 (\Delta F_i)^2}{F \sum_{i=0}^{i=F} \gamma_i c_{1i} \Delta F_i}. \quad (3\text{-}29)$$

To obtain the axial and circumferential components of velocity c_1 equation (3-29) will take the following form

$$\left.\begin{aligned} c_{1a} &= c_1 \sin \alpha_{1f} \frac{\cos \alpha_{1f}}{\cos \alpha_{1f}} = \\ &= \frac{\sum_{i=0}^{i=F} \gamma_i c_{1i} \sin \alpha_{1fi} c_{1i} \cos \alpha_{1fi} (\Delta F_i)^2}{F \sum_{i=0}^{i=F} \gamma_i c_{1i} \cos \alpha_{1fi} \Delta F_i} = \\ &= \frac{\sum_{i=0}^{i=F} \gamma_i c_{1ai} c_{1ui} (\Delta F_i)^2}{F \sum_{i=0}^{i=F} \gamma_i c_{1ui} \Delta F_i}; \\ c_{1u} &= c_1 \cos \alpha_{1f} \frac{\sin \alpha_{1f}}{\sin \alpha_{1f}} = \\ &= \frac{\sum_{i=0}^{i=F} \gamma_i c_{1i} \cos \alpha_{1fi} c_{1i} \sin \alpha_{1fi} (\Delta F_i)^2}{F \sum_{i=0}^{i=F} \gamma_i c_{1i} \sin \alpha_{1fi} \Delta F_i} = \\ &= \frac{\sum_{i=0}^{i=F} \gamma_i c_{1ui} c_{1ai} (\Delta F_i)^2}{F \sum_{i=0}^{i=F} \gamma_i c_{1ai} \Delta F_i}. \end{aligned}\right\} \quad (3\text{-}30)$$

Equation (3-30) permits the determination of the values of c_{1a} and c_{1u} allowing for the effect of compressibility of the fluid, i. e., for $c_1 > 0.5c_s$. When $c_1 \leq 0.5c_s$ equation (3-30) simplifies to the following

$$\left.\begin{aligned} c_{1a} &= \frac{\sum_{i=0}^{i=F} c_{1ai} c_{1ui} (\Delta F_i)^2}{F \sum_{i=0}^{i=F} c_{1ui} \Delta F_i}; \\ c_{1u} &= \frac{\sum_{i=0}^{i=F} c_{1ui} c_{1ai} (\Delta F_i)^2}{F \sum_{i=0}^{i=F} c_{1ai} \Delta F_i}. \end{aligned}\right\} \quad (3\text{-}30a)$$

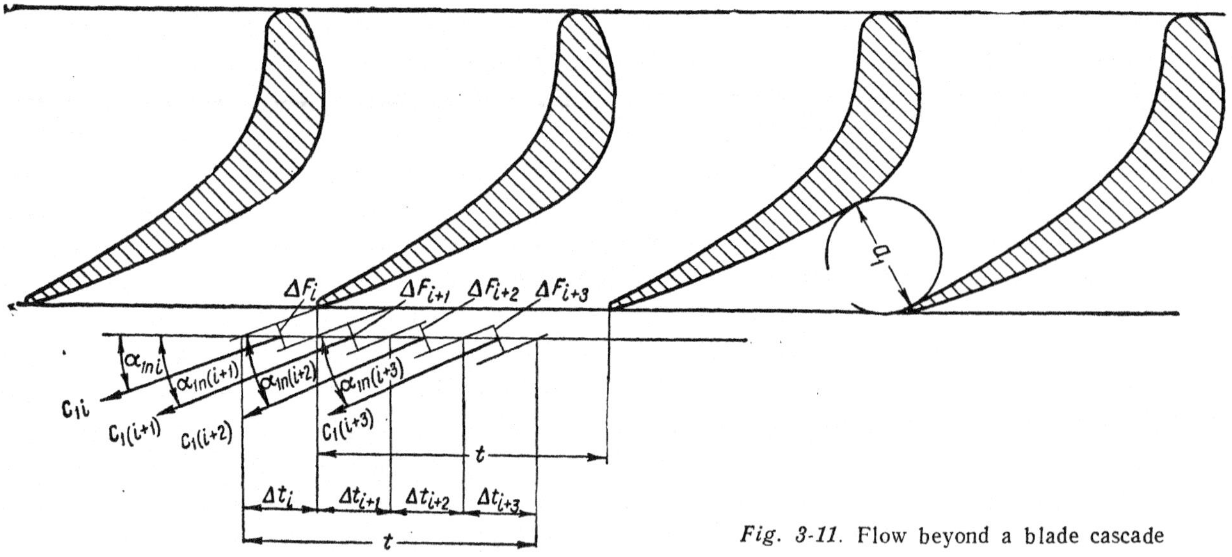

Fig. 3-11. Flow beyond a blade cascade

Similarly, for the cascades of moving blades, we obtain

$$\left.\begin{aligned} w_{2a} &= w_2 \sin\beta_{2f}\frac{\cos\beta_{2f}}{\cos\beta_{2f}} = \\ &= \frac{\sum_{i=0}^{i=F}\gamma_i w_{2ai} w_{2ui}(\Delta F_i)^2}{F\sum_{i=0}^{i=F}\gamma_i w_{2ui}\Delta F_i}; \\ w_{2u} &= w_2 \cos\beta_{2f}\frac{\sin\beta_{2f}}{\sin\beta_{2f}} = \\ &= \frac{\sum_{i=0}^{i=F}\gamma_i w_{2ui} w_{2ai}(\Delta F_i)^2}{F\sum_{i=0}^{i=F}\gamma_i w_{2ai}\Delta F_i}. \end{aligned}\right\} \quad (3\text{-}31)$$

Equation (3-31) enables the calculation of w_{2a} and w_{2u} taking into consideration the compressibility of the fluid, i. e., when $w_2 > 0.5c_s$. When $w_2 \leq 0.5c_s$ we have

$$\left.\begin{aligned} w_{2a} &= \frac{\sum_{i=0}^{i=F} w_{2ai} w_{2ui}(\Delta F_i)^2}{F\sum_{i=0}^{i=F} w_{2ui}\Delta F_i}; \\ w_{2u} &= \frac{\sum_{i=0}^{i=F} w_{2ui} w_{2ai}(\Delta F_i)^2}{F\sum_{i=0}^{i=F} w_{2ai}\Delta F_i}. \end{aligned}\right\} \quad (3\text{-}31a)$$

Velocity of flow after the blade cascade is obtained from the average values of circumferential and axial velocities derived from the equation of continuity as follows

$$\left.\begin{aligned} c_1 &= \sqrt{c_{1a}^2 + c_{1u}^2} \text{ — for guide blade cascades;} \\ w_2 &= \sqrt{w_{2a}^2 + w_{2u}^2} \text{ — for moving blade cascades.} \end{aligned}\right\} \quad (3\text{-}32)$$

The calculations carried out above enable us to determine the basic parameters for the design of blades, such as,

a) velocity coefficients

$$\left.\begin{aligned} \varphi &= \frac{c_1}{c_{1t}} \text{ — for guide blade cascades;} \\ \varphi &= \frac{w_2}{w_{2t}} \text{ — for moving blade cascades;} \end{aligned}\right\} \quad (3\text{-}33)$$

b) the energy utilised for doing useful work, per kilogram of the fluid, on the blades

$$\left.\begin{aligned} E_n &= \frac{2gc_1^2}{A} \text{ — for guide blades;} \\ E_b &= \frac{2gw_2^2}{A} \text{ — for moving blades;} \end{aligned}\right\} \quad (3\text{-}34)$$

c) the energy content of the fluid before the blade cascade per kilogram; available for doing work

$$\left.\begin{aligned} E_{on} &= \frac{2gc_{1t}^2}{A} \text{ — for guide blades;} \\ E_{ob} &= \frac{2gw_{2t}^2}{A} \text{ — for moving blades;} \end{aligned}\right\} \quad (3\text{-}35)$$

d) theoretical nozzle and blade efficiencies of he guide and moving blade cascades

$$\left.\begin{aligned}\eta_n=\frac{E_n}{E_{on}}=\frac{c_1^2}{c_{1t}^2}=\varphi^2 &\text{ — for guide blades;}\\ \eta_b=\frac{E_b}{E_{ob}}=\frac{w_2^2}{w_{2t}^2}=\psi^2 &\text{ — for moving blades;}\end{aligned}\right\} \quad (3\text{-}36)$$

e) relative energy loss in the blade cascades

$$\left.\begin{aligned}\xi_n=\frac{E_{on}-E_n}{E_{on}}=1-\varphi^2 &\text{ — for guide blades;}\\ \xi_b=\frac{E_{ob}-E_b}{E_{ob}}=1-\psi^2 &\text{ — for moving blades.}\end{aligned}\right\} \quad (3\text{-}37)$$

The actual mass flow of the fluid through the blade cascades is obtained from modified equation (3-26) as follows

$$\left.\begin{aligned}G_{an}=\mu_n G_t=\frac{\gamma_1 c_1^* \sin\alpha_{1f}}{\gamma_{1t}c_{1t}\sin\alpha_1}\pi dl\,\gamma_{1t}\sin\alpha_1=\\ =\pi dl\gamma_1 c_1^*\sin\alpha_{1f} \text{ — for guide blades;}\\ G_{ab}=\mu_b G_t=\frac{\gamma_2 w_2^*\sin\beta_{2f}}{\gamma_{2t}w_{2t}\sin\beta_2}\pi dl'' w_{2t}\gamma_{2t}\sin\beta_2=\\ =\pi dl''\gamma_2 w_2^*\sin\beta_{2f} \text{ — for moving blades.}\end{aligned}\right\} \quad (3\text{-}38)$$

It may be supposed with sufficient accuracy that $c_1^*\approx c_1$, $\alpha_{1f}\approx\alpha_1$, $w_2^*\approx w_2$ and $\beta_{2f}\approx\beta_2$. Substituting these values in equation (3-38) finally we obtain

$$\left.\begin{aligned}G_{an}=\pi dl\gamma_1 c_1\sin\alpha_1 &\text{ — for guide blades;}\\ G_{ab}=\pi dl''\gamma_2 w_2\sin\beta_2 &\text{ — for moving blades.}\end{aligned}\right\} \quad (3\text{-}38a)$$

Chapter Four

DETERMINATION OF NOZZLE AND BLADE DIMENSIONS

4-1. DETERMINATION OF NOZZLE SIZE

If steam is admitted to a turbine through nozzles placed along the full length of its circumference, i.e., steam is delivered to all the revolving blades at one and the same time, then such a turbine is known as a full admission turbine. On the other hand, if steam is admitted to only some fraction of the circumference, then such an arrangement is generally known as partial admission.

The relation between the length of the arc m occupied by the nozzles and the total circumference πd is known as the degree of partial admission

$$\varepsilon=\frac{m}{\pi d}=\frac{tz}{\pi d}, \quad (4\text{-}1)$$

where d—the mean diameter of the disc carrying the blades;

t—pitch of the blades at the mean diameter;

z—the number of blade passages.

The exit cross-section of a convergent nozzle arrangement in a direction perpendicular to the direction of velocity vector c_1 is determined as follows

$$f_1=alz, \quad (4\text{-}2)$$

where a—width of the exit (minimum) cross-section of the nozzle;

l—height of the nozzle at the exit section;

z—the number of blade passages.

From the equation of continuity we have

$$G_1 v_1=f_1 c_1, \quad (4\text{-}3)$$

where G_1—mass flow of steam through the nozzles, kg/sec;

v_1—the specific volume of steam at the exit section, m³/kg;

c_1—actual velocity of steam at the exit nozzle section.

From equations (4-1), (4-2) and (4-3) we have

$$G_1 v_1=alzc_1=tlz\sin\alpha_1=\pi d\varepsilon lc_1\sin\alpha. \quad (4\text{-}4)$$

From the last equation we obtain

$$l=\frac{G_1 v_1}{\pi d\varepsilon c_1\sin\alpha_1} \quad (4\text{-}5)$$

and

$$\varepsilon=\frac{G_1 v_1}{\pi dlc_1\sin\alpha_1}. \quad (4\text{-}6)$$

Equations (4-5) and (4-6) enable the determination of the basic dimensions of a nozzle.

Since all the quantities in equation, excepting l and ε, are known from heat drop calculations, we can determine l assuming some value of ε from equation (4-5) or assuming l, ε may be determined from equation (4-6). As will be shown later the energy losses in a nozzle increase with the decrease in nozzle height and the degree of partial admission. Hence it is recommended that the nozzle height l should be taken not less than 10 mm and ε not less than 0.2.

For turbines of smaller capacities it is found that at the normal speed of rotation of 3,000 r.p.m. the values of l and ε are small. In such cases, i.e., for turbines of capacities of up to 4,000 kW it is usual to increase the r.p.m. up to 6,000 or more, so that the mean diameter of the rotor is reduced for the same circumferential velocities, in consequence of which the values of l and ε increase. For large capacity turbines with relatively large blades the value of partial admission reaches unity, so that

$$l=\frac{G_1 v_1}{\pi d c_1 \sin \alpha_1}. \qquad (4\text{-}7)$$

For convergent-divergent nozzles the minimum cross-section is determined according to equation (2-5b) and the exit height of the nozzles according to equation (4-7), where the values of v_1 and c_1 are the specific volume and velocity of steam at the exit section of the nozzle respectively.

Example 2-7. Find the nozzle dimensions for a steam turbine operating at $p_0=16.5$ ata, $t_0=295°C$, $p_1=10$ ata and mass flow of steam $G_1=4$ kg/sec.

We shall assume $d=800$ mm[1], $\alpha_1=18°$, $\varphi=0.95$.

The relation $p_1/p_0=10/16.5=0.606$ is higher than the critical value, consequently we may assume a convergent nozzle for the heat drop; from the i-s diagram we have

$$i_0-i_{1t}=722-695=27 \text{ kcal/kg}.$$

Velocity of steam at the nozzle exit will be

$$c_1=0.95\times 91.5\times 27=452 \text{ m/sec}.$$

Heat lost in the nozzles will be

$$h_n=\left(\frac{1}{\varphi^2}-1\right)\frac{c_1^2}{8{,}378}=\left(\frac{1}{0.95^2}-1\right)\frac{452^2}{8{,}378}=$$
$$=2.63 \text{ kcal/kg}.$$

Setting off the value of this heat loss on the i-s diagram we obtain the value v_1, the specific volume of steam at the nozzle exit

$$v_1=0.232 \text{ m}^3/\text{kg}.$$

Assuming the nozzle height to be 10 mm the degree of partial admission to the turbine stage is obtained from equation (4-6)

$$\varepsilon=\frac{G_1 v_1}{\pi d l c_1 \sin \alpha_1}=\frac{4\times 0.232}{3.14\times 0.8\times 0.01\times 452\times 0.309}=0.265.$$

4-2. DETERMINATION OF THE HEIGHT OF MOVING BLADES

The height of blades at entry l_1' (Fig. 4-1) is made slightly larger than the height of the nozzles. For short blades usually l_1' is made larger by about 2 to 4 mm than l. For blades of greater length the difference between l_1' and l may be as much as 4 mm or more. The exit cross-section of the moving blades in a direction perpendicular to the direction of steam flow is determined by the equation

$$f_2=\frac{G v_2}{w_2}, \qquad (4\text{-}8)$$

where v_2—specific volume of steam at the exit from the moving blades, point A_2 in Figs (2-2) and (2-7). Exit cross-section of the blades in the plane of rotation of the disc will be

$$f_{2a}=\frac{f_2}{\sin \beta_2}=\frac{G v_2}{w_2 \sin \beta_2}. \qquad (4\text{-}9)$$

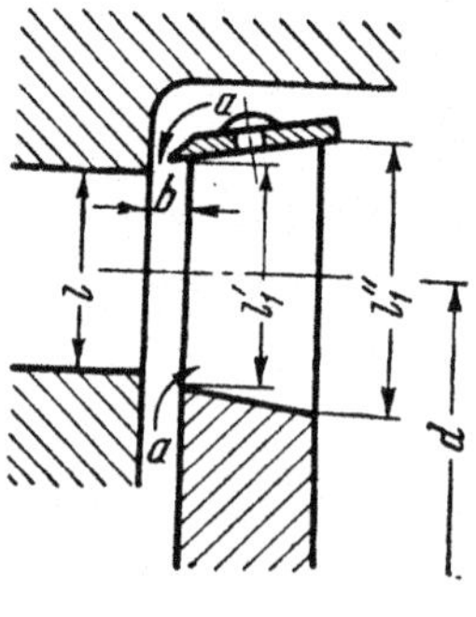

Fig. 4-1. Turbine stage

At the same time the value of f_{2a} may be expressed by

$$f_{2a}=\frac{a_1 l_1'' \varepsilon z_1}{\sin \beta_2}=l_1'' t_1 z_1 \varepsilon=\pi d l_1'' \varepsilon, \qquad (4\text{-}10)$$

where d —mean diameter of the disc carrying the moving blades;
a_1—width of the exit section of the moving blades;
t_1—blade pitch along the mean diameter;
l_1''—exit height of the moving blades;
z_1—number of blades mounted on the rotating disc.

From equations (4-9) and (4-10) we have

$$l_1''=\frac{G v_2}{\pi d \varepsilon w_2 \sin \beta_2}. \qquad (4\text{-}11)$$

When steam is delivered along the complete circumference of the disc $\varepsilon=1$.

From the velocity triangles (Fig. 2-6) we obtain

$$c_1 \sin \alpha_1=c_{1a}; \qquad w_2 \sin \beta_2=c_{2a},$$

where c_{1a} and c_{2a} are the components of the velocity c_1 and w_2 along the turbine axis.

Substituting these values in equations (4-5) and (4-11) in place of c_1 and w_2 and dividing equation (4-11) by equation (4-5) we obtain the relation

$$\frac{l_1''}{l}=\frac{v_2 c_{1a}}{v_1 c_{2a}}, \qquad (4\text{-}12)$$

from which

$$l_1''=l\frac{v_2 c_{1a}}{v_1 c_{2a}}. \qquad (4\text{-}13)$$

For purely impulse turbines we may approximate

$$l_1''=l\frac{c_{1a}}{c_{2a}}. \qquad (4\text{-}14)$$

From the point of view of streamlining the parts through which fluid (steam) flow takes place, large differences between the values l_1'' and l are not desirable.

[1] In turbine design either the stage diameter may be determined from a given heat drop or it may be assumed to suit the manufacturing conditions, in which case the heat drop is calculated according to the diameter assumed and the ratio $(u/c_1)_{opt}$.

Chapter Five

ENERGY LOSSES IN STEAM TURBINES

5-1. CLASSIFICATION OF TURBINE LOSSES

The increase in heat energy required for doing mechanical work in actual practice as compared to the theoretical value, in which the process of expansion takes place strictly according to the adiabatic process, is termed as energy loss in a steam turbine. In a turbine stage the quantity of heat drop actually converted into mechanical work on the turbine shaft is less than the theoretically calculated values for an ideal turbine stage. Mechanical losses, leakage of steam through glands, etc., contribute in addition to the decrease in useful work done on the turbine shaft. All the losses appearing in an actual turbine may be divided into two groups:

(1) internal losses, i. e., losses directly connected with the steam conditions while in its flow through the turbine;

(2) external losses, i. e., losses which do not influence the steam conditions.

The following types of losses may be classified under the first named heading:

losses in regulating valves;

losses in nozzles (guide blades);

leaving velocity losses (exit velocity);

losses due to friction of disc carrying the blades and windage losses;

losses due to clearance between the rotor and guide blade discs;

losses due to wetness of steam;

losses in the exhaust piping, etc.

Under the second heading the following losses may be grouped:

mechanical losses;

losses due to leakage of steam from the labyrinth gland seals.

5-2. LOSSES IN REGULATING VALVES

It is necessary that steam before entering the turbine proper should pass through a main stop valve and a regulating valve which without exception form an integral part of the turbine. Thus the steam conditions before the turbine are directly related to the fresh steam conditions before the stop valve (and regulating valve). The flow of steam through the stop and the regulating valve (or valves) is accompanied by loss of energy due to throttling. We may assume that during the process of throttling the total heat content of steam per kg remains constant, i. e., $i_0=\text{const}$.

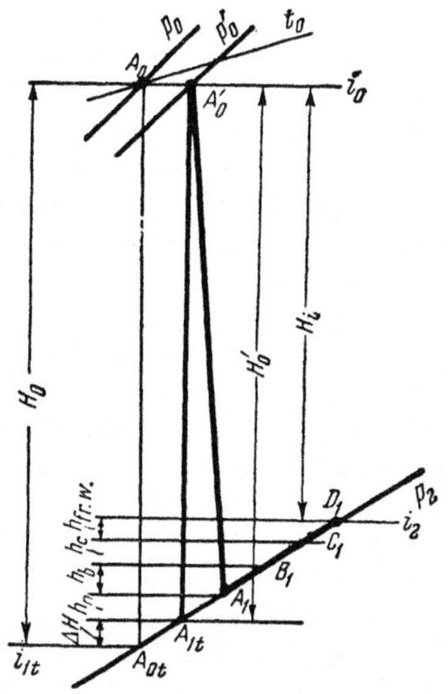

Fig. 5-1. Process of expansion of steam through regulating mechanism along with consequent losses due to throttling

Fig. 5-1 shows the process of expansion of steam through the regulating mechanism along with the consequent losses due to throttling; H_0—assumed value of heat drop in the turbine. Because of throttling process taking place in the regulating valve (or valves) the available heat drop in the turbine decreases from H_0 to H_0', i. e., there is a loss in available energy equal to $\Delta H = H_0 - H_0'$.

The magnitude of loss of pressure due to throttling with the regulating valves fully open may be as much as 5% of the pressure of fresh steam p_0.

In contemporary steam turbine practice it has been possible to reduce this pressure loss to as low as 3% and less by making use of streamlined forms for the regulating valves at those places where steam flow occurs. For design purposes a loss of pressure

$$\Delta p_v = (0.03 - 0.05)\, p_0 \qquad (5\text{-}1)$$

is recommended for accounting for the throttling effects in a regulating valve.

5-3. LOSSES IN NOZZLES

Losses of kinetic energy of steam while flowing through nozzles or guide blade passages are caused

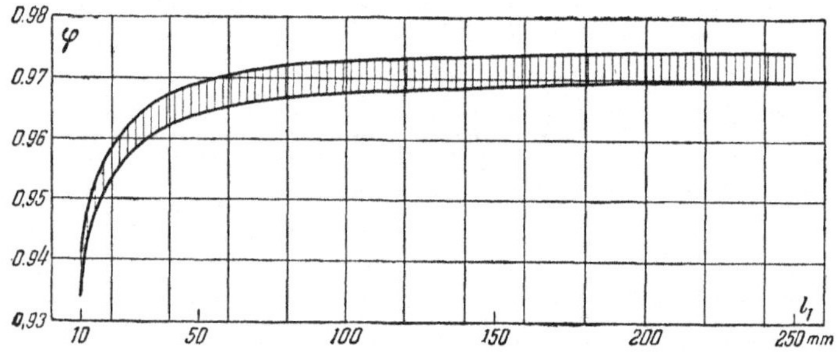

Fig. 5-2. Velocity coefficient φ for convergent nozzles as a function of nozzle height *l*

because of energy losses of steam before entering the nozzles, frictional resistance of the nozzle walls, viscous friction between particles, deflection of the flow, growth of boundary layer, turbulence in the wake (wake losses) and losses at the roof and floor of the blades (or nozzles), etc.

Losses in the velocity of steam issuing from the nozzle are accounted for by the velocity coefficient φ. The magnitude of velocity coefficient, as has been found from experimental investigations, basically depends on the nozzle dimensions (length, height and curvature), roughness of the walls and the velocity of flow, the form of the blade passages, etc. Velocity coefficient depends to a very large extent on the height of the nozzles, rapidly decreasing with the decrease in nozzle height. The velocity coefficient may be taken, for roughly cast nozzles, equal to 0.93 to 0.94, for thoroughly cast and machined nozzles—0.95 to 0.96 and for smoothly milled nozzles—0.96 to 0.97. Up to the present time complete data regarding the influence of various factors such as low steam velocities (less than 200 m/sec), high steam velocities (more than 1,000 m/sec) and the steam conditions, etc., on the velocity coefficient φ are not available.

However, for design purposes values of the velocity coefficient may be taken from graph shown in Fig. 5-2[1]. In convergent-divergent nozzles the velocity coefficient also depends on the conditions of expansion. Under conditions of back pressures higher than the design values, when shock waves appear in the flow, the velocity coefficient φ decreases with consequent increase in the magnitude of losses. Hence in such cases it is advisable to assume a value of velocity coefficient sufficiently lower than usual. The energy losses in a nozzle are determined according to equation (2-8a). All the energy expended in the process of expansion taking place in the limits of the oblique exit portion is deemed as lost energy.

5-4. LOSSES IN MOVING BLADES

Losses in moving blades are caused due to various factors. It was shown in 2-5 and 2-6 that the total losses in moving blades are accounted for by the velocity coefficient ψ. The total losses are comprised of the following:

1. Losses due to trailing edge wake. Steam issuing from the nozzle exit enters the annular space between the rows of nozzles and the moving blades. At the exit from the nozzle the flow of steam is in the form of clearly defined jets. These jets after leaving the nozzle walls mix with each other and form a homogeneous flow, with consequent formation of eddies. Thus there appear losses due to the turbulence in the steam flow while leaving the exit of the nozzles. These losses are known as the leaving losses or wake losses (Fig. 5-3). As has been remarked elsewhere these wake losses influence the magnitude of the velocity coefficient ψ and, further, are instrumental in disturbing the uniformity of flow before entry to the moving blades leading to losses in the moving blades (losses due to the periodic nature of flow). These losses depend on the thickness of the trailing edges of nozzles (guide blades).

2. Impingement losses. Steam, before entering the moving blade passages, meets with the leading edges of the blade profiles, the presence of which causes a disturbance in the flow thus leading to energy losses. These losses depend on the shape of the blade profiles at entry.

3. Losses due to leakage of steam through the annular space *b* between the stator and the shrouding (Fig. 4-1). For any stage of a turbo-machine there

[1] This graph has been taken from the book, *Steam Turbine Operation at Variable Loads*, by G. S. Samoilovich and B. M. Troyanovsky, Moscow, 1953.

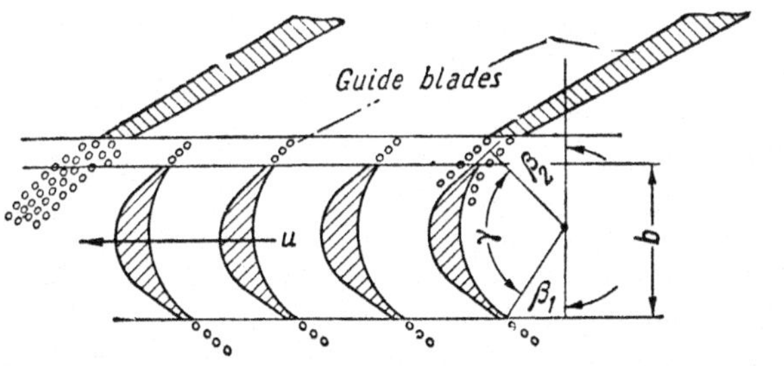

Fig. 5-3. Diagrammatic representation of losses in moving blades

is an unavoidable leakage of steam through the space *b* shown in Fig. 4-1 leading to unproductive flow as far as the creation of mechanical torque on the rotor is concerned, thus adding to the energy losses. To avoid this leakage it is usual to make the height of the moving blades slightly greater than the exit height of the nozzles (guide blades). This increase in the height of the moving blades leads to partial filling of steam in the moving blade passages, which establishing a partial vacuum in the annular space between the discs of moving and fixed blades draws the steam from the annular space into the moving blades as shown by arrows *a* in Fig 4-1. The negative effect of this suction may be mitigated by using labyrinth seals and increased degree of reaction.

However, increase in the degree of reaction causes a reversal of steam flow through the gap between the stator and the moving blade shrouding, thus leading to further leakage losses.

4. Frictional losses. Steam while flowing through nozzle or blade passages encounters frictional resistance to its flow over the nozzle walls, to overcome which some of the energy content is dissipated. Frictional losses depend on the length of the nozzle walls, i.e., the chord *b*, as also the nature of the wall surface: rough, smooth, etc.

5. Losses due to the turning of the steam jet in the blades.

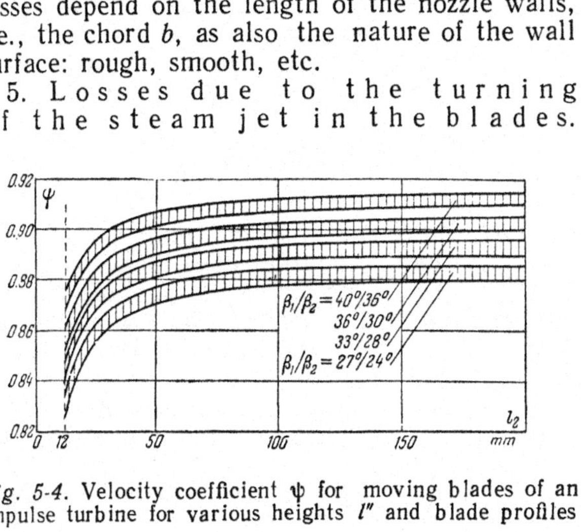

Fig. 5-4. Velocity coefficient ψ for moving blades of an impulse turbine for various heights l'' and blade profiles

These losses are caused by the mutual friction of steam particles caused by the difference in the lengths of the path described by them resulting from the curvature of the blade surface along which the steam particles flow. It has been found from experimental observations of short blades that secondary flows make their appearance, i. e., there is turbulence, which seriously affects the efficiency and adds to the losses. Losses due to turning of the steam in the blades depend on the angle of turning $\upsilon=180-(\beta_1+\beta_2)$ (Fig. 5-3). With increase in the angle υ, i. e., with the decrease in the angles β_1 and β_2, these losses grow rapidly and in the end predominate in effect over all other losses.

6. Losses due to shrouding. The difference between the heights l and l' (Fig. 4-1) affects the uniformity of flow at the entry to the nozzle and, as a rule, leads to the increase in the thickness of the boundary layer and consequent increase in losses. However, the influence of shrouding on the blade losses has not been systematically investigated till the present time.

The above enumerated losses do not profess to include all the types of losses occurring in a turbine blade or nozzle system or to completely explain the physical phenomenon occurring in a nozzle while there is a flow of steam or any fluid through the nozzle or the blade system.

The total losses occurring due to the above enumerated factors also depend on the velocity w of steam, height of blades and the degree of reaction. All these factors stated above which aid in the increase in losses are accounted for by the velocity coefficient ψ. To get an idea of the losses we may make use of graph shown in Fig. 5-4. The loss of energy, in heat units, is determined from equations (2-20a) and (2-27a). To reduce the losses in a moving blade it is imperative that the blade dimensions be chosen properly to the best advantage, i. e., pitch t_1, radius of curvature r, the blade profile itself, the angles of entry β_1 and exit β_2 (Fig. 5-3). According to Brilling the most advantageous pitch of blades is

$$t_1=\frac{r}{\sin\beta_1+\sin\beta_2}. \quad (5\text{-}2)$$

5-5. LEAVING VELOCITY (CARRY-OVER) LOSSES

Steam leaves the exit of the moving blades with an absolute velocity of c_2. In multistage turbines the velocity energy of the exit steam may be used either fully or partially in the succeeding stages. To enable the utilisation of energy equivalent to the velocity energy of the exit steam from the blades it is necessary to maintain as small a gap as possible between the blades of the previous stage and the nozzles of the following stage.

If the space between the moving blades of the previous stage and the nozzles of the following stage is large the velocity energy of the exit steam is completely lost for useful work, as is the case in the first impulse (regulating) stage of the turbine shown in Fig. 1-8, in the stage before the chamber from where steam is bled (Fig. 1-4), when the diameters of two adjacent stages differ considerably (Fig. 7-8), and in the last stage of a turbine. The magnitude of energy loss because of the leaving velocity in heat units is given by the equation

$$h_e = A\frac{c_2^2}{2g} = \frac{c^2}{8,378}\,[\text{kcal/kg}]. \qquad (5\text{-}3)$$

The leaving losses h_e increase the heat content of the exit steam. These losses are added on the *i-s* diagram after the nozzle and blade losses (Fig. 5-1) or, in other words, after the internal losses in a turbine stage if this leaving velocity is made use of in the nozzles of the following stage. Losses due to the exit velocity of the steam in the last stages of small- and medium-capacity turbines at a not very high vacuum do not exceed 1 to 2% of the heat drop in the turbine. For turbines of large capacity and for turbines operating at high vacuums these losses may reach a value as high as 3-4% or even more.

5 6. LOSSES DUE TO DISC FRICTION AND WINDAGE

Frictional forces appear between the rotating turbine disc and the steam enveloping it. The rotating disc drags the particles near its surface and imparts to them an accelerating force in the direction of its rotation. A definite amount of mechanical work is spent in overcoming the effect of friction and imparting this acceleration. The work spent in overcoming friction and accelerating the steam particles is again converted into heat, thus increasing the heat content of the steam. In the case of partial admission there is a good deal of turbulence over the arc where steam is not admitted. This turbulence causes windage losses which basically consist of the following: friction and impingement of steam on the blades, intermittent admission of steam into the moving blades in the case of partial admission. Besides, when there is partial admission of steam only that portion of the moving blade rotor disc, which is before the nozzle arrangement at the given instant, is filled with the incoming steam, all the remaining blade passages being filled with spent steam. When these blades in their turn come before the nozzles some portion of the kinetic energy of the incoming steam is spent in clearing away this spent steam. Loss of energy for the above is known as scavanging losses. All these losses lead to the increase in the heat content of steam. To determine the magnitude of these losses it is usual to make use of Stodola's empirical formula:

$$N_{\text{wind}} = \lambda\,[1.07\,d^2 + 0.61\,z\,(1-\varepsilon)\,dl_1^{1.5}] \times \times \frac{u^3}{10^6}\gamma\,[\text{kW}]^1, \qquad (5\text{-}4)$$

where N_{wind}—power lost in overcoming friction and ventilation (windage);

λ—coefficient usually taken as equal to unity for air and highly superheated steam (according to Levitsky), for ordinary superheats between 1.1 and 1.2, for saturated steam equal to 1.3;

d—diameter of the disc measured at the mean blade height, in metres;

z—number of velocity stages on the disc;

ε—degree of partial admission of steam;

l_1—height of blades, in cm;

u—circumferential velocity at the mean diameter in m/sec;

γ—specific weight of steam in which the disc rotates, in kg/m³.

Determination of N_{wind} disc friction and ventilation is quite often carried out with the help of Forner's formula as well

$$N_{\text{wind}} = \beta \times 10^{-10}\,d^4 n^3\,l_1\,\gamma\,[\text{kW}] \qquad (5\text{-}4\text{a})$$

where n—r.p.m. of the turbine,

β—coefficient equal to 1.76 for single-row discs, 2.06 for double-row discs and 2.80 for a three-row disc.

Losses due to disc friction and ventilation in heat units may be determined according to the following equation

$$h_{\text{wind}} = \frac{102\,N_{\text{wind}}}{427\,G}\,[\text{kcal/kg}], \qquad (5\text{-}5)$$

where G—mass flow of steam through the stage,

h_{wind}—thermal equivalent of the work expended in overcoming friction and windage.

Losses due to disc friction and windage are shown on the *i-s* diagram just as the other losses h_n, h_b and h_e (Fig. 5-1).

The magnitude of h_{wind} is appreciable in the case of discs carrying two or three rows of blades

---

[1] For two- or three-row discs a mean blade height is used for l_1 in equation (5-4).

with partial admission of steam at high pressures. For stages operating at low pressures and full admission of steam all along its circumference the magnitude of h_{wind} is quite insignificant, and it may be neglected for the last stages of a condensing turbine. In the case of reaction turbines in the absence of nozzle discs and with full admission of steam, losses due to disc friction and ventilation may be neglected, since due to the drum type of construction these losses are very small.

5-7. CLEARANCE LOSSES

a) Impulse Turbines

The pressure stage of an impulse turbine is shown in Fig. 5-5,*a*. There is a difference of pressure between the two sides of the nozzle disc which is fixed to the turbine stator, due to the expansion of steam in the nozzle. The diaphragms carrying the moving blades rotate whereas the nozzle discs are stationary so that there is always a small clearance space between the rotating and the stationary discs. As a consequence of the difference in the pressures between

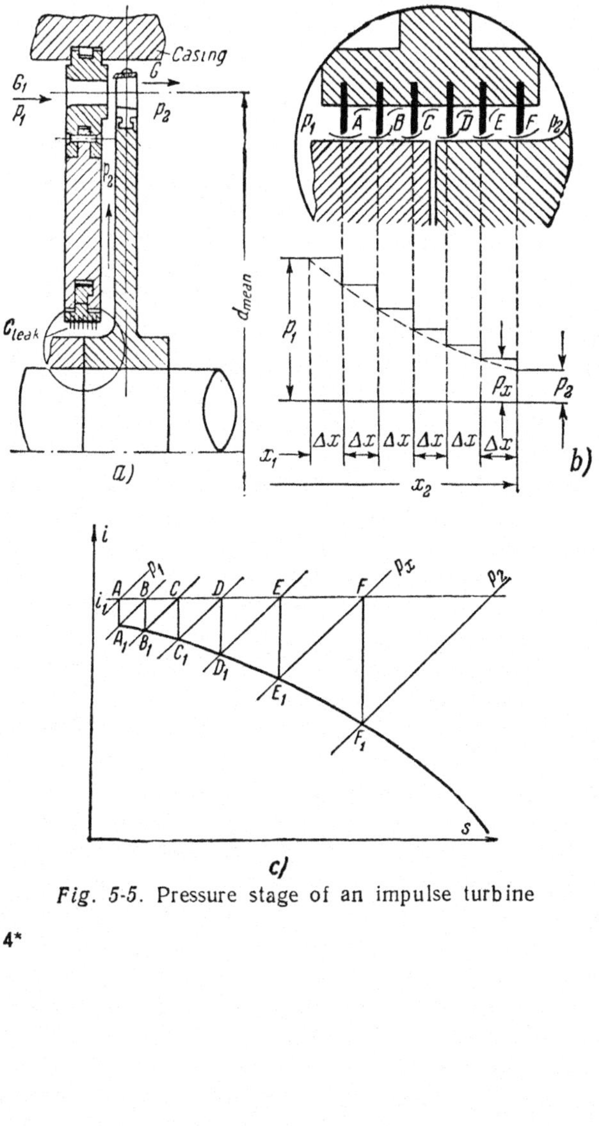

Fig. 5-5. Pressure stage of an impulse turbine

the two sides of the diaphragm carrying the nozzles there is a steam flow through this clearance space. This steam flow detracts from the available energy for conversion into mechanical work. The leakage of steam through the clearance space contributes to the increase in heat content of the steam leaving the stage and may be termed as a loss which can be accounted for on the *i-s* diagram. To reduce this steam leakage from one side of the disc to the other special labyrinth packings are employed (Fig. 5-5,*b*).

The loss of energy due to leakage of steam through clearance space may be determined from the following expression

$$G_{leak}(i_0 - i_2), \qquad (5\text{-}6)$$

where G_{leak}—mass flow of steam through the clearance space, kg/sec;
i_0—heat content of steam before the diaphragm;
i_2—heat content of steam after the moving blade diaphragm taking into consideration all the losses excepting those due to leakage.

Loss of heat content for doing useful work in a stage due to leakage is determined from the relation

$$h_{leak} = \frac{G_{leak}}{G}(i_0 - i_2)\,[\text{kcal/kg}], \qquad (5\text{-}7)$$

where G—mass flow of steam through the given turbine stage, kg/sec.

These losses, h_{leak}, as well are included in the *i-s* diagram. All the quantities appearing in equation 5-7, excepting G_{leak}, are already known from the heat drop calculations of a turbine stage. The quantity G_{leak} is found as follows. Fig. 5-6,*b* shows the process of pressure drop in a labyrinth seal for a diaphragm. The pressure before the diaphragm is equal to p_1 and after the disc carrying the moving blades it is equal to p_2. Consequently all the pressure drop taking place in the labyrinth seal from p_1 to p_2 (for a pure impulse stage) is distributed between the various compartments *A*, *B*, *C*, *D*, *E*, and *F*. Pressure drop takes place in each of the annular spaces between the labyrinth seal and the shaft. In the first gap pressure drops to p'. The velocity of steam *c* which results from the expansion of steam in the labyrinth gap is completely lost in the chamber *A* because of eddy formation and impact on the walls of the chamber. As a consequence of this the heat drop which had taken place during expansion is restored back to the steam to its original value.

In the next gap the pressure again undergoes a decrement from p' to p''. Again the velocity of steam c resulting from the expansion of steam in the labyrinth gap is lost in the next chamber B during which process the heat content is restored to its original value. A similar process continues in all the following chambers of the labyrinth seal. After the last of the labyrinth chambers the pressure drops to p_2, whereas the heat content increases to i_0 due to the presence of losses. In the above case the heat lost through the turbine shaft by conduction has been neglected.

Fig. 5-5,*c* shows a part of the *i-s* diagram for the heat drop restoration in a labyrinth seal. The adiabatic expansion taking place in the labyrinth gaps is shown by the lines AA_1, BB_1, CC_1, etc. The lines A_1B, B_1C, C_1D, etc. (constant pressure lines) show the process of heat restoration due to loss of steam velocity in the labyrinth chambers.

From the equation of continuity we may write for any labyrinth gap the following equation

$$G_{\text{leak}}\, v = f_s c, \tag{5-8}$$

where v—specific volume of steam in the labyrinth gaps (at points A_1, B_1, C_1, etc.), in m^3/sec;

c—velocity of steam in the gaps due to adiabatic expansion in m/sec;

$f_s = \pi ds$—area of the annular gap, m^2;

s—width of the gap.

Since the same mass of steam flows through the labyrinth gaps and since the areas of the annular spaces for all the labyrinth gaps are the same, we may write

$$\frac{c}{v} = \frac{G_{\text{leak}}}{f_s} = \text{const}. \tag{5-9}$$

This equation is an approximate one for labyrinth seals of different diameters.

With the expansion of steam in the gap its specific volume continuously increases. Consequently to maintain the conditions given by equation (5-9) the velocity of steam c must increase from one labyrinth gap to another, i. e., the heat drop must go on increasing for each succeeding gap (Fig. 5-5,*c*).

While designing labyrinth seals the following two cases must be borne in mind: (1) the velocity of steam in the last labyrinth gap is less than critical and (2) this velocity is equal to the critical velocity.

For a small pressure difference $p'-p''$ between the two sides of the labyrinth foil (at velocities less than the critical) the velocity of steam in the annular gap may be determined from the equation [1]

$$c = 100\sqrt{2g(p'-p'')v}, \tag{5-10}$$

where v—average specific volume of steam between the two pressures p' and p''.

The quantity of steam flowing through the gap is determined from the equation

$$G_{\text{leak}} = \frac{f_s c}{v} = 100 f_s \sqrt{2g\frac{p'-p''}{v}}. \tag{5-11}$$

This equation may be further expressed in the following manner

$$\frac{p'-p''}{v} = \frac{1}{2g}\left(\frac{G_{\text{leak}}}{100 f_s}\right)^2. \tag{5-12}$$

Since all the points characterising the steam conditions in each of the labyrinth chambers lie on the line i_1=const (Fig. 5-5,*c*), the change in the steam condition in the labyrinth can be deemed to be according to the gas law

$$p_1 v_1 = pv = \text{const} = B; \tag{5-13}$$

$$\text{whence } v = \frac{B}{p},$$

where B is a constant.

Substituting B/p in place of v in equation (5-12) we obtain

$$(p'-p'')p = \frac{B}{2g}\left(\frac{G_{\text{leak}}}{100 f_s}\right)^2. \tag{5-14}$$

If we denote the change of pressure $p'-p''$ through $-\Delta p$ ($p'-p'' = -\Delta p$) and divide both sides of this equation by Δx (Fig. 5-5), we obtain

$$-\frac{\Delta p}{\Delta x}p = \frac{BG^2_{\text{leak}}}{2g \times 100^2 f^2_s} \times \frac{1}{\Delta x} = \frac{a}{\Delta x}, \tag{5-15}$$

$$\text{where } a = \frac{BG^2_{\text{leak}}}{2g \times 100^2 f^2_s}.$$

For small differences in pressure it may be supposed with sufficient accuracy that

$$\frac{\Delta p}{\Delta x} = \frac{dp}{dx}.$$

So that equation (5-15) now becomes

$$-p\frac{dp}{dx} = \frac{a}{\Delta x}, \text{ or } -pdp = \frac{a}{\Delta x}dx. \tag{5-16}$$

Integrating the last obtained equation between the limits p_1 and p_2 and x_1 and x_2 (Fig. 5-5,*b*) we obtain

$$p_1^2 - p_2^2 = 2a\frac{x_2 - x_1}{\Delta x}. \tag{5-17}$$

---

[1] p' and p'' are expressed here in kg/cm².

From Fig. 5-5,*b* it follows that $\frac{x_2 - x_1}{\Delta x}$ represents the number of labyrinth chambers z. Hence taking the last mentioned relation into consideration and substituting the values for B and a from equations (5-13), (5-15) and (5-17) we finally obtain

$$G_{\text{leak}} = 100 f_s \sqrt{\frac{g(p_1^2 - p_2^2)}{z p_1 v_1}}. \quad (5\text{-}18)$$

If the velocity of steam reaches the critical value in the last chamber of the labyrinth seal the mass flow of steam through it (and consequently through all the chambers of the labyrinth) can be determined according to the equation

$$G_{\text{leak}} = 203 f_s \sqrt{\frac{p_x}{v_x}} = 203 f_s \sqrt{\frac{p_x^2}{p_1 v_1}}, \quad (5\text{-}19)$$

where p_x and v_x—pressure and specific volume of steam in the last chamber (kg/cm² and m³/kg).

The mass flow of steam through the preceding chambers $(z-1)$ of the labyrinth is determined in accordance with equation (5-18) as

$$G_{\text{leak}} = 100 f_s \sqrt{\frac{g(p_1^2 - p_x^2)}{(z-1) p_1 v_1}}. \quad (5\text{-}20)$$

Equating the right-hand sides of equation (5-19) and (5-20) we have

$$2.03^2 p_x^2 = \frac{g}{z-1}(p_1^2 - p_x^2),$$

whence

$$p_x^2 = \frac{g p_1^2}{2.03^2 (z-1) + g}. \quad (5\text{-}21)$$

Substituting the value of p_x^2 from equation (5-21) in equation (5-19)

$$G_{\text{leak}} = 203 f_s \sqrt{\frac{g p_1^2}{[2.03^2 (z-1) + g] p_1 v_1}} =$$

$$= 100 f_s \sqrt{\frac{g}{z+1.5} \times \frac{p_1}{v_1}}. \quad (5\text{-}22)$$

The critical pressure of steam in the last chamber of the labyrinth must be determined first so as to decide which of the two equations (5-18) and (5-22) is to be used for labyrinth designs.

For superheated steam $p_{cr} \approx 0.55 p_x$.

Substituting this value in place of p_x in equation (5-21) we obtain

$$p_{cr} = 0.55 p_1 \sqrt{\frac{\frac{g}{2.03^2}}{z - 1 + \frac{g}{2.03^2}}} = \frac{0.85 p_1}{\sqrt{z+1.5}}. \quad (5\text{-}23)$$

When the pressures p_1, p_2 and the number of labyrinth chambers z are known the leakage losses are determined as follows.

If the value of critical pressure obtained from equation (5-23) is less than p_2 then the velocity of steam in the last labyrinth is less than the critical velocity and the mass flow of leakage steam G_{leak} must be determined from equation (5-18). If on the other hand p_{cr} is greater than p_2 then the velocity of steam in the last labyrinth will be greater than critical and consequently all calculations will be based on equation (5-22).

Example 5-1. Determine the quantity of steam leakage occurring in a labyrinth seal for a turbine stage if the steam conditions before the diaphragm are $p_1 = 10$ ata and $t_1 = 260$°C. The pressure after the diaphragm $p_2 = 5$ ata, number of labyrinth chambers $z = 5$. Diameter of the disc for the given stage is $d = 350$ mm. Clearance between the disc and the labyrinth foils is $s = 0.3$ mm. The specific volume of steam for the conditions before the diaphragm $v_1 = 0.2427$ m³/kg (from steam tables).

The annular area for steam flow

$$f_s = \pi d s = \pi \times 0.35 \times 0.0003 = 0.00033 \text{ m}^2.$$

The critical pressure of steam in the fifth chamber of the labyrinth will be

$$p_{cr} = \frac{0.85 \times 10}{\sqrt{5+1.5}} = 0.32 \text{ ata}.$$

Since $p_{cr} < p_2$ the velocity of steam in the last labyrinth will be less than critical and the steam leakage will be determined using equation (5-18)

$$G_{\text{leak}} = 100 \times 0.00033 \times \sqrt{\frac{9.81(10^2 - 5^2)}{5 \times 10 \times 0.2427}} = 0.033 \times$$

$$\times 7.8 = 0.257 \text{ kg/sec}.$$

b) Losses in Reaction Turbines

Fig. 5-6 shows schematically a stage of a reaction turbine. In a reaction turbine the guide (fixed) blades are attached immediately to the stator of the turbine whereas the moving blades are attached to the drum or rotor.

Thus both between the guide blades and the rotor and the moving blades and the stator of

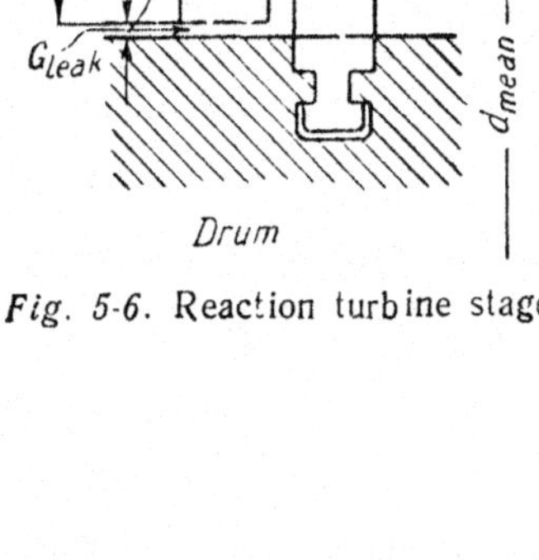

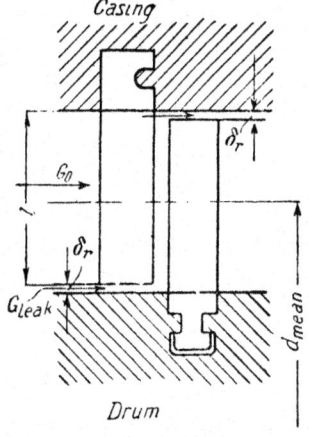

Fig. 5-6. Reaction turbine stage

the turbine a small gap will be formed. The value of the gap δ_r is chosen in such a way so that during operation of the turbine the blades of the turbine, both moving and fixed, do not scrape against the stator and rotor respectively. Since pressure drop occurs both in the moving and the guide blades there exists a leakage of steam through the radial gap δ_r. This leakage steam is a loss from the point of view of conversion of heat energy into mechanical work.

Heat losses through the radial clearance may be approximately determined from the equation[1]

$$h_{\text{leak}} = \frac{\delta_r}{l \sin \alpha_1} (i_0 - i_2) \qquad (5\text{-}24)$$

where α_1—the angle subtended by the exit steam jet to the turbine axis (for the guide blades), in degrees;

δ_r—width of the clearance, in mm;

l—height of the guide blades, in mm;

i_0—heat content of the steam before the guide blades, in kcal/kg;

i_2—heat content of steam after the moving blades considering all the losses excepting those due to leakage of steam through the radial clearance, in kcal/kg.

The heat losses may also be determined from the empirical formula of Anderkhoob

$$h_{\text{leak}} = 1.72 \frac{\delta_r^{1.4}}{l} h_0. \qquad (5\text{-}24a)$$

5-8. LOSSES DUE TO WETNESS OF STEAM

In the case of condensing turbines the last few stages usually operate under conditions at which steam is wet resulting in the formation of minute droplets of water. These droplets under the influence of centrifugal force are thrown out towards the periphery. At the same time these droplets of water receive an accelerating force from the steam particles in the direction of flow. Thus some of the kinetic energy of the flowing steam is lost in accelerating these water droplets. Since the absolute velocity of steam c_1 is considerably greater than that of the water droplets c_{1w}, the direction of the relative velocity vector w_{1w} (the angle of entry of the water droplets into the moving blade passages) differs from that of the steam jet. The water droplets are deflected onto the back of the moving blades as a result of which the moving blades experience an impact force caused by the impingement of water droplets on their backs. The impact of the water droplets

Fig. 5-7 Moving blades of a low-pressure stage

does not leave the working of the moving blades unaffected. It has been found from practical investigations that the blade tips are subjected to wear from one side. Fig. 5-7 distinctly shows such an erosion of the trailing blade tips. At the same time some of the kinetic energy of the steam passing through the moving blade passages has to be spent in overcoming the impact of the water droplets which actually acts as a braking force.

In the region where steam is wet all the quantity of steam passing through the stage under consideration does not contribute to mechanical work. Only that portion of steam which remains dry during the process of expansion does work. That fraction of steam which during the process of expansion condenses into minute water droplets basically governs the losses resulting from wetness of steam.

The losses due to wetness of steam may be determined with permissible accuracy for practical purposes from the equation

$$h_{\text{wetness}} = (1 - x) h_i \qquad (5\text{-}25)$$

where h_i—the heat drop utilised in a turbine stage considering all the losses excepting those due to wetness;

x—average dryness fraction of steam in the stage under consideration.

$$x = \frac{x_1 + x_2}{2},$$

[1] Since it is usual to have $\alpha_1 = \beta_2$ in reaction stages, equation (5-24) is valid for leakage loss calculations for moving blades as well.

where x_1 and x_2—dryness fractions of steam before the nozzle (guide blades) and after the moving blades of the stage.

5-9. EXHAUST PIPING LOSSES

Depending upon the velocity of flow in the exhaust piping there is loss of pressure of the steam while flowing in the exhaust piping to the condenser.

Exhaust piping losses may be neglected in the case of back-pressure turbines since the velocities of flow are negligible (30-50 m/sec). In condensing turbines there are considerably greater velocities of flow (100-120 m/sec) and hence the losses arising may not be overlooked.

Loss of pressure in exhaust pipings of condensing turbines may be determined from the equation

$$p_2 - p_{2k} = \lambda \left(\frac{c_s}{100}\right)^2 p_{2k}, \qquad (5\text{-}26)$$

where p_2—pressure of steam after the blades;
p_{2k}—pressure of steam in the exhaust piping;
c_s—velocity of steam in the exhaust piping;
λ—coefficient varying from 0.07 to 0.1.

When the velocity of exhaust steam $c_s = 100$ m/sec, $p_{2k} = 0.05$ ata and $\lambda = 0.1$ pressure drop will be equal to

$$p_2 - p_{2k} = 0.1 \times 0.05 = 0.005 \text{ ata},$$

i. e., the pressure after the blades would be equal to $p_2 = 0.005$ ata. Heat losses in the turbine increase the heat content of the exhaust steam, thus reducing the quantity of heat usefully utilised.

5-10. EXTERNAL LOSSES

a) Mechanical Losses

These losses are caused as a result of energy expended in overcoming the resistances of the journal and thrust bearings including the journal bearings of the generator or any other machine coupled to the turbine shaft such as the main oil pump, governor, etc. There are additional losses in the case of turbines which have water-sealed bearings or which have their shafts water-cooled at the ends to obviate undesirable resistances.

Mechanical losses in a turbine installation are accounted for by the quantity termed as the mechanical efficiency, η_m, and the total me-

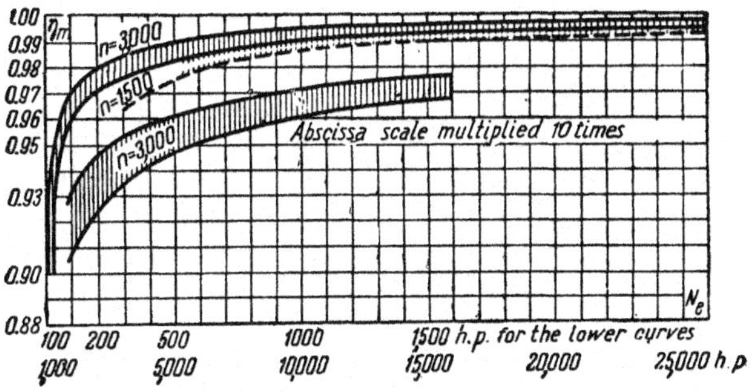

Fig. 5-8. Mechanical efficiency of a turbine

chanical losses may be determined experimentally.

For design purposes curves of mechanical efficiency such as the ones shown in Fig. 5-8 may be used. The curves in Fig. 5-8 give the mean values of mechanical efficiencies for turbines of various capacities.

b) Losses Due to Steam Leakage through the End Seals

To reduce the leakage of steam and thus the loss of useful energy from the two ends of the turbine where the turbine shaft projects out from the stator, labyrinth seals are resorted to (Figs. 1-4, 1-5, 1-6 and 1-8). In Fig. 1-4 points *5* and *7* show the forward and aft end labyrinth seals. In Fig. 1-5 the end labyrinth seals are shown by Figure 7.

Due to the difference existing between the pressures inside the stator and the outside atmosphere there exists a leakage of steam through the end labyrinth seals. The leakage of steam through the end seals does not influence the variation of steam conditions inside the turbine and therefore is classified under external losses.

Leakage of steam through the end seals is calculated according to equations (5-18) and (5-22).

While calculating the steam leakage losses through the internal clearances, it is usual to determine mass flow of steam for a given pressure drop as well as the number of labyrinth chambers required for such a pressure drop.

When designing end labyrinth seals it may be necessary to determine the number of labyrinth chambers for a given mass flow and a given pressure drop. In such cases we may make use of any of the two equations (5-18) and (5-22). Thereafter the critical pressure p_{cr} is determined from equation (5-23) and is compared with the pressure p_2 (pressure of the surroundings into which steam leaks from the seals). If these values do not agree all the calculations are repeated using the second equation.

5-11. EFFICIENCY OF A TURBINE

The relation between the work of one kilogram of steam L_u at the periphery of the disc carrying the moving blades to the theoretical work which it can accomplish is known as the relative efficiency of the blades

$$\eta_u = \frac{L_u}{L_0} = \frac{AL_u}{i_0 - i_{1t}}. \qquad (5\text{-}27)$$

The relation between the useful work done by 1 kg of steam L_i in a stage or inside a turbine to the theoretical available work L_0 is termed as the internal efficiency of a stage or turbine.

$$\eta_{0i} = \frac{L_i}{L_0} = \frac{i_0 - i_2}{i_0 - i_{1t}} = \frac{H_i}{H_0}. \qquad (5\text{-}28)$$

The relative internal efficiency is easily determined from the heat drop process of the stage or the turbine drawn on the *i-s* diagram (Figs 5-1 and 8-2).

Economic performance of a steam turbine largely depends on its internal efficiency η_{0i}.

The relation between the theoretical adiabatic heat drop in the turbine $H_0 = i_0 - i_{1t}$ (kcal/kg) and the heat available from the boiler $i_0 - q$ (kcal/kg) is known as the thermal efficiency

$$\eta_t = \frac{H_0}{i_0 - q} = \frac{i_0 - i_{1t}}{i_0 - q}, \qquad (5\text{-}29)$$

where q—sensible heat of the condensate, the temperature of which is equal to the temperature of the exhaust steam.

Power developed at the rim of the turbine discs is found from the equation

$$N_u = \frac{427\, Gh_u}{102} \text{ [kW]}, \qquad (5\text{-}30)$$

where G—mass flow of steam through the given stage, kg/sec;

$h_u = h_0 - h_n - h_b - h_e$—heat usefully utilised for doing work at the rim of the turbine disc (h_0 is the theoretical adiabatic heat drop in the stage under consideration).

Internal power of a stage

$$N_i' = \frac{427\, Gh_i}{102} \text{ [kW]}. \qquad (5\text{-}31)$$

Internal power of a turbine

$$N_i = \frac{427\, GH_i}{102} \text{ [kW]}, \qquad (5\text{-}32)$$

where G—mass flow of fresh steam through the turbine, kg/sec;

$h_i = h_0 - \Sigma h_1$—usefully utilised heat drop in a given stage (Σh_1—sum of all the heat losses in a given turbine stage);

$H_i = \Sigma h_i$—usefully utilised heat drop in all the stages of the turbine under consideration.

The power developed by an ideal turbine

$$N_0 = \frac{427\, GH_0}{102} = \frac{427\, GH_i}{102\, \eta_{0i}} = \frac{N_i}{\eta_{0i}}, \qquad (5\text{-}33)$$

from which

$$N_i = N_0 \eta_{0i}. \qquad (5\text{-}34)$$

Effective power developed at the turbine shaft will be

$$N_{\text{eff}} = N_i - \Delta N_m, \qquad (5\text{-}35)$$

where ΔN_m—power lost in overcoming mechanical resistances, in kW.

The relation between the effective power of a turbine to its internal developed power is termed as the mechanical efficiency of the turbine

$$\eta_m = \frac{N_{\text{eff}}}{N_i}. \qquad (5\text{-}36)$$

The relation between N_{eff} and N_0 is termed as the relative effective efficiency

$$\eta_{re} = \frac{N_{\text{eff}}}{N_0} = \frac{N_i \eta_m \eta_{0i}}{N_i} = \eta_m \eta_{0i}. \qquad (5\text{-}37)$$

The product of relative effective efficiency and the thermal efficiency gives the so-called absolute effective efficiency

$$\eta_e = \eta_{re}\eta_t = \eta_{0i}\eta_m\eta_t = \eta_i\eta_m. \qquad (5\text{-}38)$$

The relation between the power developed at the generator terminals N_e and N_{eff} is termed as generator efficiency

$$\eta_g = \frac{N_e}{N_{\text{eff}}}. \qquad (5\text{-}39)$$

The relative electrical efficiency of the turbo-generator is

$$\eta_{r.el} = \eta_{re}\eta_g = \eta_{0i}\eta_m\eta_g. \qquad (5\text{-}40)$$

and the absolute electrical efficiency of the turbo-generator is given by

$$\eta_{ae} = \eta_e\eta_g = \eta_{0i}\eta_t\eta_m\eta_g = \eta_i\eta_m\eta_g. \qquad (5\text{-}41)$$

The absolute electrical efficiency of the turbo-generator governs the economic performance of the set. It depends both on the mechanical construction of the set and on the thermodynamic cycle of the turbine.

5-12. DETERMINATION OF MASS FLOW OF STEAM

For the design of a turbine it is usual to specify the power developed at the generator terminals, initial and final steam conditions and the number of revolutions per minute.

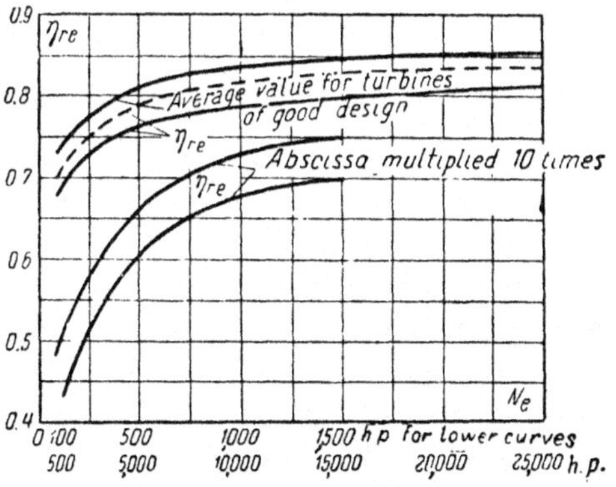

Fig. 5-9. Relative effective efficiencies of a turbine

On the basis of equations (5-32) and (5-33) we may write

$$N_{ae}=\frac{427\,GH_i}{102}\eta_m\eta_g=\frac{427\,GH_0}{102}\eta_{oi}\eta_m\eta_g\ [\mathrm{kW}] \quad (5\text{-}42)$$

or

$$N_{ae}=\frac{427\,GH_0}{102}\eta_{re}\eta_g\ [\mathrm{kW}]. \quad (5\text{-}43)$$

If a speed reduction gearing is used then we have

$$N_{ae}=\frac{427\,GH_0}{102}\eta_{re}\eta_r\eta_g\,[\mathrm{kW}], \quad (5\text{-}44)$$

where η_r—efficiency of the reduction gearing.

If we substitute mass flow of steam per hour instead of per second in equation (5-44) then

$$N_{ae}=\frac{427\,DH_0}{3{,}600\times 102}\cdot\eta_{re}\eta_g\eta_r=$$
$$=\frac{DH_0\eta_{re}\eta_r\eta_g}{860}\,[\mathrm{kW}]. \quad (5\text{-}45)$$

D—mass flow of steam per hour.

Equation (5-45) expresses the relationship between the power developed at the generator terminals and the mass flow per hour for a giv-

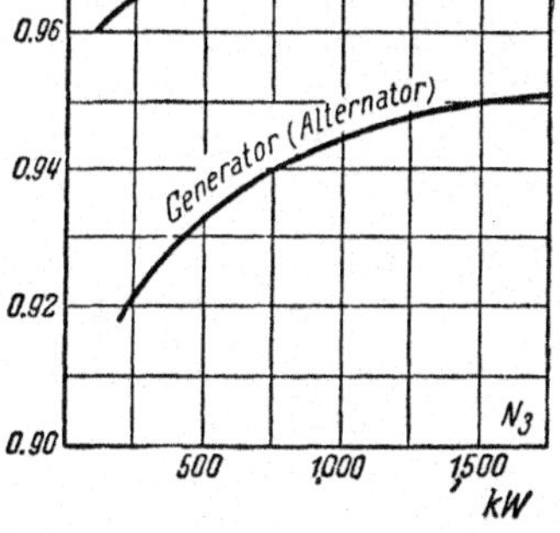

Fig. 5-10. Efficiencies of low-speed generators and toothed gearings

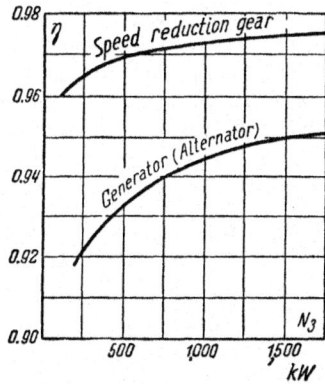

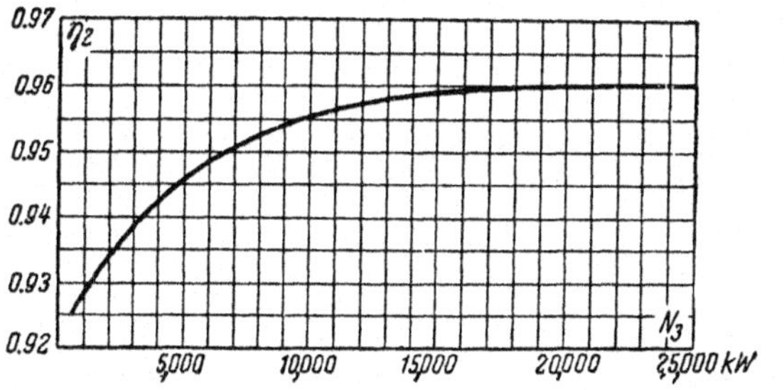

Fig. 5-11. Generator efficiencies according to Elektrosila Works data

en theoretical heat drop and efficiencies η_{re}, η_r and η_g.

The mass flow of steam through the turbine is determined from equation (5-45)

$$D=\frac{860N_{ae}}{H_0\eta_{re}\eta_r\eta_g}\,[\mathrm{kg/hr}]. \quad (5\text{-}46)$$

The specific mass flow of steam per kWh for a condensing or a back-pressure[1] turbine is found by dividing equation (5-46) by N_{ae}

$$d_{ae}=\frac{D}{N_{ae}}=\frac{860}{H_0\eta_{re}\eta_r\eta_g}\,[\mathrm{kg/kWh}]. \quad (5\text{-}47)$$

To assess the economic performance of an installation sometimes the specific heat flow per kWh is used

$$q_{ae}=d_{ae}(i_0-i_k)=\frac{860\,(i_0-i_k)}{H_0\eta_{re}\eta_r\eta_g}=$$
$$=\frac{860}{\eta_e\eta_r\eta_g}\,[\mathrm{kcal/kWh}]. \quad (5\text{-}48)$$

Here i_k—sensible heat of the condensate at the point of extraction.

For turbines without reduction gearing $\eta_r=1$. During the preliminary calculations for the design of a steam turbine it is usual to calculate the mass flow of steam using equation (5-46). The quantity H_0 is found from *i-s* diagrams and the various efficiencies are taken from the practical test results available.

These efficiencies are made precise during the detailed design.

Fig. 5-9 shows the curves of relative effective efficiency as a function of the power developed by the turbine. With the increase in turbine capacity the relative effective efficiency η_{re} increases. For turbines of good design this efficiency reaches a figure of 0.85. For turbines of lower capacities the value of η_{re} is usually considerably less. These curves are a great help during the preliminary design calculations.

[1] Here the leakage losses through the end seals have not been considered.

The efficiencies η_r and η_g may be obtained from the test results of previous designs shown in Figs 5-10 and 5-11.

Mechanical efficiency η_m may be obtained from the graphs shown in Fig. 5-8.

Heat calculations for the design of a turbine are made on the basis of mass flows obtained from equation (5-46). From heat calculations the internal efficiency η_{oi} is determined. The product of $\eta_{oi}\eta_m$ finally gives the value of η_{re}.

Chapter Six

SINGLE-STAGE TURBINES

6-1. SINGLE-STAGE IMPULSE TURBINE WITH ONE VELOCITY STAGE

The arrangement of a single-stage impulse turbine is shown in Fig. 6-1.

Fresh steam expands in the nozzles *4* from its initial pressure p_0 to the final pressure p_1. During the process of expansion the velocity of steam increases from its initial value c_0 to c_1 in the exit section of the nozzle. The velocity of steam is reduced from c_1 to c_2 while flowing through the moving blades *3*, i. e., the kinetic energy of the flowing steam is converted into mechanical work on the blades. The exhaust steam from the turbine is led away through the pipe *6*. Fig. 6-1 also shows the variations in the velocity of steam while flowing through a single-stage impulse turbine (or a stage of an impulse turbine).

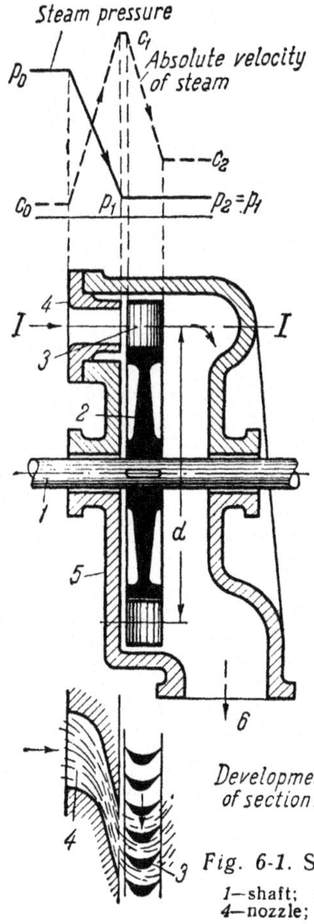

Fig. 6-1. Single-stage impulse turbine

1—shaft; *2*—disc; *3*—moving blades; *4*—nozzle; *5*—stator; *6*—exhaust pipe

Steam issuing from the stationary nozzles with an absolute velocity of c_1 enters the moving blades. In conformity with the shape of the moving blades and the losses taking place in them the direction and velocity of the steam flowing through them undergo a change, and the steam leaves the moving blades with an absolute velocity c_2. The flow of steam through the moving blades exerts a force P_u on them in the direction of rotation. Since every action has an equal and opposite reaction the moving blades also exert an equal force P_u on the steam, but in a direction opposite to the direction of rotation. We shall determine this force P_u in the direction of rotation.

From mechanics we know that the force exerted is equal to the change in momentum

$$Pt = m(c_x - c_y),$$

where t—time during which force P acts;

c_x and c_y—velocities of the mass m at the beginning and end of the interval of time under consideration.

While passing through the passages of the moving blades the momentum of the flowing steam decreases, giving rise to an impulse force on the moving blades. Since the directions of flow of steam do not coincide with the direction of rotation the force P_u in the direction of rotation u is obtained from the projections of the velocities c_1 and c_2 along u, i. e., c_{1u} and c_{2u} (Fig. 6-2) multiplied by the mass flow of steam $m=G/g$.

Thus the force exerted due to change in momentum of the flowing steam will be

$$P_u t = \frac{G}{g}(c_{1u} - c_{2u})\ [\text{kg/sec}], \qquad (6\text{-}1)$$

where G—mass flow of steam through the blades in t seconds.

When the mass flow is equal to unity and the interval of time $t=1$ sec, equation (6-1) becomes

$$P_u = \frac{1}{g}(c_{1u} - c_{2u})\ [\text{kg}]. \qquad (6\text{-}2)$$

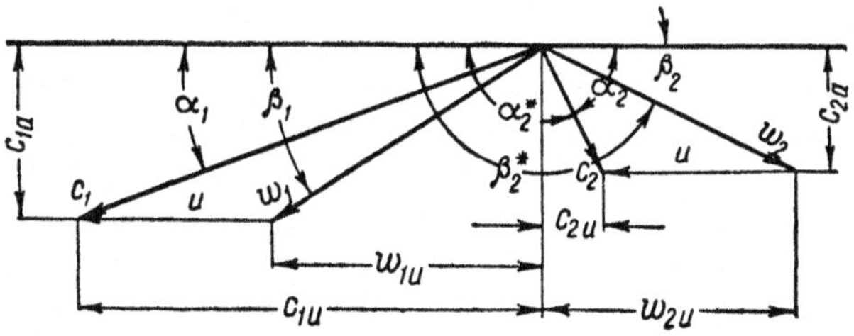

Fig. 6-2. Velocity triangles for an impulse stage

The distance travelled by the blades in one second is equal to u [m/sec] (u—circumferential velocity of the disc at its mean diameter, see Fig. 4-1). Hence the work done by one kilogram of steam at the periphery of the disc in an interval of one second is equal to

$$L_u = P_u u = \frac{u}{g}(c_{1u} - c_{2u}) \text{ [kgm]}. \qquad (6\text{-}3)$$

The theoretical work of one kg of steam at the rim of the disc for an ideal turbine, i. e., in the absence of losses in either nozzles or blades will be

$$L_0 = \frac{c_{1t}^2}{2g} \text{ [kgm]}. \qquad (6\text{-}3a)$$

The relation between L_u and L_0 is the relative efficiency of the blade disc [see also equation (5-27)]:

$$\eta_u = \frac{L_u}{L_0} = \frac{2u(c_{1u} - c_{2u})}{c_{1t}^2}. \qquad (6\text{-}4)$$

It is interesting to investigate on what factors the value of this coefficient depends. We shall carry out the following transformations.

From the diagram of velocity triangles (Fig. 6-2) we have

$$c_{1u} = c_1 \cos\alpha_1 = w_1 \cos\beta_1 + u; \qquad (6\text{-}5)$$

$$c_{2u} = c_2 \cos\alpha_2 = w_2 \cos\beta_2^* + u. \qquad (6\text{-}6)$$

From expressions (6-5) and (6-6) we obtain

$$c_{1u} - c_{2u} = w_1 \cos\beta_1 - w_2 \cos\beta_2^*. \qquad (6\text{-}7)$$

Substituting $w_2 = \psi w_1$ in equation (6-7) we have

$$c_{1u} - c_{2u} = w_1 \cos\beta_1 - \psi w_1 \cos\beta_2^* =$$

$$= \left(1 - \psi\frac{\cos\beta_2^*}{\cos\beta_1}\right) w_1 \cos\beta_1.$$

From equation (6-5) we have $w_1 \cos\beta_1 = = c_1 \cos\alpha_1 - u$, so that

$$c_{1u} - c_{2u} = \left(1 - \psi\frac{\cos\beta_2^*}{\cos\beta_1}\right)(c_1 \cos\alpha_1 - u). \qquad (6\text{-}8)$$

Since

$$c_1 = \psi c_{1t},$$

$$c_{1t}^2 = \frac{c_1^2}{\varphi^2}. \qquad (6\text{-}9)$$

Substituting expressions (6-8) and (6-9) in equation (6-4) we obtain

$$\eta_u = 2\varphi^2\left(1 - \psi\frac{\cos\beta_2^*}{\cos\beta_1}\right)\left(\cos\alpha_1 - \frac{u}{c_1}\right)\frac{u}{c_1}. \qquad (6\text{-}10)$$

Now $\beta_2^* = \pi - \beta_2$ hence $\cos\beta_2^* = -\cos\beta_2$. Substituting this expression in equation (6-10) we finally obtain

$$\eta_u = 2\varphi^2\left(1 + \psi\frac{\cos\beta^2}{\cos\beta_1}\right)\left(\cos\alpha_1 - \frac{u}{c_1}\right)\frac{u}{c_1}. \qquad (6\text{-}11)$$

For the particular case when the angles β_1 and β_2 are equal the equation simplifies to

$$\eta_u = 2\varphi^2(1 + \psi)\left(\cos\alpha_1 - \frac{u}{c_1}\right)\frac{u}{c_1}. \qquad (6\text{-}12)$$

From the above equation it follows that the quantity η_u depends upon the fraction u/c_1, the nozzle angle α_1, velocity coefficients φ and ψ and the blade angles β_1 and β_2. The value of η_u decreases with the decrease in velocity coefficients φ and ψ. The optimum value of the nozzle angle α_1 lies within the limits of 14 to 20°. A decrease in the value of angle α_1 less than 14° leads to smaller inlet angle β_1 which may not always give better efficiencies. η_u increases with the decrease in angle β_2 only up to a certain value and hence it is usual to take a value of $\beta_2 = \beta_1 - (3 \text{ to } 6)°$.

The relation u/c_1 appears as the basic characteristic of a turbine stage (or a single-stage turbine).

When the values of the angles α_1, β_1, β_2 and the velocity coefficients φ and ψ remain constant in equation (6-11) the value of η_u depends only on the following expression

$$\left(\cos\alpha_1 - \frac{u}{c_1}\right)\frac{u}{c_1} = \cos\alpha_1\frac{u}{c_1} - \left(\frac{u}{c_1}\right)^2. \qquad (6\text{-}13)$$

When $\frac{u}{c_1} = 0$ and $\frac{u}{c_1} = \cos a_1$ expression (6-13) becomes zero and consequently η_u is zero.

In order to determine the optimum value of u/c_1 the first differential coefficient of the right-hand side in equation (6-13) has to be equated to zero

$$\frac{d\left[\frac{u}{c_1}\cos\alpha_1 - \left(\frac{u}{c_1}\right)^2\right]}{d\left(\frac{u}{c_1}\right)} = 0,$$

whence

$$\cos a_1 - 2\left(\frac{u}{c_1}\right) = 0$$

and

$$\left(\frac{u}{c_1}\right)_{opt} = \frac{\cos\alpha_1}{2}. \qquad (6\text{-}14)$$

Substituting the value of u/c_1 from equation (6-14) in equation (6-12) we have

$$\eta_{u\,max} = \frac{\varphi^2}{2}\left(1 + \psi\frac{\cos\beta_2}{\cos\beta_1}\right)\cos^2\alpha_1. \qquad (6\text{-}15)$$

for $\beta_1 = \beta_2$ the equation becomes

$$\eta_{u\,max} = \frac{\varphi^2}{2}(1+\psi)\cos^2\alpha_1. \qquad (6\text{-}16)$$

Fig. 6-3 shows the variations of η_u as a function of u/c_1.

This graph is obtained in the following manner. Taking a known value of angle α_1, various values of u/c_1 from 0 to cos α_1 are substituted in equation (6-11) or (6-12), thus obtaining η_u. From the data thus obtained η_u can be plotted as a function of u/c_1.

6-2. DETERMINATION OF $\left(\frac{u}{c_1}\right)_{opt}$ FROM THE RELATIVE INTERNAL EFFICIENCY

In the previous section the most advantageous value of u/c_1 was found at which η_u has a maximum value. However, economic operation of a turbine stage (or a single-stage turbine) does not depend so much on η_u as on η_{oi}. Hence $(u/c_1)_{opt}$ has to be determined with reference to $\eta_{oi\,max}$ and not $\eta_{u\,max}$.

As has been observed before, besides loss of energy in the blades, there are additional losses due to friction between the revolving disc and the surrounding steam, windage, leakage through internal clearances, etc. Since heat drop calculations are carried out on the assumption of a constant heat drop, i.e., constant theoretical steam velocity c_1, losses resulting from leakages and wetness of steam would be constant and independent of the value u/c_1. Losses due to friction between the disc and the surrounding steam and those due to windage effects depend on the circumferential velocity as well as the diameter of the disc. For a constant steam velocity c_1 and at a given speed of rotation the diameter of the turbine disc must differ for different values of u/c_1 from which it follows that the losses due to disc friction and windage directly depend on the value of u/c_1. Hence while determining the optimum value of u/c_1 besides losses in the blades, losses due to disc friction and windage must also be accounted for.

The losses h_{leak} and $h_{wetness}$ do not influence the internal efficiency η_{oi} as a function of u/c_1, and hence while investigating the influence of u/c_1 on the variation of η_{oi} these losses may not be taken into consideration. Therefore the internal relative efficiency may be expressed as follows

$$\eta_{oi} = \frac{L_u - L_{fr.w}}{L_0} = \eta_u - \zeta_{fr.w} \qquad (6\text{-}17)$$

where $L_{fr.w}$ —losses due to disc friction and windage per kg of steam, in kgm;

$$\zeta_{fr.w.} = \frac{L_{fr.w.}}{L_0};$$

$L_{fr.w.}$ may be obtained from the relation

$$L_{fr.w.} = \frac{h_{fr.w.}}{A} = \frac{102N_{fr.w.}}{G};$$

so that

$$\zeta_{fr.w.} = \frac{L_{fr.w.}}{L_0} = \frac{102N_{fr.w.}}{Gc_{1t}^2}2g. \qquad (6\text{-}18)$$

The optimum value of (u/c_1) for $\eta_{oi\,max}$ is carried out in the following manner. For various values of (u/c_1) the quantity η_u is determined from equations (6-11) and (6-12). From the results obtained a graph of η_u as function of (u/c_1) is plotted (Fig. 6-3). For these values (u/c_1) the circumferential velocity u is determined: $u=(u/c_1)\,c_1$ from which the mean diameter of the turbine disc is calculated, viz. $d=60\;u/\pi n$.

Next velocity of steam, nozzle and blade dimensions and other quantities necessary for the calculation of $L_{fr.w.}$ are determined. The power loss due to disc friction and windage N_{wind} for each of the assumed values of u/c_1 is calculated from equation (5-4). Substituting this value in equation (6-18) disc friction and windage losses are determined. $\zeta_{fr.w.}$ as a function of u/c_1 is shown in Fig. 6-3.

Since $\eta_{oi} = \eta_u - \zeta_{fr.w.}$ we may obtain a graph of η_{oi} as a function of u/c_1 by deducting the value of $\zeta_{fr.w.}$ from η_u directly in the graphs shown in Fig. 6-3. The value of u/c_1 obtained in the above manner at $\eta_{oi\,max}$ will be the most economical value. The final design of a turbine is carried out with the value of u/c_1 obtained as mentioned above, i.e., for the optimum conditions.

The internal relative efficiency of a turbine stage also may be obtained directly from the

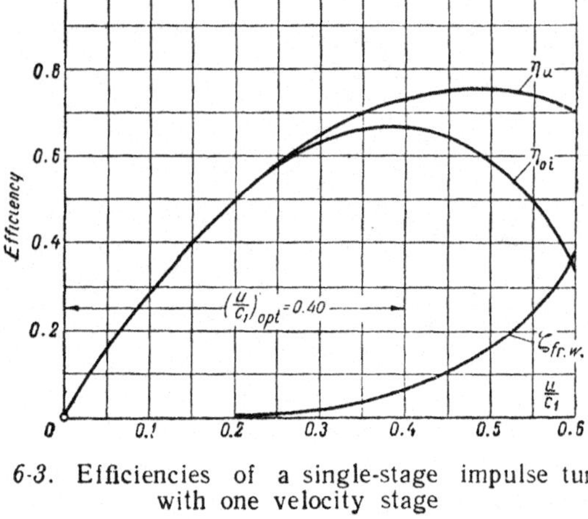

Fig. 6-3. Efficiencies of a single-stage impulse turbine with one velocity stage

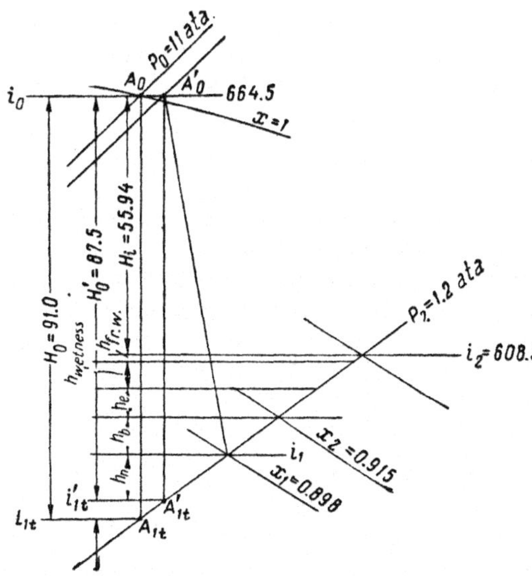

Fig 6-4. I-s diagram for a single-stage impulse turbine

i-s diagram as a relation between the actual heat drop utilised and the theoretical adiabatic heat drop.

For an impulse stage we may write (Fig. 6-4)

$$\eta_u = \frac{H_u}{H'_0} = \frac{H'_0 - h_n - h_b - h_e}{H'_0}. \quad (6\text{-}18a)$$

The relative internal efficiency of the stage is

$$\eta_{oi} = \frac{H_i}{H'_0} = \frac{H'_0 - h_n - h_b - h_e - h_{\text{wet ness}} - h_{\text{fr.w.}}}{H'_0}. \quad (6\text{-}18b)$$

As a check it is recommended that the efficiencies η_u and η_{oi} should also be evaluated from equations (6-4) and (6-17). A good agreement of the results obtained would indicate the accuracy of the calculations.

The relative internal efficiency of a single-stage turbine

$$\eta_{ol} = \frac{H_i}{H'_0} = \frac{H'_0 - h_n - h_b - h_e - h_{\text{wetness}} - h_{\text{fr.w.}}}{H_0}. \quad (6\text{-}18c)$$

6-3. DESIGN OF SINGLE-STAGE IMPULSE TURBINES (SEQUENCE OF CALCULATIONS)

For the design of a single-stage impulse steam turbine the following quantities are specified:

(1) generating capacity of the turbo-generator set at the terminals N_e, kW;

(2) number of revolutions per minute—n;

(3) pressure and temperature of fresh steam p_0 and t_0 respectively;

(4) pressure of exhaust steam (back pressure) p_2, ata.

The theoretical heat drop H'_0 is directly obtained from the *i-s* diagram. Preliminary calculations of mass flow of steam are carried out in accordance with § 5-12. Next velocity of steam issuing from the nozzles is calculated from

$$c_1 = 91.5\varphi\sqrt{H'_0}.$$

The angle of inclination α_1 of the nozzle to the turbine disc may be assumed between the limits of 14-20 degrees, since large values of angle α_1 tend to decrease the efficiency of the stage. Thereafter for various values of u/c_1 η_u and $\zeta_{\text{fr.w}}$ are evaluated. Plotting a graph with the values obtained $\eta_{oi\max}$ is found with its corresponding value of $(u/c_1)_{\text{opt}}$.

The circumferential velocity u and the diameter of the disc d are obtained from the value of $(u/c_1)_{\text{opt}}$. During the detailed design of the turbine stage velocity triangles are drawn and the various losses and the relative internal efficiency η_{oi} are determined. From the value of the relative internal efficiency mass flow of steam is recalculated and the final nozzle and blade dimensions are determined.

6-4. SINGLE-STAGE IMPULSE TURBINE WITH TWO VELOCITY STAGES

Sectional view of a single-stage impulse turbine with two velocity stages is shown in Fig. 6-5.

Pressure drop from p_0 to p_2 takes place in the nozzles *4*. As a consequence of the pressure drop in the nozzle the velocity of steam increases from c_0 in the beginning to c_1. The transformation of the kinetic energy of the flowing steam into mechanical work on the turbine shaft takes place in the two rows of moving blades mounted on the turbine disc. Thus the velocity of steam is reduced from c_1 to c_2 in the first row of moving blades and from c'_1 to c'_2 in the second row. The fixed blades *7* help only in the changing of the direction of flow of the steam issuing from the first moving blade row without contributing to the mechanical work done on the shaft. However, there is a reduction in the velocity of steam flow from c_2 to c'_1 as a result of the frictional losses occurring in the guide blades. Fig. 6-5 also shows the variations of pressure and velocity while flowing through the nozzles and the two velocity stages. The efficiency of a two-row disc is comparatively lower because of the increased losses in the moving and the fixed blades. The primary advantages of a double-velocity stage impulse turbine are the simplicity of con-

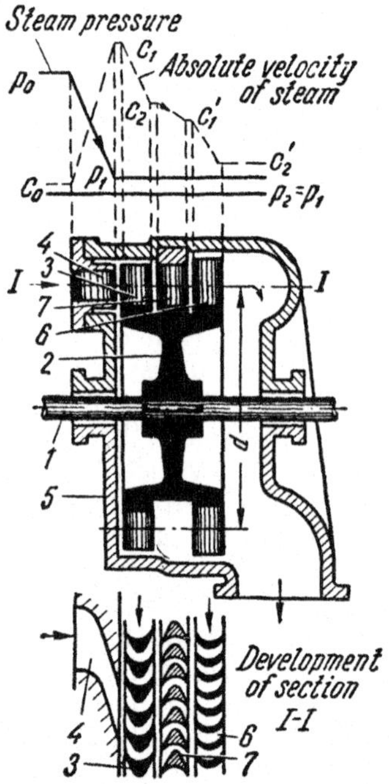

Fig. 6-5. Single-stage impulse turbine with two velocity stages

1— shaft; 2— disc; 3— first row of moving blades; 4— nozzle; 5—stator; 6— second row of moving blades; 7— guide blades

struction, compactness, lower costs of construction, reliability and easy operation. Turbines of this type are mainly used as drives for centrifugal compressors, pumps, small generators, etc., because of their inherent low efficiencies and lower capacities. However, the two-row velocity stage is widely used as the first stage (regulating stage) of present-day turbo-alternators of large capacities and high initial pressures, since large heat drops can be advantageously achieved in it without any appreciable decrement in the efficiency under varying conditions of operation. In the most recent constructions of turbines of very large capacities at supercritical pressures the single-row impulse stage is preferred because of its higher efficiency under design conditions of operation.

Fig. 1-6 shows a single-disc two-row impulse turbine of the Nevsky Works built as a prime mover for an air exhauster. Its initial pressure and temperature conditions are 11 ata and 325°C and it exhausts at a pressure of 4.5 ata.

The turbine shaft with the disc *1* mounted on it is supported by two bearings *5* and *7*. Bearing *5* acts both as a journal and thrust bearing. The thrust bearing receives the axial component of the force exerted by the steam on the moving blades and further helps to maintain the rotor in its position.

Fresh steam enters the nozzles *2* through the five regulating valves *4*. The entry of steam to the turbine is controlled by a set of regulating valves, each feeding a separate group of nozzles, thus forming the so-called nozzle governing system. The exhaust steam is led away through the pipe connecting the turbine to the condenser. The turbine is provided with labyrinth seals *6* at both ends where the turbine shaft projects out through the casing to reduce steam leakage. The main oil pump *8* provides oil under pressure for both lubrication of bearings and for governing. The bearings of the turbine are placed in special casings *9* and *10*.

Single-stage impulse turbines are built with three velocity stages as well. Increased number of velocity stages permits the utilisation of larger heat drops in a turbine with consequent increase in its capacity. However, the efficiency of such turbines would be less than that of a double-velocity stage turbine. Hence impulse turbines with three or more velocity stages are used in very rare circumstances.

The efficiency of an impulse turbine consisting of one or more velocity stages may be obtained from the formula

$$\eta_u = \frac{L_u}{L_0} = \frac{2u\,\Sigma\,(c_{1u} - c_{2u})}{c_{1t}^2}, \qquad (6\text{-}19)$$

where $\Sigma\ (c_{1u} - c_{2u})$—sum of the projections of absolute velocities at the turbine rim.

Slightly rewriting this equation we can show that the quantity η_u is a direct function of (u/c_1) and α_1. The following values of angle α_1 are recommended for impulse turbines:

(1) for two-row discs $\alpha_1 = 16$ to 22°;

(2) for three-row dics $\alpha_1 = 20$ to 24°.

Recommended values of u/c_1:

(1) for twc-row discs $u/c_1 = 0.20$ to 0.26;

(2) for three-row. discs $u/c_1 = 0.10$ to 0.18.

The optimum value of u/c_1 in each case may be obtained by preliminary calculations as has been explained before.

6-5. SEQUENCE OF CALCULATIONS FOR IMPULSE TURBINES WITH TWO VELOCITY STAGES

The following quantities are specified for the design of a turbine with two velocity stages: power generated at the generator terminals, number of revolutions per minute, pressure and temperature of fresh steam and the pressure of exhaust steam. From the *i-s* diagram (Fig. 6-6) we obtain first point A_0 which conforms to the fresh steam conditions of temperature and pressure. From this point a vertical line (adiabatic expansion) is dropped to meet the exhaust pressure isobar in the point A_{1t}.

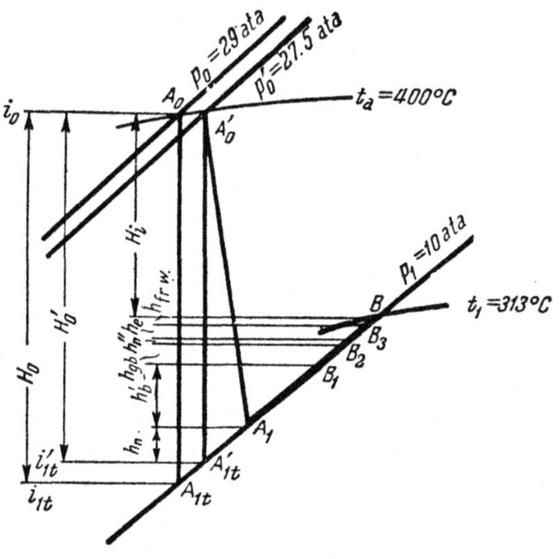

Fig. 6-6. I-s diagram for an impulse turbine with two velocity stages

Loss of pressure in the regulating valves is accounted for by assuming $\Delta p = p_0 - p_0' = (0.03$ to $0.05)p_0$. Setting off this loss Δp on the *i-s* diagram we obtain point A_0', conforming to the steam conditions just before the nozzles.

The adiabatic heat drop from the *i-s* diagram will be

$$H_0 = i_0 - i_{1t}.$$

For preliminary calculations mass flow of steam through the turbine is determined according to formula (5-46); value of efficiency is found from graphs. Adiabatic heat drop in the nozzle is determined from the equation

$$H_0' = i_0 - i_{1t}'.$$

Velocity of steam at the nozzle exit is obtained from

$$c_1 = 91.5\varphi\sqrt{H_0'}.$$

Next five or six different values of u/c_1 are assumed, viz.

a) for two-row discs from 0.10 to 0.3;

b) for three-row discs from 0.05 to 0.20.

Since velocity of steam c_1 is a constant quantity, for each assumed value of u/c_1 the circumferential velocity u is determined and velocity triangles are drawn for each value of u as shown in Fig. 6-7.

The velocity of steam c_1 is drawn from the origin 0 making an angle α_1 to the direction of rotation of the turbine. Relative velocity w_1 and inlet angle β_1 are obtained from the inlet velocity triangle of moving blades. The exit velocity triangle for the first row of moving blades is drawn by assuming an exit angle β_2 and taking $w_2 = \psi w_1$. From this triangle the absolute velocity of exit steam c_2 and its angle of inclination to the disc α_2 are determined. The velocity of steam issuing from the guide blades is obtained from

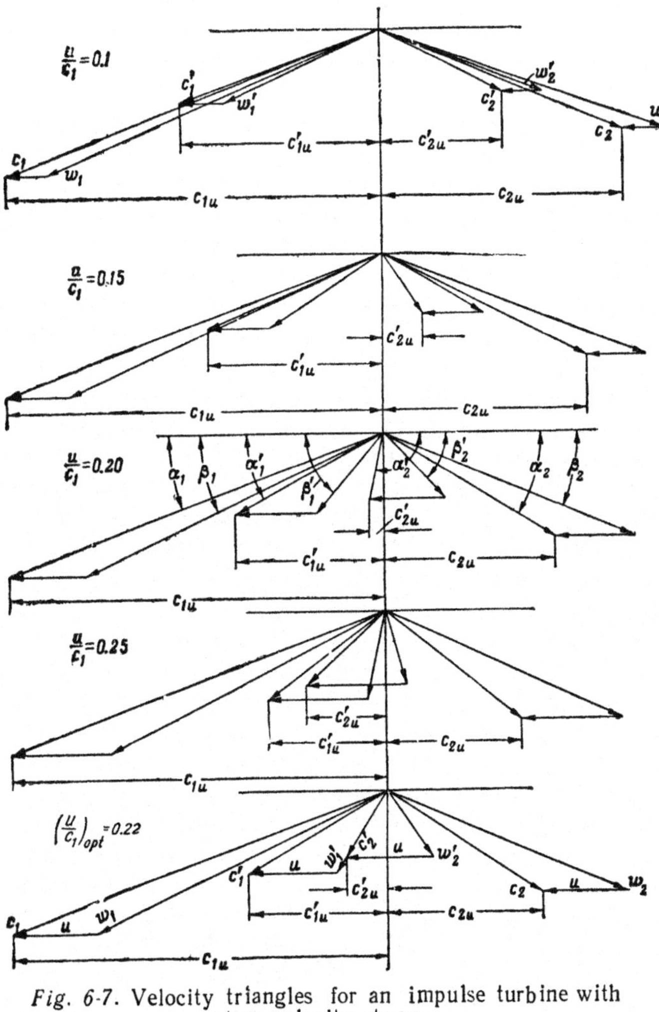

Fig. 6-7. Velocity triangles for an impulse turbine with two velocity stages

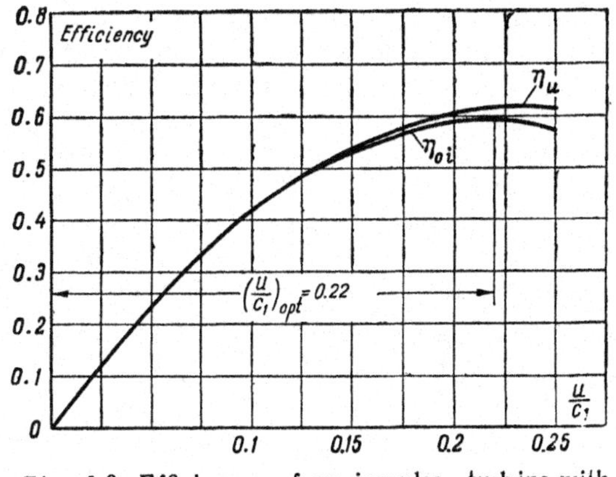

Fig. 6-8. Efficiency of an impulse turbine with two velocity stages as a function of u/c_1

the relation $c'_1 = \psi_{gb} c_2$ (ψ_{gb}—velocity coefficient for guide blades. The angle of inclination of guide blades is an assumed quantity, $\alpha'_1 = \alpha_2 - (3 \text{ to } 5°)$.

The inlet velocity triangle for the second row of moving blades is drawn from the values of c'_1 and α'_1 obtained as explained before. From this velocity triangle we have w'_1 and β'_1, i. e., the relative velocity of steam at entry to the second row of moving blades as well as the inlet blade angle for the second row. The relative velocity of steam at the exit of the second row of moving blades is obtained from the relation $w'_2 = \psi' w'_1$. The exit velocity triangle is drawn assuming the exit angle for the second row of moving blades, from which the magnitude and direction of the exit velocity c'_2 and α'_2 are determined. The diagram of velocity triangles for a two-row single-stage impulse turbine is obtained in the manner described above (Fig. 6-7).

The efficiency at the rim of the turbine disc, for the values of u/c_1 assumed for the given case, are found from equation (6-19)

$$\eta_u = \frac{2u \Sigma (c_{1u} - c_{2u})}{c_{1t}^2},$$

where c_{1u}, c_{2u}, c'_{1u} and c'_{2u} are taken from the velocity diagrams.

Exactly similar diagrams of velocities are drawn for all the other values of u/c_1 and the efficiency η_u is determined for each case. Next a graph is plotted for the efficiency η_u against the values of u/c_1 assumed for the calculation of the above efficiency (Fig. 6-8). The most economic value of u/c_1 is obtained from this graph, as was done in the case of a single-row single-stage impulse turbine. To determine values of η_{oi} for the various values of u/c_1, the power loss $N_{fr.w.}$ is determined from equation (5-4) and the coefficient $\zeta_{f.w.}$ from equation (6-18). The exit height of the nozzles from which the degree of partial admission is determined, is originally assumed within a range of 10 to 20 mm depending on the capacity of the turbine. A mean height of the blades l_{mean} may be assumed to be $1.5l$. The loss coefficient $\zeta_{fr.w.}$ is also plotted as a function of u/c_1 (not shown in Fig. 6-8). The values of the efficiency η_{oi} are directly obtained by graphically deducting $\zeta_{fr.w.}$ from η_u. The optimum value of (u/c_1) is next obtained from the graph where η_{oi} occurs. The final design of the turbine is carried out with the value of u/c_1 thus obtained. The sequence of calculations for the final design is the same as has been the case for the preliminary design calculations.

The efficiency η_u is obtained from equation (6-19) from which the energy losses h_n, h'_b, h_{gb}, h''_b, h_e, etc., are evaluated. To check the correctness of the value obtained for η_u, it is recalculated from the equation

$$\eta_u = \frac{H'_0 - h_n - h'_b - h_{gb} - h''_b - h_e}{H'_0}.$$

If the results obtained for η_u by both methods agree within allowable limits the design is deemed to be satisfactory. If otherwise, the error must be found out and corrected values obtained. Next the value of disc friction and windage losses $h_{fr.w.}$ is determined and the relative internal efficiency of the stage is obtained

$$\eta_{oi}^{st} = \frac{H'_0 - (h_n + h'_b + h_{gb} + h''_b + h_e + h_{fr.w.})}{H'_0}.$$

The efficiency of the turbine as a whole will be

$$\eta_{oi} = \frac{H'_0 - (h_n + h'_b + h_{gb} + h''_b + h_e + h_{fr.w.})}{H_0}.$$

If the product of η_{oi} and η_m sufficiently agrees with the assumed value of η_{re}, then the mass of flow of steam obtained from the preliminary calculations may be taken as the correct value. If, however, there is a divergence of more than 2% the energy loss $h_{f.w.}$ must be re-evaluated and the correct mass flow obtained.

The throat and exit height of the nozzles, degree of partial admission and the dimensions

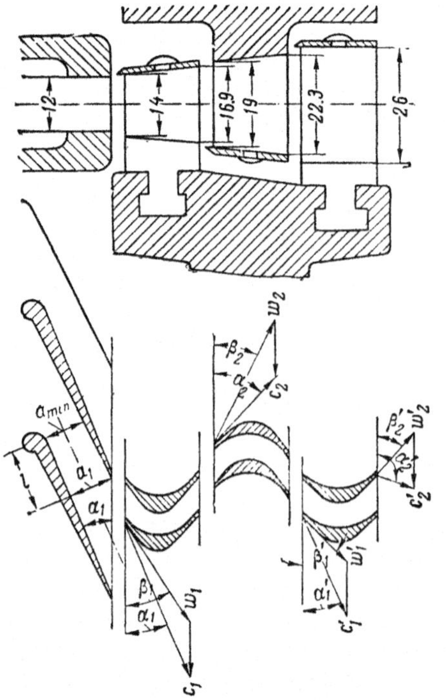

Fig. 6-9. A row with two velocity stages

of the moving and guide blades are calculated on the basis of the mass flow of steam obtained in the manner shown above.

6-6. DESIGN OF A TWO-ROW IMPULSE TURBINE

Example: Design a single-stage impulse turbine with two rows of moving blades for the following conditions. Generating capacity at the generator terminals $N_e = 1{,}000$ kW; $n = 3{,}000$ r.p.m. Pressure and temperature of supply steam $p_0 = 29$ ata and $t_0 = 400°C$. Pressure of exhaust steam $p_2 = 10$ ata.

We shall assume the pressure loss in the regulating valves to be 5% of the supply pressure p_0, so that the pressure before the nozzles will be $p_0' = 0.95 \times p_0 = 0.95 \times 29 = 27.5$ ata. The theoretical heat drop is obtained from the i-s diagram (Fig. 6-6)

$$H_0 = i_0 - i_{1t} = 773 - 706 = 67 \text{ kcal/kg}.$$

The theoretical heat drop taking place in the nozzles will be

$$H_0' = i_0 - i_{1t}' = 773 - 708.5 = 64.5 \text{ kcal/kg}.$$

Velocity of steam at the nozzle exit will be

(a) actual:

$$c_1 = 91.5\varphi\sqrt{H_0'} = 91.5 \times 0.95\sqrt{64.5} = 698.5 \text{ m/sec};$$

(b) theoretical:

$$c_{1t} = \frac{c_1}{\varphi} = \frac{698.5}{0.95} = 735 \text{ m/sec}.$$

Let us assume the nozzle angle as 20°. Exit angle for the first row of moving blades $\beta_2 = \beta_1 - 3°$.

Guide blade angle at exit $\alpha_1' = \alpha_2 - 3°$.

Exit angle for the second row of moving blades $\beta_2' = \beta_1' - 3°$.

Next we shall draw velocity triangles for various assumed values of u/c_1 (Fig. 6-7). All the quantities required for the construction of these velocity triangles are shown in Table 6-1. The values of η_u for the various values of u/c_1 are also shown in this table. A graph of η_u and u/c_1 can now be plotted.

The mass flow of steam through the turbine will be

$$G = \frac{860N_e}{3{,}600H_0\eta_{0i}\eta_m\eta_g} = \frac{860 \times 1{,}000}{3{,}600 \times 67.0 \times 0.58 \times 0.96 \times 0.928} = 6.91 \text{ kg/sec},$$

where $\eta_m = 0.96$ and $\eta_g = 0.928$ (from graphs 5-8 and 5-11); η_{0i} is assumed to be 0.58.

Disc friction and windage losses are determined from equations (5-4a) and (6-18).

We shall assume the mean height of the blades as 20 mm; coefficient $\beta = 2.06$ for a two-row disc (equation 5-4a); specific weight of steam in the blade chamber $\gamma = 3.95$ kg/m³, conforming to the given data.

The quantities $N_{\text{fr.w.}}$ and $\zeta_{\text{fr.w.}}$ are also shown for all the values of u/c_1 in Table 6-1. Graphically deducting $\zeta_{\text{fr.w.}}$ from η_u (Fig. 6-8) we obtain η_{0i} as a function of u/c_1, from which the optimum value of u/c_1 is found to be 0.22.

The final design calculations are now carried out with this value of u/c_1. The results of these calculations as well are shown in Table 6-1.

The heat losses in the turbine blades are determined from the equations

a) for nozzles:

$$h_n = \frac{c_{1t}^2 - c_1^2}{8{,}378} = \frac{735^2 - 698.5^2}{8{,}378} = 6.325 \text{ kcal/kg};$$

b) for moving blades of the first row:

$$h_b' = \frac{w_1^2 - w_2^2}{8{,}378} = \frac{555^2 - 445^2}{8{,}378} = 12.18 \text{ kcal/kg};$$

c) for guide blades:

$$h_{gb} = \frac{c_2^2 - c_1'^2}{8{,}378} = \frac{320^2 - 272^2}{8{,}378} = 3.38 \text{ kcal/kg};$$

d) for moving blades of the second row:

$$h_b'' = \frac{w_1'^2 - w_2'^2}{8{,}378} = \frac{160^2 - 141^2}{8{,}378} = 0.68 \text{ kcal/kg};$$

e) for exit velocity:

$$h_e = \frac{c_2'^2}{8{,}378} = \frac{135^2}{8{,}378} = 2.175 \text{ kcal/kg}.$$

To check the correctness of the heat losses obtained above we shall find the efficiency η_u from these values and compare it with the result obtained graphically for the optimum u/c_1

$$\eta_u = \frac{H_0' - (h_n + h_b' + h_{gb} + h_b'' + h_e)}{H_0'} = \frac{64.5 - (6.325 + 12.18 + 3.38 + 0.68 + 2.175)}{64.5} = \frac{39.76}{64.5} = 0.616.$$

The calculation error $\frac{0.616 - 0.615}{0.616} \times 100 = 0.16\%$ is permissible.

Losses due to disc friction and windage will be determined from the equation

$$h_{\text{fr.w.}} = \frac{102N_{\text{fr.w.}}}{427G} = \frac{102 \times 40.3}{427 \times 6.91} = 1.39 \text{ kcal/kg}.$$

The heat drop utilised in the turbine will be

$$H_i = H_0' - (h_n + h_b' + h_{gb} + h_b'' + h_e + h_{\text{fr.w.}}) = 64.5 - 26.13 = 38.37 \text{ kcal/kg}.$$

The relative internal efficiency of the turbine, without considering the losses in the regulating valves, will be

$$\eta_{0i}' = \frac{H_i}{H_0'} = \frac{38.37}{64.5} = 0.595,$$

which agrees well with the value of $\eta_{0i\,\max}$ obtained from the graph shown in Fig. 6-8.

The relative internal efficiency of the turbine, taking into consideration the losses occurring in the regulating valves, will be

$$\eta_{0i} = \frac{H_i}{H_0} = \frac{38.37}{67} = 0.572.$$

From the value of η_{0i} obtained we shall now find the exact mass flow of steam through the turbine

$$G_0 = \frac{860N_e}{3{,}600H_i\eta_m\eta_g} = \frac{860 \times 1{,}000}{3{,}600 \times 38.37 \times 0.96 \times 0.928} = 7.00 \text{ kg/sec}.$$

Table 6-1

$\frac{u}{c_1}$	0.10	0.15	0.20	0.25	0.22	Dimension
$u=\left(\frac{u}{c_1}\right)c_1$	69.85	104.7	139.7	174.5	153.5	m/sec
$d=\frac{60u}{\pi n}$	0.445	0.666	0.890	1.113	0.978	m
w_1	630	600	569	537.5	555	m/sec
β_1	22°20′	23°30′	25°10′	26°30′	25°20′	deg
$\beta_2=\beta_1-3°$	19°20′	20°30′	22°10′	24°30′	22°20′	deg
ψ	0.82	0.82	0.82	0.83	0.82	—
$w_2=\psi w_1$	516	491	466	446	445	m/sec
c_2	450	395	345	297	320	m/sec
α_2	22°20′	25°50′	30°40′	38°30′	32°50′	deg
$\alpha_1'=\alpha_2-3°$	19°20′	22°50′	27°40′	35°30′	29°50′	deg
ψ_{gb}	0.82	0.82	0.83	0.87	0.85	—
$c_1'=\varphi_{gb}c_2$	369	324	286	250	272	m/sec
w_1'	302	235	177.5	148.5	160	m/sec
β_1'	23°50′	32°30′	49°20′	78°20′	58°10′	deg
$\beta_2=\beta_1'-3°$	20°50′	29°30′	46°20′	75°20′	55°10′	deg
ψ_1	0.82	0.84	0.88	0.88	0.88	—
$w_2'=\psi' w_1'$	247.5	197.5	157	131	141	m/sec
c_2'	236	122.5	117.5	189	135	m/sec
α_2'	27°	54°30′	104°30′	138	121°50′	deg
c_{1u}	656	656	656	656	656	m/sec
c_{2u}	416	352	296	232.5	266	m/sec
c_{1u}'	346	297	251	203	235	m/sec
c_{2u}'	210.5	70	−30	−140	−71	m/sec
$\Sigma(c_{1u}-c_{2u})$	1,628.5	1,375	1,173	951.5	1,083	m/sec
$2u\,\Sigma(c_{1u}-c_{2u})$	227,000	287,000	326,500	331,000	332,500	m²/sec²
η_u	0.421	0.532	0.605	0.613	0.615	—
$N_{fr.w.}=\beta n^3 d^4 l\gamma\times10^{-10}$	1.73	8.65	27.6	69.0	40.3	kW
$\zeta_{fr.w.}=\frac{204g N_{fr.w.}}{G c_{1t}^2}$	0.00093	0.00465	0.0148	0.0371	0.0216	—

The difference between the mass flow of steam obtained from the preliminary calculations and the final one

$$\Delta G=\frac{7.00-6.91}{7.00}\,100=1.28\%,$$

is within the permissible limits and therefore all the calculations need not be repeated.

The front labyrinth seal consists of $z=40$ partitions, shaft diameter $d=80$ mm, annular gap between the shaft and the labyrinth partitions $s=0.3$ mm,

$$f_s=\pi\times0.08\times0.0003=0.0000755\ \text{m}^2.$$

The pressure in the last labyrinth compartment will be from equation (5-23)

$$p_{cr}=\frac{0.85\times10}{\sqrt{40+1.5}}=1.32\ \text{ata}.$$

Steam leakage through the labyrinth seal will be from equation (5-22)

$$G_{leak}=0.00755\times\sqrt{\frac{9.81}{40+1.5}\times\frac{10}{0.25}}=0.0226\ \text{kg/sec}.$$

Since $\frac{p_1}{p_0'}=\frac{10}{27.5}=0.363$ is less than critical we shall make use of convergent-divergent nozzles, the minimum or throat section for which will be

$$f_{min}=\frac{G_0+G_{leak}}{203\sqrt{\frac{p_0'}{v_0'}}}=\frac{7.023}{203\sqrt{27.5/0.1125}}=0.00221\ \text{m}^2=$$

$$=22.1\ \text{cm}^2,$$

where $v_0'=0.1125$ m³/kg — specific volume of steam before the nozzles.

The exit section of the nozzles, assuming that the continuity equation holds good, will be

$$f_{max}=\frac{(G_0+G_{leak})\,v_1}{c_1}=\frac{7.023\times 0.253}{698.5}=$$
$$=0.00254\ m^2=25.4\ cm^2,$$

where $v_1=0.253$ m³/kg — specific volume of steam at the nozzle exit (point A_1 in Fig. 6-6).

We shall assume the exit height of the nozzles to be $l=12$ mm. The degree of partial admission would then be

$$\varepsilon=\frac{f_{max}}{\pi dl\sin\alpha_1}=\frac{0.00254}{\pi\times 0.978\times 0.012\sin 20^\circ}=0.202,$$

which is within the allowable limits.

The inlet height of the first row of moving blades may be assumed as equal to

$$l_1'=l+2=12+2=14\ \text{mm}.$$

The exit height of the first row of moving blades therefore will be from equation

$$l_1''=\frac{G_0v_1'}{\pi d\varepsilon w_2\sin\beta_2}=\frac{7.00\times 0.263}{\pi\times 0.978\times 0.202\times 455\sin 22°20'}=$$
$$=0.0169\ m=16.9\ mm,$$

where $v_1'=0.263$ m³/kg — specific volume of steam at the exit of the first row of moving blades (point B_1 in Fig. 6-6).

Assuming the inlet height of the guide blades as

$$l_{gb}=l_1''+2{,}1=16.9+2{,}1=19\ \text{mm},$$

the exit height of these blades will be found from equation

$$l_{gb}''=\frac{G_0v_{gb}}{\pi d\varepsilon c_1'\sin\alpha_1'}=$$
$$=\frac{7\times 0{,}266}{\pi\times 0{,}978\times 0{,}202\times 272\sin 29°50'}=0{,}0223\ m=22{,}3\ mm,$$

where $v_{gb}=0.266$ m³/kg — specific volume of steam at the exit of the guide blades (point B_2 in Fig. 6-6).

Exit height of the second row of the moving blades will be

$$l_2''=\frac{G_0v_2}{\pi d\varepsilon w_2'\sin\beta_2'}=$$
$$=\frac{7{,}00\times 0{,}267}{\pi\times 0{,}978\times 0{,}202\times 141\sin 55°10'}=0{,}0260\ m=26\ mm,$$

where $v_2=0.267$ m³/kg — specific volume of steam at the exit of the second row of the moving blades (point B_3 in Fig. 6-6).

The inlet height of the second row of the moving blades will be assumed as equal to the exit height of the first row of the moving blades, i.e., 26 mm.

The various dimensions of the turbine are shown in the sectional diagram shown in Fig. 6-9.

We shall make use of 20 nozzles placed along the circumference of the disc, so that the throat section of each nozzle will be

$$f_{min}'=\frac{f_{min}}{20}=\frac{22.1}{20}=1.105\ \text{cm}.$$

We shall assume the height of the nozzles at the throat section to be the same as that at the exit, i.e., $l_{min}=l=12$ mm. The width of the nozzles at the throat section will therefore be

$$a_{min}=\frac{f_{min}'}{l_{min}}=\frac{1.105}{1.2}=0.92\ cm=9.2\ mm.$$

Width of the nozzles at the exit will be

$$a_1=\frac{f_{max}}{20l}=\frac{25.4}{20\times 1.2}=1.057\ cm=10.57\ mm.$$

Assuming the divergence of the nozzles to be $\gamma=6°$ the length of the divergent portion will be

$$l=\frac{a_1-a_{min}}{2\tan\frac{\gamma}{2}}=\frac{10.57-9.2}{2\tan 3°}=\frac{1.37}{2\times 0.0524}=13\ \text{mm}.$$

Chapter Seven

MULTISTAGE TURBINES

7-1. IMPULSE TURBINES WITH PRESSURE STAGES

Multistage turbines with pressure stages have found a wide field of usage in industry as prime movers for large-capacity electric generators, because of their ability to produce larger power in comparison with a single-stage turbine.

The number of pressure stages vary within a very wide range: from 4-5 to as much as 40 pressure stages. In multistage turbines expansion of steam takes place from p_0' before the nozzles of the first stage to the back pressure p_2 after the blades of the last stage in the nozzle systems of all the pressure stages consecutively. The velocity and pressure variation diagram shown in the upper part of Fig. 7-1 shows the process of transformation of potential energy into the kinetic in a three pressure-stage impulse turbine. The pressure of steam drops from p_0' to p_1 in the nozzles of the first stage, and at the same time the velocity increases from c_0 to c_1. This steam velocity decreases from c_1 to c_2 while passing through the first stage moving blades, i. e., the kinetic energy of the flowing steam gets converted into mechanical work on the turbine shaft. An exactly similar process takes place in all the consecutive stages as well. The distribution of heat drop in a large number of pressure stages enables the attainment of lower velocities for the steam flowing through the system of moving blades and consequently more advantageous values are obtained for the relation u/c_1 as well as ψ, which tend to improve the efficiency of the turbine. Multistage turbines have in addition many other advantages which will be described later.

7-2. HEAT DROP PROCESS ON THE *I-S* DIAGRAM FOR MULTISTAGE TURBINES

On the *i-s* diagram shown in Fig. 7-2 the initial steam conditions are denoted by the point A_0 (pressure p_0, temperature t^0). H_0 gives the total heat drop neglecting losses; p_{2c} is the back pressure (if the turbine is a condensing one then this would be the pressure in the condenser).

Point A_0' conforms to the steam conditions before the first group of nozzles. H_0'—heat drop occurring in the turbine taking into consideration the heat losses in the regulating valves and exhaust piping, h_0'—theoretical heat drop in the first pressure stage. Line $A_0'a_1$ shows the actual process of expansion of steam in the first pressure stage. The steam conditions at the exit from the first stage (i. e., the initial conditions for the second stage) are obtained by setting off the heat losses h_b' in the blades, $h'_{fr.w.}$ due to friction and windage and the carry-over loss or exit velocity loss h_e' occurring in the first stage on the *i-s* diagram. The heat utilised for doing mechanical work will therefore be

$$h_i' = h_0' - h_n' - h_b' - h'_{fr.w.} - h_e'. \qquad (7\text{-}1)$$

Either full or partial use is made of the carry-over velocity from the previous stage in the succeeding stage in almost all multistage turbines. However, in the following stages the carry-over velocity is not utilised: first stage (in the case of nozzle governing), last stage, all those stages after which steam is extracted for regeneration or other purposes, as well as all those stages after which there is a sufficiently large chamber which may have been designed to simplify the construction. In all the above cases because of the presence of an appreciably large space between the blades of the above-mentioned stages and the nozzles of the following ones, it is not possible to utilise the carry-over velocity.

If the carry-over velocity of a stage is made use of in the following stage then it is included in the *i-s* diagram as follows (Fig. 7-2): the heat equivalent of the carry-over velocity is added vertically to the heat drop of the following stage from the point got by setting off all the losses in the previous stage, i. e., the point where the heat drop process is completed in the first stage (e. g., points c_1, d_2, l_3 in Fig. 7-2). Points d_0^{I}, d_0^{II}, d_0^{III} give fictitious steam parameters before the nozzles of the concerned stages (stagnation parameters).

Thus the heat drop utilised in any of the stages of a multistage turbine will be

$$h_i = h_0 + \mu h_e^{pr} - h_n - h_b - h_{fr.w.} - h_{leak} - h_{wetness} - h_e, \qquad (7\text{-}1a)$$

where μ —coefficient of utilisation of the carry-over velocity; (usually from 0.7 to 0.8). Sometimes $\mu = 1$,

h_e^{pr} — carry-over loss from the previous stage.

There are no leakage losses in the first stage, since the nozzles of the first disc are directly fixed to the turbine stator.

The following losses are set off from the point a_{2t} from the second stage onwards

$$h_n'', h_b'', h''_{fr.w.}, h''_{leak}, h_e''.$$

In the case of the third and the fourth stage to the above-mentioned losses are also added losses due to wetness of steam since these stages would be operating under conditions of wet steam (Fig. 7-2 refers to a condensing turbine).

The steam conditions after the nozzles in the various stages of the turbine are given by the

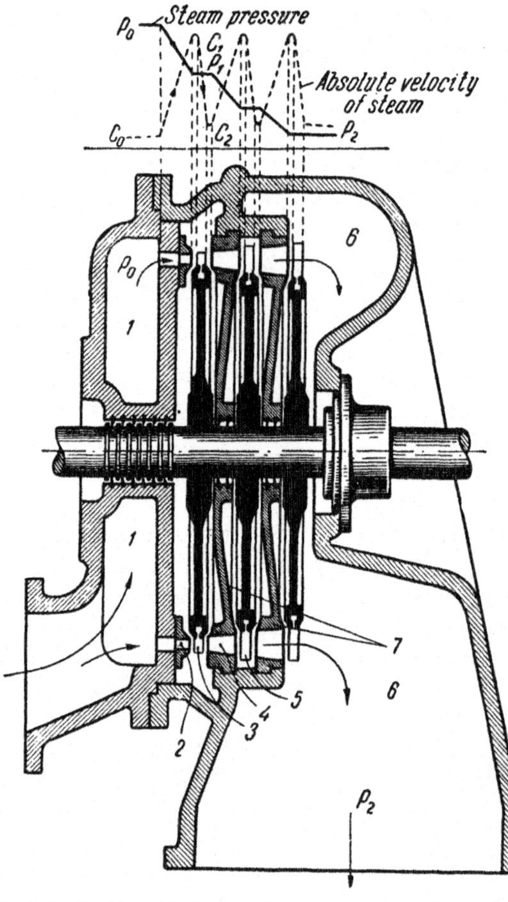

Fig. 7-1. Section through a three pressure stage impulse turbine

1 and *6*— fresh steam and exhaust steam chambers; *2* and *4*— nozzles; *3* and *5*— moving blades; *7*— diaphragms

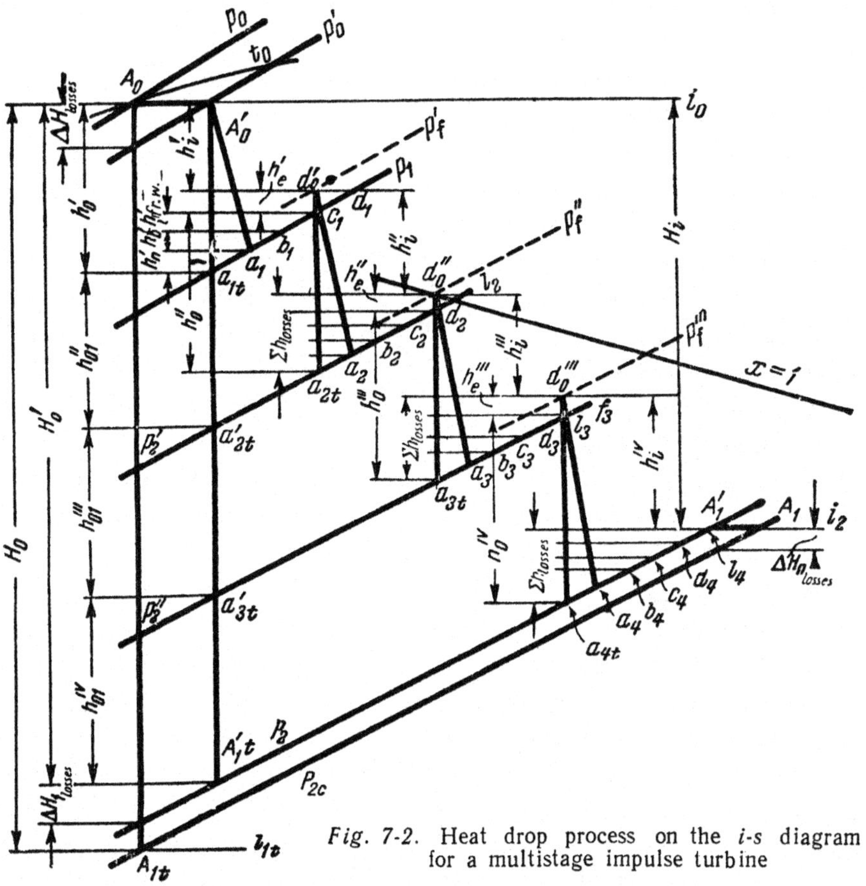

Fig. 7-2. Heat drop process on the *i-s* diagram for a multistage impulse turbine

points a_1, a_2, a_3 and a_4, and the steam conditions after the moving blades are given by points b_1, b_2, b_3 and b_4. The conditions of back-pressure steam after the last row of moving blades are given by point A_1', and point A_1 gives the steam conditions after the exhaust piping, i.e., the steam pressure and temperature in the condenser.

The heat utilised in the turbine will therefore be

$$H_i = \sum h_i = h_i^{I} + h_i^{II} + h_i^{III} + h_i^{IV}\,[1] \qquad (7\text{-}2)$$

where h_i^{I}, h_i^{II}, h_i^{III}, h_i^{IV} are the heat drops utilised in each of the turbine stages.

The relative internal efficiency of a stage for a miltistage turbine (including the carry-over velocity from the previous stage) will be

$$\eta_{oi}^{st} = \frac{h_i}{h_0 + \mu h_e^{pr}}. \qquad (7\text{-}3)$$

The relative internal efficiency of the turbine as a whole will be

$$\eta_{oi} = \frac{\sum_1^z h_i}{H_0} = \frac{H_i}{H_0}. \qquad (7\text{-}4)$$

7-3. HEAT RECOVERY COEFFICIENT

The constant pressure lines (isobars) diverge on an *i-s* diagram, and therefore with the increase in entropy, the adiabatic heat drop between two pressures increases. The heat drop process in a multistage turbine does not strictly take place along the theoretical adiabatic line but along a somewhat curved line as shown in the figure (Fig. 7-2). The actual heat drop process considering all the heat losses in each of the stages for a multistage turbine with four pressure stages is shown in Fig. 7-2, and is given by the line $A_0'a_1d_0'a_2d_0''a_3d_0'''a_4A_1'$.

Since the isobars are divergent we have $h_0^{II} > h_{01}^{II}$, $h_0^{III} > h_{01}^{III}$ and $h_0^{IV} > h_{01}^{IV}$. Consequently, the theoretical heat drop of the original adiabatic process $A_0'A_{1t}'$ will be less than the sum of the heat drop in each of the four stages for the actual process of expansion, i.e.,

$$H_0' = h_{01}^{I} + h_{01}^{II} + h_{01}^{III} + h_{01}^{IV} < \\ < h_0^{I} + h_0^{II} + h_0^{III} + h_0^{IV}.$$

For a turbine with z stages we may write

$$H_0' < \sum_1^z h_0. \qquad (7\text{-}5)$$

The difference between the right- and left-hand sides of equation 7-5 gives the heat returned to the system, in the case of multistage turbines, for utilisation in the succeeding stages.

It is usual to express the relationship between H_0' and $\sum_1^z h_0$ as

$$\sum_1^z h_0 = (1 + \alpha) H_0', \qquad (7\text{-}6)$$

where $\alpha < 1$ is known as the heat recovery

[1] Here and elsewhere the turbine stages are marked as I, II, III... or ′ ″ ‴.

c o e f f i c i e n t. From equation (7-6) we have

$$(1+\alpha)=\frac{\sum_{1}^{z} h_0}{H_0'} \quad \text{and}$$

$$\alpha=\frac{\sum_{1}^{z} h_0 - H_0'}{H_0'}=\frac{h_v}{H_0'}, \qquad (7\text{-}7)$$

where h_v— heat recovered in the various stages as a result of utilisation of the carry-over velocity. It would be erroneous, however, to conclude from the above that heat losses in a turbine are to be counted as positive advantages, since only a portion of these losses is recovered while at the same time such an increase in the heat losses leads to a notable reduction in the efficiency.

The larger the number of stages, the greater is the heat recovery, so that increase in the number of pressure stages may be deemed as a desirable factor. The heat recovery coefficient increases with the increase in the number of stages and decrease in the efficiency of the turbine.

In general the value of α is taken as 0.03 to 0.08. Often $\alpha=0.04$ to 0.06.

For a multistage turbine we have

$$H_i=h_i^{I}+h_i^{II}+h_i^{III}+\ldots+h_i^{z},$$

$$\text{or } H_0'\eta_{oi}=h_0^{I}\eta_{oi}^{I}+h_0^{II}\eta_{oi}^{II}+$$

$$+h_0^{III}\eta_{oi}^{III}+\ldots+\eta_0^{z}\eta_{oi}^{z}.$$

Assuming that the efficiency of each stage remains the same, we have

$$H_0'\eta_{oi}+\sum_{1}^{z} h_0\eta_{oi}^{st},$$

from which

$$\eta_{oi}=(1+\alpha)\,\eta_{oi}^{st},$$

where η_{oi}^{st} —efficiency of each stage.

From the last equation it follows that the efficiency of a multistage turbine as a whole is greater than the average efficiency of each of its stages.

7-4. CHARACTERISTIC COEFFICIENT FOR MULTISTAGE TURBINES

It was shown in paragraphs 6-1 and 6-2 that the most economical single-stage impulse turbine is characterised by the ratio u/c_1. The same relation also governs the design of multistage turbines.

The heat drop occurring in the nozzles of a stage [neglecting utilisation of exit velocity (carry-over velocity)] may be expressed as follows

$$h_0=\frac{Ac_1^2}{2g\varphi^2}=\frac{c_1^2}{8{,}378\varphi^2}=\frac{u^2}{8{,}378\varphi^2x^2}, \qquad (7\text{-}8)$$

where $x=(u/c_1)$, or it may be further written as

$$x^2h_0=\frac{u^2}{8{,}378\varphi^2}. \qquad (7\text{-}8a)$$

For a multistage turbine summation of the left- and right-hand sides gives

$$\sum x^2h_0=\frac{1}{8{,}378\varphi^2}\sum u^2. \qquad (7\text{-}8b)$$

If it is assumed that the ratio x is the same for all the stages then it may be taken outside the summation so that

$$8{,}378\varphi^2x^2=\frac{\sum u^2}{\sum h_0}=\frac{\sum u^2}{(1+\alpha)H_0'} \qquad (7\text{-}8c)$$

or finally

$$Y=8{,}378\varphi^2x^2=\frac{\sum u^2}{(1+\alpha)H_0'}=\frac{\sum u^2}{H_0\alpha}. \qquad (7\text{-}8d)$$

The coefficient Y was first suggested by C. A. Parsons and is known as the characteristic coefficient for multistage turbines. This coefficient is similar to the ratio u/c_1 and is the main factor governing the economical design of the turbine as a whole

An exactly similar relationship is obtained for turbines making use of the carry-over velocity as well as for reaction turbines.

The relation between the characteristic coefficient and the relative effective efficiency is shown in Fig. 7-3. From the graph it is clearly seen that η_{re} increases with the increase in the value of Y. There is a rapid increase in the value of η_{re} up to $Y=2{,}000$, beyond which the growth of η_{re} slows down.

For a given theoretical heat drop in a turbine the characteristic coefficient Y increases with increase in Σ_{u^2}. Large values of Σ_{u^2} may be

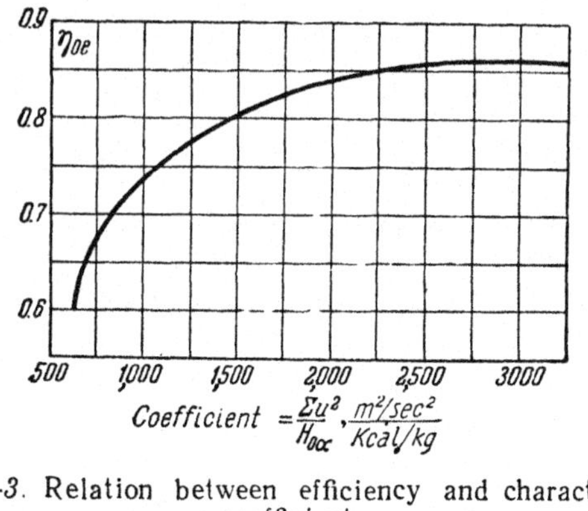

Fig. 7-3. Relation between efficiency and characteristic coefficient

obtained either by increasing the number of stages, by increasing the diameter of the blade discs, or r. p. m.

Single-cylinder turbines with relatively smaller number of stages (from 4 to 8-10) are compact, cheap, and very convenient from the operational point of view. However, their efficiency is not very high which is the main disadvantage of this type of turbine. A typical example of such a turbine [K-3 (AK-3)][1] is shown in Fig. 7-4. This turbine is made by the Kirov Turbine Building Works and has a capacity of 3,000 kW at 3,000 r. p. m. The turbine consists of a two-row Curtis stage and four pressure stages. The former practice was to build large-capacity turbines with larger number of stages at a shaft speed of 1,500 r. p. m. However, the leading factories (L. M. W., Kh. T W., etc.) have now switched over to the construction of turbines with shaft speeds of 3,000 r. p. m., thus reducing the number of pressure stages.

[1] A non-standard turbine.

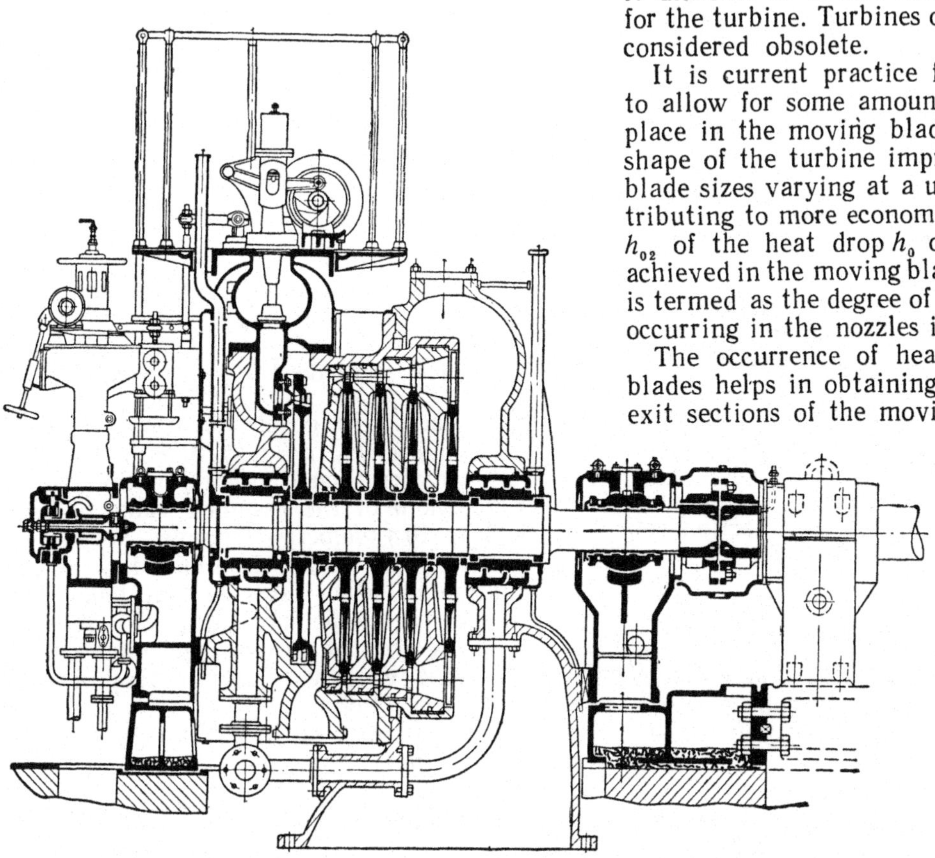

Fig. 7-4. Multistage impulse turbine built by the Kirov Turbine Works, type OK-30; capacity 3,000 kW at 3,000 r. p. m.

7-5. REACTION IN PRESSURE STAGES

Purely impulse turbines are mainly built with several stages with medium capacities of power generation. Their main advantages are simplicity of construction, low costs, reliability and convenience of operation.

The exit sections of the moving blades in turbines of the pure impulse type are usually made slightly greater than the design value, and consequently, while the turbine is operating under normal design conditions, the exit sections of the moving blade passages are not filled completely by the flowing steam. This leads to eddy formation and entrainment of the surrounding steam with the main flow as well as suction of steam through the annular gap between the crown of the moving blades and the stator. At the roof of the blades a spiral flow is established, i. e., there occur additional losses thus reducing the velocity coefficient ψ, and consequently the efficiency of the turbine.

Besides in turbines of the pure impulse type the exit heights of the moving blades rapidly increase especially in low-pressure stages making it difficult to obtain smooth streamlined shape for the turbine. Turbines of such a design are now considered obsolete.

It is current practice for multistage turbines to allow for some amount of heat drop to take place in the moving blades as well so that the shape of the turbine improves considerably, the blade sizes varying at a uniform rate, thus contributing to more economical designs. A fraction h_{02} of the heat drop h_0 occurring in a stage is achieved in the moving blades. The ratio $h_{02}/h_0=\varrho$ is termed as the degree of reaction. The heat drop occurring in the nozzles is equal to $h_{01}=h_0-h_{02}$.

The occurrence of heat drop in the moving blades helps in obtaining a full flow even in the exit sections of the moving blades thus obviating eddy formation as well as suction and entrainment. However, the salutary effect of the degree of reaction on the efficiency of the turbine is lost if the heat drop occurring in the moving blades is relatively large, since such a state leads to leakage of steam through the annular space between the moving blades and the turbine stator, i. e., the efficiency decreases. In such cases radial labyrinth seals are provided.

Besides, as a result of the heat drop occurring in the moving blades, i.e., with the increase in the degree of reaction there is a considerable amount of pressure exerted on the blade disc which is transmitted to the thrust bearing.

Hence, to avoid large axial thrusts it is usual to allow for a degree of reaction of only 4 to 5% in the high-pressure stages. In the low-pressure stages, however, the degree of reaction may reach a value as much as 20% (sometimes 30%), or even more if the pressure drop is relatively small. In steam turbines of large capacities it is now usual to have 50 to 60% degree of reaction for the low-pressure stages (these stages may very well be termed as reaction stages). To reduce the axial thrust on the blade discs of high-pressure stages it is usual to provide pressure equaliser holes for these discs, if the degree of reaction for these stages does not go beyond 5 to 15%. The low-pressure stages are not provided with such pressure equaliser holes, since their presence—if there is a considerable degree of reaction—would lead to large leakage losses and consequent decrease in turbine efficiency. The magnitude of pressure difference in the low-pressure stages being inconsiderable such pressure equalisation for low-pressure stages is deemed unnecessary.

In the case of pure impulse turbines while operating at conditions other than designed there is a possibility of existence of some degree of reaction and consequent pressure difference between the two sides of the revolving disc. Leakage of steam through the diaphragm seals (especially if the annular gap is relatively large) also enhances the degree of reaction and axial thrust. Hence pressure equaliser holes are provided in the case of pure impulse turbines as well. Turbines operating with an inconsiderable degree of reaction do not materially differ in construction from those of the pure impulse type, the required degree of reaction being provided by suitably altering the exit sections of the guide as well as the moving blades.

Design of Pressure Stages with Degree of Reaction

The theoretical velocity of steam issuing from a nozzle is obtained from the equations:

a) utilising the carry-over velocity from the previous stage

$$c_{1t}=91.5\sqrt{h_{01}+\frac{\mu c_2^2}{8{,}378}}=$$
$$=91.5\sqrt{(1-\varrho)h_0+\mu h_e^{pr}}, \qquad (7\text{-}9)$$

where

h_e^{pr}—heat equivalent of the carry-over velocity from the previous stage;

μ—coefficient of utilisation of the carry-over velocity;

b) neglecting the carry-over velocity

$$c_{1t}=91.5\sqrt{(1-\varrho)h_0}. \qquad (7\text{-}9')$$

Coefficient μ depends on the magnitude of the axial gap between the moving blades and the stationary nozzles and also on the thickness of the trailing edges of the moving blades and the inlet edges of the nozzles. With increase in this gap μ decreases. For design purposes μ may be taken equal to 0.7 to 0.8. If the axial gap is comparatively small it is usual to take μ equal to 1.

The actual velocity of steam at the nozzle exit is

$$c_1=\varphi c_{1t}. \qquad (7\text{-}10)$$

The diagram of velocity triangles is drawn with this velocity c_1 and an angle $\alpha_1=11$ to 20°.

Relative velocity of steam at entry to the moving blades and the angle β_1 are obtained directly from the velocity triangles, or analytically.

The theoretical relative velocity of steam at the exit from the moving blades is found from equation (2-25):

$$w_{2t}=91.5\sqrt{\frac{w_1^2}{8{,}378}+h_{02}}=$$
$$=91.5\sqrt{\frac{w_1^2}{8{,}378}+\varrho h_0}. \qquad (7\text{-}11)$$

The relative velocity of steam at the exit from the moving blades taking into consideration the losses will be

$$w_2=\psi w_{2t}=91.5\psi\sqrt{\frac{w_1^2}{8{,}378}+\varrho h_0}. \qquad (7\text{-}12)$$

Velocity coefficient φ for steam velocities of 300 to 400 m/sec and $\varrho=5$ to 10% is taken as 1.5 to 2% greater than that for impulse stages.

With the help of the assumed value of $\beta_2=\beta_1-(3$ to $10°)$ velocity triangles are drawn from which c_2 and α_2 are determined.

Evaluation of heat losses, actual heat drop utilised in a stage, efficiency, internal developed power and the sizes of moving and guide blades is carried out exactly as in the case of pure impulse turbine stages.

Heat losses in moving blades are determined from the formula

$$h_b=\frac{w_{2t}^2-w_2^2}{8{,}378}=(1-\psi^2)\frac{w_{2t}^2}{8{,}378}=$$
$$=(1-\psi^2)\left(\frac{w_1^2}{8{,}378}+\varrho h_0\right). \qquad (7\text{-}13)$$

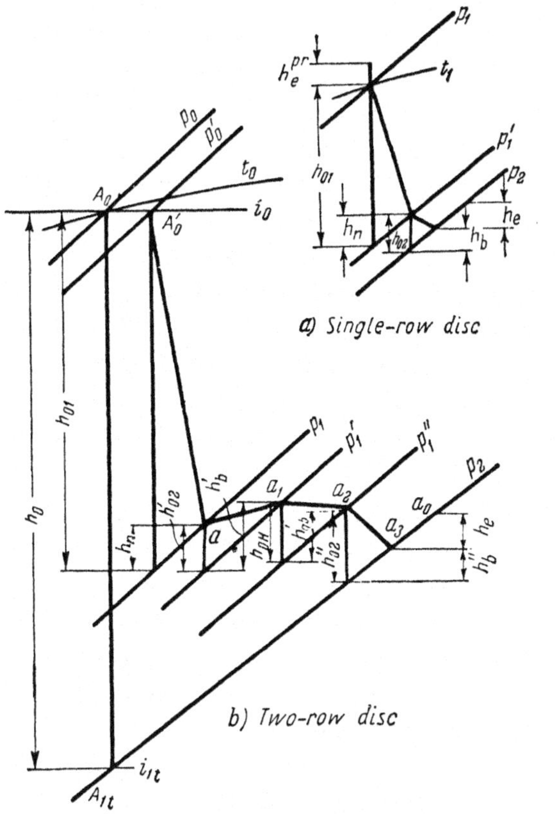

Fig. 7-5. I-s diagram for an impulse stage with some reaction

The work done by one kilogram of steam on the moving blades of a stage is determined from equation (6-3).

The theoretical work of one kg of steam on the moving blades of a stage will be

$$L'_b = \frac{c^2_{ad}}{2g},$$

where

$$c_{ad} = 91.5\sqrt{\mu h^{pr}_e + h_{01} + h_{02}} = \sqrt{c^2_{1t} + w^2_{2t} - w^2_1}$$

—theoretical velocity of steam if all the heat drop occurring in the stage is completely converted into kinetic energy.

The efficiency at the rim of the disc carrying the moving blades is obtained from the formula

$$\eta_u = \frac{2u(c_{1u} - c_{2u})}{c^2_{ad}} \qquad (7\text{-}14)$$

$$\text{or } \eta_u = \frac{2u(c_{1u} - c_{2u})}{c^2_{1t} + w^2_{2t} - w^2_1}, \qquad (7\text{-}14a)$$

which is valid for a stage with any given degree of reaction[1].

[1] See 7-10 for the derivation of a similar equation for efficiencies of reaction turbines.

The size of blades is determined in the same manner as in the case of impulse stages. The specific volume of steam at the exit of the nozzles is obtained from the pressure p'_1. The complete process of expansion for a stage with reaction is shown in Fig. 7-5, *a*.

7-6. REACTION FOR A TWO-ROW DISC

As has been mentioned before, there is every possibility of separation of the steam jet from the walls of the moving blades if there is a large difference in the blade heights at entry and exit, which leads to larger heat losses due to the appearance of eddy formation. It has been found from experience at the various Soviet factories that good results are obtained for moving blades of 40 mm width and guide blades of 30 mm width if the ratio between inlet and exit blade heights for a two-row disc is

$$\frac{l''_1}{l} = 1.20; \quad \frac{l_{gb}}{l''_1} = 1.30; \quad \frac{l''_2}{l_{gb}} = 1.30$$

and for a three-row wheel

$$\frac{l''_3}{l} = 2.6 \text{ to } 3.2.$$

Greater ratios are recommended for short blades and smaller ratios for long blades. The increase in the blade height from inlet to outlet may be taken to be the same for all the rims.

The exit height of the blades of the second and the third rows are found from the formulas

a) for a two-row wheel

$$l''_2 = \frac{Gv'_2}{\pi d e w'_2 \sin\beta'_2}; \qquad (7\text{-}15)$$

b) for a three-row wheel

$$l''_3 = \frac{Gv''_2}{\pi d e w''_2 \sin\beta''_2}; \qquad (7\text{-}15a)$$

where v'_2 and v''_2 — specific volumes of steam after the moving blades of the second and third rows;

w''_2 — relative velocity of steam at the exit of moving blades of the third row;

β''_2 — angle of exit of the relative velocity vector w''_2.

A reduction in the blade heights may be achieved by increasing the angles β'_2 and β''_2 and velocities w'_2 and w''_2. Increase in angles β'_2 and β''_2 is advisable up to a certain limit, since with an increase in these angles the exit velocities

c'_2 and c''_2 (of the moving blades of the second and third rows) increase and, simultaneously, the efficiency decreases, because of the increased exit velocities. It is therefore advisable to have blades of smaller heights notwithstanding the presence of some degree of reaction, thus obtaining a more uniform variation in the increase in blade heights without in any way detracting from the efficiency of the turbine. Allowing for some degree of reaction is highly desirable especially in the case of admission of steam along the entire circumference. In the case of partial admission, especially with small values of ε and short blades, the presence of reaction is not always desirable since due to the relatively large sizes of gaps as compared with the blade heights there is very high probability of serious leakages leading to decrease in efficiency. Hence turbines of small generating capacities with velocity stages and partial admission are usually made as pure impulse turbines. However, some degree of reaction is allowed for in the case of two-row curtiss stages (regulating stages) in the present-day turbines.

In general the degree of reaction as used in current practice is 4 to 6% (may go up to 10-12%) a large fraction of which is made use of in the moving blades of the low-pressure stages.

The steam velocities c_{1t} and c_1 are determined from equations (7-9a) and (7-10).

The heat losses in the nozzles are found from equation (2-8a).

The steam conditions after the nozzles (Fig. 7-5, *b*) are obtained from the *i-s* diagram. The heat drop occurring in the nozzles is h_{01} and the heat losses for this heat drop are h_n. Setting off h_n on the *i-s* diagram point *a* is obtained which conforms to the steam conditions after the nozzles at the pressure p_1. The exit height of the nozzles is determined from the formula (4-5). Velocity triangles are drawn for the inlet to the moving blades exactly as in the case of impulse turbines. Values of w_1 and β_1 are obtained from these. The exit velocity triangle is drawn by assuming a suitable exit angle β_2 and velocity coefficient ψ. The relative velocity at exit from the moving blades is obtained from the following equation

$$w_2 = 91.5\psi \sqrt{\frac{w_1^2}{8{,}378} + \varrho_1 h_0}, \qquad (7\text{-}16)$$

where $\varrho_1 = h'_{02}/h_0$—degree of reaction in the moving blades of the first row,

h'_{02} — heat drop occurring in the moving blades.

Theoretical relative velocity at exit is obtained from

$$w_{2t} = \frac{w_2}{\psi}. \qquad (7\text{-}17)$$

Heat losses in the moving blades are obtained from equation (2-27a). The steam conditions at the exit from the moving blades of the first row and the specific volume of steam (point a_1 in Fig. 7-5, *b*) v_2 are obtained from the *i-s* diagram with the help of heat drop h'_{02} and the heat loss in the moving blades h'_b. The pressure p'_1 after the first row of moving blades is also determined by the point a_1. The exit height of the blades of the first row is obtained from equation (4-11). The absolute velocity of steam at the exit from the moving blades as well as the angle α_2 are obtained directly from the velocity triangles.

Velocity triangles for the second row of the moving blades are drawn with the help of absolute steam velocity at entry to the second row, which is

$$c'_1 = 91.5\psi_{gb} \sqrt{\frac{c_2^2}{8{,}378} + \varrho_{gb} h_0}, \qquad (7\text{-}18)$$

and some suitably assumed values for α'_1 and ψ_{gb}.

In the above equation $\varrho_{gb} = h_{ogb}/h_0$—degree of reaction in the guide blades;

h_{ogb}— heat drop occurring in the guide blades.

The theoretical velocity of steam at the exit from the guide blades will be

$$c'_{1t} = \frac{c'_1}{\psi_{gb}}. \qquad (7\text{-}19)$$

Heat losses in the guide blades will be

$$h'_{gb} = \frac{c'^2_{1t} - c'^2_1}{8{,}378}. \qquad (7\text{-}20)$$

Steam conditions at the exit from the guide blades and the specific volume v'_{gb} are obtained from the *i-s* diagram by setting off the heat drop and heat loss h_{ogb} and h'_{gb} (point a_2 in Fig. 7-5, *b*) from point a_1. Point a_2 thus obtained also determines the pressure p''_1 after the guide blades. The exit height of the guide blades is obtained from the equation

$$l''_{gb} = \frac{Gv'_{gb}}{\pi d\varepsilon c'_1 \sin\alpha'_1}, \qquad (7\text{-}21)$$

where α'_1 — the absolute angle of exit of steam from the guide blades; $\sin\alpha'_1 = \frac{a'_1}{t_{gb}}$ (a'_1— width of the guide blade passage and t_{gb}— pitch of the guide blades).

Steam velocity w'_1 at the entry to the moving blades of the second row and the angle β'_1 are obtained from the velocity triangles. Assuming the exit angle β'_2 (from Fig. 5-4) the velocity coefficient ψ' is obtained for the second row moving blades. The exit velocity of steam would then be

$$w_2 = 91.5\psi' \sqrt{\frac{w_1'^2}{8{,}378} + \varrho_2 h_0}, \qquad (7\text{-}22)$$

where $\varrho_2 = \frac{h''_{02}}{h_0}$ — degree of reaction in the moving blades of the second row;

h''_{02}—heat drop occurring in the second row of moving blades.

The exit velocity triangles are drawn from the values thus obtained from which velocity c'_2 and angle α'_2 are determined. Heat losses in the second row of the moving blades h''_b are obtained from equation (2-27a) where w'_{2t} and w'_{2t} are substitued in place of w_{2t} and w_2.

Steam conditions at the exit from the second row of the moving blades and the specific volume v'_2 are obtained as before by adding the heat drop h''_{02} and deducting the heat losses h''_b (point a_3 on Fig. 7-5, *b*).

Exit height of the moving blades of the second row is determined from equation (7-15).

The efficiency at the periphery of the disc may be obtained from the equation

$$\eta_u = \frac{2u \sum (c_{1u} - c_{2u})}{c_{ad}^2}, \qquad (7\text{-}23)$$

where

$$c_{ad} = 91.5\sqrt{h_{01} + h'_{02} + h_{ogb} + h''_{02}}.$$

The steam conditions after the moving blades are obtained (point a_3 in Fig. 7-5, *b*) by setting off on the *i-s* diagram carry-over losses h_e and losses due to disc friction and windage.

While designing a two-row (or three-row) disc of a turbine it is usual to select such a degree of reaction that the increase in blade heights is gradual and uniform.

7-7. HEAT DROP CALCULATIONS FOR MULTISTAGE IMPULSE TURBINES

For the purpose of governing it is the current practice to make use of a single-row (or a two-row) impulse wheel in the first stage of a turbine. In the event of throttle governing, however, such a regulating stage is absent.

In order to avoid very small nozzle heights the first stage is usually provided with partial admission. If, however, the turbine has a large capacity there may be full admission of steam all along the periphery for the regulating stage.

For condensing turbines operating at high vacuum the blades of the low-pressure stages are made as large as possible, because of the very large specific volumes obtained at these pressures. Of course, the blade dimensions are such as to conform to the minimum requirements from the point of mechanical strength.

Hence during the design of multistage turbines it is usual to make preliminary calculations for the first (regulating), second and the last stages. The number of intervening stages and their details are worked out only after obtaining the basic dimensions for the first, second and the last stages.

1. Preliminary Calculations for the First (Regulating) Stage

The various dimensions for the first stage must be carried out in such a way that $l \geqslant 10$ [1] mm and $\varepsilon \geqslant 0.2$. The exit section for the nozzles is obtained from the equation of continuity, i. e.,

$$f_1 = \frac{G_0 v_1}{c_1} = \frac{\pi d \varepsilon l \sin \alpha_1}{c_1}, \qquad (7\text{-}24)$$

where c_1 — velocity of steam at the nozzle exit;

v_1 — specific volume of steam after the nozzles.

Substituting in equation (7-24) u/x in place of c_1, where $x = u/c_1$, we have

$$G_0 v_1 x = \pi d \varepsilon l u \sin \alpha_1. \qquad (7\text{-}25)$$

Since $u = \frac{\pi d n}{60}$ we shall have:

$$60 G_0 v_1 x = \pi^2 d^2 \varepsilon l n \sin \alpha_1. \qquad (7\text{-}25a)$$

Solving the above equation for the diameter we have

$$d = \sqrt{\frac{60 G_0 v_1 x}{\pi^2 \varepsilon l n \sin \alpha_1}}. \qquad (7\text{-}26)$$

The quantities x, ε, l and α_1 in equation (7-26) depend upon the mechanical design of the first stage. If suitable values are assumed for the above-mentioned quantities we find that the unknown quantities are d and v_1. Thus equation (7-26) can be solved only by assuming various values for d for which the specific volume v_1 is determined. The correctness of the assumption is now checked by comparing the assumed heat drop with the one obtained for this specific volume. For example, if for the purpose of preliminary calculations we assume the value of d then we have $u = \pi d n/60$, velocity of steam $c_1 = u/x$, adiabatic heat drop in the stage $h_0 = c_1^2/8{,}378\varphi^2$ and the heat losses in the nozzles $h_n = (1 - \varphi^2)\, c_{1t}^2/8{,}378\varphi^2$.

[1] It is the current practice to have l not less than 20 to 30 mm for large-capacity turbines.

Marking off these values of h_0 and h_n on the i-s diagram the steam conditions and the specific volume v_1 after the nozzles are determined. Substituting this value of v_1 in equation (7-26) diameter d is found. If the value of d obtained from the above equation does not agree with the assumed value then a further similar calculation has to be carried out with slight variation in the assumed value of d until a satisfactory result is obtained.

2. Preliminary Calculations for the Second Stage

Partial admission of steam is made use of for the first stage only in the case of turbines with large mass flows. If, however, the capacity of the turbine is comparatively small partial admission may be continued for some of the high-pressure stages. However, in order to maintain a high value for the efficiency of the turbine it is desirable that full admission may be effected for as many stages as possible in order to reduce losses due to disc friction and windage, even in the case of small power turbines. In such a case minimum height of the nozzles is assumed to be from 10 to 15 mm and $\varepsilon=1$, from which the diameter d is determined. The last named quantity d is obtained with the help of equation (7-26) in the manner described in the preceding paragraph.

The second stage of turbines of medium and large capacities are designed with values of $\varepsilon=1$ and nozzle heights of 15 to 20 mm, depending on the required disc diameter.

The steam conditions before the nozzles of the second stage are obtained by making detailed calculations for the first stage (as described in Chapter Six).

3. Preliminary Calculations for the Last Stage of a Condensing Turbine

From the equation of continuity we may write for the exit section of the moving blades of the last stage

$$f_2 w_2 = G_0 v_2 = \pi d l w_2 \sin\beta_2, \qquad (7\text{-}27)$$

where v_2—specific volume of steam at the exit from the moving blades.

Since $w_2 \sin\beta_2 = c_2 \sin\alpha_2$ we have

$$G_0 v_2 = \pi d l c_2 \sin\alpha_2. \qquad (7\text{-}27a)$$

Substituting $d/l=\vartheta$ in equation (7-27) we have

$$G_0 v_2 = \frac{\pi d^2}{\vartheta} c_2 \sin\alpha_2. \qquad (7\text{-}27b)$$

Solving this equation for the diameter d we obtain

$$d = \sqrt{\frac{G_0 v_2 \vartheta}{\pi c_2 \sin\alpha_2}}. \qquad (7\text{-}28)$$

The exit velocity losses may be expressed as

$$\frac{c_2^2}{8{,}378} = h_e = \zeta_e H_0. \qquad (7\text{-}29)$$

where ζ_e— coefficient of heat loss for the carry-over velocity in the last stage.

From equation (7-29) we obtain

$$c_2 = 91.5\sqrt{\zeta_e H_0}. \qquad (7\text{-}29a)$$

Velocity c_2 of the steam leaving the last stage moving blades is completely lost. The greater the carry-over velocity c_2, the greater will be the losses and consequently the lower the efficiency.

For condensing turbines the loss due to carry-over velocity reaches a value $\zeta_e=1$ to 3% of the theoretical heat drop H_0 occurring in the turbine. The quantity ζ_e is suitably assumed while carrying out the preliminary calculations.

Substituting the value of c_2 obtained from equation (7-29a) in equation (7-28) we have the diameter d as

$$d = \sqrt{\frac{G_0 v_2 \vartheta}{\pi \times 91.5\sqrt{\zeta_e H_0}\sin\alpha_2}}. \qquad (7\text{-}30)$$

For condensing turbines of medium and small capacities the ratio $d/l=\vartheta$ should be taken, as far as possible, not less than 5 to 6. For large-capacity turbines the value of ϑ, in view of unavoidable circumstances, is reduced to 3 to 3.5 or even up to 2.7.

If the ratio $\vartheta<8$ to 10 the circumferential velocity at the outer edges of the moving blades considerably differs from the velocity at the root of the blades. Hence the long blades of the low-pressure stages are made with a continuously varying entrance angle, i. e., the blades are given a certain amount of twist so as to avoid shock when the steam enters the blade passages. Blades of such twisted profiles, of course, add to the cost of manufacture and at the same time the process of manufacture also becomes more complicated.

The specific volume of steam at the exit from the moving blades is obtained from the i-s diagram assuming a suitable value for the efficiency for preliminary design purposes. It is very much desirable that the exit angle for the absolute velocity of steam from the moving blades of the last stage be 90°, since the heat loss due to carry-over velocity would be a minimum for $\alpha_2=90°$. Hence for preliminary calculation purposes we may take $\sin\alpha_2=1$ in equation (7-30).

Having obtained d we may determine $u=\pi dn$ 60. If the value obtained for u is too high then we may either reduce the ratio ϑ or increase ζ_e.

At the present time many of the large turbine building factories make use of circumferential velocities, at the mean diameter, of about 315 m/sec. In the large-capacity turbines of latest design built by L. M. W. the circumferential velocity, at the mean diameter, for the last stage is of the order of 314 m/sec. For some of the large-capacity turbines operating at supercritical parameters (under construction) the circumferential velocity at the mean diameter is as much as 350 to 360 m/sec.

If the values obtained for u are excessive even at the limiting values of ϑ and ζ then it is not possible to build the last stage as a single unit with all the steam flowing through the blades. In such cases the steam flow is divided into two streams each of which passes through separate low-pressure stages, as is shown in Fig. 7-19. Fig. 10-10 shows a turbine built by L. M. W. where from the last but one stage a portion of the steam is led away directly into the condenser by passing the last stage.

The heat drop occurring in the last stage will be

$$h_{0z}=\frac{u^2}{8{,}378\varphi^2 x^2}, \tag{7-31}$$

where $u/c=x$ may be assumed within the limits of 0.48 and 0.52 (higher values being used for stages with large reaction). For turbines with back-pressure equation (7-30) is not suitable for determining the diameter of the last stage. Equation (7-26) may be used in such cases assuming ε to be unity.

4. Distribution of Heat Drop in Various Stages

A multistage turbine may be built with different numbers of pressure stages. If the number of stages is small then the velocity of steam at the exit from the nozzles may be equal or higher than the critical value. In the case of a multistage turbine with a large number of pressure stages, excepting in the first (regulating) and one or two following stages, the steam velocity is invariably less than the critical value. After having determined the various basic dimensions for the first, second and last stages the remaining heat drop is distributed in the remaining stages. Depending upon the required efficiency the value of coefficient Y is found from the graph shown in Fig. 7-3.

On the basis of equation (7-8d) we may then write

$$YH_{0\alpha}=\Sigma u^2=z_1u_1^2+z_2u_2^2+\ldots+z_nu_n^2=zu_{av}^2,$$

where z_1, z_2,..., z_n—number of stages having velocities u_1, u_2, .., u_n.

If originally the number of stages is assumed equal to z then we may determine the mean circumferential velocity as

$$u_{\text{av}}^2=\frac{YH_{0\alpha}}{z}.$$

The theoretical heat drop in a stage for this mean or average velocity is obtained from the equation

$$h_{\text{av}}=\frac{u_{\text{av}}^2}{8{,}378\varphi^2 x_{\text{av}}^2}, \tag{7-32}$$

where

$$x_{\text{av}}=\frac{x_1+x_2+x_3+\ldots+x_z}{z}.$$

For pressure stages we may assume

$$x_2=x_3=x_4=\ldots=x_z=x,$$

and x_1 is obtained from the preliminary calculations for the first stage.

The circumferential velocity u is varied in such a way for each succeeding stage as to obtain a smooth contour for the turbine shape. The heat drop in any of the stages would then be

$$h_0=\frac{u^2}{8{,}378\varphi^2 x^2}. \tag{7-32a}$$

From equations (7-32) and (7-32a) we obtain a new expression for determining the theoretical heat drop which is

$$h_0=h_{\text{av}}\frac{u^2}{u_{\text{av}}^2}\times\frac{x_{\text{av}}^2}{x^2}. \tag{7-33}$$

If it is found that, consequent to the distribution of heat drop in all the intervening stages, $\Sigma h_0\neq(1+\alpha)H_0'$ then x should be slightly altered and h_0 recalculated for all the stages until the desirable result is achieved.

For turbines with back pressure it is usual to take the diameter of the intervening stages equal to, or slightly greater than, the diameter of the second stage, in order to obtain identical diameters at the root of the blades. A similar method is followed for determining the diameters of intervening stages for the high-pressure part of a condensing turbine as well. Final design calculations are carried out after achieving a satisfactory heat distribution in the intervening stages. Nozzle angles are usually assumed from 11 to 18° increasing up to as much as 30 to 35° for the low-pressure stages. The ratio u/c_1 for multistage turbines is assumed from 0.48 to 0.5. If the efficiency is not the criterion then the value of u/c_1 may be 0.44 or even less.

For pressure stages of equal diameters or uniformly varying blade heights, in the absence of large spacing (gap) between nozzles and moving blades, the carry-over velocity from the previous stage should be taken into consideration while calculating velocity of steam issuing from the nozzles of the succeeding stage.

While carrying out design calculations dimensions of nozzles and blades should be verified for adaptability by checking these dimensions on the scale drawing of the turbine. If it is found that these dimensions are such that there is a serious difference in diameters, etc., of the adjacent stages then angles α_1 and β_2 should be increased. Hence for the high-pressure stages it is recommended that angle α_1 should be between 11 and 14°, gradually increasing α_1 (usually only in the low-pressure stages for condensing turbines) up to 20° and in exceptional cases even higher.

In modern high-pressure turbines α_1 is usually taken equal to 11 to 13° for the first few stages.

7-8. DESIGN PROCEDURE FOR MULTISTAGE IMPULSE TURBINES.

Example: Design a multistage impulse turbine for the following given conditions. Generating capacity at the generator terminals $N_e = 5{,}600$ kW, r.p.m. = 3,000. Pressure and temperature of fresh steam $p_0 = 29$ ata, and $t_0 = 400°C$. Exhaust steam pressure (condenser pressure) $p_2 = 0.05$ ata.

On the i-s diagram (Fig. 7-6) point A_0 conforms to the fresh steam conditions. The theoretical heat drop, neglecting losses in the regulating valves, is given by the adiabatic A_0A_{1t}, i.e.,

$$H_0 = i_0 - i_{1t} = 773 - 505 = 268 \text{ kcal/kg}.$$

Assuming the losses in the regulating valves to be 5% of p_0 we have the pressure before the nozzles equal to

$$p'_0 = (1 - 0.05)\, p_0 = 0.95 \times 29 = 27.5 \text{ ata (point } A'_0).$$

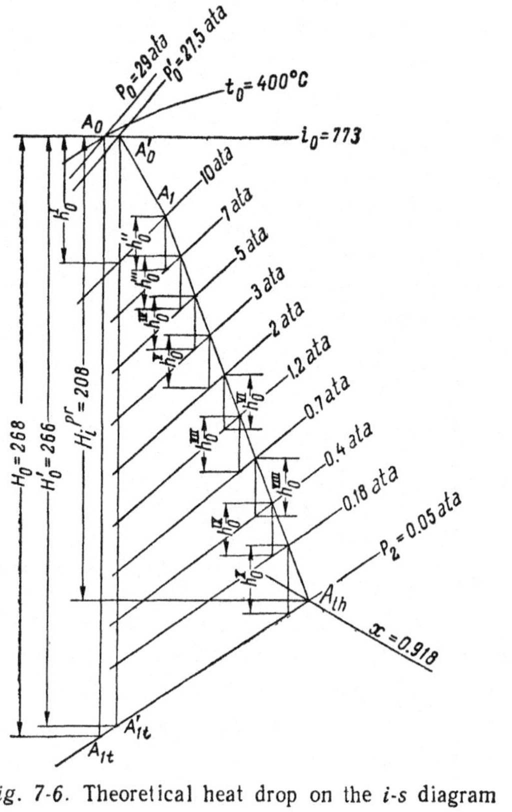

Fig. 7-6. Theoretical heat drop on the i-s diagram

Hence the adiabatic heat drop, considering the losses in the regulating valves, will be given by $A'_0A'_{1t}$, i.e.,

$$H'_0 = 773 - 507 = 266 \text{ kcal/kg}.$$

Mass flow of steam through the turbine will be

$$G_0 = \frac{860 N_e}{3{,}600 H_0 \eta_{re} \eta_g} = \frac{860 \times 5{,}600}{3{,}600 \times 268 \times 0.760 \times 0.947} = 6.94 \text{ kg/sec}.$$

where η_{re} and η_g are taken from Figs 5-9 and 5-11.

Let us now determine the basic dimensions of the first, second and last stages.

F i r s t s t a g e: We shall make use of a two-row wheel for the first stage in order to utilise a greater portion of the heat drop in it.

For small-capacity turbines as well as those with low steam consumption it is advisable that a greater portion of the heat drop be utilised in the first stage from the point of view of reducing the leakage losses (at lower pressures in the first stage chamber) as well as simplifying the construction.

Design of such a stage has been shown in 6-6. This two-row impulse wheel is utilised as the first regulating stage for the multistage turbine under consideration. Expressing the heat drop process for the first stage on the i-s diagram we obtain point A_1 which conforms to the steam conditions before the nozzles of the second stage.

S e c o n d s t a g e: Under conditions of full admission of steam all around the disc of the second stage the diameter of the disc will be given by equation (7-26)

$$d_2 = \sqrt{\frac{60 G_0 x v_1}{\pi^2 l n \sin \alpha_1}}.$$

Assuming $x = 0.42$, $l = 10$ mm and $\alpha_1 = 12°$ for the nozzles of the second stage we have

$$d_2 = \sqrt{\frac{60 \times 6.94 \times 0.42 v_1}{\pi^2 \times 0.01 \times 3{,}000 \times 0.208}} = \sqrt{2.81 v_1}.$$

For preliminary calculation purposes we shall estimate d_2 to be about 990 mm so that

$$u = \pi d_2 n/60 = \pi \times 0.990 \times 3{,}000/60 = 156 \text{ m/sec};$$

$$c_1 = u/x = 156/0.42 = 372 \text{ m/sec};$$

$$c_{1t} = c_1/\varphi = 372/0.95 = 392 \text{ m/sec};$$

$$h_0 = c_{1t}^2/8{,}378 = 392^2/8{,}378 = 18.25 \text{ kcal/kg};$$

$$h_n = (c_{1t}^2 - c_1^2)/8{,}378 = (392^2 - 372^2)/8{,}378 = 1.85 \text{ kcal/kg}.$$

Setting off the above obtained heat drop in the nozzles on the i-s diagram from the point A_1 we have the pressure p_1 and specific volume v_1 after the nozzles as

$$p_1 = 7.25 \text{ ata and } v_1 = 0.348 \text{ m}^3/\text{kg}.$$

Substituting the value of v_1 in the formula $d_2 = \sqrt{2.81 v_1}$ we have $d_2 = \sqrt{2.81 \times 0.348} = 0.99$ m = 990 mm which confirms the assumption made by us previously.

L a s t s t a g e: From equation (7-30) the diameter of the last stage will be

$$d_z = \sqrt{\frac{G_0 \vartheta v_2}{\pi \times 91.5 \sqrt{\zeta_e H_0} \sin \alpha_2}}.$$

Assuming $\vartheta = 6$, $\zeta_e = 1.0\%$ and $\alpha_2 = 90°$ we have

$$d_z = \sqrt{\frac{6.94 \times 6 \times v_2}{\pi \times 91.5 \times \sqrt{0.01 \times 268}}} = \sqrt{0.0884 \times v_2}.$$

Specific volume of steam after the moving blades of the last row is obtained as follows.

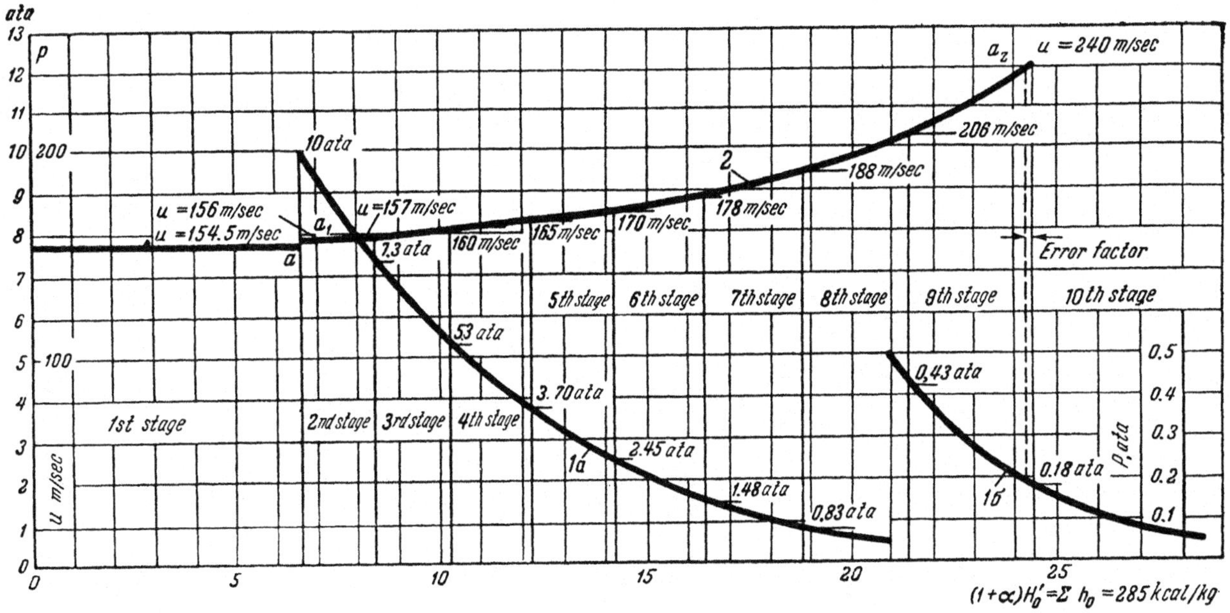

Fig. 7-7. Auxiliary graph for heat drop distribution in the various turbine stages

The theoretical internal relative efficiency of the turbine is given by

$$\eta_{oi}=\frac{\eta_{re}}{\eta_m}=\frac{0.76}{0.98}=0.776, \text{ where } \eta_m \text{ is taken from}$$

Fig. 5-8.

Hence the heat drop usefully utilised inside the turbine will be

$$H_i^{th}=H_0'\eta_{oi}=268.0\times 0.776=208 \text{ kcal/kg}.$$

Setting off this heat drop on the i-s diagram we obtain point A_{th} where the specific volume $v_2=26.4$ m³/kg. Hence diameter of the last stage will be

$$d_z=\sqrt{0.0884\times 26.4}=1.53 \text{ m}=1{,}530 \text{ mm}.$$

Circumferential velocity

$$u_z=\frac{\pi\times 1.53\times 3{,}000}{60}=240 \text{ m/sec which is permissible.}$$

To obtain a compact and a comparatively low priced turbine we shall assume the number of stages to be

$$z=10,$$

and the characteristic coefficient $Y=1{,}400$.

For the above assumed values of z and Y the efficiency expected of the turbine will be comparatively lower. Let us assume the heat recovery coefficient $\alpha=0.065$. Consequently we have

$$u_{av}^2=\frac{Y(1+\alpha)H_0'}{z}=\frac{1{,}400\times 284}{10}\approx 40{,}000;$$

$$\text{or } u_{av}=200 \text{ m/sec}.$$

We shall assume the ratio u/c_1 to be the same for all the stages. Usually, the smaller the number of stages, the lower is the value of this ratio, hence we shall assume a value 0.42 for u/c_1.

The value of $(u/c_1)_{av}$ for the turbine will be

$$x_{av}=\frac{x_1+9x}{z}=\frac{0.22+9\times 0.42}{10}=0.4.$$

The average heat drop per stage from equation (7-32) will be

$$h_{av}=\frac{40{,}000}{8{,}378\times 0.95^2\times 0.4^2}=33.1 \text{ kcal/kg}$$

where φ has been assumed to be 0.95 and constant for all the stages.

To facilitate the determination of heat drop distribution in the various turbine stages a graph as shown in Fig. 7-7 is drawn. In Fig. 7-6 the points A_0', A_1 and A_{th} are joined by straight lines. Line A_1A_{th} represents the theoretical heat drop process in the pressure stages of the turbine.

Theoretical heat drop H_0' is arbitrarily divided into h_0^I, h_0^{II}, h_0^{III}, etc. (heat drop in each stage) by arbitrarily chosen isobars. Measuring these heat drops along the abscissa and the pressures 10, 7.5, 3 ata, etc., along the ordinate we obtain the graph 7-7 (curves *Ia* and *Ib* are drawn to different scales for pressure in Fig. 7-7). These curves represent the variation of pressure in the various stages of the turbine conforming to the process of expansion of steam given by line A_1A_{th} in Fig. 7-6.

The circumferential velocities for the first, second and last stages are also plotted against heat drop in graph 7-7 (points a a_1 and a_z). Joining these points by a smooth curve 2 we obtain the graph for circumferential velocity in the various stages. From the values of u obtained from this graph for the various stages and knowing the value of x from equation (7-32a) or (7-33) the heat drop in each of the stages is determined. If it is found that the sum $\sum_1^z h_0$ obtained as described above does not equal $(1+\alpha)H_0'$ then the value of x should be altered in one or two last stages so as to attain $\sum_1^z h_0=(1+\alpha)H_0'$.

Knowing the heat drops h_0^I, h_0^{II}, h_0^{III}, etc., the determination of pressures in the stages becomes very simple (from curves *Ia* and *Ib* in Fig. 7-7).

Distribution of Heat Drop

Nomenclature \ Stage No.	1	2	3	4	5	6	7	8	9	10
u, m/sec	154.5	156	157	160	165	170	178	188	206	240
u^2, m²/sec²	23,850	24,300	24,600	25,500	27,150	28,900	31,700	35,800	42,400	57,700
u^2_{av}, m²/sec²	40,000	40,000	40,000	40,000	40,000	40,000	40,000	40,000	40,000	40,000
x^2_{av}	0.16	0.16	0.16	0.16	0.16	0.16	0.16	0.16	0.16	0.16
x	0.22	0.42	0.42	0.42	0.42	0.42	0.42	0.42	0.43*	0.431*
x^2	0.0483	0.176	0.176	0.176	0.176	0.176	0.176	0.176	0.185	0.186
h_0, kcal/kg	65.5	18.25	18.5	19.15	20.4	21.7	23.8	26.4	30.3	41.0
p_1, ata	10	7.3	5.3	3.70	2.45	1.48	0.83	0.43	0.18	0.05

* With $x=0.42$ proper distribution of heat drop could not be obtained: $\Delta h_{error}=2.3$ kcal/kg, and accordingly the ratio u/c_1 has been altered in the last two stages.

The results of preliminary heat distribution and the pressures obtained in each of the stages are shown in tabular form (Table 7-1).

Detailed heat drop calculations are carried out by redrawing the entire process on the i-s chart from the various quantities obtained from the preliminary calculations.

7-9. REACTION TURBINES

Reaction turbines are as a rule made only as multistage turbines. In the case of reaction turbines steam undergoes expansion both in the guide as well as the moving blades so that there is always a difference of pressure between the two sides of guide and moving blades. This difference of pressure exerts a thrust on the blades in the axial direction. To reduce the axial thrust, it is usual to mount the moving blades on a drum which serves the purpose of the rotor as well. The guide blades are attached to the stator of the turbine. Consequently the construction of a reaction turbine considerably differs from that of an impulse turbine. Fig. 7-8 shows the sectional view of a reaction turbine. The conversion of potential energy into kinetic energy is also shown in this figure. Curve p_0-p_2 shows the pressure variation and the dotted line indicates the variation of absolute velocity. The velocity c increases in the guide blades because of heat drop occurring in them. Work is done on the shaft of the turbine because of the reduction of kinetic energy of the flowing steam in the moving blade passages, i. e., reduction of $c_1^2/2g$, as well as on account of the heat drop h_{02} occurring in the moving blades which accounts for the increase in exit relative velocity w.

The topmost curve in Fig. 7-8 represents the variation of heat drop in each of the turbine stages. The heat content continuously goes on decreasing both in the guide and moving blades. There is a slight rise in the heat content of steam while passing over from the smaller diameter to the larger one of the rotor because of the carry-over losses from the last stage of the first group of blades.

The turbine in Fig. 7-8 is provided with a balance piston *8* to counteract the effect of axial thrust on the turbine rotor. The carry-over velocity from the preceding stage is utilised in the

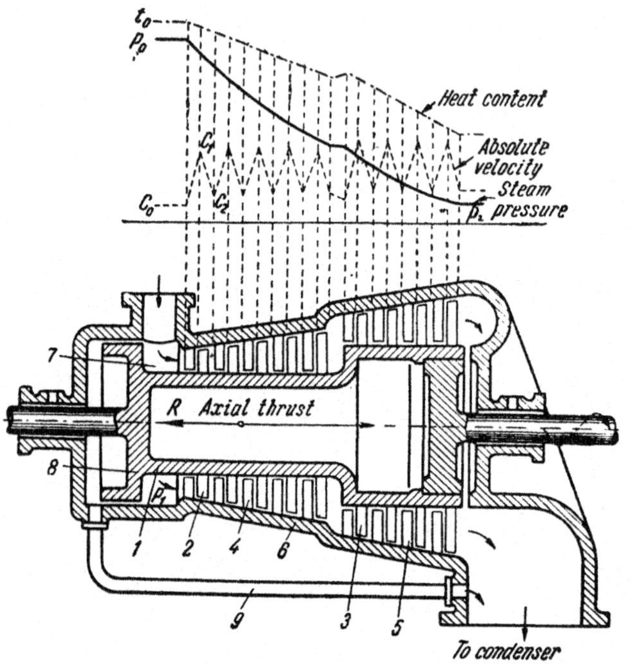

Fig. 7-8. Sectional view of a reaction turbine

1— rotor drum; *2* and *3*— moving blades; *4* and *5*— guide blades; *6*— casing; *7*— annular steam chamber; *8*— balance piston; *9*— pressure equaliser steam pipe

guide blades of the succeeding one in reaction turbines also, as was done in the case of impulse turbines. However, carry-over velocity is not available for utilisation in the guide blades of the first stage (there being no preceding stage) as well as in those groups of guide blades which are situated immediately after a sharp increase in the diameter of the rotor (or if the steam is admitted to a second cylinder from the first).

7-10. EFFICIENCY OF A REACTION TURBINE

Equation (6-3) expressing the work done by 1 kg of steam is valid for reaction turbines also, i.e.,

$$L_u = \frac{u}{g}(c_{1u} - c_{2u}).$$

The theoretical work done by 1 kg of steam on the moving blades of a stage is given by

$$L_0 = \frac{c_{ad}^2}{2g},$$

where c_{ad}— theoretical velocity of steam if all the heat drop occurring in the reaction stage is converted into kinetic energy.

Velocity c_{ad} is obtained from the equation

$$c_{ad} = 91.5\sqrt{h_{01} + h_{02}} \qquad (7\text{-}34)$$

since $h_{01} = c_{1t}^2/8{,}378$ and $h_{02} = (w_{2t}^2 - w_1^2)/8{,}378$, substituting these values in place of h_{01} and h_{02} equation (7-34) may be expressed as

$$c_{ad} = \sqrt{c_{1t}^2 + w_{2t}^2 - w_1^2} \qquad (7\text{-}34a)$$

or

$$c_{ad}^2 = c_{1t}^2 + w_{2t}^2 - w_1^2. \qquad (7\text{-}34b)$$

Efficiency at the rim, from equation (6-4), will be

$$\eta_u = \frac{L_u}{L_0} = \frac{2u(c_{1u} - c_{2u})}{c_{ad}^2} = \frac{2u(c_{1u} - c_{2u})}{c_{1t}^2 + w_{2t}^2 - w_1^2} \qquad (7\text{-}35)$$

$$\text{or} \quad \eta_u = \frac{2u(c_{1u} - c_{2u})}{\frac{c_1^2}{\varphi^2} + \frac{w_2^2}{\varphi^2} - w_1^2}. \qquad (7\text{-}35a)$$

Since $h_{01} = h_{02}$ in reaction turbines, in order to obtain blades of uniform section both for the moving and the fixed blades it is usual to have

$$w_2 = c_1 \quad \text{and} \quad \beta_2 = \alpha_1;$$

so that

$$c_2 = w_1, \quad \alpha_2 = \beta_1 \quad \text{and} \quad \psi = \varphi.$$

Substituting $c_1 = w_2$ and $\varphi = \psi$ in equation (7-35a) we obtain

$$\eta_u = \frac{2u(c_{1u} - c_{2u})}{\frac{2c_1^2}{\varphi^2} - w_1^2}. \qquad (7\text{-}35b)$$

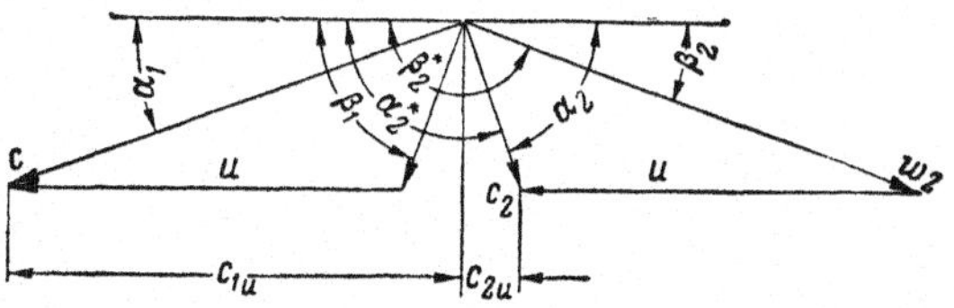

Fig. 7-9. Velocity triangles of a reaction stage

From the velocity triangles (Fig. 7-9) we therefore have $c_{1u} - c_{2u} = w_1\cos\beta_1 - w_2\cos\beta_2^* = w_1\cos\beta_1 + w_2\cos\beta_2 = c_1\cos\alpha_1 - u + c_1\cos\alpha_1 = 2c_1\cos\alpha_1 - u$ and $w_1^2 = c_1^2 + u^2 - 2uc_1\cos\alpha_1$.

Substituting the above values of $c_{1u} - c_{2u}$ and w_1 in equation (7-35b) we have

$$\eta_u = \frac{2u(2c_1\cos\alpha_1 - u)}{\frac{2c_1^2}{\varphi^2} - c_1^2 - u^2 + 2uc_1\cos\alpha_1}. \qquad (7\text{-}35c)$$

Dividing both the numerator and the denominator in equation (7-35c) by c_1^2 and rearranging we obtain

$$\eta_u = \frac{2\left(2\cos\alpha_1 - \frac{u}{c_1}\right)\frac{u}{c_1}}{\left(\frac{2}{\varphi^2} - 1\right) + \left(2\cos\alpha_1 - \frac{u}{c_1}\right)\frac{u}{c_1}}. \qquad (7\text{-}35d)$$

It follows from equation (7-35d) that, when α_1 and φ are constant, the value of η_u depends only upon the ratio u/c_1; when $u/c_1 = 0$ and $u/c_1 = 2\cos\alpha_1$, $\eta_u = 0$.

The efficiency η_u as a function of u/c_1 for $\alpha_1 = 20°$ and $\varphi = 0.92$ is shown in Fig. 7-10, from which it is seen that η_u attains a high figure, when the ratio u/c_1 has a value between the limits of 0.6 to 1.3. The maximum values of η_u seem to be at u/c_1 equal to 0.8 to 0.9. However, it is

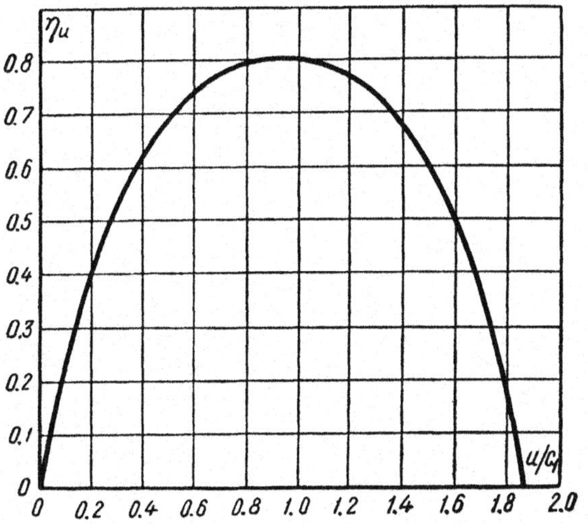

Fig. 7-10. Efficiency η_u as a function of u/c_1

found that at these ratios of u/c_1 the number of stages required becomes much too large to be convenient. Hence in order to reduce the number of stages to a reasonable figure the ratio u/c_1 may be utilised at a slightly lower value, say 0.6 to 0.7 (for the first stage of low priced turbines this value may be further decreased). The efficiency of the turbine remains practically of a high order in spite of the reduction in the value of the ratio u/c_1.

Angles of 18 to 20° are recommended for the exit of both the moving and the guide blades. In order to reduce the height of blades in the low-pressure stages the exit angle may be increased up to 35 to 40°.

7-11. DESIGN PROCEDURE FOR REACTION TURBINES

1. Auxiliary Graph for Design of Reaction Turbines

Fig. (7-11) shows the heat drop process on the i-s diagram. From the known parameters of fresh and exhaust steam we obtain H_0. Assuming the losses in the valves to be Δp the available heat drop will be H_0'.

The efficiencies η_{re}, η_g and η_m for a turbine of given capacity are obtained from the graphs shown in Figs 5-8, 5-9 and 5-11.

The efficiency $\eta_{oi}=\frac{\eta_{oe}}{\eta_m}$ and $H_i=H_0\eta_{oi}$.

Setting off this value of heat drop on the i-s diagram the conditions of exhaust steam are determined (point B in Fig. 7-11).

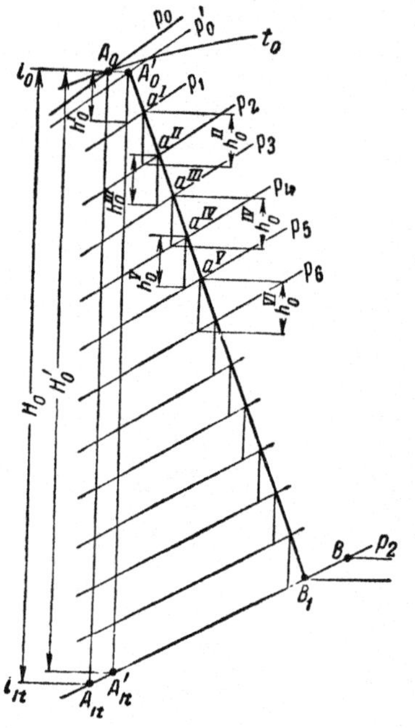

Fig. 7-11. Theoretical heat drop process of a reaction stage on the i-s diagram

The carry-over loss from the last stage will be assumed for preliminary design purposes. This value of h_e will be measured from point B downwards to B_1. Point B_1 conforms to the steam conditions immediately after the last row of the moving blades. The line joining the points A_0', B_1, and B represents the theoretical heat process in the turbine. We shall now divide the total heat drop H_0' by arbitrary isobars p_1, p_2, p_3 giving h_0^{I}, h_0^{II}, h_0^{III}, etc., in each of the stages. At every point of intersection of the straight line $A_0'B_1$ by the arbitrarily drawn isobars, the specific volume of steam is determined: v_0', v_1, v_2, etc., (at points A_0^{I}, a^{I}, a^{II}, a^{III}, etc.).

To effect a proper heat drop distribution in the various stages we shall plot a graph as shown in Fig. 7-12. $p=f\,[(1+\alpha)H_0']$ (curve *1*) and $v=\varphi\,[(1+\alpha)H_0']$ (curve *2*). The quantity $\sum h_0 = (1+\alpha)H_0'$ is measured along the abscissa and the corresponding pressures and volumes are measured along the ordinates. The scale for heat drop should be not less than 2 mm for each kcal/kg.

2. Preliminary Design Calculations for the First and Final Stages

The mass flow of steam per second through the turbine from equation (5-46) will be

$$G_0=\frac{D_0}{3{,}600}=\frac{860N_e}{3{,}600H_0\eta_{re}\eta_g}\,. \qquad (7\text{-}36)$$

Diameter of the first stage is obtained from equation (7-26) assuming full admission, i.e., $\varepsilon=1$,

$$d_1=\sqrt{\frac{60G_0x_1v_1}{\pi^2 l_1 n\sin\alpha_1}}\,.$$

The height of guide blades for the first stage of small- or medium-capacity turbine should be taken not less than 20 to 25 mm. For turbines of large capacities, however, l_{gb} should not be less than 35 to 40 mm.

The value of x_1 should be taken from 0.55 to 0.80 and the angle $\alpha_1=18$ to 20°.

The procedure of determining the diameter of the last stage is the same as was adopted in the case of impulse turbines. Thus the diameter d_z of the last stage will be

$$d_z=\sqrt{\frac{G_0v_2\vartheta}{\pi\times 91.5\sqrt{\zeta_eH_0}\sin\alpha_2}}\,,$$

where $\vartheta\geqslant5$ to 6 — for turbines of medium capacity and

$\vartheta=3.0$ to 3.2 for turbines of large capacity,

$\zeta_e=(1$ to $2\%)$ of H_0,

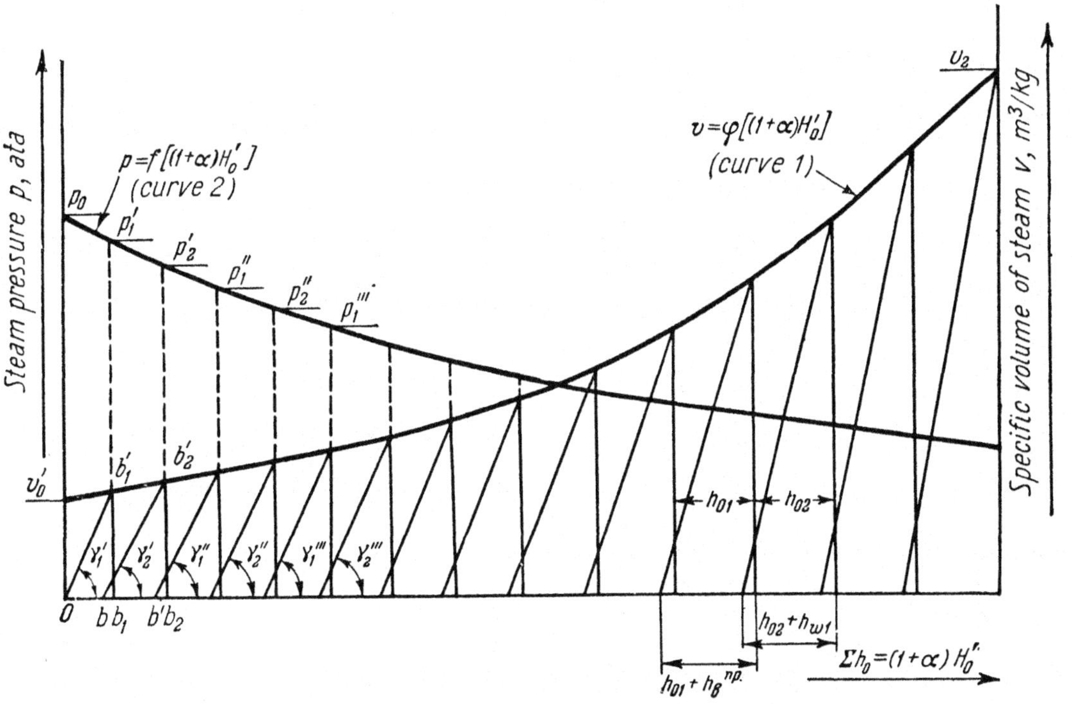

Fig. 7-12. Auxiliary graph for heat drop distribution in reaction turbines

v_2 — specific volume of steam at point B_1 from i-s diagram shown in Fig. 7-11.

3. Heat Drop in Turbine Stages

For the guide blade system of any stage we may write

$$c_1 f_{gb} = c_1 \pi d l \sin\alpha_1 = G_0 v_1. \qquad (7\text{-}37)$$

Solving this equation for c_1 and substituting $60\,u/\pi n$ for d we have

$$c_1 = \frac{G_0 v_1 n}{60 u l \sin\alpha_1}. \qquad (7\text{-}37a)$$

Multiplying both the right- and left-hand sides of this equation by $c_1/8{,}378\varphi^2$ we have

$$\frac{c_1^2}{8{,}378\varphi^2} = \frac{G_0 v_1 n}{60\times 8{,}378\varphi^2 x_1 l \sin\alpha_1}, \qquad (7\text{-}37b)$$

since $\frac{c_1^2}{8{,}378\varphi^2} = h_{01} + \frac{c_{2pr}^2}{8{,}378}$, equation (7-37b) becomes

$$h_{01} + \frac{c_{2pr}^2}{8{,}378} = \frac{G_0 v_1 n}{502{,}800\varphi^2 x_1 l \sin\alpha_1}, \qquad (7\text{-}37c)$$

where c_{2pr}—exit velocity of steam from the preceding stage.

Equation (7-37c) may further be expressed as

$$\frac{M_v v_1}{M_h (h_{01} + h_e^{pr})} = \frac{M_v \times 502{,}800\varphi^2 x_1 l \sin\alpha_1}{M_h G_0 n} = \tan\gamma_1, \qquad (7\text{-}37d)$$

where

$$\frac{c_{2pr}^2}{8{,}378} = h_e^{pr},$$

M_v and M_h — are the scales used for the volumes and heat drops while plotting the graph shown in Fig. 7-12. Consequently, the relation between the specific volume of steam at the exit of the guide blades and the theoretical available energy $h_{01} + h_e^{pr}$ is expressed by the tangent of the angle γ_1 (Fig. 7-12).

For the guide blades of the first stage as well as for those groups of guide blades which are situated immediately after a sudden variation in the drum diameters, $h_e^{pr} = 0$. For the moving blades of any given stage we may write

$$w_2 = \frac{G_0 v_2 n}{60 u l'' \sin\beta_2}. \qquad (7\text{-}37e)$$

Multiplying the above equation throughout by $\frac{w_2^2}{8{,}378\psi^2}$,

$$\frac{w_2^2}{8{,}378\psi^2} = \frac{G_0 v_2 n}{60\times 8{,}378\psi^2 x_2 l'' \sin\beta_2}, \qquad (7\text{-}37f)$$

where $x_2 = \frac{u}{w_2}$.

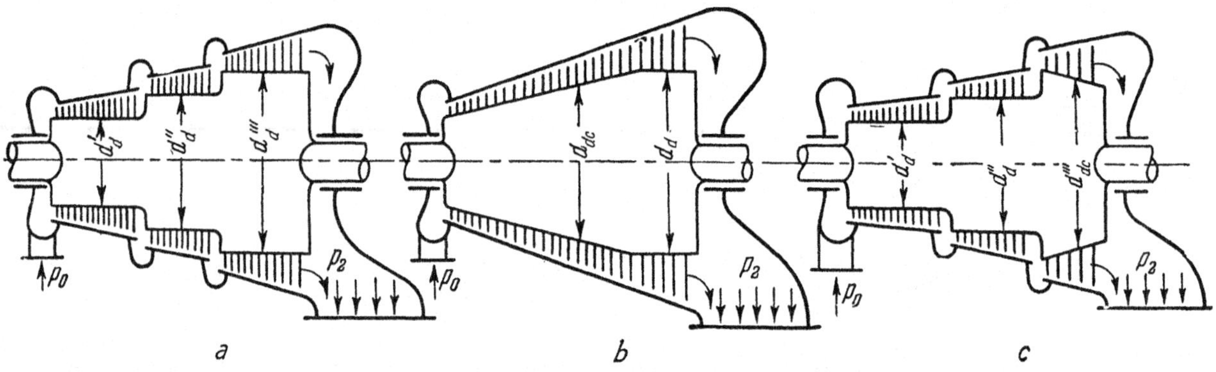

Fig. 7-13. Principal types of reaction turbines

Since $w_2^2/8{,}378\psi^2 = h_{02} + w_1^2/8{,}378$, equation (7-37f) becomes

$$h_{02} + \frac{w_1^2}{8{,}378} = \frac{G_0 v_2 n}{502{,}800\psi^2 x_2 l'' \sin\beta_2}, \quad (7\text{-}37g)$$

where $w_1^2/8{,}378 = h_{w_1}$ — energy content of 1 kg of steam entering the working blades with a velocity w_1. Equation (7-37g) may be further expressed as

$$\frac{M_v v_2}{M_h(h_{02} + h_{w_1})} = \frac{M_v 502{,}800\psi^2 x_2 l'' \sin\beta_2}{M_h G_0 n} = \tan\gamma_2. \quad (7\text{-}37h)$$

The relation between the specific volume of steam at the exit of the working blades and the theoretical heat drop $h_{02} + h_{w1}$ is expressed by the tangent of the angle γ_2 as shown in Fig. 7-12.

Since for reaction turbines $\varphi \approx \psi = 0.94$ to 0.97, equations (7-37d) and (7-37h) may be transformed into

$$\tan\gamma_1 = Ax_1 l \sin\alpha_1, \quad (7\text{-}37i)$$

$$\tan\gamma_2 = Ax_2 l'' \sin\beta_2; \quad (7\text{-}37j)$$

where $A = \frac{502{,}800\varphi^2 M_v}{G_0 n M_h} = \frac{502{,}800\psi^2 M_v}{G_0 n M_h}$.

The values of $h_{01} + h_e^{pr}$ and $h_{02} + h_{w_1}$ are determined directly from the graph 7-12 by reading the values of tangents γ_1 and γ_2.

For a proper distribution of heat drops in the various stages it is necessary to know the shape of the turbine (rough approximation). Fig. 7-13 shows a typical construction of a reaction turbine. Only after deciding the probable shape of the turbine we may proceed with the calculations for the distribution of heat drop in the various turbine stages.

From equations (7-37i) and (7-37j),

$$\tan\gamma_1 = A\frac{d'_d x_1 \sin\alpha_1}{\left(\frac{d'_d}{l}\right)}; \quad (7\text{-}38)$$

$$\tan\gamma_2 = A\frac{d''_d x_2 \sin\beta_2}{\left(\frac{d''_d}{l''}\right)}, \quad (7\text{-}38a)$$

where d'_d and d''_d — drum diameters for guide and working blades respectively.

Dividing equation (7-38a) by (7-38) we have

a) for the moving blades of the succeeding stage

$$\tan\gamma_2 = \tan\gamma_1 \frac{d''_d x_2 \sin\beta_2 \left(\frac{d'_d}{l}\right)}{d'_d x_1 \sin\alpha_1 \left(\frac{d''_d}{l''}\right)}; \quad (7\text{-}39)$$

b) for the guide blades of the succeeding stage

$$\tan\gamma_1 = \tan\gamma_2 \frac{d'_d x_1 \sin\alpha_1 \left(\frac{d''_d}{l''}\right)}{d''_d x_2 \sin\beta_2 \left(\frac{d'_d}{l}\right)}. \quad (7\text{-}39a)$$

Equations (7-39) and (7-39a) are valid for any type of construction. If instead of a gradual variation in the drum diameter, a sort of stepping up arrangement is used (Fig. 7-13,a) then for the blades fixed on the constant diameter of the drum,

since d'_d is equal to d''_d, we have

$$\tan\gamma_2 = \tan\gamma_1 \frac{x_2\left(\frac{d_d}{l}\right)\sin\beta_2}{x_1\left(\frac{d_d}{l''}\right)\sin\alpha_1}; \quad (7\text{-}39b)$$

$$\tan\gamma_1 = \tan\gamma_2 \frac{x_1\left(\frac{d_d}{l''}\right)\sin\alpha_1}{x_2\left(\frac{d_d}{l}\right)\sin\beta_2}. \quad (7\text{-}39c)$$

For the group of stages where $\alpha_1 = \beta_2$

$$\tan\gamma_2 = \tan\gamma_1 \frac{x_2\left(\frac{d_d}{l}\right)}{x_1\left(\frac{d_d}{l''}\right)}; \quad (7\text{-}39d)$$

$$\tan\gamma_1 = \tan\gamma_2 \frac{x_1\left(\frac{d_d}{l''}\right)}{x_2\left(\frac{d_d}{l}\right)}. \quad (7\text{-}39e)$$

7-12. DISTRIBUTION OF HEAT DROP IN THE TURBINE STAGES

Stage diameter and height of the guide blades of the first stage as well as the stage diameter and height of the working blades of the last stage are known quantities from the preliminary design. Thus drum diameters for the first and last stages may be determined as: $d'_d = d_1 - l$ and $d^z_d = d_z - l_z$. Depending on the diameters d'_d and d^z_d one of the types of construction shown in Fig. 7-13 is selected for the turbine drum. For turbines of small capacity having $d_z \leqslant 1.5 d_1$ the drum may be as shown at (*b*) in Fig. 7-13. Types *a* and *c* also may be used but with a single-diameter step up for the turbine drum. For back-pressure turbines it is usual to have the drum of a constant diameter; in very rare cases the drum may be conical in shape. The variations in the height of blades for the first group of stages is tabulated as follows:

Stage No.	Row No.	Rotor dia., mm	Blade height, mm	Ratio d_d/l	Mean dia., mm	Circumferential velocity u, m/sec

F i r s t s t a g e. Using the data obtained from the preliminary design, for the guide blades, we have

$$\tan \gamma'_1 = A \frac{d'_d x'_1 \sin \alpha_1}{\left(\frac{d'_d}{l_1}\right)}.$$

Line Ob'_1 is drawn from the origin O (Fig. 7-12) at an angle γ'_1 until it intersects the curve *2*. The heat drop h'_{01} occurring in the guide blades of the first stage is obtained by dropping a perpendicular from point b'_1 on to the Ox axis (intercept Ob_1). Pressure after the guide blades is obtained from curve 1 and the specific volume at point b'_1 from curve *2*.

The velocity of steam at the exit of the guide blades will be

$$c_1 = 91.5 \varphi \sqrt{h'_{01}}.$$

Velocity w_1 is obtained from the velocity triangle. The energy content at entry to the working blades will be

$$h_{w_1} = \frac{w_1^2}{8{,}378}.$$

The numerical value of h_{w_1}, to a proper scale, is measured along the Ox axis from point b_1 to the left (point b). Tangent γ'_2 is determined for the working blades according to formula (7-39). All quantities except x'_2 are known, and hence the value of x'_2 is suitably assumed. From point b a line is drawn at an angle γ'_2 to the Ox axis until it intersects curve *2* at point b'_2. The distance, along the axis, between the points b_1 and b_2 gives the heat drop h'_{02} occurring in the first row of the moving blades.

The velocity of steam issuing from the moving blades is given by

$$w_2 = 91.5 \psi \sqrt{h'_{02} + h_{w_1}}$$

from which the value of x'_2 is determined

$$x'_2 = u / w_2.$$

If the value of x'_2 thus obtained differs from the value assumed previously then a different value of x'_2 has to be taken and the calculations repeated, until a proper agreement is reached. As a check tan γ'_2 is also calculated from

$$\tan \gamma'_2 = \frac{M_v v_2}{M_h (h'_{02} + h_{w_1})},$$

where $h'_{02} + h_{w_1}$ — theoretical value of the energy entering the working blade passages; the numerical value is obtained from the graph shown in Fig. 7-12 (intercept bb_2). The specific volume v_2 is obtained from curve *2* (point b'_2).

The error for the value of tan γ'_2 should not exceed 1%.

The velocity of steam c_2 after the working blades is obtained from the velocity triangles from which the energy at exit (carry-over loss) will be

$$h_e = c_2^2 / 8{,}378.$$

This value is measured from point b_2 to the left (point b'). Simultaneously with the distribution of heat drop the height of blades for each stage is also checked:

a) for guide blades according to equation (4-7)

$$l_1 = \frac{G_0 v_1}{\pi d_1 c_1 \sin \alpha_1};$$

b) for working blades—according to equation (4-11)

$$l'' = \frac{G_0 v_2}{\pi d_1 w_2 \sin\beta_2}.$$

Second stage:

Tan γ_1'' for the second stage guide blades is obtained from the equation

$$\tan\gamma_1'' = \tan\gamma_1' \frac{d_{d2}' x_2' \sin\alpha_1'' \left(\frac{d_{d1}'}{l_1}\right)}{d_{d1}' x_1' \sin\alpha_1' \left(\frac{d_{d2}'}{l_2}\right)},$$

where the quantities with index 2 and two superscripts (") refer to the guide blades of the second stage. Similarly for the working blades we have

$$\tan\gamma_2'' = \tan\gamma_1' \frac{d_{d2}'' x_2'' \sin\beta_2'' \left(\frac{d_{d1}'}{l_1}\right)}{d_{d1}' x_1' \sin\alpha_1' \left(\frac{d_{d2}'}{l_2''}\right)}.$$

The values of x_2' and x_2'' are assumed previous to the determination of tan γ_1'' and tan γ_2''. Conforming to these assumed values of x_2' and x_2'' the theoretical heat drop occurring in the guide and moving blades is determined with the help of the graph in Fig. 7-12, i.e.,

$$(h_{01}'' + h_e^{pr}) \quad \text{and} \quad (h_{02}'' + h_{w_1}).$$

Velocities c_2 and w_2 are determined from velocity triangles and the values of x'_2 and x''_2 are checked as described before. In the event of a difference in the values of x'_2 and x''_2 between the assumed and calculated ones the process has to be repeated until a satisfactory solution is obtained. An exactly similar method is used for determining the heat drops in all the following stages. For the final stages besides varying the ratios u/c_1 and u/w_2, it is necessary to vary the angles α_1 and β_2 as well.

For the first group of high-pressure stages it is recommended that the height of blades for each succeeding stage should be taken 1 to 1.5 mm longer than the blades of the previous stage.

Fig. 7-12 shows the heat drop distribution in the various stages of a reaction turbine. If it is found that as a result of distribution of heat drop in the various stages, the sum of all these heat drops exceeds or falls short of $(1+\alpha)\ H_0'$, then the values of x_1 and x_2 should be slightly altered so that the equation $\sum_1^z h \leq (1+\alpha)\ H_0'$ is satisfied.

Details of Heat Drop Calculations

The final calculations for thermodynamic design of the various turbine stages are carried out after obtaining a preliminary idea of the distribution of heat drops in each of them.

a) Determination of losses and efficiency for a stage.

Losses in guide blades:

$$h_{gb} = (1 - \varphi^2)(h_{01} + h_e^{pr}). \qquad (7\text{-}40)$$

Losses in working blades:

$$h_b = (1 - \psi^2)(h_{02} + h_{w_1}). \qquad (7\text{-}41)$$

Carry-over loss from equation (5-3)

$$h_e = c_2^2/8{,}378.$$

Losses due to leakage of steam through the radial clearances, from equation (5-24)

$$h_{leak} = \frac{\delta_r}{l \sin\alpha_1}(i_0 - i_2)$$

or

$$h_{leak} = 1.72 \frac{\delta_r^{1.4}}{l}(h_{01} + h_{02}).$$

Neglecting the losses due to wetness the usefully utilised heat drop will be

$$h_i' = (h_{01} + h_{02} + h_e^{pr}) - (h_{gb} - h_b - h_e - h_{leak}). \qquad (7\text{-}42)$$

Losses due to wetness of steam from equation (5-25) will be

$$h_{wetness} = (1 - x)\, h_i.$$

Heat drop utilised for doing useful work, taking into account the losses due to wetness of steam, will be

$$h_i = h_{01} + h_{02} + \mu h_e^{pr} - h_{gb} - h_b - h_{leak} - h_{wetness} - h_e$$

(for reaction stages the coefficient μ may be taken equal to unity).

We shall determine the efficiency of the stage according to equation (7-3), i.e., with the assumption that all the carry-over velocity is utilised in the next stage. Hence

$$\eta_{oi}^{st} = \frac{h_i}{h_0 + h_e^{pr}}.$$

b) Determination of turbine efficiency and capacity.

The heat drop usefully utilised in the turbine is obtained from equation (7-2):

$$H_i = \sum h_i.$$

The actual mass flow of steam through the turbine will be

$$G_{act} = \frac{860 N_e}{3{,}600 H_i \eta_m \eta_r}. \qquad (7\text{-}43)$$

If it is found that $G_{act} \neq G_0$ then the heights of guide and moving blades have to be recalculated according to the relation

$$l_{act} = \frac{G_{act}}{G_0} l, \qquad (7\text{-}44)$$

where l_{act}—actual blade heights;
l — calculated blade heights.

7-13 CALCULATION OF AXIAL THRUST

1. Axial Thrust in Reaction Turbines

In a reaction turbine the axial thrust is made up of the following:

axial pressure on the rotor collars (where the rotor dia. is increased) R_{coll};

axial pressure on the conical portion of the rotor R_c;

pressure difference between the two sides of a moving blade disc R_b;

axial thrust due to the difference in momentum of inlet and exit steam R_m.

The thrust due to axial pressure on the drum drojections is determined as

$$R_{coll} = \frac{\pi}{4} \sum_{1}^{k} [d_{di}^2 - d_{d(i-1)}^2] p_i + \frac{\pi}{4}(d_{s1}^2 - d_{s2}^2) p_{atm} \quad [kg] \qquad (7\text{-}45)$$

where $d_{d(i-1)}$ and d_{di}—rotor diameters before and after the projection (Fig. 7-13);
p_i—pressure of steam in the chamber just before the projection;
k—number of projections;
d_{s1} and d_{s2}—shaft diameters at the front and rear labyrinth seals;
p_{atm}—atmospheric pressure.

The axial pressure on the conical portion of the rotor is determined from the equation

$$R_c = \frac{\pi}{4} \sum_{1}^{zz} (d_i^2 - d_{i-1}^2) \frac{p_i + p_{i-1}}{2} \quad [kg], \qquad (7\text{-}46)$$

where d_{i-1} and d_i—drum diameters at the beginning and end of the conical portion of the rotor measured at the inlet plane of the guide or moving blades as shown in Fig. 7-14;
p_{i-1} and p_i—steam pressure conforming to these diameters;
z—number of stages.

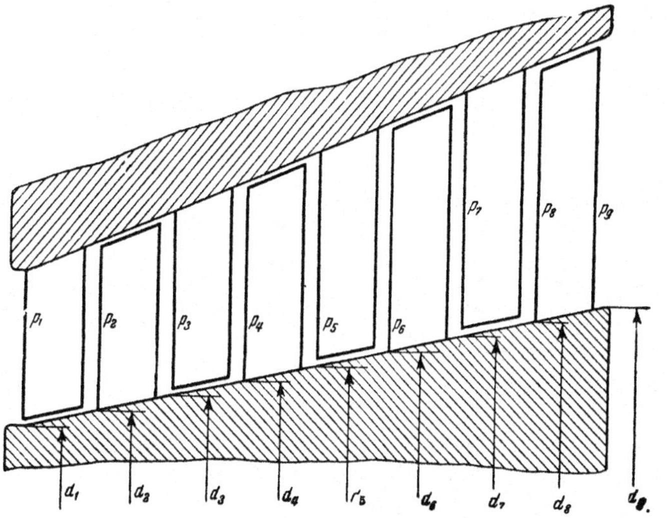

Fig. 7-14. Increase of blade heights in reaction turbines

Axial thrust on the moving blades of all the stages is calculated from the equation

$$R_b = \frac{\pi}{4} \sum_{1}^{z} (d_{ri}^2 - d_r^2)(p'_{1i} - p_{2i}) \quad [kg], \qquad (7\text{-}47)$$

where d_{ri} and d_r— diameters at the rim and root of blades;
p'_{1i} and p_{2i} — steam pressure before and after the moving blades of the stage.

Axial thrust on the rotor caused by the difference in the momentum of inlet and exit steam is determined from the equation

$$R_m = \sum_{1}^{z} \frac{G}{g} (c_{1ai} - c_{2ai}) \quad [kg] \qquad (7\text{-}48)$$

where G — mass flow of steam through the turbine, kg/sec;
g — acceleration due to gravity 9.81 m/sec²;
c_{1ai} and c_{2ai}— projections of the absolute velocities c_1 and c_2 on the turbine axis.

The total thrust on the turbine rotor is obtained by summing up the various quantities obtained from equations (7-45), (7-46), (7-47) and (7-48) so that

$$R_0 = R_{coll} + R_c + R_b + R_m = \frac{\pi}{4} \sum_{1}^{k} [d_{di}^2 - d_{d(i-1)}^2] p_i + \frac{\pi}{4}(d_{s1}^2 - d_{s2}^2) p_{ATM} + \frac{\pi}{4} \sum_{1}^{zz} [d_{di}^2 - d_{d(i-1)}^2] \frac{p_i + p_{i-1}}{2} + \frac{\pi}{4} \sum_{1}^{z} (d_{r1}^2 - d_r^2)(p'_{1i} - p_{2i}) + \sum_{1}^{z} \frac{G}{g}(c_{1ai} - c_{2ai}) \quad [kg]. \qquad (7\text{-}49)$$

While calculating for thrust care should be taken to give the proper sign for the direction in which the forces are acting: forces acting in the direction of steam flow will be deemed to be positive and vice versa.

The axial thrust R_0 acting on the rotor is balanced by a dummy piston; part of the thrust is taken up by the thrust bearing. The thrust bearing besides bearing the axial forces also helps in fixing the rotor in the proper axial direction.

The conditions for thrust balance are expressed by the equation

$$R_0 = F_{br} q = R_{coll} + R_c + R_b + R_m \text{ [kg]} \quad (7\text{-}50)$$

where F_{br} — surface of the thrust bearing, cm^2;
q — specific pressure on the thrust bearing, kg/cm^2.

For segment-type thrust bearing q is usually taken as 15 to 20 kg/cm^2 and for the collar-type no more than 5 to 8 kg/cm^2.

For turbines without balance or dummy pistons (Fig. 7-13) all the quantities except F_{br} and q in equations (7-49) and (7-50) are known. Assuming a suitable value for q F_{br} may be easily determined.

In the case of turbines with dummy pistons (Figs 1-4, 7-8, 7-18) besides F_{br} and q, diameter d_p of the dummy piston is also an unknown quantity. In such cases this diameter is also included in equation (7-45) as one of the projections. If suitable values for F_{br} and q are assumed on the basis of facility of construction and satisfactory strength of thrust bearing, dia. d_{dp} may be easily obtained from equations (7-49) and (7-50). If d_{dp} thus obtained does not suit the rotor construction one or both of the quantities F_{br} and q may have to be altered. Calculation of axial thrust is illustrated in Fig. 7-15.

2. Axial Thrust in Impulse Turbines

The axial pressure in impulse turbines may be expressed by an analogous equation

$$R_0 = R_{coll} + R_d + R_m \text{ [kg]} \quad (7\text{-}51)$$

where R_{coll} — axial pressure on rotor collars, obtained from equation (7-45);
R_d — axial pressure on the disc surfaces, caused by pressure drop as a result of reaction which may be determined from equation (7-47); d_r is taken as the diameter of the hub;
R_m — axial thrust caused by the difference in the momentum of steam at inlet and exit (equation 7-48).

It should be noted that equation (7-47) is only an approximate one, since in the absence of pressure equalising holes the actual pressure exerted by steam on the discs would be considerably greater than what is obtained from this equation. Depending upon the sizes of clearances between the guide blade diaphragms and the shaft and between the moving blades and the stator, the axial thrust can be as much as 10 to 20% greater than values obtained from equation (7-47). On the other hand, if pressure equalising holes are provided the actual axial thrust would be lower than what would be obtained from equation (7-47). Depending upon the areas of the pressure equalising holes, labyrinth clearances for guide blade, diaphragms, etc., the discs may be relieved of axial thrust to as much as 50 to 80% of the values obtained from equation (7-47) [1].

The determination of pressure drop across the discs, taking into account the effect of steam leakage through the clearances as well as through the pressure equalising holes, becomes a matter of considerable difficulty in the absence of dependable data for flow coefficients. As a result of incorrect calculations of axial thrust there have been quite a large number of serious accidents due to breaking or melting of the thrust bearings. Besides, the axial thrust on the rotor depends to a very large extent on the operation conditions, labyrinth seals both for the front and rear ends as well as between diaphragms and rotors, etc. The axial thrust thus varies within very wide limits.

Recently a new method of measuring axial pressures in turbines while under operation has been suggested. This method makes use of variation of resistance of the measuring element when deformed due to pressure variations on the membrane to which the measuring element is glued. Consequent to these investigations M. A. Trubilov has suggested a method for checking the satisfactory operation of thrust bearings on the basis of the temperature rise of white metal of the thrust bearing shoes [2]. This suggestion has been found to be of very great practical value and is now used very widely for checking the satisfactory operation of thrust bearings.

7-14. IMPULSE-REACTION TURBINES

Pure reactions of small and medium capacity are not very convenient from the constructional

[1] One of the most valuable investigations in this field is that of V. V. Zvyagintsev, "Approximate Method of Determining Pressure Drops in Steam Turbine Discs", *Sovetskoye Kotloturbostroyenie*, 1938, No. 7, and G. S. Samoilovich and B. I. Morozov, "About the Coefficient of Mass Flow Through Pressure Relief Holes in Turbine Discs", *Teploenergetika*, 1957, No. 8.

[2] Information Letter of the V. T. I., No. 1952-7, Gosenergoizdat, 1953.

point of view, since with blade heights of 25 to 30 mm and $u/c_1 \geqslant 0.65$ to 0.70 the number of stages required would be exceedingly large and at the same time the efficiency considerably decreases, on account of larger leakage losses. Hence reaction turbines in their purest form are made only for large capacities (not less than 20,000 kW) with medium values for fresh steam (i. e., with medium pressures and temperatures). If pure reaction turbines are built with very high steam pressures and temperatures (30 to 35 ata) a lot of subsidiary problems crop up. If the heat drop utilised in the first stage is not very high the pressure in the remaining portion of the turbine substantially increases which further complicates the construction of flanges, front and rear labyrinth seals, etc. The combined impulse-reaction turbine obviates many of these difficulties and hence has found ready usage with the turbine designers all over the world.

The most widely used impulse-reaction turbine consists of a first stage working on the impulse principle (more often known as the Curtis stage) followed by a number of reaction stages. The first stage may be either single-row or multirow velocity compounded. The use of an impulse (regulating) stage with velocity compounding permits the utilisation of a large heat drop in the nozzles and consequently helps in obtaining lower temperatures and pressures in the reaction stages. The use of impulse stages operating with heat drops of 40 to 60 kcal/kg and more enables the reduction of a number of reaction stages. Hence a combined impulse-reaction turbine is found to be simple in construction as well as cheaper to make.

Combined types of turbines are also made with the high-pressure portion of the turbine consisting only of impulse blading and the low-pressure stages consisting only of reaction stages.

7-15. DESIGN OF IMPULSE-REACTION TURBINES

Example: Design a combined impulse-reaction turbine with the following data: $N_e = 6{,}600$ kW capacity at generator terminals, $n = 3{,}000$ r. p. m.; $p_0 = 29$ ata; $t_0 = 400°$ C; $p_2 = 0.05$ ata.

We shall assume efficiencies from the curves given in Figs 5-8, 5-9 and 5-11.

$$\eta_{re} = 0.765;\ \eta_g = 0.948;\ \eta_m = 0.98.$$

The relative internal efficiency of the turbine will be

$$\eta_{oi} = \frac{\eta_{re}}{\eta_m} = \frac{0.765}{0.98} = 0.782.$$

Steam pressure before the nozzles of the first stage, considering the losses in the regulating valves

$$p_0' = p_0 - \Delta p = 29 - 0.05 \times 29 = 27.5 \text{ ata}.$$

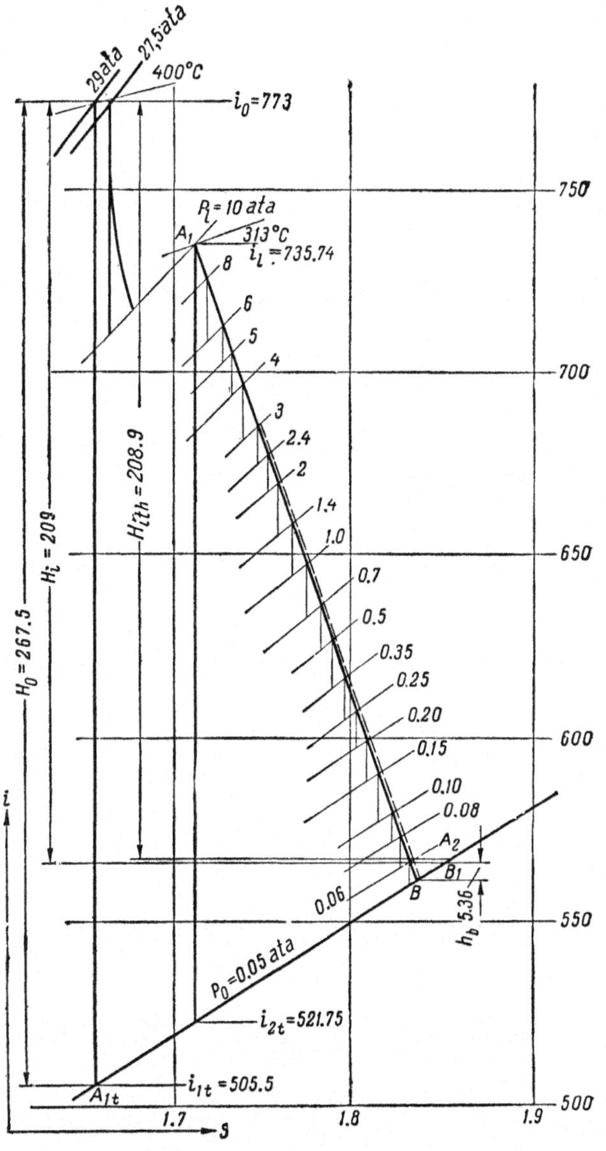

Fig. 7-15. Theoretical heat drop on the *i-s* diagram for a combined impulse-reaction turbine

From the *i-s* diagram (Fig. 7-15) the theoretical heat drops will be

$$H_0 = i_0 - i_{1t} = 773 - 505.5 = 267.5 \text{ kcal/kg};$$

$$H_0' = i_0 - i_{1t}' = 773 - 507.5 = 265.5 \text{ kcal/kg}.$$

Heat drop utilised in the turbine will be

$$H_i = H_0 \eta_{oi} = 267.5 \times 0.782 = 209 \text{ kcal/kg}.$$

Determining the mass flow of steam through the turbine we have

$$G_0 = \frac{D_0}{3{,}600} = \frac{860 N_e}{3{,}600 H_0 \eta_{re} \eta_g} =$$

$$= \frac{860 \times 6{,}600}{3{,}600 \times 267.5 \times 0.765 \times 0.948} = 8.14 \text{ kg/sec}.$$

p, ata	10	8	6	5	4	3	2.4	2.0
v, m³/kg	0.271	0.325	0.413	0.480	0.578	0.734	0.907	1.025
t or x (°C or %)	313	292	265	248	229	203	185	169
h_0, kcal/kg	13.6	17.3	10.4	11.8	15.7	10.8	9.7	15.5

In order to utilise a large heat drop in the first impulse stage we shall use a two-row disc for the regulating stage. Pressure of steam before the guide blades of the first reaction stage shall be assumed as 10 ata. Then the theoretical heat drop occurring in the nozzles of the first stage will be from the i-s diagram $h_{0th}=i_0-i_{tin}=773-708.5=64.5$ kcal/kg.

Referring to 6-6 where this example has been worked out we find that

mean diameter . . . . . . $d=983$ mm,
exit height of nozzle . . . $l=12$ mm;
nozzle angle . . . . $\alpha_1=20°$.

Since the mass flow of steam in this case is different, the degree of partial admission also will vary, which will be according to equation (7-25a)

$$\varepsilon=\frac{60G_0v_1x_1}{\pi^2d^2ln\sin\alpha_1}=$$
$$=\frac{60\times8.14\times0.253\times0.22}{\pi^2\times0.983^2\times0.012\times3{,}000\sin20°}=0.23,$$

where $v_1=0.253$ m³/kg and $x_1=0.22$ (from calculations of 6-6 and 7-8).

Windage and disc friction losses will be determined from equation

$$h_{fr.w.}=\frac{102N_{fr.w.}}{427G_0}=\frac{102N_{fr.w.}}{427\times8.14}=$$
$$=0.0294\times40.3=1.182 \text{ kcal/kg},$$

where $N_{fr.w.}$ is obtained from equation (5-4a); we shall take the mean blade height to be 20 mm.

Losses in the nozzles, guide and moving blades as well as the carry-over losses remain unchanged (see 6-6).

Heat drop utilised in the stage will be

$$h_{ith}=h_{oth}-\Sigma h_{losses}=64.5-6.325-12.18-3.38-$$
$$-0.68-2.175-1.182=38.58 \text{ kcal/kg}.$$

Marking off this heat drop on the i-s diagram we find the steam conditions before the guide blades of the first reaction stage (point A_1 in Fig. 7-15).

Marking off the heat drop usefully utilised in the turbine H_i on the i-s diagram equal to 209 kcal/kg we obtain point A_2 which gives the steam conditions at exhaust. The carry-over loss for the last stage will be assumed as 2% of H_0 so that $h_{ez}=0.02$, $H_0=0.02\times\times267.5=5.36$ kcal/kg. Marking off this loss h_{ez} on the i-s diagram downwards from point A_2 we get point B which gives the steam conditions at the exit section of the last row of the moving blades. Joining points A_1 and B by a straight line we obtain the theoretical heat drop process for the reaction stages on the i-s diagram. For the arbitrarily chosen isobars we shall now determine theoretical heat drop between each pair of isobars and the specific volume of steam at each of the points of intersection of isobars with the straight line joining points A_1 and B as also the temperatures of superheated steam or dryness fraction in the region of saturation.

The results of the above calculations are given in Table 7-2.

From the data given in Table 7-2 we now plot the auxiliary graph shown in Fig. 7-16: $v=f(\Sigma h_0)$ curves $1a$ and $1b$; $p=f_1(\Sigma h_0)$ curves $2a$ and $2b$; $t=f_2(\Sigma h_0)$ curve 3 and $x=f_3(\Sigma h_0)$ curve 4, where x is the dryness fraction.

Next we determine the dimensions of the first and last reaction stages.

The first reaction stage (the second turbine stage).

We shall take the height of the guide blades to be $l=25$ mm.

Diameter at the mean circumference from equation (7-26) will be

$$d_1=\sqrt{\frac{60G_0x_1v_1}{6\pi^2l_1\sin\alpha_1}}=\sqrt{\frac{60\times8.14\times0.715v_1}{\pi^2\times0.025\times3{,}000}}=\sin20°=$$
$$=\sqrt{1.378v_1}=\sqrt{1.378\times0.283}=0.625 \text{ m}=$$
$$=625 \text{ mm};$$

x_1 is assumed to be 0.715.

Circumferential velocity u_1 will be

$$u_1=\frac{\pi d_1n}{60}=\frac{\pi\times0.625\times3{,}000}{60}=98.2 \text{ m/sec}.$$

Velocity of steam at the exit of the guide blades will be

$$c_1=\frac{u_1}{x_1}=\frac{98.2}{0.715}=137 \text{ m/sec}.$$

$$\text{Heat drop } h_{01}=\frac{c_1^2}{8{,}378\varphi^2}=\frac{137^2}{8{,}378\times0.92^2}=2.65 \text{ kcal/kg}.$$

Pressure of steam after the guide blades from curve $2a$ (Fig. 7-16) $p_1'=9.65$ ata.

The last stage: Diameter of the last stage will be determined from equation (7-30)

$$d_z=\sqrt{\frac{G_0\vartheta v_2}{\pi\times91.5\sqrt{\zeta_eH_i}\sin\alpha_2}}=$$
$$=\sqrt{\frac{8.14\times7\times26.65}{\pi\times91.5\sqrt{0.02\times267.5\times1}}}=1.53 \text{ m}=1{,}530\text{mm},$$

where $\vartheta=7$, $\zeta_e=0.02$ and $\alpha_2=90°$ (assumed values).

Circumferential velocity at the mean diameter will be

$$u_z=\frac{\pi d_zn}{60}=\frac{\pi\times1.53\times3{,}000}{60}=241 \text{ m/sec},$$

which is not permissible for thin-walled discs and hence the last group of stages have to be accommodated on one or two discs of sufficiently large width (thickness).

Height of moving blades will be determined as

$$l_z=\frac{d_z}{\vartheta}=\frac{1{,}530}{7}=219 \text{ mm}.$$

Table 7-2

Various Stages of Reaction Turbine

1.4	1.0	0.7	0.5	0.35	0.25	0.2	0.15	0.10	0.08	0.06	0.05
1.375	1.813	2.425	3.29	4.54	6.16	7.53	9.8	14.17	17.4	22.1	26.65
141	115	90	98.5	97.5	96.3	95.6	91.6	93.3	92.6	91.7	91.2
14.4	14.8	13.8	12.7	12.0	8.0	10.8	14.0	7.2	9.5	5.9	—

Drum diameter for the second stage is $d_d = 625 - 25 = 600$ mm, and for the last stage $d_{d_z} = 1{,}530 - 219 = 1{,}311$ mm.

The rotor will be constructed, therefore, with one or more offsets (collars).

Distribution of Heat Drop in Turbine Stages

Velocity coefficients $\varphi = \psi = 0.92$ will be assumed for all the turbine stages.

For the first group of stages we shall take $\alpha_1 = \beta_2 = 20°$. Exit height of each of the succeeding stage blades shall be made longer than the previous one by 1 mm.

We shall assume that 15 stages will be located on the portion where the drum diameter is 600 mm. For this group of stages we shall calculate the mean diameter d_{mean}, the ratio d_d/l and u — the circumferential velocity. The results of all these calculations are given in Table 7-3.

The second stage:

From the preliminary calculations we have for the second stage guide blades: $c_1 = 137$ m/sec, $u = 98.2$ m/sec, $\alpha_1 = 20°$ and $h_{01} = 2.65$ kcal/kg.

From the inlet velocity triangle (Fig. 7-17) we obtain $w_1 = 55.5$ m/sec and $\beta_1 = 57°30'$.

Heat content of steam at entry to the moving blades will be

$$h_{w_1} = \frac{w_1^2}{8{,}378} = \frac{55.5^2}{8{,}378} = 0.368 \text{ kcal/kg}.$$

Tan γ_1'' (Fig. 7-16) is obtained as

$$\tan \gamma_1'' = \frac{M_v v_1}{M_h (h_{01} + h_e^{pr})} = \frac{100 \times 0.283}{2 \times 2.65} = 5.34$$

where M_v and M_h are the scales to which volumes and heat drops are plotted. With $\tan \gamma_1'' = 5.34$ we draw on Fig. 7-16 the triangle *O12*. From curve *2a* we find the pressure after the guide blades which is found to be $p_1' = 9.65$ ata.

Table 7-3

Guide Blades

Stage No.	d_d, mm	l, mm	d_{mean}, mm	d_d/l	u m/sec	u/c_1	$h_e^{pr} + h_{01}$ kcal/kg	v_1, m³/kg	p_1', ata
1a	—	12	983	—	154.5	0.220	64.5	0.25	27.5
1b	—	20.9	—	—	—	—	—	0.266	10.0
2	600	25	625	23.95	98.2	0.715	2.650	0.283	9.65
3	600	27	627	22.2	98.6	0.738	2.520	0.299	9.00
4	600	29	629	20.7	98.8	0.755	2.430	0.316	8.32
5	600	31	631	19.3	99.0	0.767	2.350	0.334	7.78
6	600	33	633	18.15	99.4	0.783	2.273	0.352	7.27
7	600	35	635	17.15	99.6	0.790	2.239	0.383	6.82
8	600	37	637	16.20	100.0	0.794	2.239	0.393	6.37
9	600	39	639	15.40	100.3	0.794	2.263	0.416	5.95
10	600	41	641	14.63	100.5	0.788	2.295	0.441	5.60
11	600	43	643	13.95	100.8	0.784	2.321	0.467	5.16
12	600	45	645	13.34	101.2	0.772	2.418	0.500	4.78
13	600	47	647	12.75	101.4	0.757	2.540	0.537	4.42
14	600	49	649	12.25	101.8	0.744	2.651	0.574	4.06
15	600	51	651	11.76	102.2	0.726	2.794	0.619	3.73
16	600	53	653	11.32	102.5	0.700	3.009	0.670	3.40
17	800	40	840	20.00	132.0	0.780	4.040	0.750	2.97
18	800	44	844	18.15	132.6	0.764	4.262	0.818	2.56
19	800	48	848	16.65	133.4	0.743	4.551	0.962	2.20
20	800	52	852	15.37	134.0	0.695	5.271	1.125	1.83
21	800	56	856	14.27	134.6	0.699	5.215	1.333	1.51
22	800	62	862	12.90	135.5	0.650	6.16	1.604	1.19
23	800	70	870	11.42	136.7	0.602	7.27	1.99	0.90
24	800	78	878	10.25	138.0	0.594	7.625	2.60	0.66
25	1,330	90	1,420	14.78	223	0.845	9.83	4.00	0.418
26	1,330	115	1,445	11.55	227	0.812	11.077	6.58	0.283
27	1,330	155	1,485	8.60	233	0.840	10.865	10.25	0.143
28	1,330	195	1,525	6.82	239	0.777	13.41	18.45	0.075

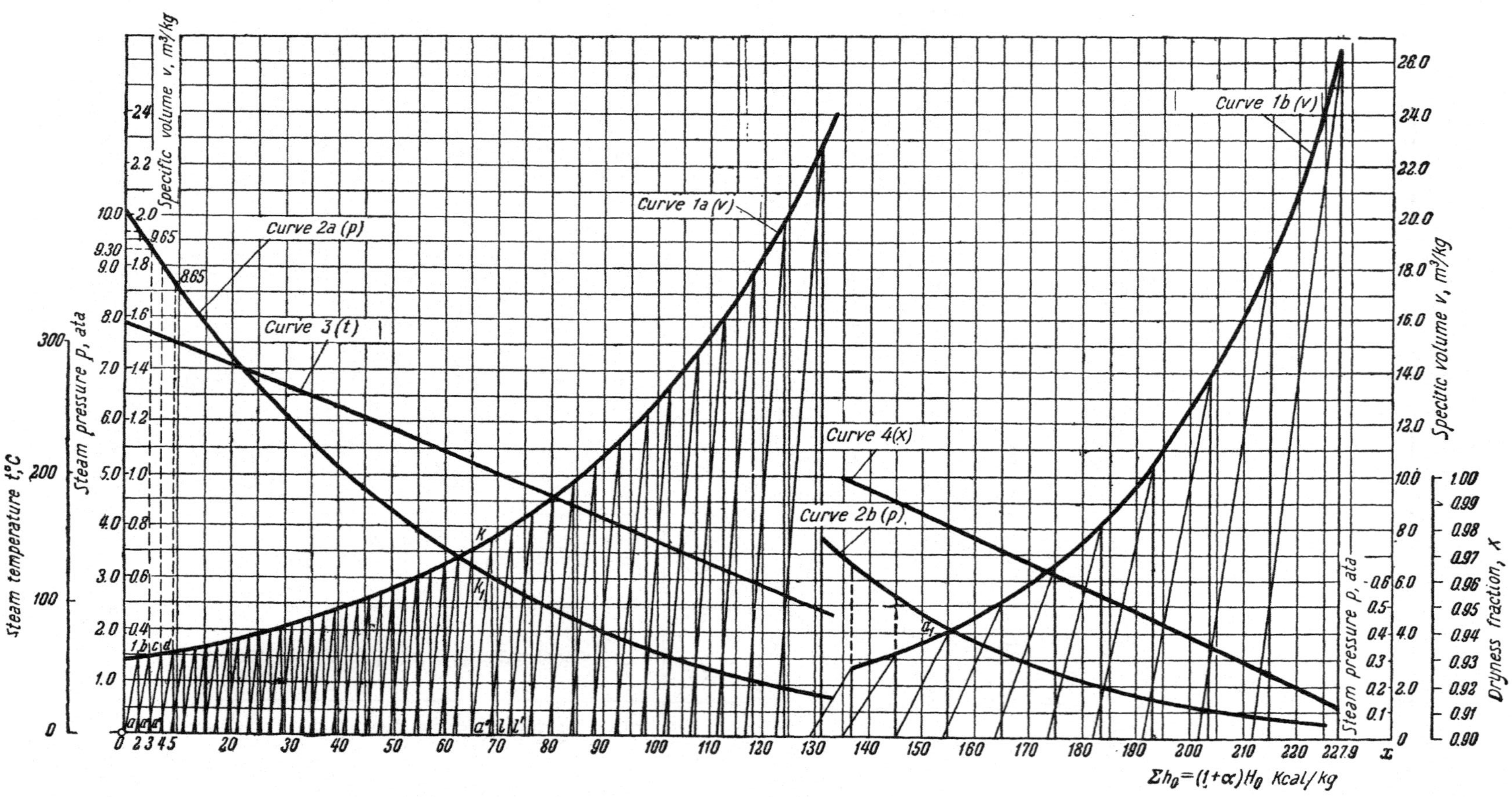

Fig. 7-16. Distribution of heat drop in the various stages of a reaction turbine

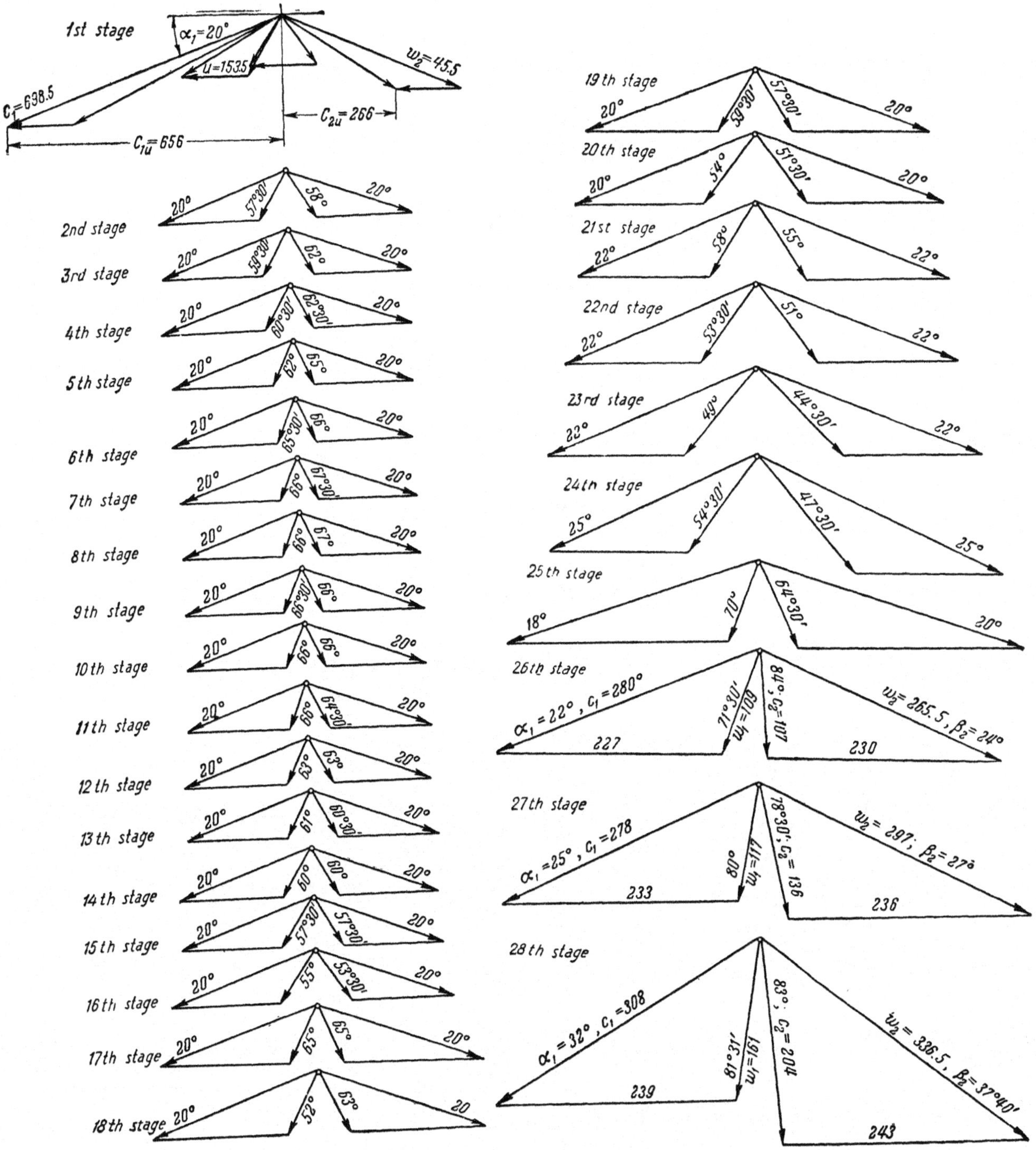

Fig. 7-17. Velocity triangles for an impulse-reaction turbine

From equation (7-39d) we have $\tan\gamma_2''$ for the working blades as

$$\tan\gamma_2''=\tan\gamma_1''\frac{\left(\frac{d_d}{l}\right)x_2''}{\left(\frac{d_d''}{l''}\right)x_1''}=5.34\frac{23.95}{23.05}\times\frac{x_2''}{0.715}=$$

$$=7.75\,x_2''=7.75\times0.728=5.63,$$

where x_2'' is previously assumed.

For this value of $\tan\gamma_2''$ drawing a similar triangle *ab3* on Fig. 7-16 we obtain from curves *1a* and *2a* specific volume and pressure after the moving blades as $v_2=0.29$ m³/kg and $p_2=9.30$ ata. The theoretical available energy for doing work in the moving blade passages is measured along the *Ox* axis as *a3*, i. e.,

$$h_{02}+h_{w_1}=a3=03-h_{01}+h_{w_1}=4.86-2.65+0.368=$$
$$=2.578\text{ kcal/kg}.$$

Velocity $w_2=91.5\psi\sqrt{h_{02}+h_{w_1}}=91.5\times0.92\times$
$\times\sqrt{2.578}=135$ m/sec.

$x_2''=\frac{u}{w_2}=\frac{98.4}{135}=0.728$ which is equal to the value assumed previously.

From equation (4-11) we obtain the exit height of the working blades, which is

$$l''=\frac{8.14\times0.29\times10^3}{\pi\times0.626\times135\sin20^\circ}=26\text{ mm}.$$

This value also agrees with the value assumed previously. From exit velocity triangle we have

$$c_2=52.5\text{ m/sec and }\alpha_2=58^\circ.$$

Carry-over losses

$$h_e=\frac{52.5^2}{8{,}378}=0.33\text{ kcal/kg}.$$

Third stage:

Tan γ_1''' for the guide blades will be

$$\tan\gamma_1'''=\tan\gamma_1''\frac{\left(\frac{d_d''}{l''}\right)x_1'''}{\left(\frac{d_d}{l}\right)x_1''}=$$

$$=5.34\frac{23.95x_1'''}{22.2\times0.715}=8.05x_1'''=8.05\times0.738=5.93,$$

where x_1''' is assumed to be 0.738.

With this value of $\tan\gamma_1'''$ the triangle *a'c4* is drawn (Fig. 7-16). The theoretical available energy in the guide blades is obtained from the relation

$$h_{01}+h_e^{pr}=a'4=04-03+h_e^{pr}=7.05-4.86+0.33=$$
$$=2.52\text{ kcal/kg}.$$

Velocity of steam

$$c_1=91.5\times0.92\sqrt{2.52}=133.7\text{ m/sec}.$$

$x_1'''=\frac{u}{c_1}=\frac{98.6}{133.7}=0.738$, which coincides with the assumed figure; *u* here is taken from Table 7-3.

From curves 1*a* and 2*a* we have

$$v_1=0.299\text{ m}^3\text{/kg and }p_1'=9.0\text{ ata}.$$

To check the accuracy of the assumed value for the exit height of the guide blades we calculate the same according to equation (4-7)

$$l=\frac{8.14\times0.299\times10^3}{\pi\times0.627\times133.7\sin20^\circ}=27\text{ mm},$$

from which we find that the value assumed is satisfactory.

From velocity triangle we obtain $w_1=53$ m/sec and $\beta_1=59^\circ30'$.

The available energy at entry to the moving blades is

$$h_{w_1}=\frac{53^2}{8{,}378}=0.335\text{ kcal/kg}.$$

Tan γ_2''' for the moving blades will be

$$\tan\gamma_2'''=\tan\gamma_1''\frac{\left(\frac{d_d}{l}\right)x_2'''}{\left(\frac{d_d''}{l''}\right)x_1''}=5.35\frac{23.95x_2'''}{21.4\times0.715}=$$

$$=8.35\times0.743=6.2,$$

where x_2''' is assumed to be 0.743.

With this value for γ_2''' triangle *a''d5* is drawn in Fig. 7-16. The theoretical available energy in the moving blades will be $h_{02}+h_{w_1}=a''5=05-04+h_{w_1}=9.2-$ $-7.05+0.335=2.485$ kcal/kg.

Velocity $w_2=91.5\times0.92\sqrt{2.485}=132.7$ m/sec from which $x_2'''=\frac{u}{w_2}=\frac{98.7}{132.7}=0.743$ which is equal to the assumed value.

From curves *a* and *2a* we obtain $v_2=0.308$ m³/kg and $p_2=8.65$ ata.

Height of blades

$$l''=\frac{8.14\times0.308\times10^3}{\pi\times0.628\times132\times7\sin20^\circ}=28\text{ mm}.$$

From velocity triangle we have

$$c_2=51\text{ m/sec and }\alpha_2=62^\circ.$$

Carry-over loss

$$h_e=\frac{51^2}{8{,}378}=0.31\text{ kcal/kg}.$$

All the succeeding stages on the drum dial of 600 mm are designed in exactly the same manner. We shall accommodate only 15 stages on the constant drum dia. of 600 mm, since if the low-pressure stages are also designed for the same diameter, as a result of smaller values for the ratios u/c_1 and u/w_2, the efficiencies obtained would be lower. Hence we shall have a drum dia. of 800 mm for the low-pressure stages.

The basic data obtained as a result of the above calculations is shown in Tables 7-3 and 7-4.

Seventeenth stage:

We shall assume the exit height of the guide blades to be

$$l=40\text{ mm}.$$

The circumferential velocity *u* will then be (at the mean dia. of $800+40=840$ mm)

$$u=\frac{\pi\times0.84\times3{,}000}{60}=132\text{ m/sec}.$$

For the guide blades we shall assume the ratio u/c_1 equal to 0.78.

Table 7-4

Moving Blades

Stage No.	d'_d, mm	l'', mm	d''_{mean}, mm	$\frac{d''_d}{l''}$	u, m/sec	u/w_2	$h_{w_1}+h_{02}$, kcal/kg	v^2, m³/kg	p^2 ata
1a	—	16.9	983	—	154.5	—	—	0.263	10.0
1b	—	22.7		—		—	—	0.267	
2	600	26	626	23.05	98.4	0.728	2.578	0.290	9.30
3	600	28	628	21.40	98.7	0.743	2.485	0.308	8.65
4	600	30	630	20.00	98.9	0.762	2.390	0.325	8.05
5	600	32	632	18.70	99.2	0.715	2.318	0.343	7.53
6	600	34	634	17.60	99.5	0.786	2.239	0.372	7.03
7	600	36	636	16.70	99.8	0.792	2.239	0.383	6.62
8	600	38	638	15.80	100.1	0.795	2.239	0.404	6.16
9	600	40	640	15.00	100.4	0.792	2.275	0.429	5.75
10	600	42	642	14.30	100.6	0.786	2.305	0.454	5.38
11	600	44	644	13.63	101.0	0.777	2.397	0.484	4.96
12	600	46	646	13.05	101.3	0.767	2.454	0.517	4.60
13	600	48	648	12.50	101.6	0.750	2.587	0.555	4.25
14	600	50	650	12.00	102.0	0.735	2.720	0.596	3.91
15	600	52	652	11.53	102.4	0.712	2.928	0.645	3.56
16	600	54	654	11.12	102.7	0.687	3.151	0.700	3.23
17	800	42	842	19.05	132.3	0.775	4.114	0.795	2.78
18	800	46	846	17.40	133.0	0.758	4.351	0.90	2.38
19	800	50	850	15.97	133.7	0.723	4.794	1.035	2.03
20	800	54	854	14.80	134.3	0.665	5.743	1.228	1.65
21	800	58	858	13.78	134.9	0.663	5.842	1.46	1.35
22	800	66	866	12.11	136.1	0.632	6.586	1.778	1.05
23	800	74	874	10.82	137.4	0.561	8.50	2.28	0.76
24	800	82	882	9.75	138 6	0.523	9.80	3.12	0.535
25	1,330	100	1,430	13.30	225	0.803	11.042	5.17	0.314
26	1,330	135	1,465	9.85	230	0.866	9.92	8.08	0.188
27	1,330	175	1,505	7.60	236	0.795	12.435	13.5	0.106
28	1,330	220	1,550	6.05	243	0.724	15.99	28.65	0.05

The theoretical heat drop in the guide blades will be

$$h_{01}=\frac{u^2}{8{,}378x_1^2\varphi^2}=\frac{132^2}{8{,}378\times0.78^2\times0.92^2}=4.04\ \text{kcal/kg}.$$

In Fig. 7-16 heat drop h_{01} is represented by the intercept $a°l$. The pressure and specific volume of steam after the guide blades are obtained by drawing a perpendicular at the point l intersecting curves *1a* and *2a* at points k and k_1. Points k and k_1 give the pressure and volume after the guide blades, as $p=2.97$ ata and $v=0.75$ m³/kg. We shall assume angle $\alpha_1=20°$. Height of the guide blades from equation (7-26) will be

$$l=\frac{60G_0x_1v_1}{\pi^2d^2n\sin\alpha_1}=\frac{60\times8.14\times0.78\times0.75}{\pi^2\times0.84^2\times3{,}000\times0.342}=$$
$$=0.04\ \text{m}=40\ \text{mm},$$

which agrees with the assumed figure.

The height of blades for each succeeding row shall be made longer than that of the preceding one by 2 mm.

Tangent γ_1 for guide blades will be

$$\tan\gamma_1=50\frac{v_1}{h_{01}}=50\frac{0.75}{4.04}=9.28.$$

(Here the coefficient 50 is the scale to which graph 7-16 is drawn).

Velocity of steam

$$c_1=91.5\times0.92\sqrt{4.04}=169\ \text{m/sec}.$$

From velocity triangle (Fig. 7-17) we have

$$w_1=65\ \text{m/sec and}\ \beta_1=65°.$$

Energy of steam at entry to the working blades will be

$$h_{w_1}=\frac{65^2}{8{,}378}=0.504\ \text{kcal/kg}.$$

The initial steam conditions before the moving blades are obtained from the difference

$$Ol-h_{w_1}=69.24-0.504=68.736\ \text{kcal/kg}.$$

For moving blades

$$\tan\gamma_2=\tan\gamma_1\frac{\left(\frac{d_d}{l}\right)x_2}{\left(\frac{d''_d}{l''}\right)x_1}=$$

$$=9.28\frac{20.00x_2}{19.05\times0.78}=12.47x_2=12.47\times0.775=9.67,$$

where x_2 is assumed to be equal to 0.775.

From the graph given in Fig. 7-16 we obtain $v_2=0.795$ m³/kg and $p_2=2.78$ ata. The theoretical available energy will be,

$$h_{02}+h_{w_1}=Ol'=Ol+h_{w_1}=72.85-69.24+0.504=$$
$$=4.114\ \text{kcal/kg}.$$

Velocity of steam w_2 will therefore be,

$$w_2=91.5\times0.92\sqrt{4.114}=170.7\ \text{m/sec}.$$

$x_2=\frac{u}{w_2}=\frac{132}{170.7}=0.775$ which agrees with the value assumed previously.

From the velocity triangle we obtain
$c_2 = 65.5$ m/sec and $\alpha_2 = 65°$.

The energy leaving the blades due to carry-over velocity will be equal to

$$h_e = \frac{65.5^2}{8,378} = 0.512 \text{ kcal/kg.}$$

All the succeeding stages up to the 24th are designed exactly in a like manner. The angles α_1 and β_1 for the stages from the 18th to the 20th are assumed as $\alpha_1 = \beta_1 = 20°$, for 21st to 23rd $\alpha_1 = \beta_2 = 22°$ and for the 24th stage $\alpha_1 = \beta_2 = 25°$.

Since the specific volume of steam increases rapidly in the second group of stages we shall increase the diameter to 1,330 mm. In practice this group of stages has to be mounted on one or two separate forged steel discs. To obtain a gradual increase in the height of blades for the latter stages the angles α_1 and β_2 are increased gradually.

The twenty-fifth stage:

For the guide blades we shall assume $\alpha_1 = 18°$, $u/c_1 = 0.845$ and $l = 90$ mm.

Mean diameter

$$d_{mean} = 1,330 + 90 = 1,420 \text{ mm.}$$

Circumferential velocity

$$u = \frac{\pi \times 1.42 \times 3,000}{60} = 223 \text{ m/sec.}$$

Theoretical heat drop in the guide blades will be,

$$h_{01} = \frac{u^2}{8,378\varphi^2 x_1^2} = \frac{223^2}{8,378 \times 0.92^2 \times 0,845} = 9.83 \text{ kcal/kg.}$$

Specific volume and pressure of steam after the guide blades (point a_1 in Fig. 7-16) are $v_1 = 3.95$ m³/kg and $p_1 = 0.418$ ata. The exit height of the guide blades from equation (7-26) is

$$l = \frac{60 G_0 x_1 v_1}{\pi^2 d^2 n \sin \alpha_1} = \frac{60 \times 8.14 \times 0.845 \times 4.00}{\pi^2 \times 1.42^2 \times 3,000 \sin 18°} =$$
$$= 0.09 \text{ m} = 90 \text{ mm which agrees with}$$

our assumption.

$$\text{Tan } \gamma_1 = \frac{M_v v_1}{M_h h_{01}} = \frac{10 \times 3.95}{2 \times 9.83} = 2.01.$$

Velocity of steam

$$c_1 = u/x_1 = \frac{223}{0.845} = 264 \text{ m/sec.}$$

From velocity triangle we have,

$$w_1 = 85.5 \text{ m/sec and } \beta_1 = 70°.$$

Energy content at entry to the moving blades is equal to

$$h_{w_1} = \frac{85.5^2}{8,378} = 0,872 \text{ kcal/kg.}$$

For the moving blades we shall assume $\beta_2 = 20°$, $l'' = 100$ mm and $x_2 = 0.803$, so that

$$\tan \gamma_2 = \tan \gamma_1 \frac{\sin \beta_2 \left(\frac{d_d}{l}\right) x_2}{\sin \alpha_1 \left(\frac{d_d''}{l''}\right) x_1} =$$
$$= 2.01 \frac{\sin 20° \times 14.78 \times 0.803}{\sin 18° \times 13.3 \times 0.845} = 2.34.$$

Theoretical available energy for doing work in the moving blades will be

$$h_{02} + h_{w_1} = 165.00 - 154.83 + 0.782 = 11.042 \text{ kcal/kg.}$$

Velocity of steam

$$w_2 = 91.5 \times 0.92 \sqrt{11.042} = 280 \text{ m/sec;}$$

$x_2 = u/w_2 = 225/280 = 0.803$ (as was assumed).

From velocity triangle we have

$$c_2 = 95 \text{ m/sec and } \alpha_2 = 64°30'.$$

Energy at exit (carry-over velocity) will be

$$h_e = 95^2/8,378 = 1.077 \text{ kcal/kg.}$$

The blades of the 26th, 27th and 28th stages are designed in exactly the same way as those of the 25th stage. Results of all these above calculations are tabulated in Tables 7-3 and 7-4.

After having obtained a satisfactory distribution of heat drop in the various turbine stages we shall proceed to the detailed design.

The second stage (the first reaction stage).

We shall first determine the heat losses in the turbine. Losses in the guide blades: from equation (7-40) we have

$$h_{gb} = (1 - \varphi^2)(h_{01} + h_e^{pr}) = (1 - 0.92^2)\, 2.65 =$$
$$= 0.408 \text{ kcal/kg,}$$

in the moving blades: from equation (7-41)

$$h_b = (1 - \psi^2)(h_{02} + h_{w_1}) = (1 - 0.92^2)\, 2.578 =$$
$$= 0.397 \text{ kcal/kg.}$$

The carry-over loss $h_e = 0.33$ kcal/kg.

Leakage losses through the radial clearances from equation (5-24a) will be

$$h_{leak} = 1.72 \frac{\delta_r^{1.4}}{l}(h_{01} + h_{02}') = 1.72 \frac{1^{1.4}}{25}(2.65 + 2.21) =$$
$$= 0.335 \text{ kcal/kg,}$$

where δ_r has been assumed to be equal to unity.

Heat drop usefully utilised in the stage will be

$$h_i = h_0' - \sum h_{losses} = h_{01} + h_{02} - h_{gb} - h_b - h_e - h_{leak} =$$
$$= 2.65 + 2.21 - 0.408 - 0.397 - 0.33 - 0.335 =$$
$$= 3.39 \text{ kcal/kg.}$$

For the succeeding stages the theoretical available energy is obtained from the relation

$$h_0 = h_{01} + h_{02} + h_e^{pr} \text{ [kcal/kg.]}$$

and heat drop actually utilised is $h_i = h_{01} + h_{02} + h_e^{pr} - \sum h_{losses}$ [kcal/kg].

The results of the above calculations are given in Table 7-5.

The total of heat drops usefully utilised for doing work in the turbine (from Table 7-5) $H_i = 208.9$ kcal/kg.

Actual mass flow through the turbine

$$G_{act} = \frac{860 \times 6,600}{3,600 \times 208.9 \times 0.98 \times 0.948} = 8.13 \text{ kg/sec.}$$

The difference between the value obtained and the theoretical one is within the allowable limits and hence blade heights need not be recalculated. Sectional view of the turbine is shown in Fig. 7-18. To counterbalance axial thrust a dummy piston is provided at the front end of the turbine. Some quantity of steam leaks out through

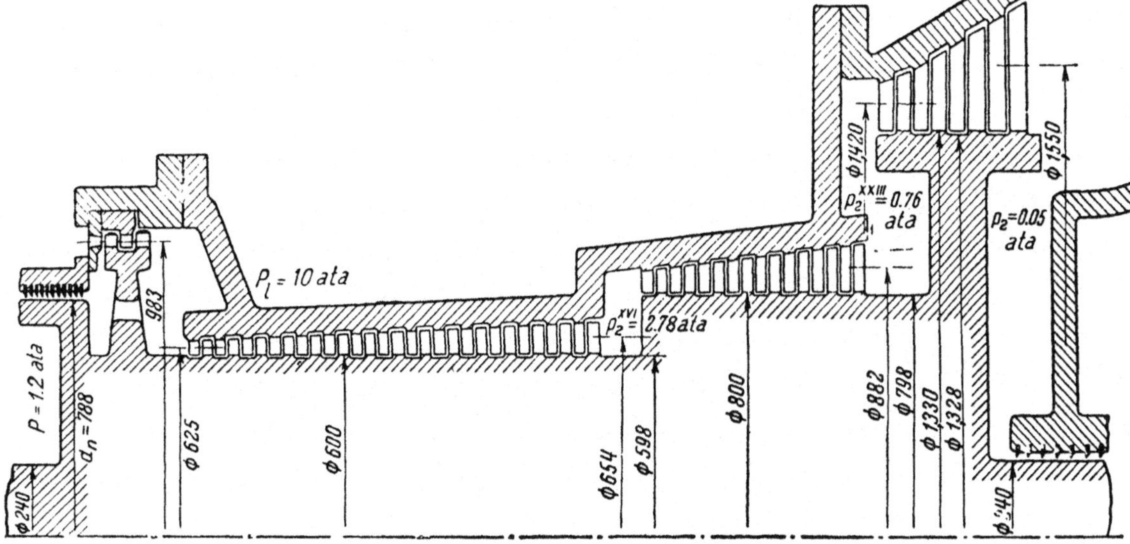

Fig. 7-18. Section through an impulse-reaction turbine

the front labyrinth seal, without doing any useful work. We shall determine this leakage loss. Steam pressure on the pressure side of the dummy piston is 10 ata (from design calculations). Pressure on the rearside of the dummy piston will be assumed as 1.2 ata. Diameter of piston $d_{dp} = 788$ mm. Number of labyrinth foils will be assumed as $z = 36$ and clearance $\delta = 0.4$ mm.

The annular area of clearances will then be

$$f_{cl} = \pi \times 0.788 \times 0.0004 = 0.00099 \text{ m}^2.$$

From equation (5-23) the pressure in the last labyrinth compartment will be

$$p'_{cr} = \frac{0.85 \times 10}{\sqrt{36 + 1.5}} = 1.39 \text{ ata}.$$

Since $p'_{cr} > 1.2$ the steam leakage through the labyrinth seal will be calculated from equation (5-22)

$$G_{leak} = 0.099 \sqrt{\frac{9.81 \times 10}{(36 + 1.5)\, 0.25}} = 0.36 \text{ kg/sec}.$$

Mass flow of steam through the nozzles of the first regulating (Curtis) stage will therefore increase by the amount G_{leak}, the leakage through the front labyrinth seal, i. e.,

$$G'_0 = G_{act} + G_{leak} = 8.13 + 0.36 = 8.49 \text{ kg/sec}.$$

The nozzle height will therefore be increased proportionally, i. e.,

$$l_n = lG'_0/G_0 = 12\,(8.49/8.13) = 12.5 \text{ mm}.$$

Determination of Axial Thrust and Dimensions of Balance Piston

The thrust bearing will be assumed to be of the segment type. The axial components of pressures will be determined according to equations (7-45), (7-47) and (7-48).

We shall assume the thrust in the direction of steam flow to be positive and vice versa. The axial pressures on the rotor collars have been calculated as per equation 7-45) and are given in Table 7-6.

Thrust due to reaction in the working blades as well as due to difference in the momentum of entering and leaving steam has been calculated according to equations (7-47) and (7-48) and are given in Table 7-7.

We shall assume the bearing surface of the thrust bearing pads to be

$$F_{br} = 300 \text{ cm}^2.$$

Specific pressure on the bearing will be assumed as

$$q = 10 \text{ kg/cm}^2.$$

Hence the axial pressure borne by the thrust bearing would be

$$F_{br}q = 10 \times 300 = 3{,}000 \text{ kg}.$$

Substituting the above in equation (7-50) we have

$$+\,39{,}600 + 6{,}279 - 72 - 3{,}000 = 6.909 d^2_{dp}$$

$$\text{or } 6.909 d^2_{dp} = 42{,}800 \text{ kg},$$

$$\text{from which } d_{dp} = \sqrt{\frac{42{,}800}{6.909}} = \sqrt{6{,}180} =$$

$$= 78.8 \text{ cm} = 788 \text{ mm}.$$

7-16. TURBINES OF OPTIMUM CAPACITY

The optimum capacity of a condensing turbine is that capacity which is the maximum attainable under stipulated conditions of initial steam pressure, temperature and revolutions per minute.

The equation for the capacity of a turbine, without extraction, may be expressed as

$$N_e = \frac{D\, H_0 \eta_{oi} \eta_m \eta_g}{860}. \qquad (7\text{-}52)$$

From the above equation it follows that the capacity of the turbine depends mainly upon the

Table 7-5

Stage No.	h_{01}, kcal/kg	c_1, m/sec	w_1, m/sec	h_{02}, kcal/kg	w_2, m/sec	c_2, m/sec	α_1/β_2, degr.	$\frac{\beta_1 \text{ calc.}}{\beta_1 \text{ assumed}}$, degr.	$\frac{\alpha_2 \text{ calc.}}{\alpha_2 \text{ assumed}}$, degr.	h_{gb}, kcal/kg	h_b, kcal/kg	h_e, kcal/kg	h_{leak}, kcal/kg	h_i, kcal/kg	t, °C or x, %	$h_{wetness}$, kcal/kg	h_i, kcal/kg	N_i, kW
1	64.5	698.5	555	0.0	455	320	20°/22°20′	25°20′	34°	6.68	13.10	2.62	1.25	38.51	313	0.00	38.580	1.196
1	—	272	160	0.0	141	135	29°50′/55°10′	58°10′	116°30′	2.90	0.44	—	—	—	—	—	—	—
2	2.65	137.0	55.5	2.21	135	52.5	20°/20°	57°30′	58°/57°30′	0.408	0.397	0.330	0.335	3.400	306	0.00	3.400	106
3	2.19	133.7	53.0	2.15	132.7	51.0	20°/20°	59°30′/60°	62°	0.388	0.383	0.310	0.276	3.313	299	0.00	3.313	103
4	2.12	131.0	51.0	2.08	130.0	51.0	20°/20°	60°30′/60°	62°30′/62°	0.314	0.368	0.310	0.249	3.209	292	0.00	3.209	100
5	2.04	129.0	50.0	2.02	128.0	49.5	20°/20°	62°	65°	0.362	0.357	0.293	0.225	3.137	286	0.00	3.137	98
6	1.98	127.0	48.0	1.98	126.5	47.5	20°/20°	65°30′/65°	66°	0.350	0.347	0.269	0.206	3.081	279	0.00	3.081	96
7	1.97	126.0	47.5	1.97	126.0	47.5	20°/20°	66°	67°30′/66°	0.345	0.345	0.269	0.194	3.056	273	0.00	3.056	95
8	1.97	126.0	47.5	1.97	126.0	47.0	20°/20°	66°	67°/66°	0.345	0.345	0.263	0.183	3.073	268	0.00	3.073	95
9	2.00	126.5	48.0	2.00	127.0	48.0	20°/20°	66°30′/66°	66°	0.349	0.351	0.275	0.176	3.112	262	0.00	3.112	97
10	2.02	127.5	48.0	2.03	128.0	48.5	20°/20°	66°	66°	0.354	0.355	0.281	0.170	3.165	255	0.00	3.165	98
11	2.04	128.5	49.0	2.11	130.0	50.0	20°/20°	66°	64°30′/65°	0.357	0.369	0.298	0.166	3.241	248	0.00	3.241	101
12	2.12	131.0	50.5	2.15	132.0	51.0	20°/20°	63°/62°	63°/62°	0.373	0.378	0.310	0.163	3.344	241	0.00	3.344	104
13	2.23	134.0	51.5	2.27	135.5	53.5	20°/20°	61°/60°	60°30′/60°	0.391	0.397	0.341	0.165	3.510	234	0.00	3.510	109
14	2.31	137.0	55.0	2.36	139.0	56.0	20°/20°	60°	60°	0.408	0.419	0.374	0.164	3.646	226	0.00	3.646	113
15	2.42	141.0	57.0	2.54	144.0	58.5	20°/20°	57°30′	57°30′	0.430	0.451	0.409	0.167	3.877	217	0.00	3.877	120
16	2.60	146.5	61.5	2.70	149.5	63.0	20°/20°	55°/54°	53°30′/54°	0.464	0.485	0.474	0.172	4.114	208	0.00	4.114	126
17	4.04	169.3	65.0	3.61	170.7	65.5	20°/20°	65°	65°	0.623	0.633	0.512	0.329	5.553	195	0.00	5.553	172
18	3.75	173.5	68.0	3.80	175.5	68.0	20°/20°	62°	63°/62°	0.656	0.670	0.551	0.295	5.890	183	0.00	5.890	183
19	4.00	179.5	70.5	4.20	185.0	75.0	20°/20°	59°30′/60°	57°30′	0.702	0.739	0.671	0.294	6.349	168	0.00	6.343	197
20	4.60	193.0	81.5	4.95	202.0	87.5	20°/20°	54°/53°	51°30′/53°	0.811	0.885	0.914	0.316	7.295	152	0.00	7.295	246
21	4.30	192.5	84.0	5.00	203.5	92.0	22°/22°	58°/56°30′	65°/56°30′	0.803	0.900	1.010	0.286	7.215	136	0.00	7.215	224
22	5.15	208.5	95.5	5.50	215.5	103	22°/22°	53°30′/52°	51°/52°	0.949	1.014	1.270	0.296	8.131	118	0.00	8.131	252
23	6.00	227.0	112.0	7.00	245.0	127.0	22°/22°	49°/47°	44°30′/47°	1.120	1.310	1.925	0.320	9.595	96	0 00	9.595	298
24	5.70	232.5	119.0	8.20	265.0	151.0	25°/25°	54°30′	47°30′	1.175	1.524	2.72	0.307	10.099	99.0	0.055	10.044	312
25	9.83	264.0	85.5	10.17	280.0	95.0	18°/20°	70°	64°30′/65°	1.514	1.700	1.077	0.382	15.327	97.1	0.306	15.021	466
26	10.00	280.0	109.0	8.50	265.5	107.0	22°/24°	71°30′	84°	1.705	1.528	1.365	0.277	14.702	95.3	0.558	14.144	440
27	9.50	278.0	117.0	10.80	297.0	136.0	25°/27°	80°	78°30′	1.670	1.915	2.210	0.225	15.645	93.4	0.855	14.790	460
28	11.20	308.0	161.0	12.90	336.5	204.0	32°/37°40′	80°30′	83°	2.065	2.465	4.97	0.212	16.598	91.2	1.280	15.318	476

$$H_i = \sum h_i = 208.927$$

Table 7-6

Nomenclature / Collars	d_d, cm	d'_d, cm	$\frac{\pi}{4}(d'^2_d - d^2_d)$, cm²	p_{coll}, kg/cm²	$\pm \Delta R_{coll} = \frac{\pi}{4}(d'^2_d - d^2_d)\, p_{coll}$, kg
Lateral surface at the front end	0.0	24	452	1.0	+452
1	24	d_{dp}	$0.785 d^2_{dp}$—452	1.2*	$+0.941 d^2_{dp} - 542$
2	d_{dp}**	59.8	2,800—$0.785\, d^2_{dp}$	10	+28,000—$7.85 d^2_{dp}$
3	59.8	79.8	2,200	2.78	+6,110
4	79.8	132.8	8,830	0.76	+6,710
5	132.8	24	13,350	0.05	—668
Lateral surface at the rear end	24	0.0	452	1.0	—452
					$R_{coll} = +39,600 - 6.909\, d^2_{dp}$

* The steam pressure in the balance piston chamber is assumed to be the same as that of steam supplied to labyrinth seals.
** d_{dp} – dummy piston dia.

Table 7-7

Stage No.	d, cm	d_0, cm	$\frac{\pi}{4}(d^2 - d_0^2)$, cm²	$\Delta p = p'_1 - p_2$, kg/cm²	$\Delta R_{reaction} = \frac{\pi}{4} \times (d^2 - d_0^2)\, \Delta p$, kg	c_{1a}, m/sec	c_{2a}, m/sec	$\Delta R_m = \frac{G}{g}(c_{1a} - c_{2a})$, kg
2	65.2	59.8	534	0.35	+187	46.5	45.0	+ 1.13
3	65.6	59.8	581	0.35	+203	46.5	45.0	+ 1.13
4	66.0	59.8	613	0.27	+165	45.0	45.0	0.0
5	66.4	59.8	660	0.25	+165	45.0	45.0	0.0
6	66.8	59.8	700	0.24	+168	43.5	42.0	+ 1.13
7	67.2	59.8	739	0.20	+148	43.5	42.5	+ 0.76
8	67.6	59.8	777	0.21	+163	42.0	42.0	0.0
9	68.0	59.8	819	0.20	+164	44.0	44.0	0.0
10	68.4	59.8	865	0.22	+190	44.0	44.0	0.0
11	68.8	59.8	904	0.20	+181	45.0	45.0	0.0
12	69.2	59.8	962	0.18	+173	45.0	45.0	0.0
13	69.6	59.8	998	0.17	+170	46.5	48.0	— 1.13
14	70.0	59.8	1,052	0.15	+158	48.0	48.0	0.0
15	70.4	59.8	1,092	0.17	+186	48.0	49.0	— 0.76
16	70.8	59.8	1,130	0.17	+192	51.0	51.0	0.0
17	88.4	79.8	1,140	0.19	+217	59.0	59.0	0.0
18	89.2	79.8	1,257	0.18	+226	60.0	60.0	0.0
19	90.0	79.8	1,377	0.17	+234	61.5	63.0	— 1.13
20	90.8	79.8	1,470	0.18	+265	66.0	69.0	— 2.27
21	91.6	79.8	1,595	0.16	+255	72.0	75.0	— 2.27
22	93.2	79.8	1,882	0.14	+264	78.0	81.0	— 2.27
23	94.8	79.8	2,060	0.14	+288	85.5	90.0	— 3.40
24	96.4	79.8	2,300	0.125	+288	97.0	112	—11.30
25	153.0	132.8	4,480	0.104	+466	81.0	85.5	— 3.40
26	160.0	132.8	6,200	0.095	+588	103	106	— 2.27
27	168.0	132.8	8,300	0.037	+307	115	134	—14.37
28	177.0	132.8	10,730	0.025	+268	160	202	—31.80
					$R_{reaction} = +6,279$ kg			$R_m = -72$ kg

mass flow of steam, since the magnitude of H_0 is determined by the steam parameters whereas the quantities η_{0i}, η_m and η_g vary within a very narrow range.

However, in the case of condensing turbines the mass flow of steam through the turbine is governed by the size of moving blades in the last stage. The length of blades for the low-pressure stages may be increased only up to certain finite lengths depending upon the strength and stiffness of the material used. Thus we find that the maximum attainable capacity of a steam turbine depends upon the steam handling capacity of its last stage.

From the equation of continuity we have, for the last stage moving blades

$$G_0 v_2 = f_b w_2 = \pi d_z l_z w_2 \sin\beta_2. \qquad (7\text{-}53)$$

Solving this equation for G_0 we have

$$G_0 = \frac{\pi d_z l_z w_2 \sin\beta_2}{v_2} = \frac{\pi d_z^2 c_2 \sin\alpha_2}{v_2 \left(\frac{d_z}{l_z}\right)}. \qquad (7\text{-}54)$$

When $\alpha_2 = 90°$; $d_z = 60u/\pi n$; and $d_z/l_z = \vartheta$ equation (7-54) becomes

$$G_0 = \frac{3{,}600 u^2 c_2}{\pi \vartheta n^2 v_2}. \qquad (7\text{-}54a)$$

Velocity c_2 of steam may be expressed as

$$c_2 = 91.5\sqrt{h_e} = 91.5\sqrt{\zeta_e H_0}, \qquad (7\text{-}55)$$

so that equation (7-54a) becomes

$$G_0 = \frac{3{,}600 u^2\, 91.5\sqrt{\zeta_e H_0}}{\pi \vartheta n^2 v_2} \qquad (7\text{-}56)$$

$$\text{or } D_0 = 3{,}600 G_0 = \frac{3{,}600 \times 3{,}600 u^2 91.5\sqrt{\zeta_e H_0}}{\pi \vartheta n^2 v_2}.$$

Substituting the value of D_0 in equation (7-52) we obtain

$$N_{e\,opt} = \frac{439{,}000 u^2 \sqrt{\zeta_e H_0} H_0 \eta_{oi} \eta_m \eta_g}{\vartheta n^2 v_2}. \qquad (7\text{-}57)$$

Equation (7-57) gives an expression for the optimum capacity for a condensing turbine, without extraction, if the limiting values of u, ϑ and ζ_e are substituted in it.

From equation (7-57) it follows that $N_{e\,opt}$ is a function of u, n, ϑ and v_2.

The limiting values for these quantities are given below:

$u = 315$ to 340 m/sec;
$\vartheta = 3.0$ to 2.8;
$\zeta_e = (0.015$ to $0.03)\, H_0$.

The value of n, in present-day practice, is invariably 3,000 r.p.m.; v_2 depends upon the condenser vacuum. If the condenser vacuum increases v_2 increases resulting in a decrease in the optimum capacity for the turbine. If, however, the steam flow in the low-pressure stages is divided into two branches it is found that all other conditions remaining the same, the optimum capacity increases to as much as 1.5 to 2 times, the capacity that would have been obtained if the flow were not branched. If steam flow is branched into four canals the optimum capacity becomes 3 to 4 times greater.

Example 7-1. Find the optimum capacity for a condensing turbine without extractions for the following given conditions $p_0 = 29$ ata; $t_0 = 400°$ C; $p_2 = 0.04$ ata; and $n = 3{,}000$ r. p. m. The steam flow in the low-pressure stages is not branched.

We shall assume

$$u = 315 \text{ m/sec.},\ \zeta_e = 0.03;\ \vartheta = 3;\ \eta_{oi} = 0.8;\ \eta_m = 0.99;$$
$$\eta_g = 0.96.$$

From the i-s diagram we have

$$H_0 = 273 \text{ kcal/kg and } v_2 = 33.2 \text{ m}^3\text{/kg}.$$

Optimum capacity of the turbine would therefore be

$$N_{e\,opt} = \frac{439{,}000 \times 315^2 \times \sqrt{0.03 \times 273} \times 273 \times 0.8 \times}{3.0 \times 3{,}000^2 \times 33.2}$$
$$\frac{\times 0.99 \times 0.96}{3.0 \times 3{,}000^2 \times 33.2} \approx 28{,}800 \text{ kW}.$$

In the U.S.S.R. single-cylinder steam turbines have been built up to a capacity of 50,000 kW at speeds of 3,000 r.p.m. Such high capacities could be attained only with high initial steam pressures and temperatures along with extraction of steam from intermediate stages for feedheating. Circumferential velocity at the mean diameter for these turbines is $u = 314$ m/sec (diameter $d_z = 2$ metres) with a slight increase in $\zeta_e H_0$ the carry-over losses for the last stage.

The single-cylinder condensing steam turbine with a capacity of 50,000 kW at 1,500 r.p.m. (built by Kh.T.W.) (Fig. 10-5) is one such example of an optimum capacity turbine. Here the ratio dia./blade height $\vartheta = 3.67$. A detailed description of this turbine is given in Chapter Ten.

If the capacity of a turbine is to be increased beyond the optimum value then one of the methods shown in Fig. 7-19 is employed. In the case of the method shown in Fig. 7-19, *c*, 30 to 40% of the steam flow is directly led away to the condenser through the outer annular exhaust. Thus the upper half of the blades experiences a pressure drop directly up to the condenser pressure. The remaining portion of steam flows through the lower half of the blades where it undergoes only partial expansion since the major portion of heat drop occurs in the last stage of the turbine. The different heat drops required for the above case both in nozzles as well as in the moving blades are brought about by designing the upper and lower halves of the blades with different blade profiles.

In the present-day practice steam turbines of large capacities are invariably built with extraction from intermediate stages for feedheating which reaches a figure as high as 30 to 35% of the total steam fed to the turbine, thus leading to a decrease in the steam led to the condenser, which in its own turn increases the optimum capacity of the turbine.

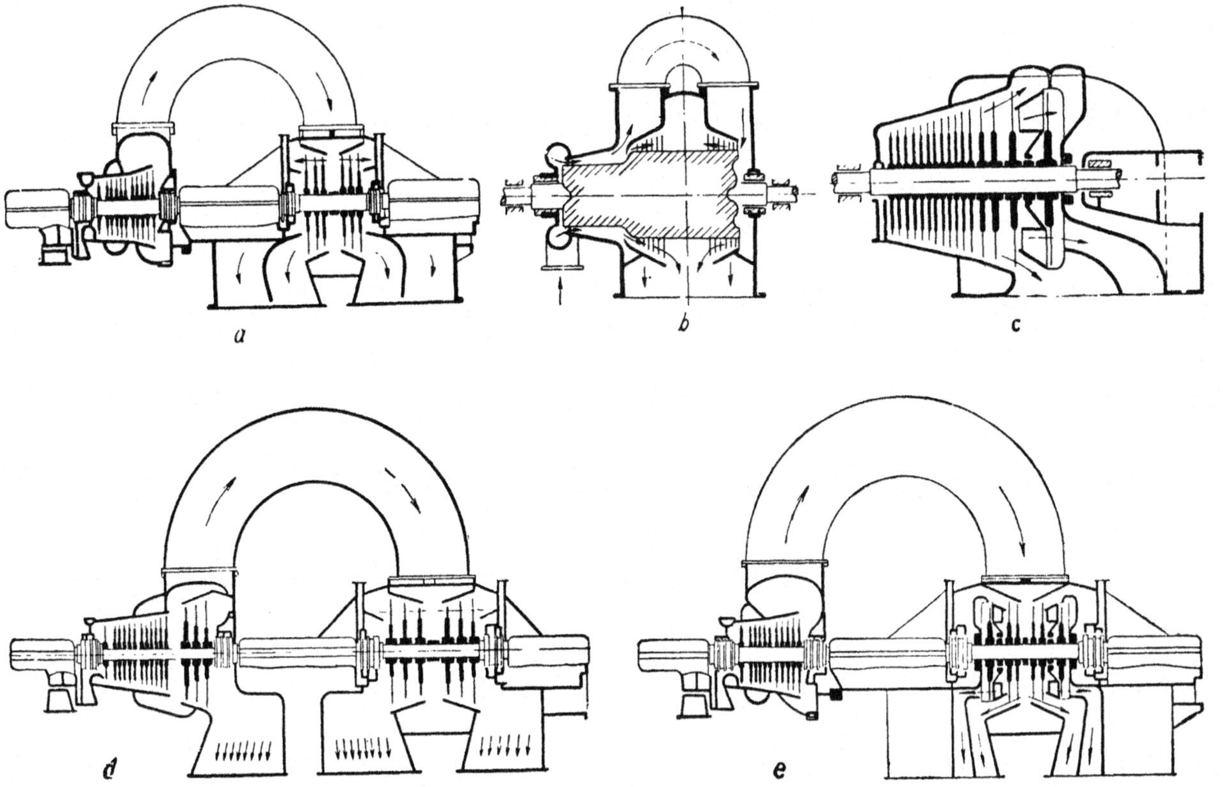

Fig. 7-19. Methods of branching steam in the low pressure stages of high-capacity condensing turbines

7-17. TURBINES WITH EXTRACTION FOR REGENERATION

In the current practice steam turbines of high initial pressures are usually constructed with extractions (from 5 to 7) from intermediate stages. For turbines of supercritical parameters the number of extractions may be as much as 8 to 9. Steam bled from the intermediate stages is usually utilised in feed heaters where it condenses giving up its latent heat to the feedwater. For medium-pressure steam turbines the number of extractions is limited to 2 to 4. Let us consider the principle of working of a turbine with three extractions (Fig. 7-20).

Steam is extracted in quantities D_{ex}^{I}, D_{ex}^{II} and D_{ex}^{III} from the three points 1, 2 and 3 and is led to three feed heaters 1, 2 and 3. Thus the mass flow of steam will be different for the various stages. Up to the first point of extraction the mass flow of steam would be D_0, from first to second point of extraction it would be $D_0 - D_{ex}^{I}$, from second to third point of extraction $D_0 - D_{ex}^{I} - D_{ex}^{II}$ and finally from the third point onwards it would be $D_0 - D_{ex}^{I} - D_{ex}^{II} - D_{ex}^{III}$. These differences in the mass flows should be taken into consideration when designing the various stages of the turbine. The capacity of the turbine having the extraction system as shown in Fig. 7-20 would be expressed by the equation

$$N_e = \frac{[D_0 h_0^{I} \eta_{0i}^{I} + (D_0 - D_{ex}^{I}) h_0^{II} \eta_{0i}^{II} + (D_0 - D_{ex}^{I} - D_{ex}^{II}) \times}{860}$$

$$\frac{\times h_0^{III} \eta_{0i}^{III} + (D_0 - D_{ex}^{I} - D_{ex}^{II} - D_{ex}^{III}) h_0^{IV} \eta_{0i}^{IV}] \eta_m \eta_g}{860}, \quad (7\text{-}58)$$

where h_0^{I}, h_0^{II}, h_0^{III} and h_0^{IV} — theoretical heat drops between the points of extraction (Fig. 7-21),

η_{0i}^{I}, η_{0i}^{II}, η_{0i}^{III}, η_{0i}^{IV} — relative internal efficiencies for the sections between points of extraction.

Capacity may also be expressed as

$$N_e = \frac{D_0 [h_0^{I} \eta_{0i}^{I} + (1 - \alpha_1) h_0^{II} \eta_{0i}^{II} + (1 - \alpha_1 - \alpha_2) h_0^{III} \eta_{0i}^{III} +}{860}$$

$$\frac{+ (1 - \alpha_1 - \alpha_2 - \alpha_3) h_0^{IV} \eta_{0i}^{IV}] \eta_m \eta_g}{860},$$

where $\alpha_1 = D_{ex}^{I}/D_0$; $\alpha_2 = D_{ex}^{II}/D_0$; $\alpha_3 = D_{ex}^{III}/D_0$.

While designing a turbine with extractions, the values for α_1, α_2 and α_3 and extraction pressures p_{ex}^{I}, p_{ex}^{II} and p_{ex}^{III} are taken from the preliminary design calculations for the regeneration scheme.

After obtaining a satisfactory heat drop distribution in the various stages of the turbine the values assumed for α_1, α_2 and α_3 as well as p_{ex}^{I}, p_{ex}^{II} and p_{ex}^{III} are made more precise.

With the values of α_1, α_2 and α_3 and p_{ex}^{I}, p_{ex}^{II}, and p_{ex}^{III} obtained as above and with the stipulated capacity N_e the mass flow of steam is determined

$$D_0 = \frac{860 N_e}{[h_0^{I}\eta_{oi}^{I} + (1-\alpha_1) h_0^{II}\eta_{oi}^{II} + (1-\alpha_1-\alpha_2) h_0^{III}\eta_{oi}^{III} + (1-\alpha_1-\alpha_2-\alpha_3) h_0^{IV}\eta_{oi}^{IV}]\,\eta_m\eta_g} . \quad (7\text{-}59)$$

Values of η_{oi}^{I}, η_{oi}^{II}, η_{oi}^{III} and η_{oi}^{IV} are suitably assumed and those for η_m and η_g are taken from graphs (Figs 5-8 and 5-11).

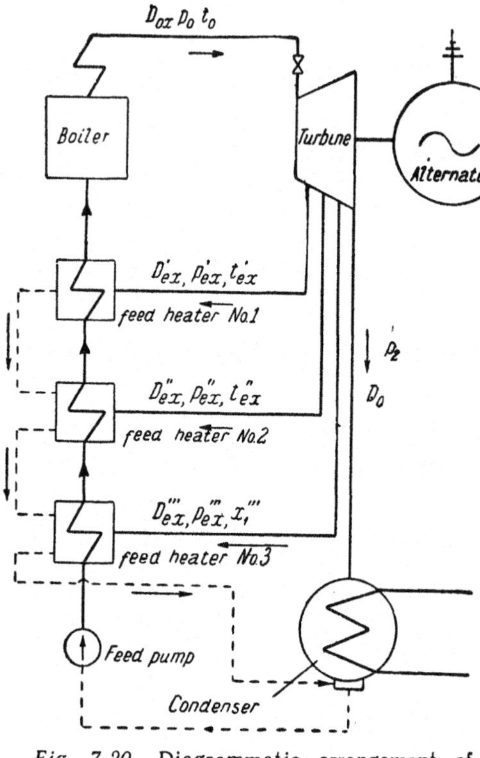

Fig. 7-20. Diagrammatic arrangement of regenerative feed heating with three extractions

7-18. DESIGN PROCEDURE FOR TURBINES WITH EXTRACTIONS

a) Heat Drop Process on the *i-s* Diagram

For the given conditions of fresh and exhaust steam the theoretical heat drop process is plotted on the *i-s* diagram as shown in Fig. 7-22, from which we find that $H_i = H_0\,\eta_{oi}$ where η_{oi} is assumed in conformity with the efficiency obtainable for such types of turbines or from the relation $\eta_{oi} = \eta_{re}/\eta_m$.

From the temperature of feedwater t_{fw}, number of extractions and the temperature of condensate the extraction pressures are determined. In order to simplify calculations we shall suppose that the feedwater is heated through the same number of degrees in all the heaters. In which case the increase in temperature of the feedwater in the heater can be obtained as

$$\Delta t = \frac{t_{fw} - t_{con}}{z},$$

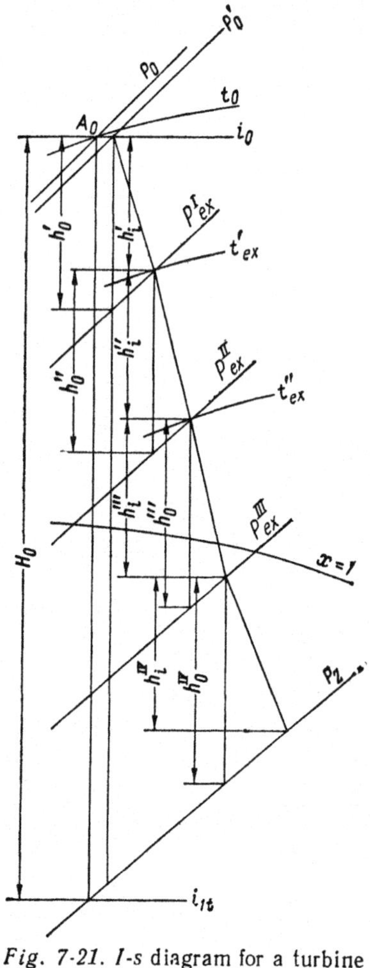

Fig. 7-21. *I-s* diagram for a turbine with three extractions

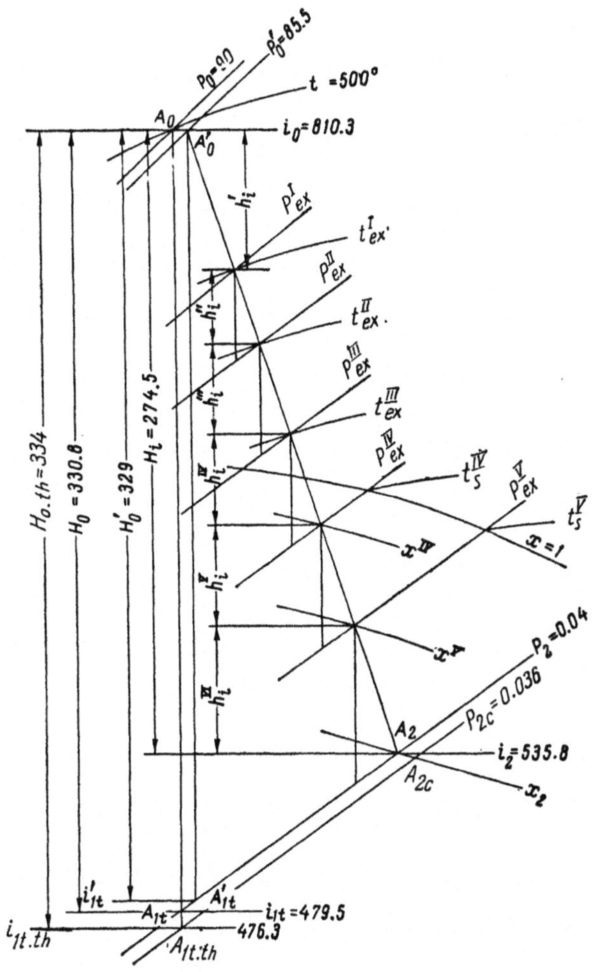

Fig. 7-22. Theoretical heat drop for a turbine with five extractions

where t_{fw}—temperature of feedwater; its value is known from thermal design calculations for the turbine;

t_{con}—temperature of condensate, which may be assumed as 1 to 2°C lower than the saturation temperature of exhaust steam;

z—number of extractions (for the case under consideration five in number).

Feedwater temperatures after the heaters will therefore be after low-pressure feed heater No. 1

$$t^{I}_{fw}=t_{con}+\Delta t;$$

after low-pressure feed heater No. 2

$$t^{II}_{fw}=t^{I}_{fw}+\Delta t;$$

after low-pressure feed heater No. 3

$$t^{III}_{fw}=t^{II}_{fw}+\Delta t;$$

after high-pressure heater No. 4

$$t^{IV}_{fw}=t^{III}_{fw}+\Delta t;$$

after high-pressure heater No. 5

$$t^{V}_{fw}=t_{fw}=t^{IV}_{fw}+\Delta t.$$

Saturation temperature of the heating steam in the feed heaters is obtained from the following equations

in heater No. 1 $t^{V}_{s}=t^{I}_{fw}+\delta t;$

in heater No. 2 $t^{IV}_{s}=t^{II}_{fw}+\delta t;$

in heater No. 3 $t^{III}_{s}=t^{III}_{fw}+\delta t;$

in heater No. 4 $t^{II}_{s}=t^{IV}_{fw}+\delta t;$

in heater No. 5 $t^{I}_{s}=t_{fw}+\delta t;$

Table 7-8

Nomenclature	Before turbine	First extr.	Second extr.	Third extr.	Fourth extr.	Fifth extr.	Condenser
Steam pressure, ata	p_0	p^{I}_{ex}	p^{II}_{ex}	p^{III}_{ex}	p^{IV}_{ex}	p^{V}_{ex}	p_2
Steam temperature or wetness (t_0 or $100-x$), °C or %	t_0	t^{I}_{ex}	t^{II}_{ex}	t^{III}_{ex}	$100-x^{IV}$	$100-x^{V}$	$100-x_2$
Heat content of steam i, kcal/kg	i_0	i^{I}_{ex}	i^{II}_{ex}	i^{III}_{ex}	i^{IV}_{ex}	i^{V}_{ex}	i_2
Saturation temperature t_s, °C	t_s	t^{I}_{s}	t^{II}_{s}	t^{III}_{s}	t^{IV}_{s}	t^{V}_{s}	t^{con}_{s}
Heat content of condensate i_{con} kcal/kg	i_{con}	i^{I}_{s}	i^{II}_{s}	i^{III}_{s}	i^{IV}_{s}	i^{V}_{s}	i^{con}_{s}
Feedwater temp. after the heater (t_{fw}), °C	—	t_{fw}	t^{IV}_{fw}	t^{III}_{fw}	t^{II}_{fw}	t^{I}_{fw}	t_{con}
Heat content of feedwater after the heater ($\bar{t}_{fw}$), kcal/kg	—	$\bar{t}_{fw}$	$\bar{t}^{IV}_{fw}$	$\bar{t}^{III}_{fw}$	$\bar{t}^{II}_{fw}$	$\bar{t}^{I}_{fw}$	$\bar{t}_{con}$

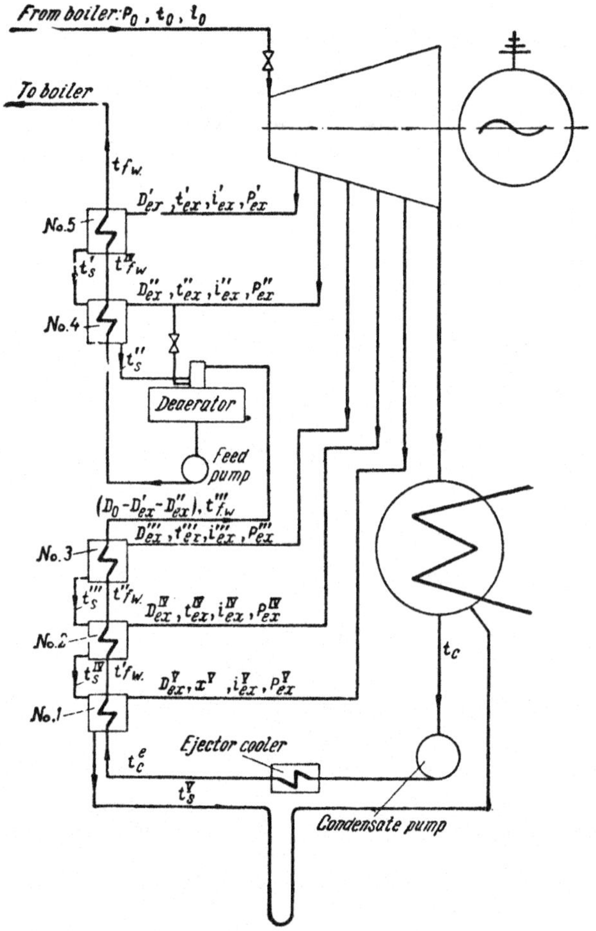

Fig. 7-23. General arrangement of regenerative feedheating — turbine VK-50

where δ_t —temperature differential between the saturation temperature of heating steam in the feed heaters and the feedwater temperature at its exit from the feed heater usually taken as 5 to 7°C.

The pressure of steam at points of extraction is determined from standard tables for saturated steam in conformity with the saturation temperatures prevailing in the feed heaters:

pressure at 1st extraction p_{ex}^{I};
pressure at 2nd extraction p_{ex}^{II};
pressure at 3rd extraction p_{ex}^{III};
pressure at 4th extraction p_{ex}^{IV};
pressure at 5th extraction p_{ex}^{V}.

These pressures are indicated on the i-s diagram (Fig. 7-22).

b) General Arrangement of Turbine Installation for Regenerative Feed Heating System

The general arrangement for regenerative feed heating system for a steam turbine installation is shown in Fig. 7-23; the various pertinent parameters are tabulated as shown in Table 7-8. From the equation of heat balance for the feed heaters the quantity of bled steam is determined for each of the extraction points as a fraction α of the total mass flow D_0, which are

1) from the firs point of extraction

$$\alpha_1 = \frac{\bar{t}_{fw} - \bar{t}_{fw}^{IV}}{(i_{ex}^{I} - \bar{t}_{s}^{I})\,\eta_5}; \qquad (7\text{-}60)^{1}$$

2) from the second point of extraction,

$$\alpha_2 = \frac{\left(\frac{1}{\eta_4}\bar{t}_{fw}^{IV} - \bar{t}_{fw}^{III}\right) - \alpha_1(\bar{t}_{s}^{I} - \bar{t}_{fw}^{III})}{i_{ex}^{II} - \bar{t}_{fw}^{III}}. \qquad (7\text{-}61)$$

The fraction of extracted steam which goes to feed heater No. 4 and the deaerator is evaluated here;

3) from the third point of extraction

$$\alpha_3 = \frac{(1 - \alpha_1 - \alpha_2)(\bar{t}_{fw}^{III} - \bar{t}_{fw}^{II})}{(i_{ex}^{III} - \bar{t}_{s}^{III})\,\eta_3}; \qquad (7\text{-}62)$$

4) from the fourth point of extraction

$$\alpha_4 = \frac{(1 - \alpha_1 - \alpha_2)(\bar{t}_{fw}^{II} - \bar{t}_{fw}^{I}) - \alpha_3(\bar{t}_{s}^{III} - \bar{t}_{s}^{IV})\,\eta_2}{(i_{ex}^{IV} - \bar{t}_{s}^{IV})\,\eta_2}, \qquad (7\text{-}63)$$

5) from the fifth point of extraction

$$\alpha_5 = \frac{(1 - \alpha_1 - \alpha_2)(\bar{t}_{fw}^{I} - \bar{t}_{con}) - (\alpha_3 + \alpha_4)(\bar{t}_{s}^{IV} - \bar{t}_{s}^{V})\,\eta_1}{(i_{ex}^{V} - \bar{t}_{s}^{V})\,\eta_1}, \qquad (7\text{-}64)$$

where η_1, η_2, η_3, η_4 and η_5— feed heater efficiencies accounting for the heat lost to the surrounding medium.

c) Quantity of Steam Flowing Through the Turbine and the Extractions

The quantity of fresh steam flowing through the turbine is determined from the following equation

$$D_0 = \frac{860\,N_e}{[h_i^{I} + (1 + \alpha_1)h_i^{II} + (1 - \alpha_1 - \alpha_2)h_i^{III} - (1 - \alpha_1 - \alpha_2 - \alpha_3)h_i^{IV} + (1 - \alpha_1 - \alpha_2 - \alpha_3 - \alpha_4)h_i^{V} + (1 - \alpha_1 - \alpha_2 - \alpha_3 - \alpha_4 - \alpha_5)h_i^{VI}]\,\eta_m\eta_g}, \qquad (7\text{-}65)$$

[1] Equations (7-60) to (7-64) are given here in their final form.

where $h_i^{I}, h_i^{II}, h_i^{III}, h_i^{IV}, h_i^{V}$ and h_i^{VI} — heat drops usefully utilised in the turbine between each of the extraction points (Fig. 7-22).

The quantities of steam bled from each of the extraction points are equal to

$D_{ex}^{I} = \alpha_1 D_0$ — from the first point;
$D_{ex}^{II} = \alpha_2 D_0$ — from the second point;
$D_{ex}^{III} = \alpha_3 D_0$ — from the third point;
$D_{ex}^{IV} = \alpha_4 D_0$ — from the fourth point;
$D_{ex}^{V} = \alpha_5 D_0$ — from the fifth point of extraction.

The quantity of steam flowing through the turbine between the various points of extraction will therefore be D_0 — mass flow through the first chamber up to the first point of extraction;

$D_1 = D_0 - D_{ex}^{I}$ between the first and the second points of extraction;

$D_2 = D_0 - D_{ex}^{I} - D_{ex}^{II}$ between the second and the third points;

$D_3 = D_0 - D_{ex}^{I} - D_{ex}^{II} - D_{ex}^{III}$ between the fourth and the fifth points;

$D_4 = D_0 - D_{ex}^{I} - D_{ex}^{II} - D_{ex}^{III} - D_{ex}^{IV}$—be ween the fourth and the fifth points;

$D_5 = D_0 - D_{ex}^{I} - D_{ex}^{II} - D_{ex}^{III} - D_{ex}^{IV} - D_{ex}^{V}$— after the fifth point of extraction.

d) Heat Drop Distribution for Turbines with Extraction

Preliminary heat drop distribution for turbines with extraction is carried out in exactly the same manner as was done in the case of turbines without extraction, i.e., preliminary calculations are carried out for the thermal design of the first, second and last stages of the turbine. Calculations of the regulating and the second stage are carried out according to equation (7-26), for which either heat drop h_0 or the diameter is assumed, the ratio u/c_1 having the optimum value in either case. The last stage dimensions are determined according to equation (7-30). For turbines of large capacities the ratio d/l is taken within the limits of 2.7 to 3 and the absolute pressure after the last stage moving blades between 0.03 and 0.035 ata.

Detailed design of the regulating stage is carried out after having obtained satisfactory results from the preliminary design. The heat losses are plotted on the i-s diagram as shown in Fig. 7-24 and the steam conditions just before the second stage nozzles, point a_1 and the theoretical heat drop h_{01} (from point a_1 to a_{1t}) are determined.

The theoretical heat drop in the second stage is obtained from the following relation

$$h_0^{II} = \frac{u^2}{8{,}378\,\varphi^2 x^2} .$$

Comparing the magnitudes of heat drops h_0^{II} and h_{01} the number of stages which may be accommodated up to the first extraction is determined. Depending upon the fresh steam conditions, the type of turbine, and the number of extractions for regenerative feedheating, the number of stages up to the first point of extraction may vary as much as from two-three to five-six. For the first group of stages up to the first extraction the heat drop in each of the stages may be taken equal in each of them or with a slight increase in successive stages. Having obtained a satisfactory figure for the number of stages up to the first point of extraction along with the heat drop occurring in each of them, the pressures after the second, third and fourth stages, i. e., p_2^{II}, p_2^{III} and p_{ex}^{I} are evaluated on the basis of the basic, adiabatic $a_1 a_{1t}$ (Fig. 7-24).

Detailed calculations for all the stages of the first group up to the first bleeding are then carried out (in the case under consideration, three stages) and the various heat losses are plotted on the i-s diagram. Results of the second stage calculations enable us to obtain the fresh steam conditions before the third stage nozzles (point k_1), and similarly from the second stage heat calculations steam conditions after the third stage are found (point k_2) and so on.

For the particular case under consideration the theoretical heat drop in the second stage is h_0^{II} and those for the third and fourth stages are given by the intercepts $k_1 k_{1t}$ and $k_2 k_{2t}$.

The theoretical heat drop from the first to the second point of extraction, h_{0II}, is now distributed over the stages of the second group. Depending upon the theoretical available energy, given by $k_1' a_{2t}$ (Fig. 7-24), as well as the diameters of the first and last stages of the second group, the number of stages to be accommodated between the first and the second extractions is decided and the available heat drop is distributed among them. Supposing that for the section between the first and second extractions the number of stages decided upon is three with heat drops of h_0^{V}, h_0^{VI} and h_0^{VII} the sum of which is equal to h_{0II}. The pressures p_2^{V}, p_2^{VI} and p_{ex}^{II} conforming to these heat drops are obtained from the i-s diagram (Fig. 7-24).

The detailed heat calculations considering all the losses occurring in the various stages (leading to an increase in entropy) are now carried out for the stages between these pressures. All the succeeding stages are evaluated in the same way as has been done for the cases shown above. While distributing the heat drop in the various stages the following points should be borne in mind:

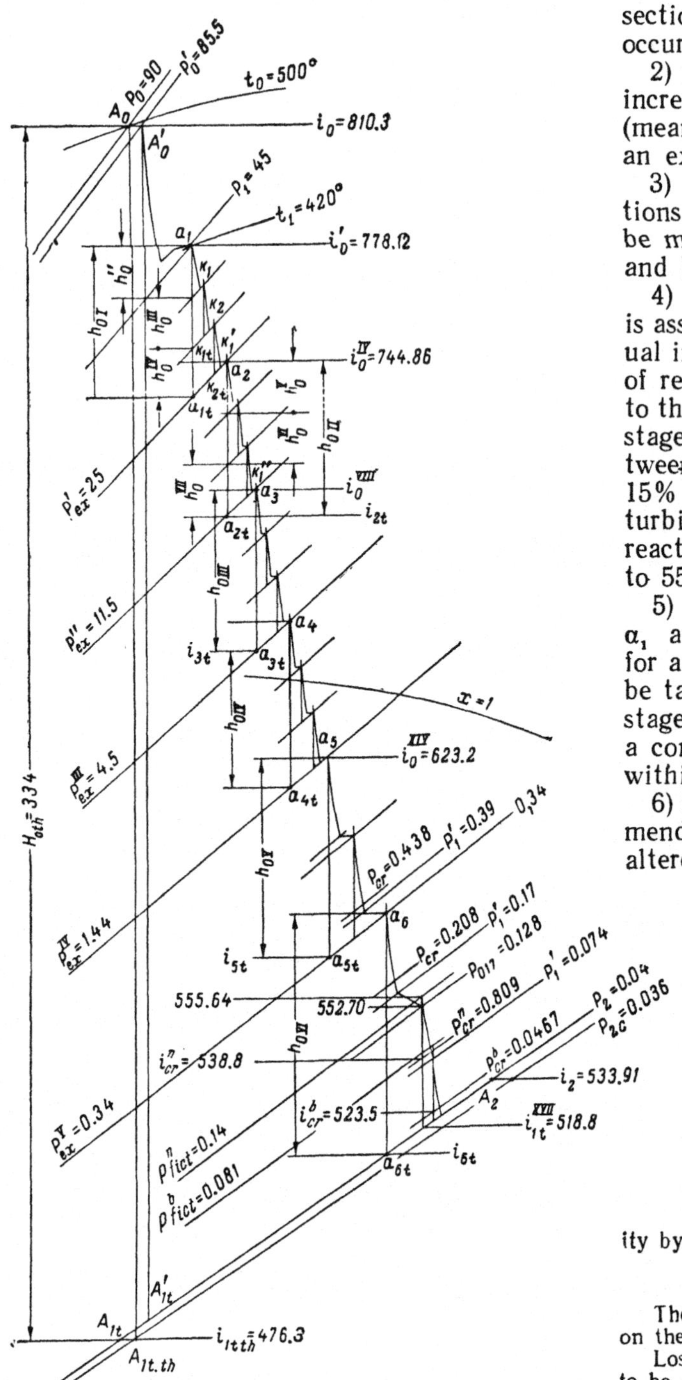

Fig. 7-24. I-s diagram for an impulse turbine with regenerative feed heating type VK-50

1) to obtain $(u/c_1)_{opt}$ and stage diameters for the section under consideration the preliminary heat calculations for the stage situated immediately after an extraction point should be carried out in such a way that the sum of the heat drops in the various stages of the particular section under consideration equals the heat drop occurring between two adjacent extractions;

2) the mean diameters of the blade discs should increase gradually from one stage to another (mean diameters may be abruptly increased after an extraction);

3) simultaneously with the detailed calculations the detailed drawings for the turbine should be made using the mean disc diameters, nozzle and blade heights, etc.;

4) the degree of reaction for the moving blades is assumed keeping in view the smooth and gradual increase in mean disc diameters, i. e., degree of reaction is increased gradually from the first to the last row of the moving blades. For the first stages the degree of reaction (ϱ) may be taken between 4 to 8%, for the central stages from 10 to 15% and for the last stages from 20 to 30%. For turbines of very large capacities the degree of reaction for the last stages may be as high as 50 to 55%;

5) besides the degree of reaction (ϱ), angles α_1 and β_2 also govern the streamlined shape for a turbine: for the first pressure stages α_1 may be taken from 11 to 13°, in the central pressure stages from 14 to 15° and for the last stages of a condensing turbine from 25 to 30°. β_2 varies within the limits of 35 to 40°;

6) in order to unify blade profiles it is recommended that the angles α_1 and β_2 should not be altered from stage to stage but kept at the same values, if only for a group of stages of the same type.

7-19. HEAT CALCULATIONS FOR TURBINE TYPE K-50-90 (VK-50)

Data: Nominal turbine capacity at the generator terminals $N_e^n = 50{,}000$ kW, $n = 3{,}000$ r.p.m., pressure and temperature of fresh steam $p_0 = 90$ ata and $t_0 = 500°$ C, steam pressure at entry to the condenser $p_{2c} = 0.036$ ata, final temperature of the feedwater at exit from the feed heaters $t_{fw} = 217°$ C.

We shall suppose that the economic capacity of the turbine will be less than the nominal capacity by 10%, i.e.,

$$N_e^{ec} = 0.90 \times 50{,}000 = 45{,}000 \text{ kW}.$$

The fresh steam conditions are denoted by the point A_0 on the i-s diagram shown in Fig. 7-22.

Losses in the stop and regulating valves are assumed to be of the order of 5% of the fresh steam pressure, so that the pressure in front of the nozzles of the first stage will be

$$p_0' = (1 - 0.05)\,90 = 85.5 \text{ ata (point } A_0').$$

Pressure losses in the exhaust piping will be determined from the equation (5-26), i.e.,

$$\Delta p = p_2 - p_{2c} = \lambda (c_{ex}/100)^2 p_{2c} =$$
$$= 0.092 (110/100)^2 \times 0.036 = 0.004 \text{ ata}$$

where $\lambda = 0.092$ and $c_{ex} = 110$ m/sec (assumed from experience). Pressure of steam after the last row of the moving blades

$$p_2 = 0.036 + 0.004 = 0.04 \text{ ata}.$$

Theoretical heat drop occurring in the turbine neglecting losses in the stop and regulating valves as well as exhaust piping will be

$$H_{max} = 810.3 - 476.3 = 334 \text{ kcal/kg (adiabatic } A_0A_{1t\,max}).$$

Theoretical heat drop neglecting losses in the regulating and stop valves but taking into account the losses in the exhaust piping will be given by

$$H_0 = 810.3 - 479.5 = 330.8 \text{ kcal/kg (adiabatic } A_0A_{1t}).$$

Adiabatic heat drop in the turbine taking into consideration losses both in the regulating valves as well as the exhaust piping would be

$$H_0' = 810.3 - 481.3 = 329 \text{ kcal/kg (adiabatic } A_0'A_{1t}').$$

The relative internal efficiency of the turbine will be assumed to be $\eta_{0i} = 0.822$, so that heat drop usefully utilised in the turbine will be

$$H_i = H_{max}\eta_{0i} = 334 \times 0.822 = 274.5 \text{ kcal/kg}.$$

Plotting this value on the *i-s* diagram (Fig. 7-22) we obtain the line $A_0'A_2$ representing the heat drop process in the turbine. From the tables of thermodynamic properties of water and saturated steam we find the saturation temperature of exhaust steam at the stipulated pressure of $p_{2c} = 0.036$ ata —

$$t_s = 26.7^\circ \text{ C}.$$

We shall take the condensate temperature to be $26.7 - 0.7 = 26^\circ$ C. For the turbine under consideration with the stipulated fresh steam parameters we shall choose five extractions for feedheating with the assumption that the feed is heated in each of the heaters through the same temperature range. Further we shall assume that the temperature of condensate at the exit from the condenser increases by one degree as a result of its being used as coolant for the air ejectors ($t_{con} = 27^\circ$ C).

Increase in temperature of feedwater in each of the feed heaters would be

$$\Delta t = \frac{t_{fw} - t_{con}}{5} = \frac{217 - 27}{5} = 38^\circ \text{ C}.$$

Feedwater temperatures after the heaters would be obtained from equations given in 7-18; these results are tabulated in Table 7-9. Sensible heat of feedwater after the heaters as well as steam pressures at points of extraction are obtained from steam tables (also tabulated in Table 7-9). With the help of these extraction pressures from the *i-s* diagram (Fig. 7-22) we next obtain theoretical available heat drop in each of the turbine sections between points of extraction (also shown in Table 7-9). For the regenerative feedheating cycle (Fig. 7-23) with the help of equations (7-60) (7-64) we obtain the relative mass flows of steam from the extractions into the heaters (assuming an efficiency of 0.98 for the heaters) as

1) from the first extraction

$$\alpha_1 = \frac{222 - 181.8}{(749.3 - 228.6)\,0.98} = 0.0789;$$

2) from the second extraction

$$\alpha_2 = \frac{\left(\frac{1}{0.98}\,181.2 - 141.9\right) - 0.0789\,(228.6 - 141.9)}{715 - 141.9} = 0.0628;$$

3) from the third extraction

$$\alpha_3 = \frac{(1 - 0.0789 - 0.0628)\,(141.9 - 103)}{(676.5 - 148.1)\,0.98} = 0.0645;$$

4) from the fourth extraction

$$\alpha_4 = \frac{(1 - 0.0789 - 0.0628)\,(103 - 65) - 0.0645\,(148.1 - 110)\,0.98}{(637 - 110)\,0.98} = 0.0585;$$

and 5) from the fifth extraction

$$\alpha_5 = \frac{(1 - 0.0789 - 0.0628)\,(65 - 27) - (0.0645 + 0.0585)\,(110 - 71.6)\,0.98}{(592.5 - 71.6)\,0.98} = 0.0548.$$

Table 7-9

Parameters	Before turbine	1st extr.	2nd extr.	3rd extr.	4th extr.	5th extr.	Condenser
Steam pressure, ata	90	25	11.5	4.5	1.44	0.34	0.04
Steam temp. or wetness (t_0 or $100 - x$), °C or %	500	354	275	187	1.1	6.4	12.7
Total heat of steam i, kcal/kg	810.3	749.3	715	676.5	637	592.5	535.8
Saturation temp. t_s, °C	301.9	222.9	185.2	147.2	110	71.6	28.6
Sensible heat of condensate i_c, kcal/kg	323.8	228.6	187.8	148.1	110	71.6	28.6
Temp. of feedwater at inlet to the feed heaters t_{fw}, °C	—	217	179	141.0	103	65	27
Sensible heat of feedwater at inlet to heaters i_{fw}, kcal/kg		222	181.2	141.9	103	65	27
Heat drop usefully utilised in each of the turbine sections between points of extraction h_i, kcal/kg	61.0	34.3	38.5	39.5	44.5	56.7	—

Total quantity of fresh steam entering the turbine from equation (7-65) will be

$$D_0 = \frac{860 \times 45{,}000}{(61 + 0.9211 \times 34.3 + 0.8583 \times 38.5 + 0.7838 \times 39.5 + 0.7251 \times 44.5 + 0.6705 \times 56.7)\,0.995 \times 0.96} = 178.5 \text{ tons/hr}$$

assuming $\eta_m = 0.995$ and $\eta_g = 0.96$.

The quantity of steam flowing through the various turbine sections and that extracted from each of the points are tabulated in Tables 7-10 and 7-11.

Table 7-10

Nomenclature	1st extr.	2nd extr.	3rd extr.	4th extr.	5th extr.
α	0.0789	0.0628	0.0645	0.0585	0.0548
D_{ex}, tons/hr	14.1	11.2	11.5	10.44	9.8
G_{ex}, kg/sec	3.92	3.12	3.20	2.9	2.72

Table 7-11

Quantity of steam flowing through a stage	Up to 1st extr.	1st to 2nd extr.	2nd to 3rd extr.	3rd to 4th extr.	4th to 5th extr.	after 5th extr.
D, tons/hr	178.5	164.4	153.2	141.7	131.26	121.46
G, kg/sec	49.6	45.68	42.56	39.36	36.46	33.74

We shall carry out the detailed stage calculations assuming the turbine to be of type VK-50-1 made by L.M.W.

Regulating stage:

We shall make the regulating stage with two blade rows (two velocity stages). We shall take the heat drop in the first stage to be 46.5 kcal/kg on the basis of available data for the construction of such turbines. From the *i-s* diagram (Fig. 7-24) steam pressure in the regulating stage steam chest will be $p_r = 45$ ata. Assuming $(u/c_1)_{opt.}$ to be 0.246, where $c_1 = 91.5\sqrt{46.5} = 625$ m/sec we have circumferential velocity $u = 0.246 \times 625 = 153.7$ m/sec and mean diameter $d_1 = 60\,u/\pi n = 60 \times 153.7/\pi \times 3{,}000 = 0.98$ m $= 980$ mm.

Detailed calculations for the first stage with heat drop of 46.5 kcal/kg and mean diameter of 980 mm will be carried out as explained in 6-6. Results of these calculations are shown in Table 7-12 and the velocity triangles are shown in Fig. 7-25.

The degrees of reaction utilised in the moving and the guide blades are as follows:

for the moving blades of the first row 2%;
for the guide blades 5%;
for the moving blades of the second row 3%.

Nozzle heights have been assumed to be 15 mm.

The degree of partial admission has been found to be 0.444 (taking into consideration steam leakage through end labyrinth seals). Losses due to disc friction and windage have been determined according to Stodola's formula. Steam leakage through end seals has been calculated on the basis of the following conditions: shaft dia. 500 mm, number of labyrinths $z = 91$, clearance $\Delta s = 0.4$ mm. Steam from labyrinth seals is exhausted into the third extraction steam chest where the prevailing pressure is 4.5 ata; pressure after nozzles of the regulating stage $p_1' = 48$ ata.

Critical pressure in labyrinth seals just before the compartment from which it is exhausted will be

$$p_{cr} = \frac{0.85\,p_1^{\mathrm{I}}}{\sqrt{z + 1.5}} = \frac{0.85 \times 48}{\sqrt{91 + 1.5}} = 4.24 \text{ ata.}$$

Steam leakage through the labyrinth seals from equation

$$G'_{leak} = 100 f_s \sqrt{\frac{g\,(p_1^{\mathrm{I}2} - p_{ex}^{\mathrm{III}2})}{z p_1^{\mathrm{I}} v_1}} = 100 \times 0.628 \times 10^{-3} \times \sqrt{\frac{9.81\,(48^2 - 4.5^2)}{91 \times 48 \times 0.063}} = 0.566 \text{ kg/sec,}$$

where $f_s = \pi d \Delta s = \pi \times 0.5 \times 0.4 \times 10^{-3} = 0.628 \times 10^{-3}$ m²;
$v_1 = 0.063$ m³/kg — specific volume of steam after the nozzles.

Total heat of steam before the nozzles of the second stage will be obtained from the relation

$$i_0' = i_0 - (h_0 - \sum h_{losses}) = i_0 - h_i = 810.3 - 32.18 = 778.12 \text{ kcal/kg.}$$

Steam conditions before the nozzles of the second stage are determined by pressure 45 ata and temp. 420°C (point a_1 in Fig. 7-24).

Making preliminary calculations for the dimensions of the last stage we have

$$d_z = \sqrt{\frac{G v_2 \vartheta}{\pi \times 91.5\sqrt{\xi_e H_0}\sin\alpha_2}} = \sqrt{\frac{33.74 \times 32.5 \times 3.1}{\pi \times 91.5\sqrt{0.022 \times 330.8}}} = 2.09 \text{ m} = 2{,}090 \text{ mm,}$$

where $G = 33.74$ kg/sec — mass flow of steam through the last stage (Table 7-11);
$v_2 = 32.5$ m³/kg — specific volume of steam after the moving blades of the last stage;
$\vartheta = 3.1$ — ratio of d_z to l_z (assumed);
$\xi_e = 2.2\%$ — carry-over losses from the last stage as percentage of H_0;
$\alpha_2 = 90°$ — exit angle of steam from the last row of the moving blades for minimum carry-over losses.

Circumferential velocity at the mean diameter will be,

$$u = \frac{\pi \times 2.09 \times 3{,}000}{60} = 328 \text{ m/sec,}$$

which is within allowable limits.

Height of blades

$$l_z = \frac{2{,}090}{3.1} = 674 \text{ mm,}$$

which is also permissible.

Detailed Heat Calculations for Pressure Stages up to the First Extraction

Theoretical heat drop from pressure 45 ata and 420°C (point a_1, Fig. 7-24) to pressure at the first extraction is

$$h_{01} = 778.12 - 737.6 = 40.52 \text{ kcal/kg.}$$

Approximate calculations show that three stages may be accommodated in the interval up to the first point of extraction. Assuming equal heat drops in each of the stages we have

$$h_0 = \frac{40.52}{3} \approx 13.5 \text{ kcal/kg}.$$

Hence we shall assume the following heat drops for the three successive stages:

13.6 kcal/kg — in the second stage;
13.45 kcal/kg — in the third stage;
13.47 kcal/kg — in the fourth stage.

Steam pressures after each stage, from *i-s* diagram, are $p_2^{II} = 37.3$ ata after the second stage; $p_2^{III} = 30.5$ ata after the third stage; $p_{ex}^{I} = 25$ ata after the fourth stage.

In order to reduce entrainment losses we shall assume 5% reaction in each of the blade rows.

For the second row we have, therefore, $u/c_1 = 0.462$; $c_1 = 91.5\sqrt{13.6} = 337$ m/sec; $u = 0.462 \times 337 = 155.5$ m/sec and $d = 60 \times 155.5/\pi \times 3{,}000 = 0.99$ m or 990 mm.

On the basis of this mean diameter of 990 mm we shall now carry out detailed calculations.

Heat drop in the nozzles $h_{01} = 0.95 \times 13.6 = 12.92$ kcal/kg and in the moving blades heat drop $h_{02} = 13.6 - 12.92 = 0.68$ kcal/kg. Steam pressure after the nozzles is 37.4 ata. Ratio of pressures $p_1^I/p_0 = 37.4/45 = 0.832 > \nu_{cr}$, i.e., steam velocity is less than the critical value. The actual velocity of steam $c_1 = 91.5 \times 0.95\sqrt{12.92} = 312.5$ m/sec and $c_{1t} = 328.5$ m/sec where $\varphi = 0.95$ (taken from the graph shown in Fig. 5-2).

Angle $\alpha_1 = 12°$ is assumed in such a way that when $\varepsilon = 1$ the nozzle heights obtained would be within the permissible range. From velocity triangles we obtain $w_1 = 164$ m/sec and $\beta_1 = 23°50'$ (Fig. 7-25); β_2 shall be assumed as 21° from which we have $\psi = 0.86$ (from graph shown in Fig. 5-4)

$w_2 = 0.86\sqrt{164^2 + 8{,}378 \times 0.68} = 155$ m/sec and $w_{2t} = 180.5$ m/sec.

Again from velocity triangles we have

$c_2 = 56.1$ m/sec; $\alpha_2 = 101°10'$; $c_{1u} = 312.5 \cos 12° = 305.5$ m/sec;

$c_{2u} = 56.1 \sin 11°10' = 10.8$ m/sec;

$$\eta_u' = \frac{2u(c_{1u} - c_{2u})}{c_{ad}^2} = \frac{2 \times 155.5(305.5 - 10.8)}{337^2} = 0.805.$$

Determining the losses in the blade passages of each stage we have

$$h_n = (1/\varphi^2 - 1)\, c_1^2/8{,}378 = \left(\frac{1}{0.95^2} - 1\right)\frac{312.5^2}{8{,}378} = 1.24 \text{ kcal/kg};$$

$$h_b = \left(\frac{1}{\varphi^2} - 1\right)\frac{w_2^2}{8{,}378} = \left(\frac{1}{0.86^2} - 1\right)\frac{155^2}{8{,}378} = 1.02 \text{ kcal/kg};$$

$$h_e = \frac{56.1^2}{8{,}378} = 0.38 \text{ kcal/kg}.$$

To check the accuracy of the results obtained above we shall find η_u as follows and compare it with η_u' calculated previously

$$\eta_u = \frac{h_0 - (h_n + h_b + h_e)}{h_0} = \frac{13.6 - (1.24 + 1.02 + 0.38)}{13.6} = \frac{10.96}{13.6} = 0.806.$$

The values of η_u and η_u' give a good agreement.

Losses due to friction and windage

$$N_{fr.w.} = \lambda\, 1.07 d^2 \frac{u^3}{10^6}\gamma = 1.07 \times 0.987^2 \frac{155.5^3}{10.6} \times 12.42 = 48.2 \text{ kW},$$

where $\gamma = 1/v_1 = 1/0.0805 = 12.42$ kg/m³; $v_1 = 0.0805$ m³/kg is the specific volume of steam after the nozzles;
$\lambda = 1$ — coefficient assumed to be unity for high superheat steam.

$$h_{fr.w.} = \frac{102 N_{fr.w.}}{427 G} = \frac{102 \times 48.2}{427 \times 49.6} = 0.23 \text{ kcal/kg}.$$

Total heat of steam after the blades with losses considered

$$i_2' = 778.12 - 13.6 + (1.24 + 1.02 + 0.38 + 0.23) = 767.30 \text{ kcal/kg}.$$

Steam leakage through diaphragm seals from equation (5-18)

$$G_{leak} = 100 f_s \sqrt{\frac{9.81\,(45^2 - 37.4^2)}{8 \times 45 \times 0.07}} = 100 \times 0.628 \times 10^{-3}\sqrt{244} = 0.98 \text{ kg/sec},$$

where $f_s = \pi d_{leak} \Delta s = \pi\, 0.5 \times 0.4 \times 10^{-3} = 0.628 \times 10^{-3}$ m²;
$d_{leak} = 500$ mm — diameter of rotor seals;
$\Delta s = 0.4$ mm — clearance in the labyrinth seals.

Losses due to steam leakage through clearances from equation (5-7) will be

$$h_{leak} = \frac{0.98}{49.6} \times 10.73 = 0.21 \text{ kcal/kg}.$$

Summation of all the losses in a stage $\sum h_{losses} = 1.24 + 1.02 + 0.38 + 0.23 + 0.21 = 3.08$ kcal/kg.

Heat drop usefully utilised in a stage $h_i = h_0 - \sum h_{losses} = 13.6 - 3.08 = 10.52$ kcal/kg.

Stage efficiency $\eta_{0i}^{st} = \frac{10.52}{13.6} = 0.774.$

Power delivered by a stage

$$N_i = \frac{427 \times 49.6 \times 10.52}{102} = 2{,}180 \text{ kW}.$$

Exit height of nozzles

$$l_n = \frac{(G - G_{leak})\, v_1 \times 10^3}{\pi d c_1 \sin\alpha_1} = \frac{(49.6 - 0.98)\, 0.0805 \times 10^3}{\pi \times 0.99 \times 312.5 \sin 12°} = 19.3 \text{ mm}.$$

Exit height of blades

$$l_2'' = \frac{G v_2 \times 10^3}{\pi d w_2 \sin\beta_2} = \frac{49.6 \times 0.081 \times 10^3}{\pi \times 0.99 \times 155 \sin 21°} = 23.2 \text{ mm}.$$

The overlap being only 3.9 mm, we shall have straight shrouding for the blade tips.

For the third stage we have: steam pressure before the nozzles 37.3 ata, total heat of steam before the nozzles $i_0^{III} + h_e^{pr} = i_0^{II} + h_e^{pr} - h_i^{II} = 778.12 - 10.52 = 767.6$ kcal/kg, $i_0^{III} = 767.6 - 0.38 = 767.22$ kcal/kg, temperature of steam before the nozzles 396°C.

We shall assume the mean diameter $d = 990 + 4 = 994$ mm, i.e., 4 mm larger than that for the second stage.

Theoretical heat drop occurring in the stage is

$$h_0 = i_0^{III} - i_{1t}^{III} = 767.22 - 753.6 = 13.62 \text{ kcal/kg}.$$

Heat drop in the nozzles $h_{01} = 0.95 \times 13.62 = 12.93$ kcal/kg and in the moving blades $h_{02} = 13.62 - 12.93 = 0.69$ kcal/kg.

Assuming that the carry-over velocity from the exit of the second stage is utilised in the third one we have the available energy in the third stage nozzles equal to

$$h_{01} + h_e^{pr} = 12.93 + 0.38 = 13.31 \text{ kcal/kg}.$$

Velocity of steam at exit from the nozzles $c_1 = 91.5 \times 0.955\sqrt{13.31} = 318.5$ m/sec and $c_{1t} = 333.5$ m/sec.

Total available energy in the stage will be $h_0 + h_e^{pr} = 13.62 + 0.38 = 14.00$ kcal/kg.

Velocity $c_{ad} = 91.5\sqrt{h_0 + h_e^{pr}} = 91.5\sqrt{14} = 342$ m/sec. All the succeeding stages are designed on exactly similar lines the results of which are tabulated in Table 7-12.

To carry out further calculations for stages situated between the first and the second points of extraction heat drop distribution and determination of the number of stages to be accommodated between these points of extraction must be evaluated first.

Theoretical available energy for doing work between the first and the second extractions ($k_1'a_{2t}$, Fig. 7-24) is equal to

$$h_{0II} + h_e^{pr} = i_0^{IV} + h_e^{pr} - i_{2t} = 745.27 - 699.2 = 46.07 \text{ kcal/kg}.$$

Theoretical heat drop occurring between the section from the first to the second point of extraction is

$$h_{0II} = i_0^{IV} - i_{2t} = 744.86 - 699.2 = 45.66 \text{ kcal/kg}.$$

Three stages would be a suitable number for this heat drop. Average heat drop in a stage will therefore be

$$h_{0\,mean} = 45.66/3 = 15.22 \text{ kcal/kg}.$$

Since the diameters are to be in a steadily increasing order we shall finally have

$$h_0^V = 15.08 \text{ kcal/kg}; \ h_0^{VI} = 15.12 \text{ kcal/kg and}$$
$$h_0^{VII} = 15.46 \text{ kcal/kg}.$$

Drawing these heat drops h_0^V, h_0^{VI} and h_0^{VII} on the i-s diagram (Fig. 7-24) we obtain the pressures after each of these stages, viz., $p_2^V = 19.7$ ata; $p_2^{VI} = 15.25$ ata and $p_{ex}^{II} = 11.5$ ata. Final calculations for these stages are carried out for heat drops occurring between the limits of the pressures as indicated above duly taking into account all the heat losses that occur during an actual process of expansion.

The diaphragms are fixed onto special rings (Fig. 7-26) so as to reduce clearance spaces between nozzles and the moving blades. Thus the carry-over velocity from the fourth stage moving blades is utilised in the nozzles of the fifth stage. Design calculations for the group of stages between the first and the second extractions are similar to those for the previous stages. Results of these calculations are shown in Table 7-12.

Steam conditions before the nozzles of the eighth stage are $p_{ex}^{II} = p_0^{VIII} = 11.5$ ata and $t_0^{VIII} = 261°$C conforming to $i_0^{VIII} = 706.52$ kcal/kg (point a_3) and $i_0^{VIII} + h_e^{pr} = 706.52 + 0.6 = 707.12$ kcal/kg. The theoretical heat drop in the next group of stages between the second and the third points of extraction

$$h_{0III} = i_0^{VIII} - i_{3t} = 706.52 - 657.7 = 48.82 \text{ kcal/kg}.$$

In order to maintain the mean diameters of the following stages of the same order as that of the seventh stage we shall take the heat drops occurring in each of these stages to be about that which is occurring in the seventh stage. To satisfy these assumptions we shall have three stages for this part of the turbine (between the second and the third points of extraction).

Mean heat drop in each of the stages would be

$$h_{0\,mean} = 48.82/3 = 16.27 \text{ kcal/kg}$$

so that finally we may assume $h_0^{VIII} = 16.15$ kcal/kg; $h_0^{IX} = 16.15$ kcal/kg and $h_0^X = 16.52$ kcal/kg.

Plotting these heat drops along the adiabatic a_3a_{3t} (Fig. 7-24) we shall determine steam pressures for each of the stages, i.e., $p_2^{VIII} = 8.7$ ata, $p_2^{IX} = 6.3$ ata. Pressure after the tenth stage is the same as pressure in the third extraction which is $p_{ex}^{III} = 4.5$ ata.

The detailed heat calculations for these stages will be conducted on the basis of the actual process of expansion of steam for the heat drops governed by the above pressures.

For the determination of friction and windage losses we shall assume λ equal to unity up to the ninth stage. For the tenth stage $\lambda = 1.2$.

Results of these calculations are tabulated in Table 7-12.

Theoretical heat drop for the group of stages between the third and the fourth point of extraction (Fig. 7-24) along the adiabatic a_4a_{4t}, $h_{0\,IV} = 663.92 - 615.9 = 48.02$ kcal/kg.

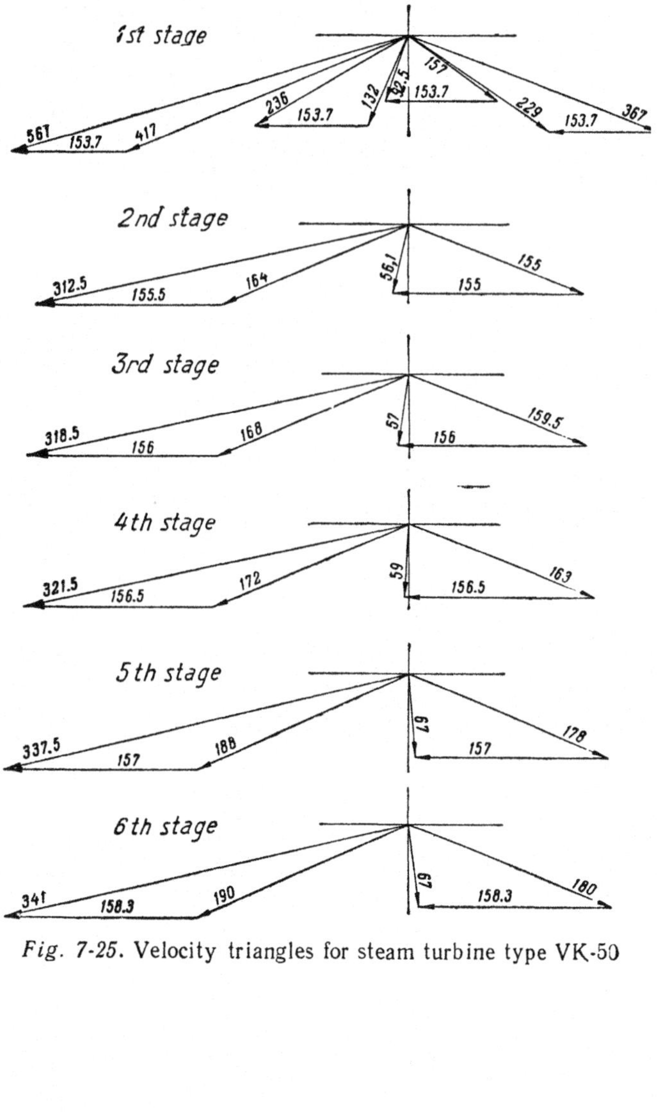

Fig. 7-25. Velocity triangles for steam turbine type VK-50

Although heat drop for this group is found to be slightly lower than that for the previous group we shall continue to have three stages for this group as well. This is desirable since for the preceding stages u/c_1 was found to be lower than the optimum value.

Average heat drop for each of the stages

$$h_{0\ mean} = 48.02/3 = 16.01 \text{ kcal/kg}.$$

We may therefore assume

$$h_0^{XI} = h_0^{XII} = 16.00 \text{ kcal/kg and}$$

$$h_0^{XIII} = 16.02 \text{ kcal/kg}.$$

Pressures conforming to these heat drops would be

$$p_2^{XI} = 3.18 \text{ ata}; \; p_2^{XII} = 2.15 \text{ ata and}$$

$$p_2^{XIII} = p_{ex}^{IV} = 1.44 \text{ ata}.$$

Detailed calculations for these stages are carried out just as has been done for the preceding groups of stages. Results of the above calculations are given in Table 7-12. For stages 11 to 13 we shall determine the exact quantity of steam flowing through the group of stages. From the high-pressure gland seal (front end) part of the leakage steam is brought into the third extraction steam chest so that mass flow $G_{11} = G_{10} - G_{ex}^{III} + \triangle G_{seal}$, where $G_{10} = 42.56$ kg/sec — mass flow through the tenth stage; $G_{ex}^{III} = 3.2$ kg/sec — extraction steam to feed heater No. 3; $\triangle G_{seal}$ — steam from front gland seal into the third extraction steam chest:

$$\triangle G_{seal} = G_{leak}^{I} - G_{leak}^{II}$$

where G_{leak}^{I} — 0.566 kg/sec — mass flow of steam through the gland seal up to the point of bleeding; G_{leak}^{II} — mass flow of steam from the first to the second point of bleeding from the high-pressure labyrinth seal. The second point of bleeding delivers into the steam space of the fifth extraction from the turbine. We shall assume for the portion of labyrinth seal between the first and the second points of bleeding forty partition foils, i.e.,

$$z_2 = 40.$$

Pressure in the labyrinth compartment just before the second bleeding from equation (5-23) is

$$p_{cr} = \frac{0.85 \times 4.5}{\sqrt{40 + 1.5}} = 0.595 \text{ ata}.$$

Steam pressure in the compartment from where steam is bled to the fifth extraction steam chest would be equal to $p_{ex}^{V} = 0.34$ ata.

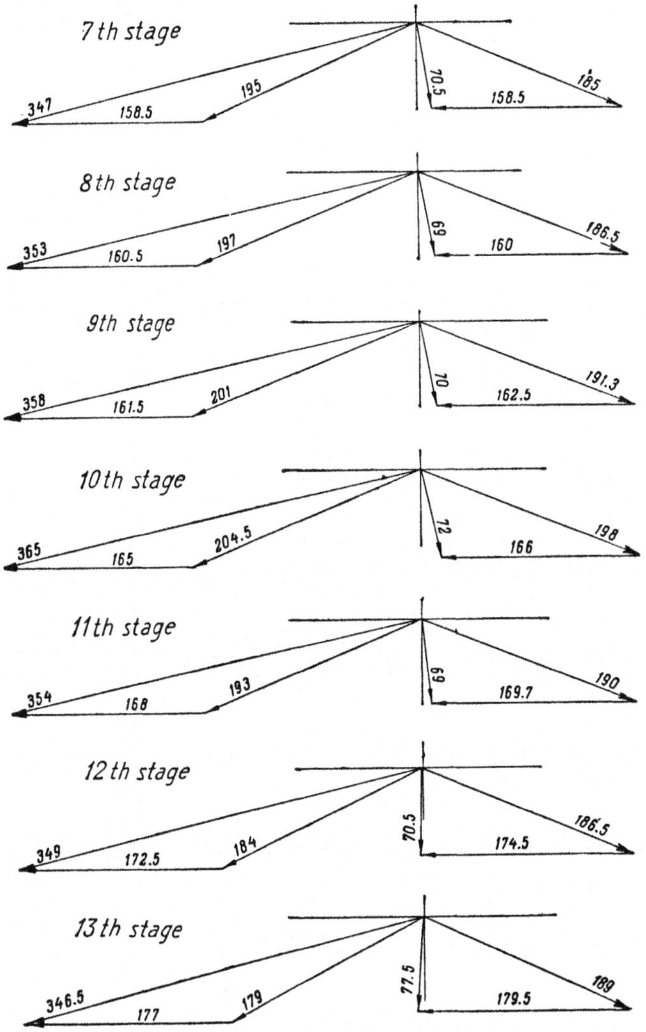

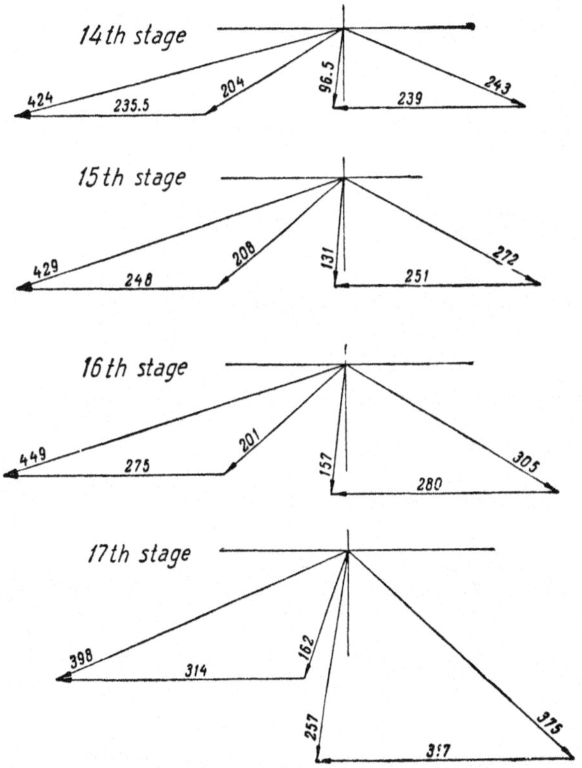

To *Fig. 7-25.*

Stage No. / Designation	Dimension	I	II	2	3	4	5	6	7
G_0	kg/sec	49.6		49.6	49.6	49.6	45.68	45.68	45.68
p_0	ata	85 5		45	37.3	30.5	25	19.7	15.25
t_0 or x	°C or %	488°		420°	396°	370°	346°	319°	289°
i_0	kcal/kg	810.3		778.12	767.22	756.12	744.86	732.37	719.68
$i_0+h_e^{pr}$	kcal/kg	810.3		778.12	767.60	756.51	745.27	732.91	720.22
$p_{fict.}$	ata	85.5		45	37.5	30.7	25.2	19.9	15.35
p_1'/p_2	ata	48/47.6	46/45	37.4/37.3	30.7/30.5	25.3/25	20/19.7	15.5/15.25	11.7/1.5
p_{cr}	ata	—		—	—	—	—	—	—
i_{cr}	kcal/kg	—		—	—	—	—	—	—
i_{1t}	kcal/kg	763.8		764.52	753.6	742.4	729.78	717.20	704.00
$h_{0\,cr}$	kcal/kg	—		—	—	—	—	—	—
h_0	kcal/kg	46.5		13.6	13.62	13.72	15.08	15.17	15.68
ρ	%	2+5+3=10		5	5	5	5	5	5
h_{01}	kcal/kg	41.8		12.92	12.93	13.02	14.30	14.40	14.87
h_{02}	kcal/kg	0.93+2.33+1.39		0.68	0.69	0.70	0.78	0.77	0.81
h_e^{pr}	kcal/kg	—		0	0.38	0.39	0.41	0.54	0.54
$h_{0\,cr}+h_e^{pr}$	kcal/kg	—		—	—	—	—	—	—
$h_{01}+h_e^{pr}$	kcal/kg	41.85		12.92	13.31	13.41	14.71	14.94	15.41
c_{cr}/w_{cr}	m/sec	—		—	—	—	—	—	—
c_{1t}	m/sec	591	268	328.5	333.5	335	350.5	353	359
φ	—	0.95	0.88	0.95	0.955	0.96	0.962	0.966	0.966
c_1	m/sec	561	236	312.5	318.5	321.5	337.5	341	347
$h_0+h_e^{pr}$	kcal/kg	46.5		13.60	14.00	14.11	15.49	15.71	16.22
c_{ad}	m/sec	625		337	342	343.5	360	362.5	368.5
u/c_{ad}	—	0.246		0.462	0.456	0.456	0.436	0.437	0.431
u/c_1	—	0.274		0.498	0.490	0.487	0.466	0.465	0.457
u	m/sec	153.7		155.5	156	156.5	157.0	158.3	158.5
d	mm	980		990	994	998	1,000	1,008	1,010
α_1/α_1'	deg	16°	30°	12°	12°	12°	13°	13°	13°
w_1	m/sec	417	132	164	168	172	188	190	195
β_1	deg	22°	66°20′	23°50′	23°20′	22°50′	24°	23°50′	23°50′
β_2/β_2'	deg	20°	35°	21°	21°	21°	22°	22°	22°
ψ	—	0.86	0.90	0.86	0.865	0 865	0.87	0.872	0.873
w_2	m/sec	367	151	155	159.5	163	178	180	185
c_2	m/sec	229	92.5	56.1	57	59	67	67	70.6
α_2	deg	33°30′	109°20′	101°10′	97°30′	92°40′	84°10′	82°40′	80°20′
c_{1u}	m/sec	539	204	305.5	311.5	314	329.5	332	338
c_{2u}	m/sec	191	−30.5	−10.8	−7.5	−2.5	6.5	8	12
$\eta_u=\frac{2u\sum(c_{1u}-c_{2u})}{c_{ad}^2}$	—	0.713		0.805	0.812	0.827	0.816	0.828	0.817
h_n	kcal/kg	4.18	1.92	1.24	1.14	1.06	1.11	0.99	1.02
h_b	kcal/kg	5.67	0.64	1.02	1.02	1.07	1.15	1.22	1.27
h_e	kcal/kg	1.02		0.38	0.39	0.41	0.54	0.54	0.60
η_{iu}	—	0.712		0.805	0.816	0.82	0.82	0.825	0.822
v_{cr}^n/v_{cr}^b	m³/kg	—	—	—	—	—	—	—	—
v_1	m³/kg	0.063	0.067	0.0805	0.096	0.112	0.136	0.169	0.209
v_2	m³/kg	0.064	0.069	0.0810	0.0975	0.114	0.138	0.171	0.210
γ_1	kg/m³	15.9		12.42	10.43	8.93	7.35	5.92	4.79
$N_{fr.w.}$ *	kW	184		48.2	42	36	30.4	25.7	20.8
$h_{fr.w.}=\frac{102\times N_{fr.w.}}{427\times G_0}$	kcal/kg	0.89		0.23	0 19	0.17	0.16	0 13	0.11
h_i'' **	kcal/kg	32.18		10.73	11.23	11.4	12.53	12 83	13.22
$f_s=\pi d_y \Delta s$	cm²	—	—	6.28	6.28	6.28	6.28	6.28	6.28
G_{leak}	kg/sec	—	—	0.98	0.78	0.69	0.62	0.51	0.42

8	9	10	11	12	13	14	15	16	17
42.56	42.56	42.56	39.85	39.85	39.85	36.95	36.95	34.23	34.23
11.5	8.70	6.3	4.5	3.18	2.15	1.44	0.74	0.34	0.128
259°	229°	196°	164°	99.7	97.9	96.3	93.9	91.3	88.4
706.52	692.87	678.79	663.92	650.06	636.21	623.2	600.88	578.16	552.70
707.12	693.45	679.38	664.78	650.63	636.80	632.2	601.99	580.21	555.64
11.7	8.85	6.4	4.5	3.20	2.18	1.44	0.76	0.36	0.14
8.9/8.7	6.4/6.3	4.6/4.5	3.20/3.18	2.18/2.15	1.5/1.44	0.81/0.74	0.39/0.34	0.17/0.128	0.074/0.040
—	—	—	—	—	—	—	0.438	0.208	0.809/0.0467
—	—	—	—	—	—	—	582.2	561.9	538.8/523.5
690.37	676.3	661.5	647.92	634.0	620.1	596.6	573.3	546.2	518.8
—	—	—	—	—	—	—	18.68	16.26	13.90/13.18
16.15	16.57	17.29	16.0	16.06	16.11	26.6	27.58	31.96	33.90
5	5	6	7	8	10	15	20	26	50
15.33	15.70	16.24	14.88	14.77	14.5	22.6	22.0	23.3	17.0
0.82	0.87	1.05	1.12	1.29	1.61	4.0	5.58	8.66	16.9
0.60	0.58	0.59	0.86	0.57	0.59	0	1.11	2.05	2.94
—	—	—	—	—	—	—	19.79	18.31	16.84 16.32
15.93	16.28	16.83	15.74	15.34	15.09	22.6	23.11	25.35	19.94
—	—	—	—	—	—	—	407	391	375.5/370
365	369.5	375	363	358.5	355.5	435	440	461	409
0.968	0.97	0.973	0.975	0.975	0.975	0.975	0.975	0.975	0.975
353	358	365	354	349	346.5	424	429	449	398
16.75	17.15	17.88	16.86	16.63	16.70	26.6	28.69	34.01	36.84
374.5	378.5	387	375.5	373	374	472	490	534	555
0.430	0.429	0.427	0.448	0.462	0.476	0.50	0.506	0.515	0.566
0.457	0.453	0.453	0.475	0.495	0.514	0.556	0.578	0.612	0.789
160.5/161	161.5/162.5	165/166	168/169.7	172.5/174,5	177/179.5	235.5/239	248/251	275/280	314/317
1,020/1,024	1,030/1,036	1,050/1,058	1,070/1,080	1,098/1,110	1,126/1,144	1,500/1,520	1,580/1,600	1,750/1,780	2,000/2,030
12°	12°	12°	12°	13°	14°	14°	16°/17°30′	16°/17°	22°/22°30′
197	201	204.5	193	184	179	204	208	201	162
22°	22°	22°	22°40′	25°15′	27°50′	27°30′	39°	40°20′	70°20′
21°	21°	21°	21°	22°	24°	23°	28°30′	31°	36°/42°30′
0.872	0.875	0.88	0.88	0.882	0.887	0.887	0.907	0.907	0.915
186.5	191.3	198	190	186.5	189	243	272	305	375
69.5	70	72	69	70.5	77.5	96.5	131	157	257
79°30′	79°	76°	83°	90°	94°30′	107°20′	95°20′	96°30′	99°
345	350	357	346	340	336	411	409	429.5	367.5
12.5	14.5	17.5	8	−1	−6	−12	−11.5	−18	−41
0.822	0.825	0.827	0.843	0.844	0.84	0.847	0.825	0.803	0.67
0.99	0.98	0.91	0.77	0.75	0.74	1.11	1.13	1.24	0.98
1.30	1.34	1.36	1.25	1.19	1.16	1.91	1.91	2.40	3.24
0.58	0.59	0.86	0.57	0.59	0.72	1.11	2.05	2.94	7.89
0.828	0.826	0.826	0.843	0.847	0.843	0.847	0.825	0.807	0.672
—	—	—	—	—	—	—	3.52	7.0	16.8/27.5
0.265	0.33	0.442	0.60	0.835	1.18	2.05	4.02	8.5	18.1
0.27	0.335	0.45	0.61	0.85	1.20	2.26	4.50	11.1	32.0
3.78	3.03	2.37	1.67	1.20	0.847	0.488	0.248	0.143	0.059
17.7	15.9	15.1	11.7	10.3	8.3	20.6	13.3	12.6	10.2
0.10	0.09	0.08	0.07	0.06	0.05	0.13	0.09	0.09	0.07
13.78	14.14	14.67	14.20	14.04	14.03	22.34	23.51	27.34	24.66
6.28	6.28	6.28	6.28	6.28	6.28	8.81	8.81	10.05	10.05
0.34	0.257	0.20	0.15	0.11	0.097	0.089	0.049	0.029	0.01

Stage No. Designation	Dimension	I	II	2	3	4	5	6	7
$h_{leak}=\frac{G_{leak}}{G_0}h_i''$	kcal/kg	—	—	0.21	0.17	0.16	0.17	0.14	0.12
$h_i'=h_i''-h_{leak}$	kcal/kg	32.18		10.52	11.09	11.24	12.36	12.69	13.10
$h_{wetness}$	kcal/kg	—	—	—	—	—	—	—	—
$h_i=h_i'-h_{wetness}$ ***	kcal/kg	32.18		10.52	11.09	11.24	12.36	12.69	13.10
$\sum h_{losses}$	kcal/kg	14.32		3.08	2.91	2.87	3.13	3.02	3.12
$\eta_{oi}^{st}=\frac{h_i}{h_0+h_e^{pr}}$	—	0.691		0.774	0.792	0.797	0.798	0.808	0.809
$N_i^{st}=\frac{427\times G_0 h_i}{102}$	kW	6,660		2,180	2,300	2,335	2,360	2,420	2,500
l_n	mm	15	20.6	19.3	22.7	26.1	25.8	31.5	38.8
l_2''	mm	18.5	28.8	23.2	27.8	30.8	30.2	36.7	44.5

* Calculated according to equation (5-4).

** Calculated according to equation $h_i''=h_0+h_e^{pr}-(h_n+h_b+h_e+h_{fr.w.})$.

*** Accuracy checked by calculating from equation $h_i=h_0+h_e^{pr}-\sum h_{losses}$.

The quantity of steam flowing between the first and the second points of bleeding from equation (5-22) would be

$$G_{leak}^{II}=100 f_s \sqrt{\frac{g}{(z_2+1.5)}\frac{p_{ex}^{III}}{v'}}=$$

$$=0.0628\sqrt{\frac{9.81}{41.5}\times\frac{4.5}{0.7}}=0.078 \text{ kg/sec},$$

where $v'=0.7$ m³/kg — specific volume of steam before the first bleeding (in the labyrinth seal).

The quantity of steam bled to the third extraction steam space

$$\Delta G_{seal}=0.566-0.078=0.49 \text{ kg/sec}.$$

The quantity of steam flowing through the 11th stage

$$G_{11}=42.56-3.2+0.49=39.85 \text{ kg/sec}.$$

From the i-s diagram we find that steam while expanding in the 12th stage becomes wet. Hence losses due to wetness must be accounted for. From equation (5-25) we have

$$h_{wetness}=(1-x_{av})h_i=\left(1-\frac{x_1+x_2}{2}\right)14.0=$$

$$=\left(1-\frac{0.997+0.979}{2}\right)14.0=0.17 \text{ kcal/kg}.$$

Calculations for turbine stages after the 4th extraction.

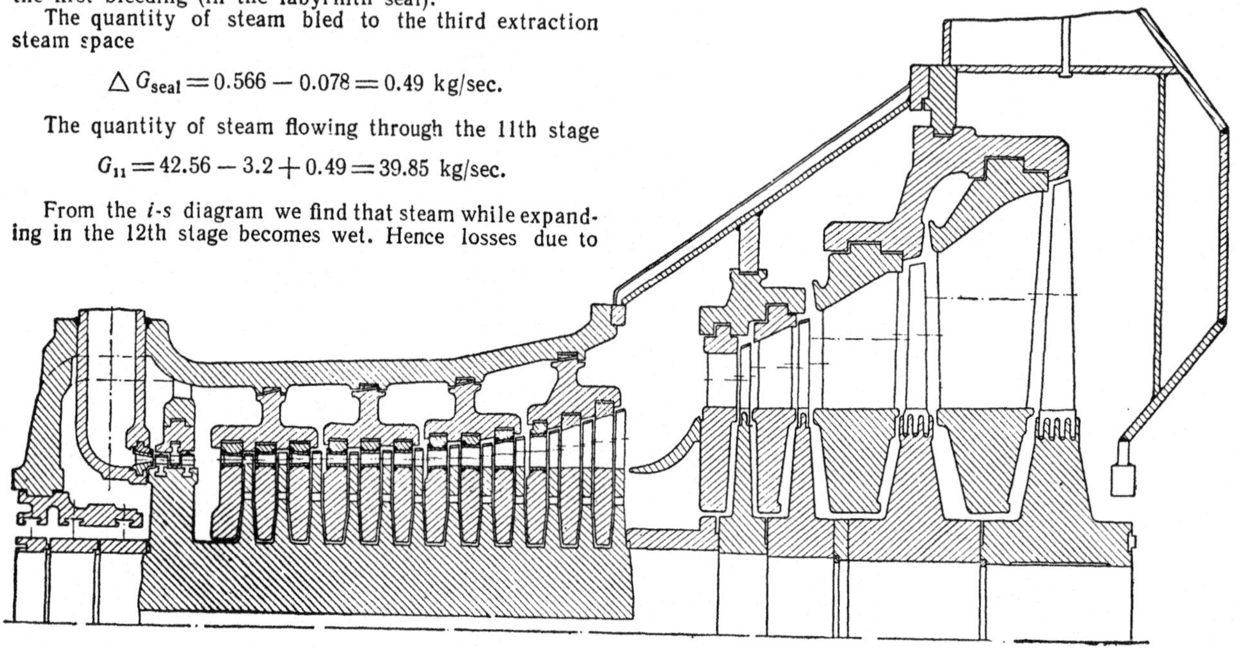

Fig. 7-26. Sectional view of impulse turbine VK-50

8	9	10	11	12	13	14	15	16	17
0.11	0.08	0.07	0.05	0.04	0.03	0.05	0.03	0.03	0.01
13.67	14.07	14.60	14.15	14.00	14.00	22.29	23.48	27.31	24.65
—	—	—	—	0.17	0.40	1.08	1.70	2.74	2.92
13.67	14.07	14.60	14.15	13.83	13.60	21.21	21.78	24.57	21.73
3.08	3.08	3.28	2.71	2.80	3.10	5.39	6.91	9.44	15.11
0.817	0.82	0.817	0.838	0.831	0.815	0.797	0.76	0.722	0.591
2,430	2,500	2,600	2,360	2,300	2,270	3,275	3,360	3,510	3,100
47.5	58.8	75	96	122	156	157	234	400	643
53.5	64.0	81.5	105	139	173.6	184	255	430	680

The quantity of steam flowing through the fourteenth stage

$$G_{14}=G_{13}-G_{ex}^{IV}=39.85-2.9=36.95 \text{ kg/sec.}$$

Total heat of steam before the nozzles of the fourteenth stage

$$i_0^{XIV}=i_0^{XIII}+h_e^{pr}-h_i^{XIII}=636.80-13.60=$$
$$=623.20 \text{ kcal/kg.}$$

On the i-s diagram the total heat at point a_{5t} is

$$i_{5t}=570 \text{ kcal/kg.}$$

Theoretical heat drop between the fourth and the fifth extractions (intercept a_5a_{5t})

$$h_{0V}=i^{XIV}-i_{5t}=623.20-570=53.20 \text{ kcal/kg.}$$

We shall have two stages between the fourth and the fifth extractions so that the average heat drop for each stage would be

$$h_{0\,mean}=53.2/2=26.6 \text{ kcal/kg.}$$

Since steam expands very rapidly in the low-pressure regions we shall have to make use of larger mean diameters as well as greater degrees of reaction in order to maintain a smooth curvature for the turbine shape.

For the fourteenth stage we shall finally choose a heat drop of 26.6 kcal/kg.

Preliminary estimates have shown that we may have mean diameters for the nozzles and blades as 1,500 and 1,520 mm respectively. The fourteenth stage is calculated in detail as was done in the case of the preceding thirteenth stage. Results of calculations are shown in Table 7-12.

The fifteenth stage:

Steam conditions after the blades of the fourteenth stage are

$$i_0^{XV}+h_e^{pr}=i_0^{XIV}-h_i^{XIV}=623.2-21.21=$$
$$=601.99 \text{ kcal/kg.}$$

Total heat of steam before the nozzles of the fifteenth stage is

$$i^{XV}=(i_0^{XV}+h_e^{XIV})-h_e^{XIV}=601.99-1.11=$$
$$=600.88 \text{ kcal/kg.}$$

Pressure before nozzles will be 0.74 ata and dryness fraction $x_1=0.939$.

If the carry-over velocity of the fourteenth stage were to be utilised in the next stage the pressure before the fifteenth stage would have been

$$p_{fict.}^{XV}=0.76 \text{ ata.}$$

Theoretical heat drop from 0.74 to 0.34 ata from the i-s diagram is

$$h_0^{XV}=i_0^{XV}-i_{1t}^{XV}=600.88-573.3=27.58 \text{ kcal/kg.}$$

Degree of reaction will be assumed as $\varrho=20\%$. Heat drop in the nozzles $h_{01}^{XV}=0.8\times 27.58=22.00$ kcal/kg.

Heat drop in the moving blades $h_{02}^{XV}=27.58-22.00=$ $=5.58$ kcal/kg.

Theoretical available energy in the nozzles is

$$h_{01}^{XV}+h_e^{XIV}=22.00+1.11=23.11 \text{ kcal/kg.}$$

From the i-s diagram pressure after the nozzles

$$p_1'=0.39 \text{ ata.}$$

Critical pressure $p_{cr}=0.577\times 0.76=0.438$ ata which indicates that the velocities obtained would be above the critical values. We shall therefore have a convergent nozzle with expansion above the critical taking place in the oblique exit portion of the nozzle.

Heat content of steam up to the critical pressure (from i-s diagram) is

$$i_{cr}=582.2 \text{ kcal/kg.}$$

Theoretical heat drop up to the critical pressure is

$$h_{0\,cr}=i_0^{XV}-i_{cr}=600.88-582.20=18.68 \text{ kcal/kg.}$$

Theoretical available energy up to the critical pressure is $h_{0\,cr}+h_e^{pr}=18.68+1.11=19.79$ kcal/kg.

Critical velocity in the nozzle throat

$$c_{cr}=91.5\sqrt{h_{0\,cr}+h_e^{pr}}=91.5\sqrt{19.79}=407 \text{ m/sec.}$$

Specific volume of steam at the throat $v_{cr}=3.52$ m³/kg.

Steam velocity at the nozzle exit

$$c_1=91.5\varphi\sqrt{h_{01}+h_e^{pr}}=91.5\times 0.975\sqrt{23.11}=429 \text{ m/sec,}$$

where $\varphi=0.975$ (from Fig. 5-2).

Theoretical steam velocity at nozzle exit

$$c_{1t} = c_1/\varphi = 429/0.975 = 440 \text{ m/sec.}$$

Heat losses in the nozzles

$$h_n = (1 - \varphi^2)\, c_{1t}^2/8{,}378 = (1 - 0.975^2)\ 440/8{,}378 = 1.13 \text{ kcal/kg.}$$

Specific volume of steam after nozzles $v = 4.02$ m³/kg.
Deflection of the jet in the oblique exit region

$$\sin\alpha_1' = \frac{c_{cr}}{c_1}\frac{v_1}{v_{cr}}\sin\alpha_1 = \frac{407}{429} \times \frac{4.02}{3.52}\sin 16° = 0.299;$$

$$\alpha_1' = 17°30' \text{ and}$$

$$\omega = \alpha_1' - \alpha_1 = 17°30' - 16° = 1°30'.$$

Exit nozzle heights

$$l_n = \frac{G_{XV}v_{cr}}{\pi \times d_{av}c_{cr}\sin 16°} = \frac{36.95 \times 3.52 \times 10^3}{\pi \times 1.58 \times 407 \times 0.2756} = 234 \text{ mm.}$$

All the remaining quantities are determined as was done for the preceding stages (Table 7-12).

Stages after the fifth extraction:

$$G_{XVI} = G_{XVII} = G_{XV} - G_{ex}^{V} = 36.95 - 2.72 = 34.23 \text{ kg/sec.}$$

$$p_{ex}^{V} = p_0^{XVI} = 0.34 \text{ ata;}$$

$$x_0^{XVI} = 0.913,\ i_0^{XVI} = 578.16 \text{ kcal/kg and}$$

$$i_0^{XVI} + h_e^{pr} = 580.21 \text{ kcal/kg.}$$

Total heat of steam at point i_{6t} is

$$i_{6t} = 512.70 \text{ kcal/kg.}$$

Theoretical heat drop in the last stage

$$h_{0VI} = i_0^{XVI} - i_{6t} = 578.16 - 512.70 = 65.46 \text{ kcal/kg.}$$

Assuming two stages with increasing heat drops we have discs (both for nozzles and moving blades) of large mean diameters and sufficiently large degrees of reaction.

For the sixteenth stage we shall assume a heat drop of $h_0^{XVI} = 31.96$ kcal/kg, mean dia. for nozzles 1,750 mm and mean dia. for blades as 1,780 mm.

Calculations for the sixteenth stage are carried out as done previously the results of which are given in Table 7-12.

T h e s e v e n t e e t h s t a g e:

Preliminary calculations have shown that in the nozzles and the moving blades of the last stage steam flows with a velocity greater than the critical value. Hence this stage will be calculated in greater detail.

Steam conditions before the stage are given by pressure 0.128 ata, dryness fraction 0.884 and total heat

$$i_0^{XVII} = (i_0^{XVI} + h_e^{XV}) - (h_i^{XVI} + h_e^{XVI}) = 580.21 - (24.57 + 2.94) = 552.70 \text{ kcal/kg,}$$

so that

$$i_0^{XVII} + h_e^{XVI} = 552.7 + 2.94 = 555.64 \text{ kcal/kg.}$$

Fictitious pressure $p_{fict} = 0.14$ ata.

Theoretical heat drop in the stage $h_0^{XVII} = i_0^{XVII} - i_{1t}^{XVII} = 552.7 - 518.8 = 33.9$ kcal/kg, where $i_{1t}^{XVII} = 518.8$ — total heat of steam after the moving blades (from the i-s diagram).

Assuming degree of reaction to be 50% heat drop in the nozzles will be

$$h_{01}^{XVII} = 0.50 \times 33.90 \approx 17.0 \text{ kcal/kg}$$

and in the moving blades

$$h_{02}^{XVII} = h_0^{XVII} - h_{01}^{XVII} = 33.9 - 17.0 = 16.9 \text{ kcal/kg.}$$

Available energy in the nozzles is equal to

$$h_{01}^{XVII} + h_e^{pr} = 17.0 + 2.94 = 19.94 \text{ kcal/kg.}$$

From the i-s diagram pressure after the nozzles

$$p_1' = 0.074 \text{ ata.}$$

Critical pressure for this group of nozzles

$$p_{cr}^{n} = 0.577 \times 0.14 = 0.809 \text{ ata.}$$

Total heat at critical pressure

$$i_{cr}^{n} = 538.8 \text{ kcal/kg.}$$

Heat drop in the nozzles up to the critical pressure

$$h_{0\,cr}^{XVII} = i_0^{XVII} - i_{cr}^{n} = 552.7 - 538.8 = 13.9 \text{ kcal/kg.}$$

Theoretical available energy for pressure drop up to critical pressure is equal to

$$h_{0\,cr}^{XVII} + h_e^{pr} = 13.9 + 2.94 = 16.84 \text{ kcal/kg.}$$

Critical velocity

$$c_{cr} = 91.5\sqrt{16.84} = 375.5 \text{ m/sec.}$$

Velocity of steam at the nozzle exit

$$c_1 = 91.5 \times 0.975\sqrt{19.94} = 398 \text{ m/sec}$$

and $c_{1t} = 398/0.975 = 409$ m/sec.
Heat losses in the nozzles

$$h_n = (1 - 0.975^2)\ 409^2/8{,}378 = 0.98 \text{ kcal/kg.}$$

Specific volume of steam after the nozzles

$$v_1 = 18.1 \text{ m}^3\text{/kg.}$$

Specific volume of steam at the critical pressure

$$v_{cr} = 16.8 \text{ m}^3\text{/kg.}$$

Jet deflection in the region of the oblique exit:

$$\sin\alpha_1' = \frac{375.5}{398} \times \frac{18.1}{16.8}\sin 22° = 0.383;$$

$$\alpha_1' = 22°30' \text{ and } \omega = 22°30' - 22° = 30'.$$

Nozzle heights at the throat section

$$l_n = \frac{34.23 \times 16.8 \times 10^3}{\pi \times 2.0 \times 375.5\ \sin 22°} = 650 \text{ mm,}$$

which excellently suits the turbine shape.

From the velocity triangles we have

$$w_1 = 162 \text{ m/sec; } \beta_1 = 70°20'.$$

Velocity energy of 1 kg of steam at entry to the moving blades expressed in heat units is equal to

$$h_{w_1} = w_1^2/8{,}378 = 162^2/8{,}378 = 3.14 \text{ kcal/kg.}$$

Plotting this heat equivalent of velocity energy on the i-s diagram (Fig. 7-24) we obtain the fictitious pressure $p^b_{fict.}$ before the moving blades

$$p^b_{fict.} = 0.081 \text{ ata}.$$

Critical pressure and specific volume of steam in the throat section of the moving blades are

$$p^b_{cr} = 0.577 \times 0.081 = 0.0467 \text{ ata}; \quad v^b_{cr} = 27.5 \text{ m}^3/\text{kg}.$$

Total heat of steam at critical pressure in the throat section of blades

$$i^b_{cr} = 523.5 \text{ kcal/kg (from the } i\text{-}s \text{ diagram)}.$$

Theoretical heat drop up to the critical pressure

$$h^b_{0\,cr} = i^{XVII}_0 - h^{XVII}_{01} + h^{XVII}_n - i^b_{cr} = 552.7 - 17.0 + 0.98 - 523.5 = 13.18 \text{ kcal/kg}.$$

Theoretical available energy in the moving blades up to the critical pressure

$$h^b_{0\,cr} + h_{w_1} = 13.18 + 3.14 = 16.32 \text{ kcal/kg}.$$

Critical velocity in the throat section of the moving blades $w_{cr} = 91.5\sqrt{16.32} = 370$ m/sec.

Height of the moving blades

$$l''_2 = \frac{34.23 \times 27.5 \times 10^3}{\pi \times 2.03 \times 370 \sin 36^\circ} = 680 \text{ mm},$$

which suits the requirements of smooth curvature for turbine shape.

The theoretical relative velocity at the exit of the moving blades $w_{2t} = \sqrt{162^2 + 16.9 \times 8{,}378} = 409$ m/sec. Actual velocity $w_2 = \psi w_{2t} = 0.915 \times 409 = 375$ m/sec. Assuming $\beta_2 = 36^\circ$ we shall find the jet deflection in the oblique exit region of the moving blades:

$$\sin \beta'_2 = \frac{370}{375} \times \frac{32}{27.5} \sin 36^\circ = 0.675;$$

$$\beta'_2 = 42^\circ 30'; \quad \omega = 42^\circ 30' - 36^\circ = 6^\circ 30'.$$

From velocity triangles we have

$$c_2 = 257 \text{ m/sec and } \alpha_2 = 99^\circ.$$

The ratio $d/l''_2 = 2{,}030/680 = 2.98$ which is permissible.

The remaining calculations are carried out as was done for the previous cases. Results of calculations are tabulated in Table 7-12.

From Table 7-12 we obtain the heat drop usefully utilised for doing mechanical work $\sum h_i = 276.39$ kcal/kg, and the internal energy of the turbine $\sum N_i = 48{,}460$ kW.

The relative internal efficiency of the turbine

$$\eta_{0i\,th} = \frac{\sum h_i}{H_{0\,th}} = 276 \times 39/334 = 0.828,$$

which is 0.6% higher than what had been assumed earlier.

While carrying out heat calculations the bleeding of steam from the high-pressure end labyrinth seal into the third extraction steam space has been taken into consideration. For all the stages starting from the 11th up to the seventeenth (both inclusive) the quantity of steam flowing through the stages has been increased by 0.49 kg/sec which has resulted in additional contribution to the power developed by the turbine, equal to

$$\Delta N_i = \frac{427 \, \Delta G_{seal} \sum\limits_{XI}^{XVII} h_i}{102} = \frac{427 \times 0.49 \times 130.43}{102} = 280 \text{ kW},$$

where

$$\sum_{XI}^{XVII} h_i = 14.15 + 13.83 + 13.6 + 21.21 + 21.78 + 24.57 + 21.73 = 130.43 \text{ kcal/kg}.$$

The internal power developed by the turbine neglecting the additional power from gland seal bled steam

$$N^T_i = \sum N_i - \Delta N_i = 48{,}460 - 280 = 48{,}180 \text{ kW}.$$

Power developed at the alternator terminals

$$N_e = N^T_i \eta_m \eta_g = 48{,}180 \times 0.995 \times 0.96 = 45{,}900 \text{ kW}.$$

The disparity between the calculated value of power at the alternator terminals with the assumed value is 900 kW which is 2% error.

Since the calculated internal efficiency for the turbine has been found to be 0.6% greater than the assumed value (0.828 — 0.822) the total design error is 2 — 0.6 = = 1.4% which is wholly permissible. Thus the above design would be deemed a satisfactory one for a turbine working at the stipulated steam parameters.

Chapter Eight

TURBINE PERFORMANCE AT VARYING LOADS

8-1. OPERATING CONDITIONS

The capacity at which a turbine operates with minimum specific heat per kW, and consequently, with maximum absolute efficiency is known as the most *economic* capacity for the turbine.

The limiting capacity at which a turbine can operate is known as the *nominal* capacity of turbine. Depending on the type of turbine the nominal capacity may be equal or greater than the most economic capacity by 10 to 25%.

It is usual to carry out thermal design of turbines on the basis of the most economic capacity desired for the turbine under consideration. For purposes of satisfactory heat drop distribution the ratio $(u/c_1)_{opt}$ is utilised.

The capacity of a turbine may, while under operation, vary between the wide limits of no load to full load. Varying power developed de-

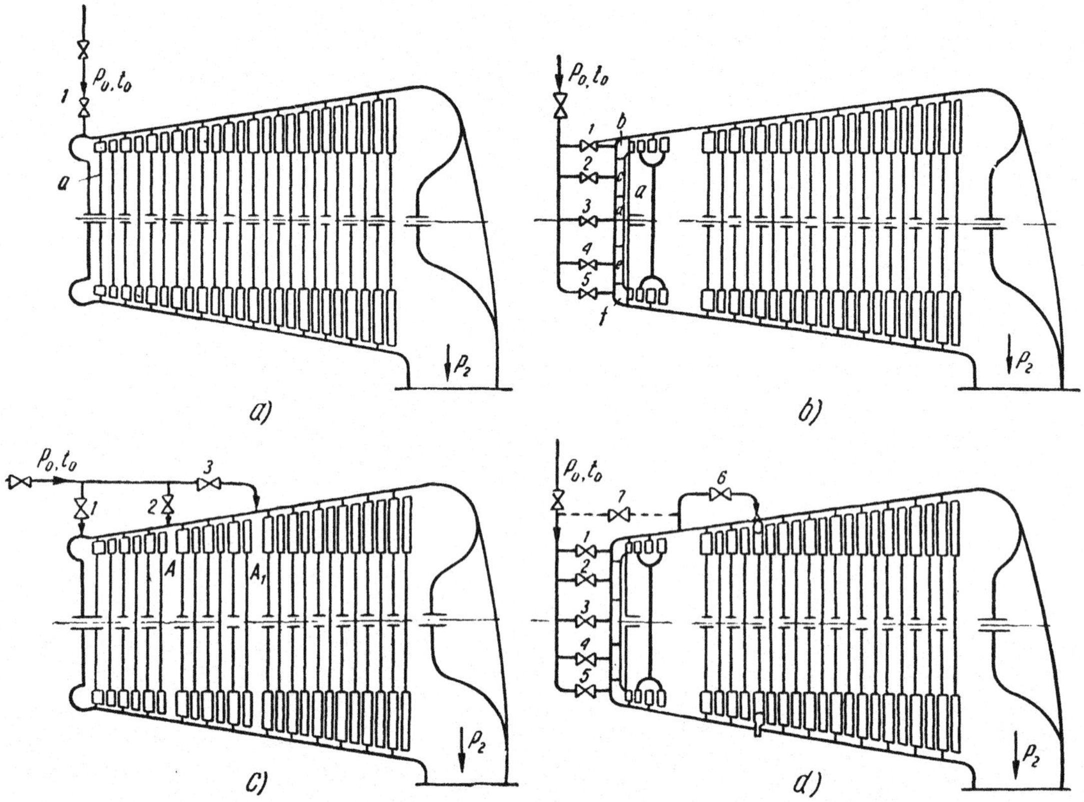

Fig. 8-1. Methods of steam supply to a turbine

pending on the varying loads on the turbine is brought about mainly by varying the quantity of steam passing through the turbine, D_0, and heat drop H_0^1 [equations (7-52) and (7-58)].

The principal methods of governing are as follows:

1) Throttling of fresh steam at entry to the turbine (throttle governing).

2) Varying the number of valve openings to the regulating stage of the turbine (nozzle control governing).

3) Supplying fresh steam directly to one or more intermediate stages (external bypass governing).

4) Bypassing steam from one intermediate stage to another (internal bypass governing).

Besides regulating valves one or two automatic stop valves are also provided before the turbine. If steam is delivered to the turbine through a single steam main one automatic stop valve is sufficient. However, if the quantity of steam flowing to the turbine is greater than 200 to 250 tons/hour then two stop valves are desirable. Under operating conditions, irrespective of the load on the turbine, these valves are kept completely open. Some of the methods of steam supply to a turbine are shown in Fig. 8-1.

8-2. THROTTLE GOVERNING

In throttle governing (Fig. 8-1,*a*) steam is delivered to all the nozzles placed along the periphery at one and the same time. The delivery of steam to the nozzles may be effected through one or two simultaneously opening throttle valves. In the high-pressure large-capacity turbines operating at very high initial temperatures built in recent times steam is fed to the turbine through as many as four separate mains having each a throttling valve mounted on it.

If fresh steam enters the turbine only through the throttle valve *1* then it would open to its maximum travel only if the turbine is operating at full load. In such cases steam consumption for nominal as well as the most economic capacities coincide. During partial loading of turbine the throttle valve opens only a fraction of its travel; thus at partial loads all the quantity of fresh steam fed to the turbine undergoes throttling accompanied by heat losses leading to a reduction of turbine efficiency.

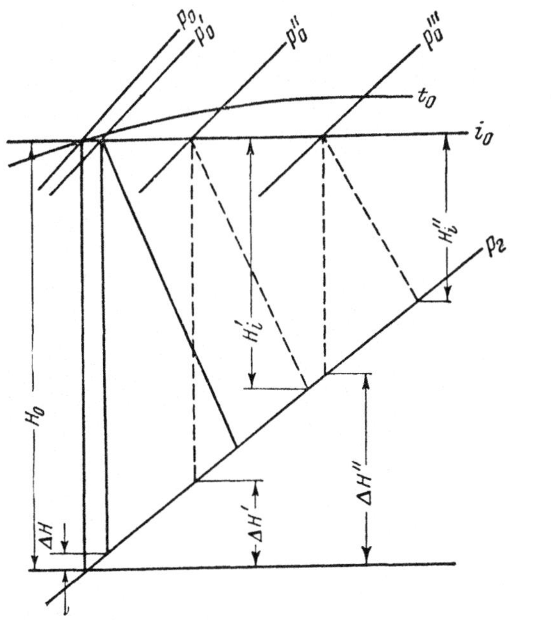

Fig. 8-2. Throttle governing of a steam turbine on the *i-s* diagram

Fig. 8-2 shows the throttling process on an *i-s* diagram. Full line indicates the heat drop process at nominal load and the broken lines indicate heat drop process at partial loads. With increased throttling the turbine capacity decreases and losses due to throttling increase in direct proportion with the reduction of load. The theoretical heat drop at low loads sharply decreases so that efficiencies at light loads are very low and uneconomical.

8-3. NOZZLE CONTROL GOVERNING

In the system of nozzle control governing fresh steam enters the first stage nozzles through several (3 to 10) valves known as either n o z z l e or r e g u l a t i n g v a l v e s. Diagrammatic arrangement of such a system with five valves is shown in Fig. 8-1, *b*. Each nozzle control valve regulates the supply of steam to its own group. Under conditions of operation at full load all the regulating valves are fully open, the degree of partial admission being usually less than unity ($\varepsilon \leq 1$).

When the load on the turbine varies the nozzle valves open or close in a definite consecutive order and hence the degree of partial admission varies with the load carried by the turbine. If these regulating valves are partially open throttling does take place as in the case of throttle governing, but, since through each of the regulating valves only a certain fraction of the total steam flows, the losses due to throttling are smaller than in the case of throttle governing where all the quantity of steam has to undergo throttling to the same extent. The efficiency of a turbine with nozzle control governing has been found to be much more stable at varying loads than if the turbine were to operate with throttle governing.

8-4. BYPASS GOVERNING

As a rule for turbines with throttle governing it is usual to make use of the system of bypass governing (Fig. 8-1, *c*), all the more so in the case of reaction turbines. Main throttle valve delivers steam to the nozzles of the first stage ($\varepsilon=1$). Usually the turbine would be developing the most economic capacity when this main throttle valve is fully open. The pressure before the nozzles of the first stage in this case would reach its maximum value, remaining practically constant for any further increase in load up to the nominal one. Increased supply of steam to cope with loads greater than the economical and up to the nominal is brought about by feeding fresh steam directly to one or more intermediate stages of the turbine (Fig. 8-1,*c*; chambers *A*, A_1).

When fresh steam enters the steam space at *A* there is an increase in the steam pressure which immediately results in an increased mass flow of steam through all the succeeding stages. The turbine capacity increases in spite of the fact that the quantity of steam flowing in the first few stages up to the bypass chamber decreases. If the turbine is provided with a single bypass valve *2*, the power developed by the turbine reaches the nominal value only when the valve is at its maximum opening position (the pressure in bypass chamber *A* reaches some limiting value). If the turbine is provided with two bypass valves (Fig. 8-1,*c*) the nominal capacity is reached only when the second bypass valve is opened fully. The mass flow of steam through all the stages situated after the bypass chamber A_1 increases and at the same time flow through stages up to the bypass chamber decreases. However, the increase of power developed in the stages after the bypass chamber takes place at a much faster rate than the decrease occurring in the stages up to chamber A_1, so that as an overall picture we have an increase in capacity up to the nominal value. At nominal loads steam pressure in chamber *A* must be less than the pressure ahead of the nozzles of the first stage; and the pressure in chamber A_1 must be less than that prevailing in chamber *A*, so that a sufficient quantity of steam continues to flow through the first few stages carrying away with it the heat generated in the stages while running idle (if there is a back pressure, as will be the case when using bypass governing, there is a tendency for the steam flow to stagnate resulting in almost idle running of stages ahead of the bypass chamber). However, in practice the initial stages do contribute to the developed power though only in a very small way.

With external bypass governing maximum efficiency is attained at economical loading, since at this load throttling losses are a minimum. At other loads, however, the efficiency sharply falls especially at partial loads. For turbines with throttle governing it is usual to have a single bypass unit. However, in rare cases two or even three bypass units may be made use of.

Although rare, turbines with nozzle control governing are also provided with bypass governing (Fig. 8-1,*d*). Here steam may be admitted simultaneously to both the first regulating as well as the third or even the fourth stage through valve *7*. Sometimes turbines with nozzle control governing are provided with bypassing of steam from one intermediate stage to another (internal bypassing), usually from the first stage steam chamber to an intermediate stage (valve *6* in Fig. 8-1,*d*). In Fig. 8-1,*d* external bypass governing is shown by the dotted line (valve *7*). Both internal and external bypass governing are, as a rule, not used simultaneously.

8-5. RELATION BETWEEN PRESSURE AND MASS FLOW OF STEAM IN A TURBINE STAGE UNDER VARYING LOAD CONDITIONS

Variations in the mass flow of steam affect the heat drop distribution as well as pressures prevailing in the turbine stages. Flugel analytically established a relation between the mass flow and pressures in a turbine stage, which for steam velocities less than critical at nozzle exits may be expressed by the equation

$$\frac{D}{D_0}=\sqrt{\frac{T_0}{T}}\sqrt{\frac{p_1^2-p_2^2}{p_{10}^2-p_{20}^2}}, \qquad (8\text{-}1)$$

where D_0 and D—mass flows of steam through the turbine conforming to the design conditions and the load under consideration;

T_0 and T—temperature in absolute degrees (°K) at design and prevailing loads;

p_{10} and p_1—steam pressures before the nozzles of the first or any other stage for design load and the load under consideration;

p_{20} and p_2—steam pressure after the moving blades of the last or any other stage conforming to the design load and the load under consideration.

The ratio $\sqrt{T_0/T}$ is usually very nearly unity and hence the following modification of equation 8-1 is used

$$\frac{D}{D_0}=\sqrt{\frac{p_1^2-p_2^2}{p_{10}^2-p_{20}^2}}. \qquad (8\text{-}2)$$

For turbines operating with high vacuum the quantities p_{20} and p_2, in view of their being very small, may be neglected.

Hence for condensing turbines we may write

$$\frac{D}{D_0}=\frac{p_1}{p_{10}}, \qquad (8\text{-}2a)$$

or

$$p_1=\frac{D}{D_0}p_{10}. \qquad (8\text{-}2b)$$

From equation (8-2b) it follows that for condensing turbines the steam pressure before the nozzles of any stage is a straight line function of the mass flow of steam. Equation (8-2b) can be utilised for determining steam pressures in any given turbine stage irrespective of whether the steam flows are subsonic or supersonic (in the last case not only for condensing but also for back-pressure turbines and in general for any given group of turbine stages). Equations (8-1), (8-2) and (8-2b) may be used with sufficient accuracy for groups of stages consisting of not less than three stages.

It may be, however, noted that the above equations may be used only when the areas of steam flow for all the stages of the group under consideration remain unaltered.

8-6. OPERATION OF TURBINES AT VARYING LOADS

Equation (8-2b) shows that steam pressure in a stage varies in a direct proportion with the variation of mass flow of steam. Thus the variations of pressures in a stage of condensing turbine operating at high vacuum as a function of mass flow of steam may be represented by a pencil of rays emanating from the origin. Fig. 8-3,*a* shows the lines of pressure variation for four intermediate stages of a condensing turbine. Points a_1, a_2, a_3 and a_4 show the pressures for design conditions (economic load) whereas points a'_1, a'_2, a'_3 and a'_4 represent pressures at maximum load (maximum continuous rating).

Pressure variations for a turbine with back pressure or deteriorated vacuum as a function of mass flow of steam may be determined from equation (8-1) or approximated from equation (8-2). If these equations are solved relative to p_1, for various values of D with $p_2=p_{20}$=const, we have

$$p_1=\sqrt{\frac{T}{T_0}\left(\frac{D}{D_0}\right)^2(p_{10}^2-p_{20}^2)+p_2^2} \quad (8\text{-}2c)$$

or

$$p_1=\sqrt{\left(\frac{D}{D_0}\right)^2(p_{10}^2-p_{20}^2)+p_2^2}. \quad (8\text{-}2d)$$

If for various values of D pressures in various stages are calculated from one of the above equations and graphs are plotted for p_1 as a function of D we then have a family of curves with their origin at point b (Fig. 8-3,*b*). This figure shows the pressure variation curves only for four of the intermediate stages where points b_1, b_2, b_3 and b_4 represent design pressures and b'_1, b'_2, b'_3 and b'_4—pressures at D_{max}.

Very often it is required to determine the mass flow of steam through a turbine or a group of stages for a constant initial pressure and a varying back pressure. In such cases solving equation (8-1) or (8-2) for D we have

$$D=D_0\sqrt{\frac{T_0}{T}\,\frac{p_1^2-p_2^2}{p_{10}^2-p_{20}^2}} \quad (8\text{-}2e)$$

or

$$D=D_0\sqrt{\frac{p_1^2-p_2^2}{p_{10}^2-p_{20}^2}}. \quad (8\text{-}2f)$$

If it is assumed that $p_1=p_{10}$=const then for various back pressures p_2 D may be determined from one of the above equations. Curve a_0c_0 represents D as a function of p_2 for condensing turbines and a_0c_0c for back pressure turbines (Fig. 8-3, *a* and *b*).

The straight line c_0c in Fig. 8-3,*b* shows that the limiting mass flow of steam D_{lim} through a turbine occurs when the back pressure is reduced. With such a flow through the turbine steam at the exit sections of nozzles as well as moving blades attains a pressure equal to the critical value p_{cr}. Hence for all back pressures less than p_{cr} the mass flow through the turbine remains unchanged.

From equations (8-1) and (8-2) any of the quantities comprising these equations such as pressure in a stage, mass flow through the turbine or group of stages may be determined for various operating conditions. As an example let us consider the operation of a condensing turbine at varying loads, with throttle governing coupled with a single bypass unit connected to the fourth stage steam space. We shall suppose that the quantities D_0, D_{max}, p_{20}, steam pressure in the bypass chamber p_{bpo} and steam pressure before the nozzles of first stage $p_{10}=p_1=p'_0$ are known from previous calculations. It is required to determine steam pressure in the bypass chamber and mass flow of steam through the bypass valve for all operating loads.

Pressure variations in the bypass chamber for mass flow D between the limits of zero and D_{max} with $p_{20}=p_2$ will be obtained from equation (8-2d)

$$p_{bp}=\sqrt{\left(\frac{D}{D_0}\right)^2(p_{bpo}^2-p_{20}^2)+p_2^2}.$$

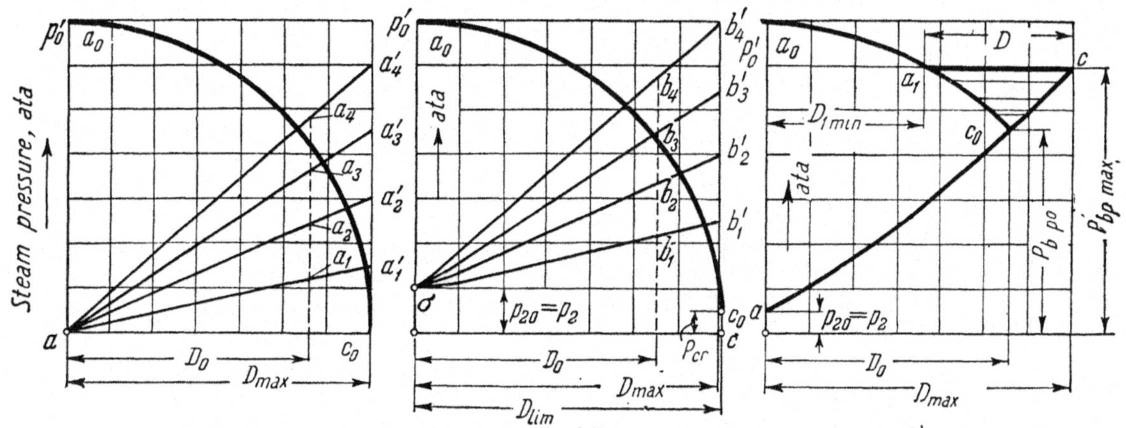

Fig. 8-3. Pressure variation in a turbine stage as a function of mass flow

p_{bp} as a function of D is shown in Fig. 8-3, c (curve ac_0c). Point c_0 represents the conditions of economic operation. Line ac_0 shows the variation of pressures in the bypass chamber for steam flows from 0 to D_0. Line c_0c shows the character of pressure variation in the bypass chamber for steam flows from D_0 to D_{max}. For the first three stages when the bypass chamber pressure $p_{bp} > p_{bp_0}$ the mass flow of steam $D_1 < D_0$. The magnitude of this flow quantity is obtained from equation (8-2f)

$$D_1 = D_0 \sqrt{\frac{(p_0')^2 - p_{bp}^2}{(p_0')^2 - p_{bp0}^2}},$$

from which it follows that when $p_{bp} = p_0'$ (point a_0) $D_1 = 0$ and $D_1 = D_0$ when $p_{bp} = p_{bp_0}$ (point c_0). Determining D_1 for several of the pressures p_{bp} lying between p_{bp_0} and p_0' we may plot the curve $a_0a_1c_0$. From the graph it is seen that the quantity of steam flowing through the first three stages, when p_{bp} is a maximum ($p_{bp\ max}$), (bypass valve completely open) equals D_{1min}. Quantity of steam flowing through the stages situated after the bypass chamber equals D_{max}. Flow of steam through the bypass valve, fully open, is obtained as the difference $D = D_{max} - D_{1min}$. When the load carried by the turbine is less than the nominal value the quantity of steam flowing through the bypass valve will be given by the intercepts between lines a_1c_0 and cc_0.

8-7. HEAT CALCULATIONS FOR VARYING LOADS

Turbine design for varying load conditions is usually carried out for a known value of D. Assuming some value for the relative internal efficiency η_{oi} the theoretical heat drop process is outlined on the i-s diagram. Steam velocities in the various stages may be either higher or lower than critical. Depending on the velocity of steam flow from the nozzles or guide blades different methods are used for the detailed calculations of a turbine stage.

1. Steam Velocity Less Than Critical

This method is based on the equation of continuity applied at the exit sections of nozzles and moving blades. The various stage dimensions are calculated in reverse order, i. e., starting with the last stage. For the steam conditions at the point A_2' (Fig. 8-4, a, b) friction and windage losses $h_{fr.w.}$ are calculated and plotted downwards from point A_2' (point B). For the last stage of a condensing turbine, friction and windage losses are insignificant and hence are usually neglected.

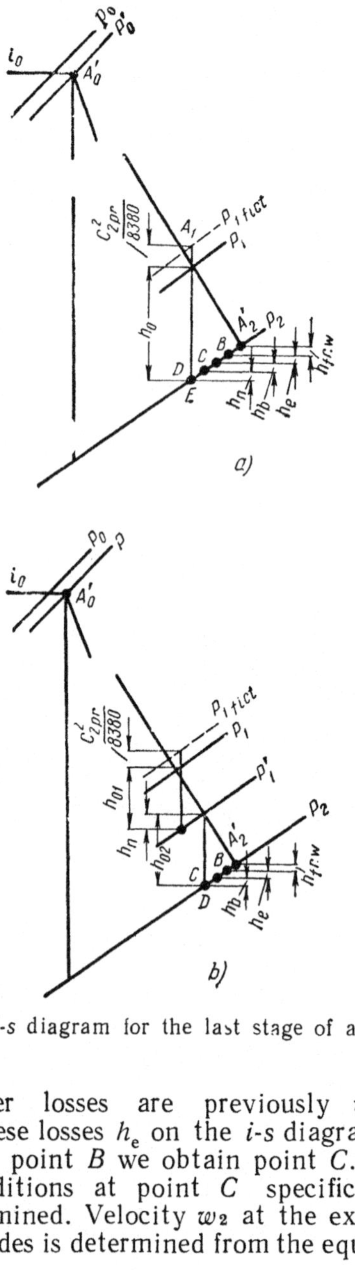

Fig. 8-4. I-s diagram for the last stage of a turbine

Carry-over losses are previously assumed. Marking these losses h_e on the i-s diagram downwards from point B we obtain point C. For the steam conditions at point C specific volume v_2 is determined. Velocity w_2 at the exit of the moving blades is determined from the equation

$$w_2 = \frac{Gv_2}{\pi dl'' \sin \beta_2}.$$

Drawing velocity triangles from the values of w_2 and β_2 velocity c_2 and losses h_e' are determined. If h_e' does not equal h_e recalculations have to be carried out. For a pure impulse turbine heat losses in the moving blades are determined as h_b and are further added from point C to obtain point D.

For the steam conditions at point D specific volume v_1 is determined for the exit sections of the nozzles. Velocity c_1 is obtained from the equation

$$c_1 = \frac{Gv_1}{\pi dl \sin \alpha_1}.$$

Next heat losses in the nozzles h_n are determined; setting off these losses downwards from the point D point E is obtained (Fig. 8-4,*a*). Numerical value of the theoretical available energy $c_1^2/8{,}378\varphi^2 = h_0 + c_{2pr}^2/8{,}378$ is then set off vertically from point E to A_1 as shown in Fig. 8-4,*a*.

Velocity w_1 is obtained from the velocity triangle. If it is found that $w_2^1 = \psi w_1$ then stage calculations are deemed to be satisfactory. If the value of w_2 obtained from continuity conditions is less than ψw_1 then the exit section of the moving blades would not be completely filled by the flowing steam. In such cases it is necessary to determine the actual exit velocity of steam $w_{2act} = \psi w_1$ and drawing velocity triangles with this value w_{2act} obtain c_2, efficiency and all the losses occurring in the stage.

Calculations for all the subsequent stages are carried out in an analogous manner. If it is supposed that the carry-over velocity from the preceding stage is not utilised in the succeeding stage then point A_1 must be placed on the actual pressure line instead of on the fictitious one. If, however, the carry-over velocity is used in the next stage point A_1 must be on the fictitious isobar, p_{1fict}., as shown in Fig. 8-4,*a*. In such cases if the actual pressure before the nozzles is to be determined the carry-over velocity must be presupposed so that measuring off the amount $c_{2pr}^2/8{,}378$ downwards from point A_1 we obtain the actual pressures prevailing before the nozzles. The accuracy of the supposition may be verified when calculating the succeeding stage.

Most of the present-day impulse turbines have some amount of reaction in their moving blades. Here the moving blade calculations would be carried out in a manner similar to the one described above.

For the calculation of nozzles it is necessary to assume the heat drop that would occur in the moving blades h_{02} and set off this value from the point D, as shown in Fig. 8-4,*b*, upwards to meet the isobar p_1'. For the steam conditions conforming to the point of intersection of h_{02} and p_1' specific volume v_1 is obtained from which velocity c_1 is calculated. Velocity w_1 is obtained from the velocity triangle constructed from the data c_1 and α_1.

The actual heat drop occurring in the moving blades is obtained from the equation

$$h_{02} = \frac{1}{8{,}378}\left[\left(\frac{w_2}{\psi}\right)^2 - w_1^2\right].$$

If the value of h_{02} obtained from the above equation differs from the assumed value it is necessary to obtain a second approximation or even a third approximation if found necessary.

Heat drop in the nozzles is obtained from the equation

$$h_{01} = \frac{1}{8{,}378}\left[\left(\frac{c_1}{\varphi}\right)^2 - c_{2pr}^2\right],$$

where c_{2pr}—velocity of steam at the exit of the preceding stage (carry-over velocity).

Next the nozzle losses h_n are calculated and plotted on the *i-s* diagram along the isobar p_1' (downwards).

The theoretical available energy is plotted upwards from the constant pressure line p_1'. The magnitude of $c_{2pr}^2/8{,}378$ is previously assumed from which steam pressure before the nozzles of the last stage is determined. Calculations for all the intermediate stages are carried out in a manner similar to that of the last stage.

2. Steam Velocities above the Critical Value

At critical and supercritical velocities heat calculations for a turbine stage may be carried out in the following manner.

Mass flow of steam as a function of pressure may be expressed by the following equation

$$G = 203 f_{min} \sqrt{\frac{p_{1f}}{v_{1f}}}, \qquad (8\text{-}3)$$

where p_{1f} and v_{1f}— actual pressure and specific volume of steam before the nozzles of any given stage if the energy content at entry to nozzles is equal to zero or fictitious pressure and specific volume of steam if energy at entry to nozzles is not equal to zero (p in kg/cm² and v in m³/kg).

From equation (8-3) it follows that with the nozzle sections remaining unaltered a decrease in pressure p_{1f} is accompanied by an increase in v_{1f} which leads to a decrease in the mass flow G. If p_{1f} increases the reverse process occurs, i.e., v_{1f} decreases and there is a consequent increase in the mass flow of steam. Equation (8-3) may be further expressed as

$$\frac{p_{1f}}{v_{1f}} = \left(\frac{G}{203 f_{min}}\right)^2. \qquad (8\text{-}4)$$

Knowing the throat area for the nozzles (f_{min}) and the mass flow G through the turbine the ratio p_{1f}/v_{1f} may be determined for all the turbine stages.

Pressure before the nozzles is determined with the help of an auxiliary graph. This auxiliary graph is plotted by presupposing some efficiency for the turbine η_{0i} and representing the heat drop process on the *i-s* diagram (Fig. 8-5). A

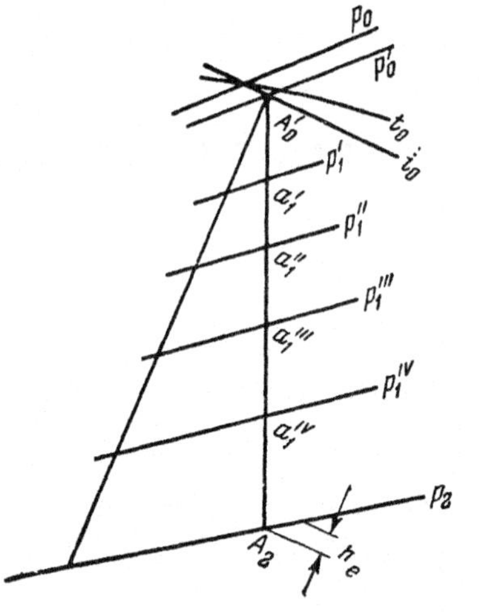

Fig. 8-5. Theoretical heat drop process on the i-s diagram

number of arbitrary isobars is marked off on the line $A_0'A_2$ as shown in Fig. (8-5) by points p_0^I, p_1^I, p_1^{II}, p_1^{III}, etc. For the state points A_0', a_1^I, a_1^{II}, a_1^{III}, etc., the specific volume of steam v_0^I, v_1^I, v_1^{II}, etc., are determined, as also the ratios p_0^I/v_0^I, p_1^I/v_1^I, p_1^{II}/v_1^{II}, p_1^{III}/v_1^{III}, etc. From the last named ratios and the values of the arbitrary pressures an auxiliary graph is plotted as shown in Fig. 8-6 (curve *a*). The ratio p_{1f}/v_{1f} is set off along the ordinates (obtained from equation 8-4) and pressures before nozzles are obtained along the abscissa for any given stage, e.g., p_{1f}^I, p_{1f}^{II}, p_{1f}^{III}, etc.

For the stage pressures thus obtained detailed heat calculations are carried out to determine the various stage dimensions, and in the process heat losses and efficiency are also determined.

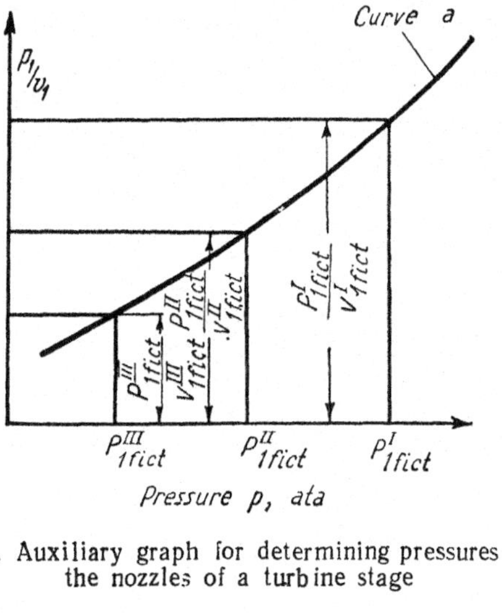

Fig. 8-6. Auxiliary graph for determining pressures before the nozzles of a turbine stage

3. Some Special Features of Heat Drop Calculations

The thermal design of a turbine is carried out on the basis of steam velocities prevailing in the various stages with the help of one of the methods mentioned above. For turbines with throttle governing, without any additional supply of steam directly to one or more of the intermediate stages, i.e., without bypass governing coupled with it, heat calculations for any of the stages are not difficult. However, with bypass governing heat calculations for the first few stages have some special features associated with them; so also for the heat calculations for the first regulating stage of a turbine with nozzle control governing. For turbines with nozzle control governing at off-design loads some of the control valves may be partially open (with varying degrees of opening for the different valves) so that there is a throttling effect on the steam passing through the control valves. These special features will be considered for some particular cases as follows:

a) Turbine with Throttle Governing Coupled with Bypass to One of the Intermediate Stages

We shall take a concrete case of a turbine where the nominal capacity is attained by supplying fresh steam to the nozzles of the fifth stage. We shall assume that the conditions of operation under consideration are such that the mass flow of steam through the turbine is D_1. Heat calculations for all the stages situated after the bypass chamber are carried out on the basis of the approximate equation.

Quantity of steam flowing through the first four stages is given by the equation

$$D_2 = D_0 \sqrt{\frac{p_1^2 - (p_2^{IV})^2}{p_{10}^2 - (p_{02}^{IV})^2}}, \qquad (8\text{-}5)$$

where p_1, p_{10}—steam pressures before the nozzles of the first stage at the design load and the partial load under consideration (in the present case they are nearly equal).

D_0—calculated value of mass flow of steam through the turbine at economic load.

Quantity of steam flowing through the bypass valve will be equal to

$$D_{bp} = D_1 - D_2.$$

Fig. 8-7 shows the heat drop process on the i-s diagram for the partial load conditions considered above (dotted lines indicate the heat drop process for design load and full lines the process at nominal load with bypass valve fully open).

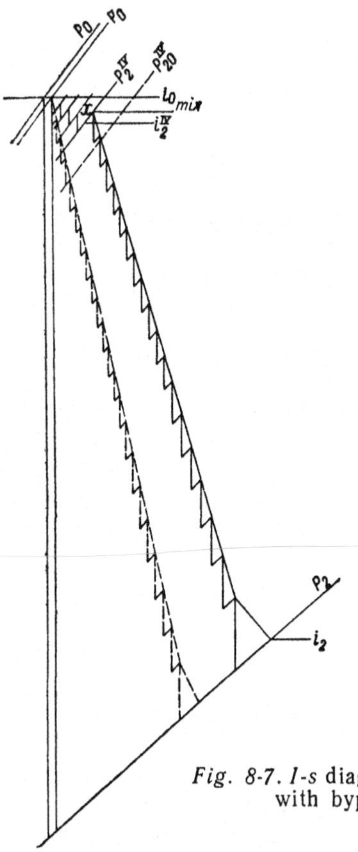

Fig. 8-7. I-s diagram for varying loads with bypass governing

Heat content (enthalpy) of the steam mix before the nozzles of the fifth stage (point x) is given by the equation

$$i_{\text{mix}} = \frac{D_2 i_2^{\text{IV}} + D_{\text{bp}} i_0}{D_1}. \qquad (8\text{-}6)$$

Detailed calculations of the first four stages must be carried out on the basis of equation of continuity and those after the bypass chamber by any one of the above described methods in conformity with the prevailing steam velocities.

The power developed by the turbine for the conditions of operation under consideration is given by the equation

$$N_{e1} = \frac{D_2 (i_0 - i_2^{\text{IV}}) + D_1 (i_{\text{mix}} - i_2)}{860} \eta_m \eta_g \qquad (8\text{-}7)$$

where η_m, η_g—mechanical and generator efficiencies.

b) Turbines with Nozzle Control Governing

Let us consider a turbine operating at partial loads, assuming that the mass flow of steam at this load is known and equal to D_1. Calculations for the various pressure stages are carried out by one of the methods already described. Thus the steam pressure in the regulating stage chamber p_r is determined from pressure stage calculations (Fig. 8-8). The dotted lines indicate the heat drop process for design load and full lines for the partial load under consideration (mass flow of D_1).

The quantity of steam flowing through the turbine is adjusted in such a way that two of the regulating valves are fully open, the third one being open partially. Quantity of steam flowing through the two fully open valves, assuming critical velocity of steam in the nozzles, will be given by

$$D_{1,2} = 3{,}600 \times 203 f_{\min 1,2} \sqrt{\frac{p_0'}{v_0'}},$$

where $f_{\min 1,2}$—throat sections of nozzles receiving steam from the first two valves.

The quantity of steam flowing through the partially open valve will, therefore, be

$$\Delta D = D_1 - D_{1,2}.$$

Thus regulating stage calculations are divided into two parts: calculations for the group of nozzles served by the two fully open control valves, and the group of nozzles served by the partially open control valve. Theoretical heat drop for the group of nozzles receiving steam from the first two valves is equal to $h_{0(1,2)}$ (Fig. 8-8). On the basis of this heat drop calculations are carried out for that group of nozzles and moving blades which are served by the first two fully open control valves. Steam conditions after the moving blades for the arc segment over which these nozzles are situated are obtained from the above calculations (point a_1' Fig. 8-8); so also the enthalpy i_2'.

For the second group of nozzles steam pressure before the nozzles must be determined first. Depending upon the degree of throttling steam velocity at the exit of the nozzles may be less, equal or more than the critical value. If $c_1 \geqslant c_{cr}$, p_0'' is obtained from the constant enthalpy line i_0=const.

This pressure must satisfy the equation

$$p_0'' = \left(\frac{\Delta G}{203 f_{\min 3}}\right)^2 v_0'',$$

where $\Delta G = \frac{\Delta D}{3{,}600}$ — mass flow per second;

$f_{\min 3}$—throat section of nozzles served by the partially open control valve;

v_0''—specific volume conforming to pressure p_0''.

Solution of this equation with two unknowns is carried out by the method of successive approximations. Presupposing a certain value for p_0'', v_0'' is determined (from steam tables). These

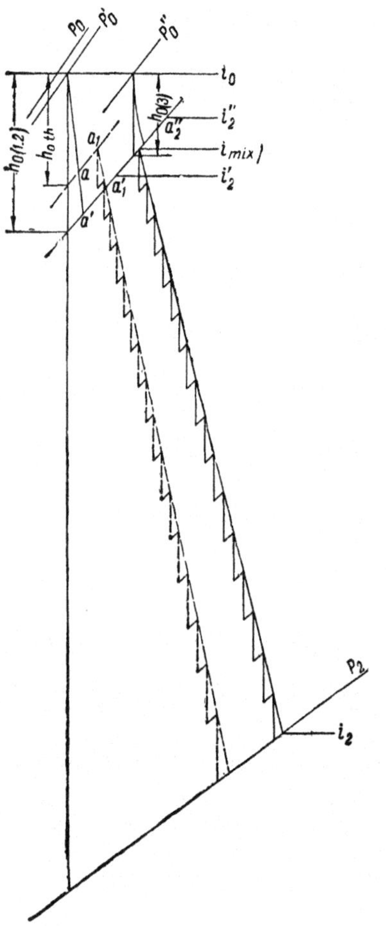

Fig. 8-8. I-s diagram for varying loads with nozzle control governing

values are next substituted in the above equation. If the equation is satisfied the value assumed for p_0'' is taken as the correct one. If not, a different value for p_0'' has to be assumed and the process repeated until a satisfactory solution is obtained.

If $c_1 < c_{cr}$ calculations are made on the basis of the equation of continuity. Steam conditions after the moving blades are presupposed. If after calculations it is found that the steam conditions before the nozzles do not agree with the assumed figure (point a_0 on the line $i_0=\text{const}$) successive approximations have to be carried out as has been described above until a satisfactory solution is obtained.

Steam conditions after the moving blades for the arc segment, where the nozzles of the third group are situated, are given by point a_2'' (Fig. 8-8); enthalpy here being i_2''.

Enthalpy of steam mixture before the nozzles of the second stage is obtained from the equation

$$i_{\text{mix}} = \frac{D_{1,2} i_2' + \Delta D i_2''}{D_1}. \qquad (8\text{-}8)$$

Power developed by the turbine

$$N_{e1} = \frac{D_{1,2}(i_0 - i_2') + \Delta D\,(i_0 - i_2'') + D_1\,(i_{\text{mix}} - i_2)}{860} \times \eta_m \eta_g, \qquad (8\text{-}9)$$

where i_2—enthalpy of steam after the moving blades of the last stage.

Chapter Nine
GOVERNORS AND GOVERNOR GEARS

9-1. BASIC CONCEPTS

The mechanical work done on the turbine shaft is converted into electrical energy in the alternator from the terminals of which electrical power is supplied to the consumer. Thus a change in the demand of the consumer, i.e., load on the alternator terminals, must influence the mechanical work done on the turbine shaft. If the turbine shaft is directly or through transmission gearings connected to a driven mechanism (pump, blower, etc.), a change in the load of these mechanisms must be a direct consequence of the change in the work done on the turbine shaft. A turbine should have the ability to operate with sufficient stability in the very wide range of no load to full load. Since there is a direct relation between the power developed by a turbine and the mass flow of steam through it, it immediately follows that any variation of load at the alternator terminals would directly affect the rate of steam flow, increasing or decreasing as the case may be depending on whether the load builds up or decreases. Under conditions of constant load there is a definite and constant relation between the turning moment exerted by the moving blades and the quantity of steam flowing through the turbine. When the load changes this relation does not hold good since the turning moment exerted exceeds the load (in case of reduction of load) so that there is an immediate increase in the speed of rotation of the turbine shaft. This process would continue until the mechanism controlling steam supply to the turbine does not come into play. Once the steam supply comes under control the turning moment again is equated to the required load and the speed is brought back to the normal figure. In general the equation of moments for the rotors of a turbo-alternator may be expressed as follows

$$M_c = M_e + M_{losses} + (I_t + I_g)\frac{d\omega}{dt}, \quad (9\text{-}1)$$

where M_c—turning moment at the turbine coupling, in kgm;

M_e— turning moment converted into electrical energy at the alternator terminals, in kgm;

M_{losses} — braking moment in bearings arising out of friction as well as heat losses in the alternator as windage, etc., in kgm;

I_t, I_g—moment of inertia of turbine and alternator rotors, in kgm sec^2;

$\frac{d\omega}{dt}$—angular acceleration of rotors ($d\omega$—an infinitely small increase in the angular velocity), in 1/sec^2.

When the turbine is operating at constant load (constant r.p.m.) angular acceleration $d\omega/dt=0$ and equation (9-1) becomes

$$M_c = M_e + M_{losses} \quad (9\text{-}2)$$

which may also be written as

$$\frac{M_c\omega}{102} = \frac{M_e\omega}{102} + \frac{M_{losses}\omega}{102}$$

$$\text{or } N_c = N_e + N_{losses} \text{ (kW)} \quad (9\text{-}3)$$

or as a generalisation

$$N_c = N_e + N_{losses} + (I_t + I_g)\frac{\omega}{102}\frac{d\omega}{dt}, \quad (9\text{-}3a)$$

where N_c—effective power developed at the turbine coupling, kW;

N_e—useful electrical power developed at the alternator terminals, kW;

N_{losses} —frictional losses in bearings and windage losses in alternator, kW;

N_e is governed by the load required by the consumer and

N_c is a direct function of D and H.

At the instant when load N_e on the alternator terminals changes an inequality is established $N_e + N_{losses} \gtreqless N_c$ which either speeds up or slows down the turbine shaft. An increase in N_e leads to a reduction of shaft speed n and vice versa. Thus any load variation on the alternator is accompanied by a change of speed (r.p.m.). It is therefore the function of the speed regulator to automatically effect a balance for equation (9-3) at any given load N_e.

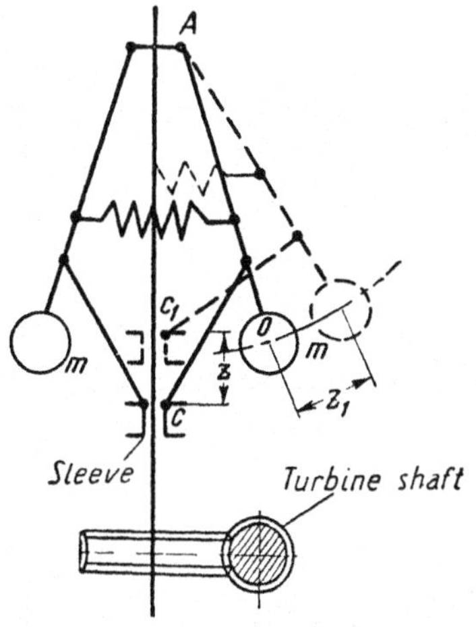

Fig. 9-1 Centrifugal governor

In turbines with automatic speed governing the steam regulating mechanisms are connected to the speed regulators. Transmission of speed variation impulse to the steam regulators is effected by various means. Operation of steam regulators is controlled by speed governors, which are themselves controlled by the centrifugal force of rotation which again varies directly as the r.p.m. of the turbine shaft.

Fig. 9-1 shows the principal parts for a centrifugal governor. With increase in shaft speed the weights m fly apart under the influence of centrifugal force, changing their position along the arc of radius AO through a distance z_1. The governor sleeve is displaced from its original position c to a new one c_1, i.e., through a distance z. If, on the other hand, speed of rotation diminishes the weights m would draw closer and the sleeve would be pushed down along the governor spindle. This displacement of the governor sleeve with variation of shaft speeds is utilised for effecting the required change in the steam supply both qualitatively and quantitatively. Speed governors are usually coupled to the turbine shaft through a toothed gearing.

The sleeve displacement of the speed regulator may be transmitted to the control valves in various ways, mechanical (levers and links, etc.) or hydraulic (with pressurised oil).

The principle of operation of a hydraulic- or oil-regulated speed governor is based on the ratio between oil pressure (oil under pressure for regulation purposes is supplied by a centrifugal pump mounted on the turbine shaft) and the speed of the turbine shaft.

Some of the methods of speed governing for steam turbines are described below.

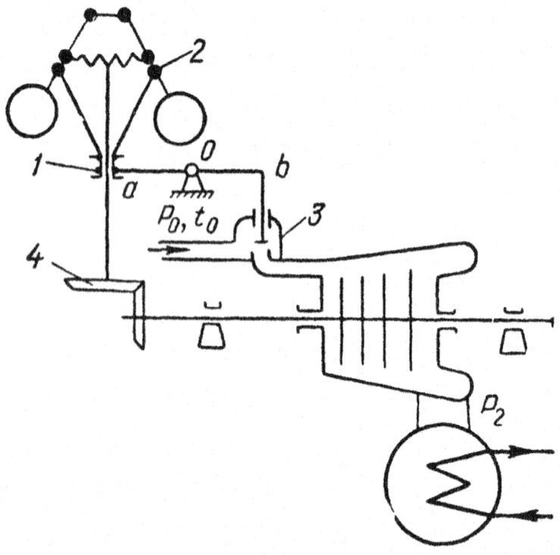

Fig. 9-2. Direct regulation

9-2. DIRECT REGULATION

General arrangement of direct speed regulation is shown in Fig. 9-2. Sleeve *1* of the centrifugal governor *2* is directly connected to the turbine throttle valve *3* by means of a lever arrangement having its fulcrum at *O*. When at the alternator terminals load increases, balance between the turning moment of the turbine shaft and the external load is disturbed leading to a reduction of the shaft speed. The centrifugal speed governor is connected to the turbine shaft through toothed gearings *4*. When the shaft speed falls the weights of the centrifugal regulator draw closer and sleeve *1* is displaced downwards and in the process displaces the end of the lever *a* connected with it. The lever *ab* is therefore rotated about its fulcrum *O*. Throttle valve *3* connected to the other end of the lever gets opened and the turbine develops greater power to cope with the increased load, however, operating at a much lower speed. Similarly in case of a reduction of load on the turbine shaft speed increases and the throttle valve starts closing. The advantage of the above system of regulation is the simplicity of its construction. However, the centrifugal force the weights of the speed governor could exert is very small (a few kilograms), so that simple direct speed governing may be used only for turbines of small capacity (from 50 to 60 kW) with light regulating valves not requiring a large power for their operation by the centrifugal speed governor. For turbines of mèdium and large capacities the force required to operate the regulating valves is substantially great, indispensable, since large resistances such as valve weights and spring resistances have to be overcome. Use of servomotors enables the generation of the required force for the operation of large regulating valves of medium and large capacity turbines.

9-3. INDIRECT SYSTEM OF GOVERNING

Figure 9-3 shows one of the indirect methods of governing using a servomotor of the piston type.

Under conditions of constant operation piston *6* of pilot valve *5* of servomotor *7* occupies a position midway between its travel, in which both the inlet and exit ports of the pilot valve connecting it to the servomotor are closed. Regulating valve *9*, for these conditions, also occupies a certain fixed position.

Any displacement of sleeve *2* of centrifugal speed regulator *1* causes a displacement of regulating valve piston *6*. In conformity with the direction of displacement of piston *6* oil under pressure from the oil pump *4* enters either chamber K or K_1 of the servomotor *7*. If oil enters the upper half portion, i. e., chamber K, valve *9* starts closing reducing the quantity of steam flow through the turbine (power developed by the turbine reduces). At the same time oil from chamber K_1 starts flowing out through the pilot valve port to drain. If oil under pressure enters chamber K_1 an exactly opposite process is effected opening the regulating valve *9*.

For the particular case under consideration, displacement of piston *6* does not require a large amount of force since it is balanced by oil pressures in the central chamber K_0 of the pilot valve. The amount of displacing force required for opening or closing of the regulating valve largely depends upon the size of piston *8* of servomotor *7* and oil pressures used.

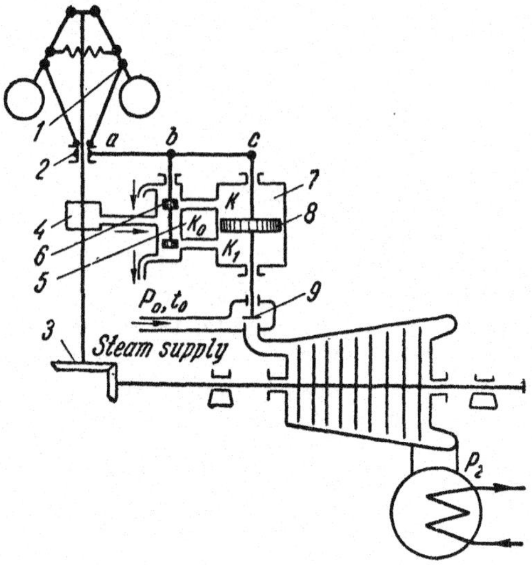

Fig. 9-3. Indirect regulation

Usually oil pressures used for servomotor operation are from 3 to 7 atm. In the current practice oil pressures of 12 to 20 atm and more have also been used.

Let us consider the regulator operation when the load on the turbine falls. As soon as the load on the turbine falls the speed of rotation of the turbine shaft increases, the governor weights fly apart because of the increase in centrifugal forces, sleeve *2* will get displaced upwards displacing with it point *b*, coupled with the servomotor piston *6*, relative to fulcrum point *c* of the lever *ac*. Servomotor space *K* now connects with the central chamber K_0 of the pilot valve and oil under pressure starts entering into the upper portion *K* of the main cylinder. Valve *9* begins to close and at the same time oil from the lower portion of the main cylinder is discharged to drain. Point *c* of lever *ac* now starts moving downwards, lever *ac* operating with respect to point *a* as fulcrum, and in the process displaces the servomotor piston *6* downwards along with it. As soon as the servomotor piston *6* occupies its original central position entry of oil to space *K* of the main cylinder is cut off, and the regulating valve occupies a new position. Quantity of steam flowing through the turbine reduces and consequently the power developed diminishes. Speed of rotation of the turbine shaft slightly increases. Lever *ac* is known as a differential lever since with its help the pilot valve piston can always be brought back to its central position.

9-4. REGULATION WITH A ROTARY SERVOMOTOR

Very often for turbines with nozzle control governing a rotary servomotor is used for transmitting the impulse for speed regulation. Such an arrangement is shown in Fig. 9-4.

Oil is delivered from the gear pump under two different pressures: 5 ata for the speed regulating servomotor, and 1.4 ata for lubrication of journal and thrust bearings. Oil under pressure enters servomotor and rotates the servomotor piston (flap) about its axis. The servomotor spindle carries a system of cams, the rotation of which causes the nozzle valves to open or close. As in the previous case servomotor piston is rotated by the oil supplied to it through a pilot valve. At the end of the process of speed

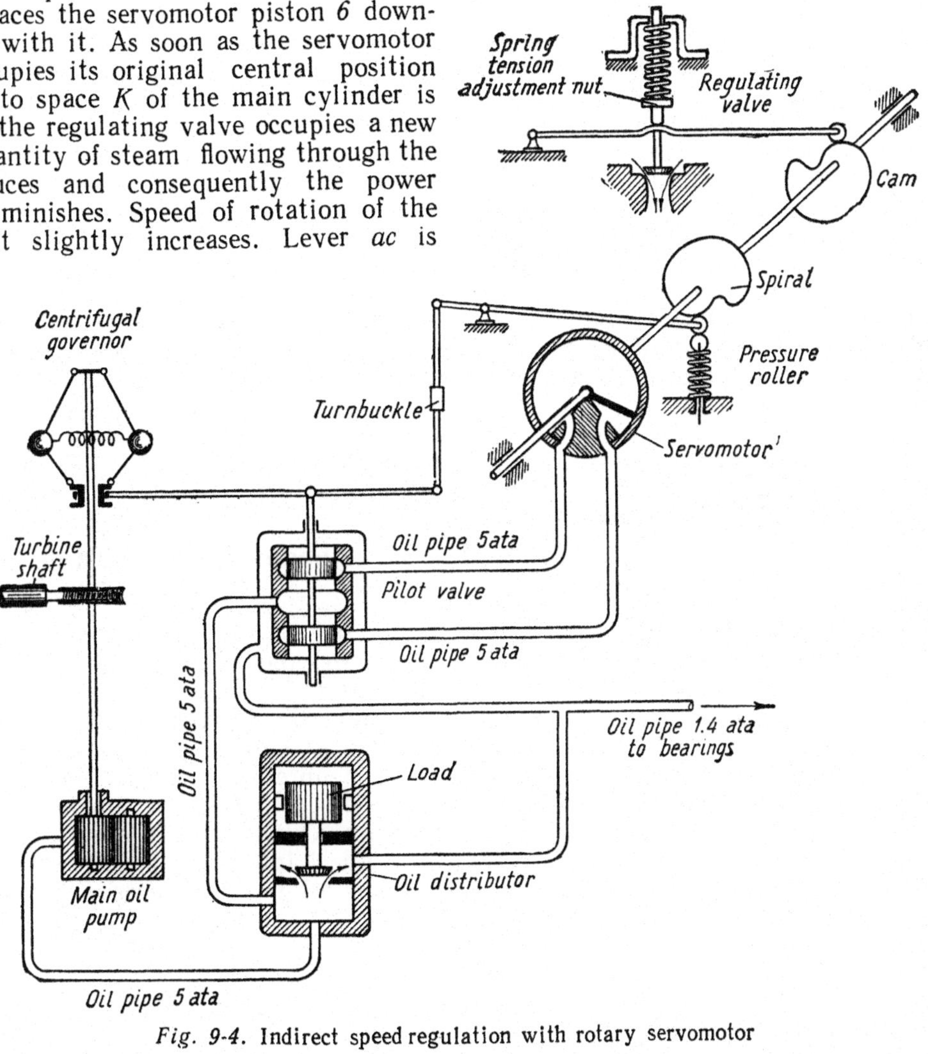

Fig. 9-4. Indirect speed regulation with rotary servomotor

regulation the pilot valve is brought back to its original central position with the help of another cam also mounted on the servomotor spindle, operating through the differential lever as was done in the previous case.

Proper orientation of a required number of cams enables the opening and closing of nozzle control valves in a definite order when the load on the turbine increases or decreases.

9-5. SPEED REGULATION WITH HYDRAULIC CONTROLS

In hydraulically operated regulating devices there are no differential lever connections. Fig. 9-5 shows the general arrangement of such a system of speed governing.

Pilot valve *1*, with the free end cut off at an angle, is connected to the sleeve of the centrifugal governor *2* through a speed reduction gearing *3*. When the turbine speed undergoes a change pilot valve *1* is displaced either upwards or downwards decreasing or increasing the orifice opening in oil box *12* thus controlling oil flow to drain through tube *13*. Thus the pressure of oil under piston *10* of the servomotor mechanism operating the steam valve increases or decreases, opening or closing the steam regulating valve *14*. Oil from oil reservoir *5* is fed to the gear pump *4* through an oil filter and inlet oilduct *6*. From the gear pump oil is directed into the regulating system through a stop valve *7* and to the lubricating system through diaphragm *9*. Oil for the regulation system is divided into two streams, one of which is directly led to oil box *12* and

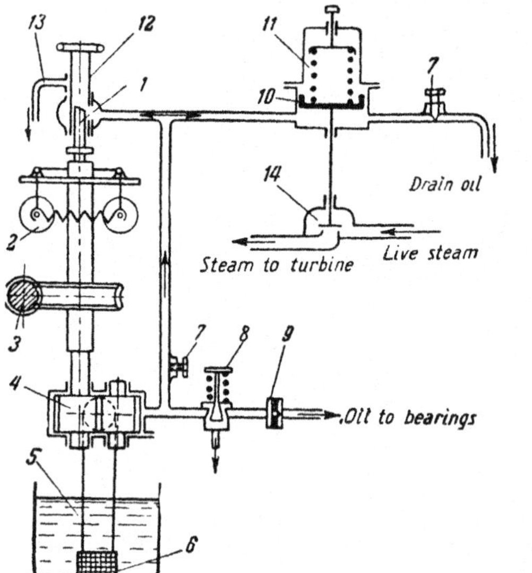

Fig. 9-5. Regulation with hydraulic servocontrols

thence to the drain and the second to the underside of the servomotor piston also joining the drain oil through the second stop valve *7*. The position of both stop valves *7* is fixed at the time of installation of the regulating system.

A spring loaded safety valve *8* permits the drainage of oil to sump in the event of excessive pressures. Spring *11* of the regulating valve *14* is always under compression and thus the valve remains open only under oil pressure, which has to overcome the force exerted by the spring. With variation of load on the turbine oil pressure under piston *10* varies, thus suitably operating the stop valve to admit the required quantity of steam.

Let us consider the sequence of operation if the load on the turbine increases.

When the load on the turbine increases the sleeve of the centrifugal governor and the pilot valve *1* are displaced upwards so that the delivery section for oil flow in oil box *12* diminishes. Oil pressure under piston *10* increases and valve *14* is opened further. There is an increased flow of steam through the turbine and as a consequence the developed power also grows. The above system of speed regulation with hydraulic controls is used by the B.B.C. (Brown Boveri & Company). Many other systems of speed regulation make use of hydraulic servomotor controls (often with differential levers).

For the system of nozzle control governing, with hydraulic servocontrols, the various nozzle control valves are adjusted for different spring tensions to effect a definite order of opening and closing of valves. The basic advantage of hydraulic servocontrols is the complete absence of connecting levers and the resulting friction and play which are responsible for the lowering of operational efficiencies.

9-6. HYDRODYNAMIC SYSTEM OF REGULATION OF V.T.I.

Fig. 9-6 shows the general arrangement of the hydrodynamic system of regulation made by the V.T.I.

In the above-given system the centrifugal governor is replaced by an impeller.[1] The impeller is mounted at the front end of the turbine shaft. Oil is fed to the impeller from the oil chamber *K*. Oil from the exit of the impeller is divided into four paths. Pipe *A* connects pressure oil from chamber K_0 to the pressure side of the diaphragm control *2*. Pipe *B* carries

[1] See "Hydrodynamic Regulation of V.T.I. for Turbines of 44,000 kW" by A. V. Shcheglayev, I. I. Galperin and G. F. Prikazchikov, *Izvestiya V.T.I.*, No. 3, 1951.

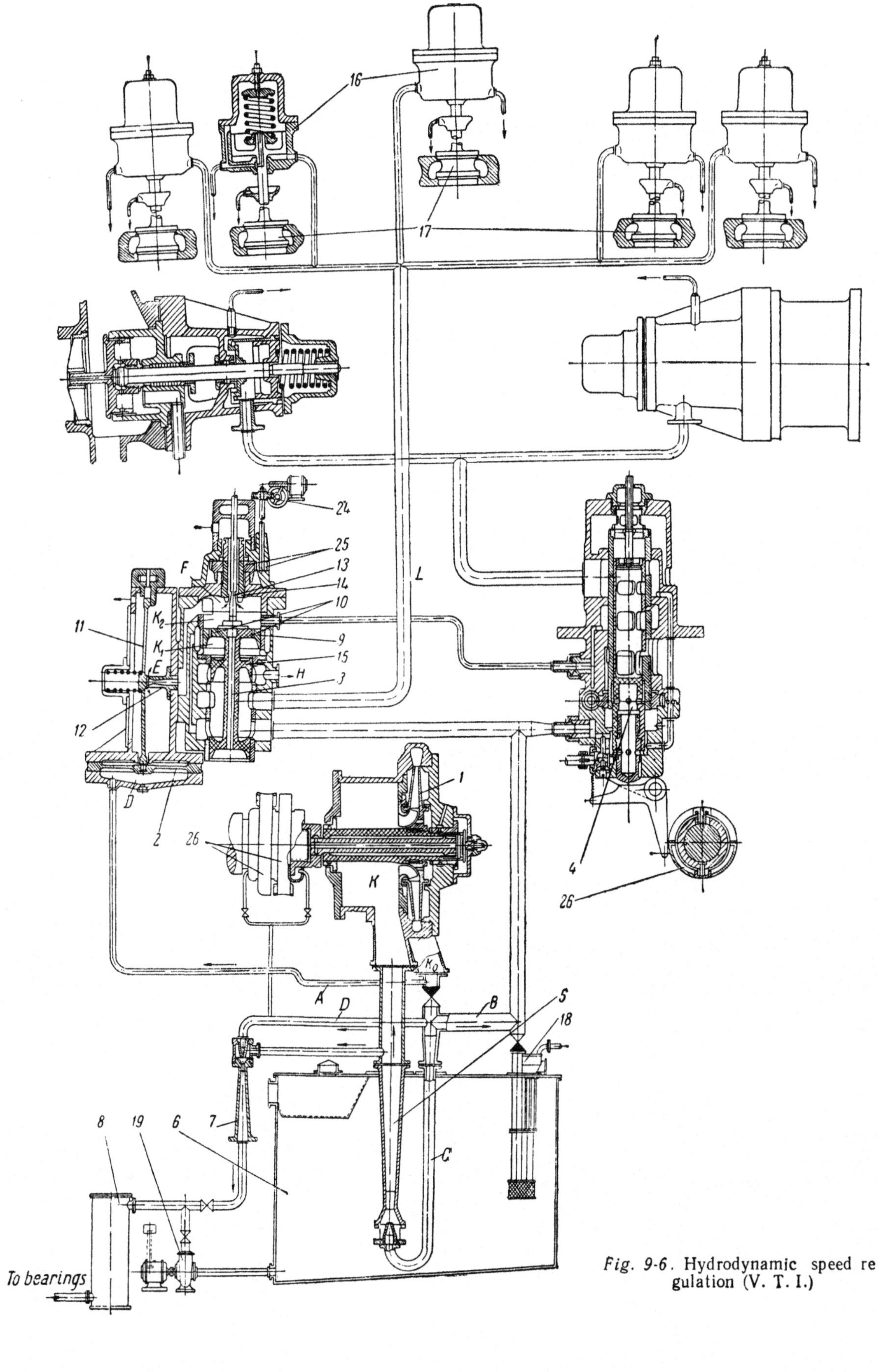

Fig. 9-6. Hydrodynamic speed regulation (V. T. I.)

pressure oil to the pilot valve *3* of the main servomotor and feed controller *4*. Pipe *C* supplies oil to the main oil ejector *5*, which sucks oil from the main oil reservoir *6* and compresses it to the inlet pressure of the impeller *1* in chamber *K*. Pipe *D* leads oil to ejector *7* which removes oil for bearing lubrication from ejector *5*. The oil cooler *8* is also connected to the delivery side of ejector *7*.

Under constant conditions of operation full working pressure is exerted on the underside of the intermediate servomotor piston *9* whereas the pressure exerted by oil from the upper side is half of the working pressure (in chamber K_2). Such a pressure differential is obtained by throttling the pressure oil through orifice *10* and drain oil through pressure relieving orifices *E*, and annular space *F* between stock *13* and sleeve *14*. For any given condition of operation the sum total of oil flowing to the drain through these two orifices *E* and *F* is always constant and equal to the amount of inflow through the orifice 10. The area ratio between the upper working surface of servomotor piston *9* and lower face of servomotor piston *15* is kept as 2 : 1. A proper selection of working pressure and areas for pistons *9* and *15* enables piston *15* and stock *13* acting as a hydraulic cut off to occupy a central position under any given steady load conditions. When the speed of rotation of the turbine shaft alters the pressure at which oil is delivered from the impeller also undergoes a change. The impulse of this change of pressure is felt in the lower compartment of the diaphragm control leading to an alteration of its curvature, which deflects the disc spring *11* at its centre as a consequence of which orifice area *E* changes and the amount of oil flowing to drain also changes. Thus the intermediate servomotor piston *9* is actually a differential piston attached to pilot valve *3* of main servomotor *16* which regulates the steam valve *17*.

The complete process of regulation is carried out with two pressure stages operated hydraulically.

Let us consider the operation of the regulating system when the load on the turbine falls. Shaft speed increases, and oil pressure at delivery from impeller *1* increases giving rise to deflection of diaphragm *2* and disc spring *11*. Orifice *E* increases in size and more oil flows out to drain from chamber K_2. Oil pressure above piston *9* decreases and pilot valve *3* starts moving upwards increasing the amount of oil discharged through *H* so that there is a drop of pressure in the oil main *L* connecting the main servomotor *16* to the steam regulating valve *17*. This decrease in pressure allows the valve to close under the pressure from the valve spring. The upward displacement of the conical spindle *13* decreases orifice *F* and reduces oil flow to drain from chamber K_2. Pilot valve *3* will continue to move upwards until in this chamber pressure does not build up to its original value of half the working pressure prevailing in oil main *L*. The total amount of oil flowing through orifices *E* and *F* throughout the operation remains equal to the inflow through orifice *10*. For the above system of speed regulation two auxiliary oil pumps, a turbo-pump *18* and an electropump *19*, are provided.

9-7. SPEEDER GEARS

From what has been said about the working of speed regulators it follows that a change in

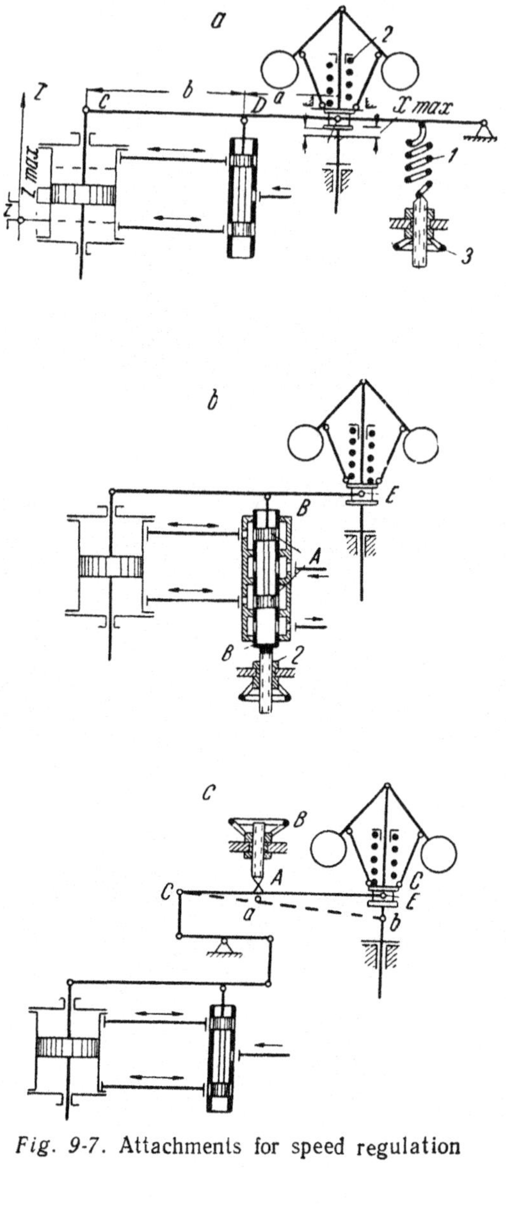

Fig. 9-7. Attachments for speed regulation

the load on the turbine is immediately followed by a change in the speed of rotation of the turbine shaft. Thus for a turbo-alternator operating in isolation its r.p.m. is a direct function of the load carried, i. e., for each and every load there is a definite particular speed of rotation.

In such cases the limits between which r.p.m. varies depends on the degree of speed governor fluctuations (usually not more than 6% of the rated r.p.m.).

If a turbo-alternator operates in conjunction with other machines in a network system delivering substantial electrical power in parallel with them its r.p.m. is then determined by the frequency at which the network operates. In this case the turbo-alternator must operate always at the rated speed throughout its range of operation, i. e., from no load to full load. This is brought about by the use of special mechanisms known as speeder gears. The mechanical designs of such gears are numerous and differ greatly in constructional details. Fig. 9-7, *a* shows the additional parts required for synchronising the speed regulation device shown in Fig. 9-3. There is an additional spring *1* with a handwheel *3*, for manually increasing or decreasing spring tension. The revolving weights of the centrifugal regulator, when there is a change in r.p.m., act on both springs *1* and *2*. For example, if there is an increase in the shaft speed the weights of the centrifugal governor compress spring *2* and extend spring *1* while displacing sleeve *E* upwards. The initial tension of spring *1* can be adjusted by rotating handwheel *3*. A change in the tension of spring *1* gives rise to a change in the tension of spring *2* at the same time displacing sleeve *E* to a new position. If, say, tension of spring *1* is increased, sleeve *E* is displaced downwards and tension of spring *2* reduces.

Lowering down of sleeve *E* at a new steady position of servomotor the pilot valve increases the opening of regulating valve and thus the power developed at practically constant r.p.m. Thus by changing the tension of spring *1* position of sleeve *E* can be varied to obtain the required developed power.

In Fig. 9-7,*b* we see that bush *B* of the pilot valve may be displaced with the help of handwheel *2*. By arbitrarily displacing bush *B* either up or down the central position of piston *A* of the pilot valve will be different for different positions of governor sleeve. Every position of bush *B*, when the pilot valve is at the centre of its travel, conforms to a definite position of sleeve *E* and, consequently, the opening of the regulating valve. Thus the amount by which the regulating valve is open and hence the amount of steam flowing through the turbine may be directly regulated by displacing bush *B*. In Fig. 9-7, *c*, the fulcrum *A* of lever *bc* may be displaced by means of the handwheel *B*. Displacement of fulcrum *A* of lever *bc* permits the displacement of the pilot valve from its central position leading to a change in the regulating valve opening and thus the steam flow. Changing the mass flow of steam through the turbine results in a change in the power developed by the turbine.

In the arrangement shown in Fig. 9-5 we see that the regulation of r.p.m. and load on the turbine is brought about by displacing bush *12*. Displacement of the latter allows the increase or decrease in oil to drain through tube *13* and thus vary oil pressure on piston *10*. Depending on the pressure of oil on piston *10* regulating valve *14* will either open or close increasing or decreasing the quantity of steam flowing through the turbine.

In the hydrodynamic speed regulation system of V.T.I. (Fig. 9-6) mechanism *24* functions as a speed regulating device. It consists of a small electric motor, by operating which bush *14* may be displaced in either direction through worm gearing *25*. Displacement of bush *14* permits the arbitrary variation of the amount of oil from the chamber over piston *9* drained through orifice *F*. A change in the pressure of oil in this chamber leads to the displacement of pilot valve *3* as well as the amount of oil drained through orifice *H*. Pressure of oil in oil main *L* and main servomotors *16* changes and steam regulating valve *17* comes to a new position thus varying the quantity of steam flow to suit the load requirements. This speeder gear device enables the operation of a turbo-alternator set at constant speeds (rated speed), when working in parallel with other machines, throughout its working range of no load to full load.

9-8. REGULATION CHARACTERISTICS

When the speeder gear is in a fixed position (Figs 9-3 to 9-6) every shaft speed conforms to a definite position of the governor sleeve and the power developed at the alternator terminals. From the observations of power developed and the r.p.m. of the turbine it is possible to plot a graph of N_e as a function of n. Such a graph is known as the graph of static regulation characteristics; Fig. 9-8, *a* shows such a graph. Fig. 9-8, *a* shows the variation of N_e with n for a given fixed position of the speeder gear. Curve *ab* shows the variation of N_e as a function of n

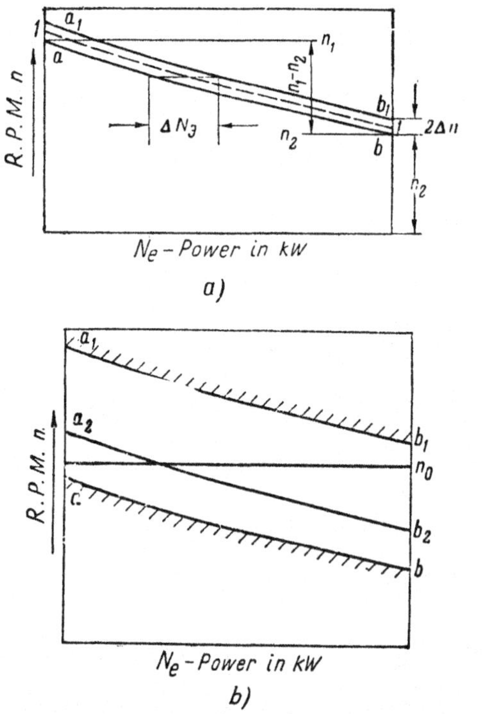

Fig. 9-8. Static regulation characteristics

for increasing loads (from no load to N_e^{nom}). The relation

$$2\frac{n_1 - n_2}{n_1 + n_2} \times 100\% = \delta \qquad (9\text{-}4)$$

is known as the degree of non-uniformity of regulation, where n_1—r.p.m. of turbine at no load,

n_2—r.p.m. of turbine at nominal load (max. cont. rating),

$\frac{n_1 + n_2}{2}$ — average turbine speed for a fixed position of the speeder gear.

If instead of average speed, rated speed is assumed then equation (9-4) becomes

$$\delta = \frac{n_1 - n_2}{n_0} \times 100\%. \qquad (9\text{-}4a)$$

Regulation of the turbine depends upon the nature of curve of variation of N_e as a function of n.

For satisfactory operation of a turbine the degree of non-uniformity must lie between the limits of 4 to 6% (i.e., for a smooth variation of N_e as a function of n).

If the characteristic of power developed versus speed is plotted for decreasing power from N_e^{nom} to no load condition, we have a slightly different curve from the one that was obtained for increasing loads, i. e., N_e versus n. This curve is shown in Fig. 9-8, *a* by a_1b_1. For one and the same value of power developed the difference in r. p. m. given by the two curves ab and a_1b_1 is $2\Delta n$ and is known as spontaneous speed variation. The value $2\Delta n = +\Delta n - (-\Delta n)$ signifies such a speed variation between the limits of which the speed governor is in its equilibrium position 1-1 and is unable to overcome the frictional forces of the speed regulator mechanism, servomotor coupling as well as the servomotor itself. Positive value of Δn gives the increase in r. p. m., from steady load conditions, necessary for overcoming the above-named frictional forces before the speed regulator can move from its equilibrium position. Similarly a negative value of Δn stands for the decrease in r. p. m. before the speed regulator can come into operation. In short, these are the r. p. m. intervals necessary for overcoming the inertia forces before the regulator can move from its equilibrium position. The exact numerical value of $2\Delta n$ entirely depends on the sensitivity of the speed regulating mechanism. The relation $2\Delta n/n_0$ is usually known as the degree of insensitivity of the regulating system:

$$\varepsilon = \frac{2\Delta n}{n_0} \times 100\%. \qquad (9\text{-}5)$$

The maximum permissible value for ε for steam turbines should in no case be more than 0.5%. In modern practice the value of ε is kept, as far as possible, less than 0.1% ($\varepsilon \leqslant 0.1\%$).

The so-called degree of insensitivity depends upon a large number of factors, such as back lash in connecting levers, inherent inertia of the regulator mechanism, friction in servomotor, pilot valve, couplings, etc.

Spontaneous speed variation for a given degree of insensitivity of regulation may be obtained from equation (9-5):

$$2\Delta n = \frac{\varepsilon n_0}{100}. \qquad (9\text{-}5a)$$

The static regulation characteristic ab may be displaced with the help of speeder gears. Fig. 9-8, *b* shows the regulation characteristics for the two extreme positions of the speeder gear (curves ab and a_1b_1). Curve a_2b_2 shows the regulation characteristics for an intermediate position of the speeder gear.

9-9. PARALLEL OPERATION OF STEAM TURBINES

It is now the established practice to operate thermal power stations and hydroelectric power stations in parallel supplying power to one and the same network system operating at a given frequency which is maintained at the required value by operating the turbines at the necessary r.p. m.

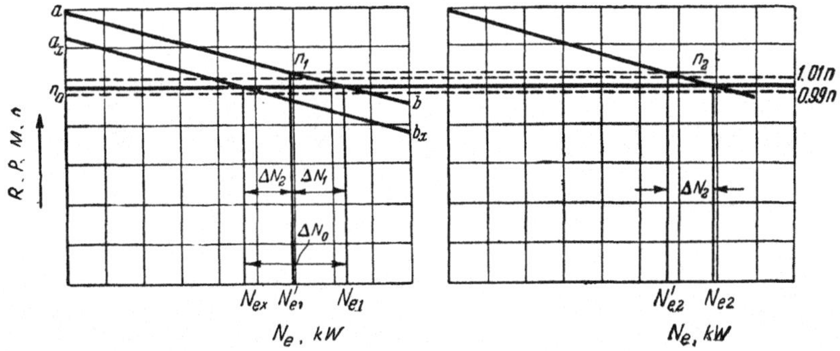

Fig. 9-9. Regulation characteristics for parallel operation

Fig. 9-9 shows the static regulation characteristics for two turbines operating in parallel. At a given moment the total load on the two turbines is $N_e^{total} = N_{e1} + N_{e2}$, where N_{e1} is the load carried by the first alternator and N_{e2} is the load on the second one. Supposing that the load on the grid has decreased by an amount ΔN_0. If now this drop in load is not controlled by the speeder gears to maintain the turbines at constant speeds of rotation, the load having fallen equally for both the turbines, their r. p. m. would increase. We shall suppose that the decrease in load for turbine No. 1 is ΔN_1 and for turbine No. 2 it is ΔN_2, the total drop in load being $\Delta N_0 = \Delta N_1 + \Delta N_2$. It is possible to transfer the complete drop in load onto turbo-alternator No. 1 by suitably regulating its speeder gear so that its regulation characteristic is displaced (line $a_x b_x$ in Fig. 9-9) downwards. The machine now operates at the same r. p. m. but at a lower load; turbine No. 2 operates at the same load as before. Load fluctuations in a network system, unless controlled by the speeder gears, invariably lead to a variation of load on the turbines and consequent change of r. p. m. (the frequency of the network system alters). The load variation on a particular turbine depends upon its static characteristics. For a turbine having gently sloping static characteristic the load variations will be quite generous whereas for a turbine having a steep static characteristic the load variations will be much less. The load on a turbine may vary spontaneously, if its speed regulating mechanism has an unsatisfactory degree of insensitivity and if the static characteristic slopes gently. Under the above cited conditions the turbine would operate under unstable conditions. Thus static regulation characteristic with a gentle slope and a large degree of insensitivity to load fluctuations is deemed to be unsatisfactory, and hence operation of a turbine with such a characteristic would be unreliable if not dangerous.

In power system terminology it is usual to subdivide turbines under the headings of "base load turbines" and "peak load turbines". High-efficiency large-capacity turbines are, as a rule, utilised as base load turbines. Turbines having low efficiencies come under the category of peak load turbines. In order to maintain base load turbines at thermal power stations operating at their maximum efficiencies it is necessary that they carry full load throughout their period of operation, whereas peak load turbines are basically meant to take up the additional loads which appear in the grid at peak hours. If the base load turbines are to operate at their maximum economic load even when there is a fluctuation of load on the network system, it is absolutely necessary that their static characteristics should be quite steep, especially at the portion where the maximum economic load occurs. However, increasing the slope of the static regulation characteristic is not very desirable from the point of view of load trippings, when the r.p.m. would rise very dangerously; the more so for the full load tripping of generators at times of accidents. The difference between the nominal operating speed and the maximum speed attained at load tripping is known as the dynamic trip speed. The dynamic trip speed depends, to a large degree, on the sensitivity of regulation, and besides on the amount of steam locked up in the steam chests, etc., after the regulating valves. The factor

$$\Delta n = \frac{\delta n_0}{100} \qquad (9\text{-}6)$$

expresses the dynamic trip speed in terms of the nominal speed as well as the degree of non-uniformity. From the above equation it follows that Δn is directly proportional to the degree of non-uniformity and hence even for base load turbines the degree of non-uniformity is kept within 6%.

9-10. OIL SUPPLY SYSTEM OF TURBINES

The principal scheme of oil supply to a steam turbine is shown in Fig. 9-10. This is one of the most widely used systems for the supply of oil to turbines.

Oil from the reservoir is sucked by the main oil pump (gear pump) through non-return valve *2*. In turbines of large capacities it is usual to provide helical gear pumps as main oil pumps. Helical gear oil pumps have certain constructional advantages over the usual toothed gear oil pumps. They require comparatively smaller power for

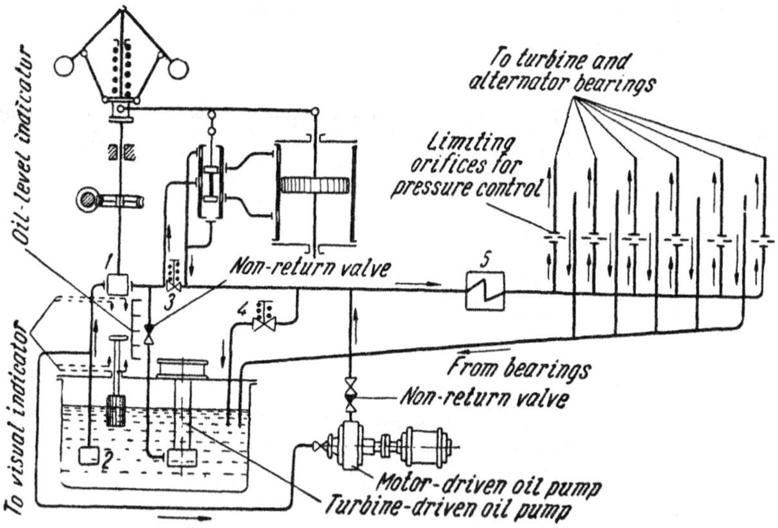

Fig. 9-10. Oil supply system of a turbine

their operation, are simpler to assemble and have a much longer useful life. The main oil pump supplies oil under pressure to the servomotor mechanism of the speed governor and to the journal and thrust bearings of the turbine, through pressure reducer *3*. Pressure of oil for lubrication purposes is maintained within the limits of 0,4 to 0.8 atm gauge. Before entering the bearings the oil is passed through oil cooler *5* where it is cooled. For a proper distribution of oil between the various bearings the oil piping system is provided with flow control diaphragms of varying orifice sizes. Used oil from the bearings is collected into a single drain oil piping and is returned to the oil reservoir. The lubricating system is provided with safety valve *4* which in case of excessive oil pressures permits some of the oil to be returned to the oil reservoir, thus maintaining the lubricating oil pressure at the required value. The drain oil from the servomotor exit is connected to the supply line of the lubricating system. In this case the total oil delivered by the oil pump will be the quantity required for lubrication. The above-given system of oil distribution reduces the pressure differential on the servomotor piston.

In some of the turbines drain oil from the servomotor and speed governor system is directly delivered into the oil reservoir. For such systems of oil supply the main oil pump is designed to handle oil in excess of the lubrication requirement, thus accounting for the drain oil from the servomotor which is directly delivered back into the oil reservoir. Sectional view of a main oil pump of the toothed gear type is shown in Fig. 9-11. The driving wheel (pinion) is mounted either directly on the turbine shaft or is driven through a toothed gearing system. The driving pinion is rigidly fixed to the shaft with the help of a key. At the suction side the hollow gaps between the teeth of the two pinions are filled with oil. When the pinions rotate at the pressure side oil is expelled from between the teeth and simultaneously some amount of oil is enmeshed between the rotating teeth at the suction side. The capacity of a gear oil pump is determined from the equation:

$$Q=\frac{2vnz}{1{,}000}\eta \text{ [l/min]}, \qquad (9\text{-}7)$$

where v—volume of the hollow gap between the teeth, cm³;
n—r.p.m. of the pinions;
z—number of teeth per pinion;
$\eta=0.7$ to 0.9—volumetric coefficient.

The power required to drive the gear pump is obtained from the equation

$$N_p=\frac{10.33P_{atm}Q}{60\times 102\eta_p} \text{ (kW)} \qquad (9\text{-}8)$$

where η_p—efficiency of the gear pump which may be supposed to be between 0.8 and 0.9.

Working oil pressure is developed by the main oil pump usually at about half the normal speed of rotation of the turbine shaft. At the time of starting and stopping the turbine the main oil pump cannot supply oil at the required pressure and in the required quantity, and hence auxiliary oil pumps are provided to supply oil under the required pressure to the bearings of the turbine and generator. These may be either turbine driven or electrically driven (Fig. 9-12). Before starting the turbine the auxiliary oil pumps are started which maintain supply of oil at the required pressure both for lubrication and for the speed governing system.

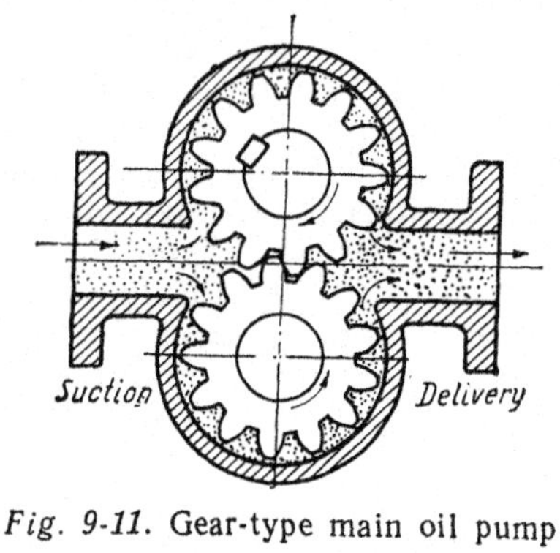

Fig. 9-11. Gear-type main oil pump

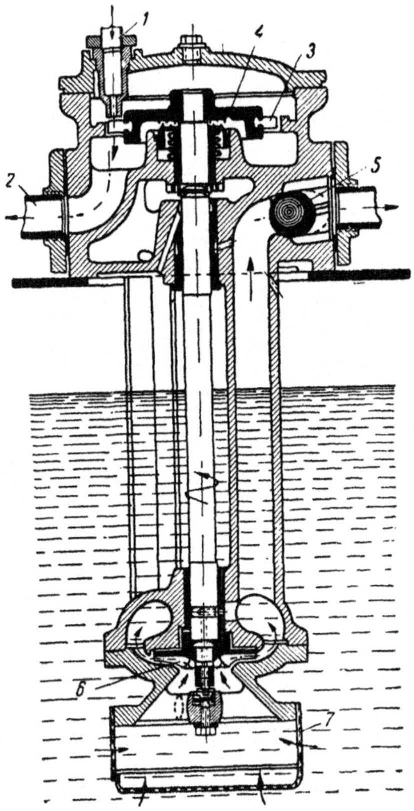

Fig. 9-12. Auxiliary oil pump (turbine driven)

Fresh steam is supplied to the turbine of the auxiliary oil pump through nozzle *1*. The rotor of the oil pump *6* is mounted on the extension of the turbine shaft carrying the turbine disc *4*. The oil pump rotor is mounted in such a way that it is always submerged in the oil of the reservoir, which helps in keeping the pump rotor always primed. From the oil pump supply oil under pressure is led to the main piping through a non-return valve *5*. When the main oil pump picks up its normal pressure the auxiliary oil pump is automatically shut off, and the non-return valve shuts off under the pressure of oil from the main oil pump.

At the time of stopping the turbine the auxiliary oil pump is started once again. Besides the auxiliary oil pump (either turbine or motor driven) it is usual to have a stand-by electrically-driven oil pump to supply oil to the various bearings (Fig. 9-10) to safeguard them from damage in case of failure of the auxiliary oil pump at the time of starting or stopping the turbine. The oil cooler is one of the most important components of the lubricating system, since lubricating oil has to be maintained at a predetermined temperature to maintain it at the required viscosity. Lubricating oil temperature at entry to the bearings, in conformity with the standard regulations, should not be lower than 35° C or higher than 40° C.

9-11. OVERSPEED TRIPPING SYSTEM

If a turbine has a not too satisfactory speed regulating system the sudden increase in the shaft speed at times of load tripping may reach dangerous figures. The usual overspeed limit is taken as about 10 to 12% of the normal operating speed. Hence every turbine is provided with one or two overspeed trips which shut off the supply of steam to the turbine if the r. p. m. exceeds a certain limit. The overspeed tripping device consists of an unstable centrifugal governor (astatic regulator). Fig. 9-13 shows the main details of construction of such a ring-type astatic regulator. The eccentric ring *1* is directly mounted on the turbine shaft. The eccentric ring is held in the position shown in Fig. 9-13 by spring *2*. The eccentricity *e* of the regulator is given by the distance between the axis of the turbine shaft and the centre of gravity of the regulator ring *1*. Distance *a* shows the regulator travel. The regulator ring *1* is displaced through *a* when the shaft speed exceeds the limiting speed. Overspeed regulators are made of various different constructions as well. One of the most widely used overspeed tripping devices is shown in Fig. 9-14. Regulator pin *1*, when overspeeding, trips lever *7* disconnecting the interlock *8*. The tensile force of the helical spring rotates lever *2* and segment *3* in the clockwise direction. Segment *3* now leaves its notch in sleeve *4* and allows the valve to close under the pressure of its own spring *5*, thus shutting off steam supply to the turbine.

If the stop valve is to be re-opened the following operations have to be carried out. The handwheel of the stop valve is rotated in the direction of closing. During this operation sleeve *4* is displaced upwards. Levers *7*, *2* and segments *3* are now brought back to their original position. The stop valve is now opened by means of the handwheel. For turbines of medium and large

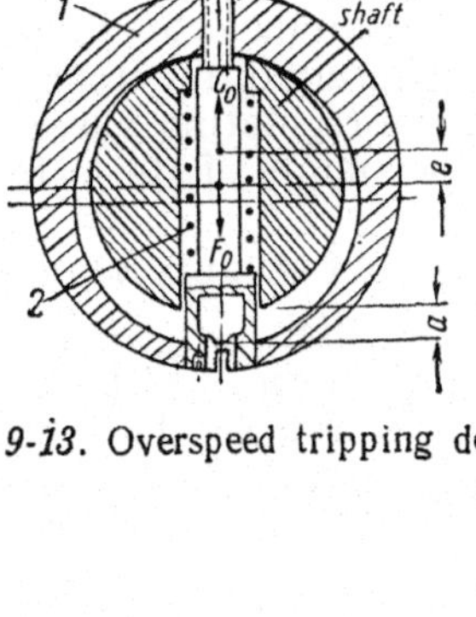

Fig. 9-13. Overspeed tripping device

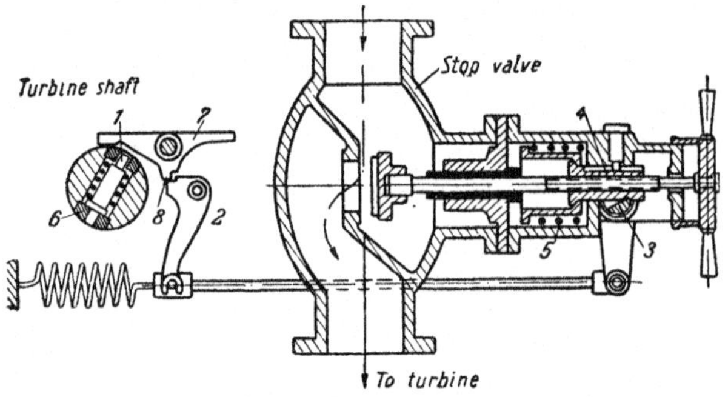

Fig. 9-14. Overspeed trip with lever control

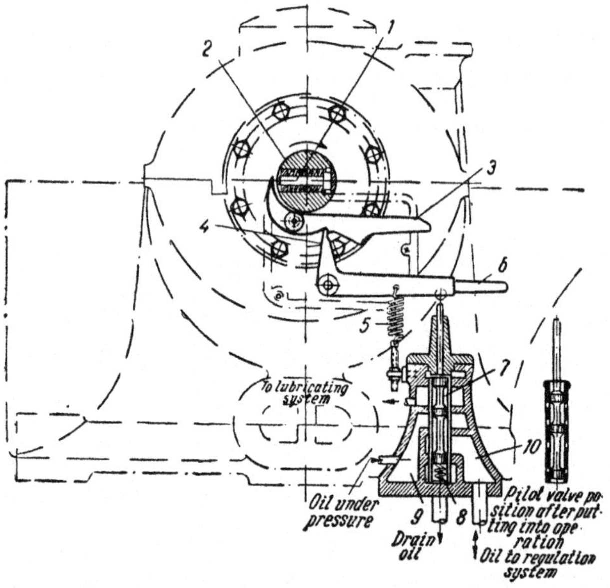

Fig. 9-15. Overspeed tripping relay with hydraulic control

capacities closing of the stop valve while overspeeding is brought about by means of hydraulic controls (Fig. 9-15). In case of overspeeding beyond the limiting speed pin *1* of the overspeed trip strikes against the left arm of lever *3* and thus releases lock *4*. Because of tension of spring *5* lever *6* pushes the servomotor piston *7* downwards. With the displacement of the servomotor piston oil normally flowing from chamber *9* to chamber *10* is directed into the lubricating system of the turbine, and at the same time the oil from chamber *10*, i. e., regulating oil, finds its way into the drain oil piping through chamber *8*. This drain of oil from the regulating system causes the regulating valves to close and thus shut off the supply of steam to the turbine.

9-12. DESIGN OF OVERSPEED TRIPPING DEVICES

Let us suppose that

G—weight of the overspeed trip;

e—the distance between the centre of gravity of the regulator tripping pin and the axis of rotation of the turbine shaft at normal speed of rotation;

a—maximum travel of the tripping pin;

$e+a$—distance between the centre of gravity of the tripping pin and the axis of turbine shaft when the overspeed tripping occurs;

F_0—initial spring tension;

f_0—initial spring sag;

k—coefficient of spring tension;

R—radius of spring.

At the normal speed of operation the centrifugal force of the overspeed trip pin must be lower than the spring tension, i. e., $C<F_0$. At the limiting speed of the turbine shaft these forces must balance each other, i. e., $C_0=F_0$ (where C_0 — centrifugal force of the tripping pin at the limiting shaft speed).

The force exerted by the spring

$$F_0=f_0k. \tag{9-9}$$

Centrifugal force of the tripping pin

$$C_0=\frac{G}{g}e\omega_{max}^2, \tag{9-10}$$

where ω_{max} — maximum allowable angular velocity of turbine shaft.

The centrifugal force of the tripping pin when the spring is under compression (when the tripping device operates) must increase at a greater rate than the increase in spring tension, i. e., the tripping device must be in an unstable condition.

The centrifugal force of the tripping pin at the limiting displacement of the trip ring is equal to

$$C_{max}=\frac{G}{g}(e+a)\omega_{max}^2. \tag{9-11}$$

For this condition the spring tension will be

$$F_{max}=F_0+ka \tag{9-12}$$

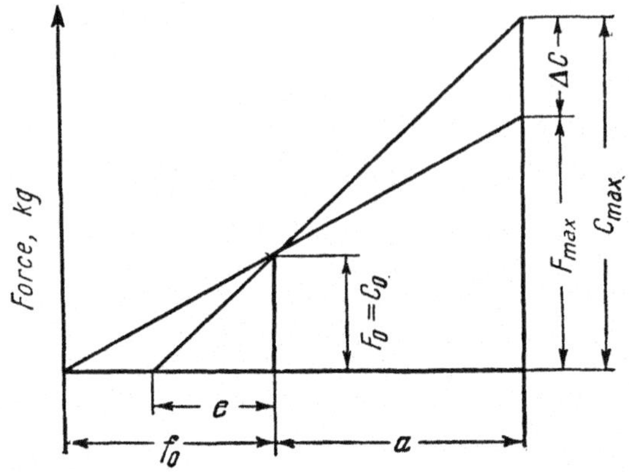

Fig. 9-16. Force variation diagram for overspeed tripping gear

It is usual to assume while designing trip devices

$$F_{max}=0.65C_{max}. \qquad (9\text{-}13)$$

Substituting values from equations (9-11) and (9-12) in equation (9-13) we have

$$F_0+ka=0{,}65\frac{G}{g}(e+a)\,\omega^2_{max}. \qquad (9\text{-}14)$$

The quantities G, e, a, and ω_{max} have to be initially determined to proceed with the design; e is usually assumed between 4 to 8 mm, and a from 3 to 7 mm. From these assumptions we can determine the values of C_0 and F_0 using equation (9-10), and k is obtained from equation (9-14):

$$k=\frac{0.65\frac{G}{g}(e+a)\,\omega^2_{max}-F_0}{a}. \qquad (9\text{-}14a)$$

Diameter d of the spring wire will be

$$d=\sqrt[3]{\frac{16RF_{max}}{\pi\sigma_d}}, \qquad (9\text{-}15)$$

where σ_d—maximum permissible spring tension for twist which may be up to 4,000 kg/cm^2.

Fig. 9-16 shows the diagram of variation of spring tension and centrifugal force exerted by the tripping pin.

Force $\Delta C=C_{max}-F_{max}$ is known as the displacing force of the tripping device.

Chapter Ten

CONSTRUCTIONAL DETAILS OF MULTISTAGE CONDENSING TURBINES

10-1. PRINCIPLES OF CONSTRUCTION OF CONDENSING TURBINES

A turbine must be dependable, economic and easy to operate, simple in construction and assembly and give easy access for repairs and maintenance. Experience shows that the construction of such a turbine is very difficult in reality. For example, the construction of a high-efficiency turbine, as a rule, always leads to a complicated design, requires more metal and in general leads to higher costs of manufacture, assembly, repairs and maintenance. On the other hand, a turbine of simple design which may be very easy to operate, cheap from the point of view of manufacture, and easy to assemble, as a rule, proves less efficient.

The turbine designer usually tries his best to satisfy all the above conditions taking into consideration their importance in a definite order. Studying the various types of new constructions, comparing each variant for economic operation and initial costs, facility for assembly, maintenance, etc., he always tries to give the customer a dependable machine. Thus dependability of operation appears as the most important factor and must always be fully satisfied independent of the remaining contributing factors. At the same time the selection of type of construction depends upon many factors, for example, turbine capacity, steam conditions, efficiency, cost of manufacture, useful life, water resources of the power station in question, etc.

In power station practice it is usual to differentiate turbines as base load turbines and peak load turbines. The base load turbines are always operated at full load and hence they should be designed for the highest possible efficiencies at rated load. Peak load turbines, since they are used only for short intervals of time for the peak periods, may have lower efficiencies. The most economic load for such turbines is about 75 to 80% of their nominal capacity. The characteristic coefficient Y appears as one of the most important parameters governing the efficiency of a turbine. (See 7-4.) For given initial and final steam conditions (p_0, t_0 and p_2) this coefficient completely depends upon $\sum_1^z u^2$, i.e., on the diameter of disc d, number of stages z and r. p. m. n. The most economic value of $\sum_1^z u^2$, and consequently, high efficiency and the best construction of a turbine may be achieved by a proper selection of the quantities d, z, and n.

a) Principles of Construction for Turbines of Small and Medium Capacities

For turbines of small and medium capacities selection of large diameters is to be favoured from the point of view of obtaining smaller number of stages, compactness of construction and lower manufacturing costs. This, however, leads to smaller heights for nozzles and blades in the first few stages, smaller degree of partial admission in the first stage and consequent continuation of partial admission in the succeeding stages. The use of short blades and nozzles as well as low degrees of partial admission leads to a considerable decrease in the efficiency of these stages and the turbine as a whole. And hence from the point of view of maintaining proper efficiency for turbines of small and medium capacities it is advisable to use not too large diameters with increasing heights for blades, increased degree of partial admission in the first stage obviating the necessity to make use of partial admission in the second and following pressure stages. However, for turbines with shaft speeds of 3,000 r. p. m., the use of smaller diameters, although efficient, requires larger number of stages which may be as many as to require the turbine to be made in two or even three units. Such a construction is again to be avoided since it would require too much metal, demand too much expense for both construction and maintenance, being complicated to operate and above all not very dependable.

In order to maintain the value of $\sum_{1}^{z} u^2$ at its optimum figure and at the same time construct a compact turbine with small diameters and a small number of stages, the r. p. m. of the turbine is increased to 5,000, 9,000 or even up to 12,000. However, high figures for r. p. m. necessitate the use of reduction gearings while coupling the turbine to an alternator, which involves additional expenditure as well as increases losses due to friction. All the above-cited factors stress some of the difficulties which a turbine designer has to tackle when designing small- and medium-capacity turbines. To properly balance all these peculiarities of design it is usual to carry out comparative designs based on adaptability for manufacture with sufficiently high efficiencies. Comparison of different variations and analysis of the performance data of the already operating turbines enable the proper selection of the most optimum variant for the turbine under consideration.

Two-row stages have found a wide usage as the regulating or governing stage for turbines of small and medium capacity. The use of two-row regulating stage is quite desirable from the designer's point of view since it simplifies the construction of such turbines. Further, turbines having two-row regulating stages are found to have stable efficiencies even when operated at conditions other than those stipulated while designing.

Two-row regulating stages usually account for large-pressure drops (40 to 50 kcal/kg or even more), which simplifies the construction of the turbine casing and labyrinth seals because of the low pressures immediately after the first stage, enabling the selection of sufficiently long blades in the low-pressure stages (in view of the greater specific volumes to be handled by the last stages).

The use of single-row regulating stages with small pressure drops across them (15 to 20 kcal/kg) complicates the construction increasing its manufacturing costs which is not very desirable. Design and construction of low-pressure stages for turbines of small and medium capacity (at normal speed of rotation 3,000 r. p. m.) do not create any difficulties. For small flow quantities branching of low-pressure steam is not at all necessary and similarly blades of limiting length need not be used. The carry-over losses in the low-pressure stages for such turbines are usually insignificant.

b) Selection of Type of Construction for Large-Capacity Turbines

The selection of the most suitable and optimum construction, in the case of condensing turbines of large capacities, poses a difficult problem. In this case a good deal of attention has to be paid not only to the construction of the first regulating, but also to the last few low-pressure stages. The reliability of operation of such a turbine depends to a large extent on the proper selection of construction type for the low-pressure stages, efficiency, cost of manufacture, etc.

The constructional details of the existing turbines are exceedingly diverse differing in efficiency, initial costs, maintenance costs, etc. Turbines with throttle governing as well as bypass governing are now obsolete. The efficiency of such turbines is seriously impaired at low loads because of throttling losses. Even at nominal loads their efficiency is not very high. In spite of all the above disadvantages turbines, with throttle and bypass governing were widely used in the past for capacities ranging from 12,000 to 50,000 kW for steam parameters of 30 to 35 ata. Even at the present time turbines with throttle governing are employed for large-capacity machines designed to operate at very high steam pressures and temperatures, if they are to be used as base load turbines, i. e., operating at their maximum economic rating (type VKT-100 of Kh. T.W.). The L. M. W. and Kh.T.W. are now building turbines of 300 MW at initial steam conditions of 240 ata and 580°C with nozzle control governing. These turbines must operate at the maximum continuous rating for maximum efficiencies.

Another type of turbine which has been used very widely is one with nozzle control governing combined with supply of steam to an intermediate stage; the efficiency of such turbines is found to be quite stable even at lower loads. However, at overloads these turbines are inefficient, which is a serious drawback. This disadvantage is brought into sharp focus especially in the case of large-capacity turbines operating at the peak of their capacities. Turbines, in which the nominal capacity is reached by supplying fresh steam to one or more of the intermediate stages because of smaller pressure drops in the regulating stage, are found to have very large number of stages leading to increased initial costs. Even at the nominal loads it is found that their efficiencies are not very high. In order to provide for economic working of large-capacity turbines in the region of their nominal loads nozzle control governing is now very extensively used. A single-row or a double-row Curtis disc is used as the first regulating stage in such turbines. Although the two-row disc considerably simplifies the construction, its use may not always give sufficiently high efficiencies. The use of single-row regulating discs gives enhanced values for efficiencies and hence their use seems to be very desirable. However, the smaller the pressure drop across the regulating stage, the greater will be the pressure in the turbine casing, the number of stages will be correspondingly more, shell thickness will be greater and the front end gland seals will have to be of quite a complicated construction, especially in the case of turbines with pressures equal to or higher than critical. Until recently it was the usual practice to have two-row Curtis disc as the regulating stage for turbines of large capacities, but now more and more turbines are built with a single-row regulating stage with the aim of increasing their efficiencies. Turbines with very high initial pressures, however, are built with two-row regulating stages, since it leads to considerable simplification of design.

The design and size of blades as well as the steam channelisation in the low-pressure stages of large-capacity condensing turbines basically depend on the shaft speed and absolute steam pressure at the blade exit. High vacuums are utilised to achieve the maximum possible efficiencies for such turbines. Increase in vacuum is accompanied by larger steam volumes to be handled by the low-pressure stages thus necessitating the use of comparatively larger sizes of blades for the last stages. In order to effectively handle such large volumes of steam, in the case of large-capacity turbines (more than 50 MW), two or even three low-pressure cylinders are utilised. The limiting capacity of a condensing turbine is governed by the handling capacities of the low-pressure stages (see 7-16).

The exit blade sections for slow-speed turbines are usually found to be considerably larger than those for high-speed turbines. Thus it is found that for turbines having equal capacities and operating at the same vacuum, the slow-speed turbines have a much simpler steam distribution in their low-pressure stages than the high-speed ones. This is an advantage. However, slow-speed turbines require a much larger number of stages or diameters than the high-speed ones thus considerably increasing their weight and initial cost. The L. M. W. and the Kh. T. W. in the 30's were building condensing turbines of 50,000 and 100,000 kW at initial steam conditions of 29 ata and temperatures of 400° C with shaft speeds of 1,500 r. p. m. L. M. W. eventually switched over to the now universally accepted speeds of 3,000 r. p. m. The 3,000 r. p. m. turbines built by L. M. W. are found to be much more compact and less in weight when compared with turbines of the same capacity and operating at the same steam conditions manufactured by the Kh.T.W. Types AK-50-2[1] and AK-100-1[1] turbine rotors built by L.M.W. are less than half the weight of the Kh.T.W. rotors, and consequently cheaper.

However, L. M. W. had to make use of double-gallery blades for the low-pressure stage (Bauman stage) which considerably complicates their design (see 10-4). The increase in r. p. m. from 1,500 to 3,000 permits the reduction of gross weight of turbines by as much as 30%. This is one of the main advantages of high-speed turbines. In the case of turbines operating at very high pressures the increase in r. p. m. enables the increase in nozzle and blade heights for the first pressure stages leading to higher efficiencies. In the Soviet Union the construction of steam turbines is carried out in accordance with the GOST stipulations for steam pressures, turbine ratings, type of constructions, etc. This systematic classification of turbines is a prerequisite for the standardisation of turbine parts permitting the interchange of parts for turbines of a particular series.

The experience of the Soviet turbine builders shows that a large number of turbine details such as bearings, labyrinth seals, regulating devices and especially dimensions and profiles of nozzles and moving blades, may be standardised. Current trends in turbine building are directed mostly towards the use of high pressures and temperatures for the supply steam. The efficiency of a turbine increases with the increase in pressure and temperature of fresh steam as well as with the use of regenerative feedheating system. It is therefore highly desirable that the maximum permissible initial steam parameters be coupled with as many extractions as possible for feedheating; as also to use the bled steam for the purposes of industrial heating, etc. The highest temperatures which may be allowable in a steam turbine are governed by the quality of the material used, its cost and cost of manufacture. Increase in initial pressures leads to the increase in wet-

[1] Non-standard turbines.

ness of exhaust steam, thus complicating the construction of the low-pressure stages, labyrinth seals and high-pressure cylinder flanges.

In the high-pressure steam turbines now being built, the wetness (theoretically) is found to be 12 to 14% for the exhaust steam. It is possible to build turbines without intermediate reheating if the initial pressure does not exceed 90 ata and initial temperature is not lower than 500°C. The use of very high pressures even with very high initial superheats always leads to increased wetness of exhaust steam and thus to increased losses due to wetness, decrease in overall efficiency and erosion of the low-pressure stage blading. Hence for turbines operating at pressures higher than 100 ata intermediate reheating is invariably practised. The quality of material to be used for stator and rotor as well as for blades, nozzles and diaphragms of the first stages depends upon the initial temperatures as well as the intermediate reheating temperatures. The experience of the past few years has shown that for fresh and reheat steam with temperatures not exceeding 565 to 580°C cheaper variety of pearlite steels can be used with full reliance. For turbines operating at temperatures higher than 580°C both for fresh and reheat steam the use of costlier variety of austenite steels becomes necessary.

Turbines now being built at pressures of 130 ata and 565°C with intermediate reheat up to 565° C along with the use of a highly developed system of regenerative feedheating permit the enhancement of efficiencies through 25 to 30% as compared with the turbo-generator sets of medium pressures (29 ata and 400°C). Large-capacity turbines operating at critical and supercritical initial parameters are being built either with one or two shafts. In the United States it is the usual practice to use two shafts for turbines of 250 to 340 MW. Here in the Soviet Union, the L. M. W. and Kh. T. W. are building 300 MW turbines in a single unit at 580 and 565°C for fresh and reheat steam respectively and 240 ata on a single shaft. Two-shaft units require two alternators. Their design is more complicated requiring much more steel and consequently costlier to manufacture as well as assemble. However, two-shaft units have comparatively smaller carry-over losses in their last stages even with very high vacuums, which may be advantageously utilised for building turbines of high efficiencies.

At present in the Soviet Union large-capacity turbines have been built at critical and supercritical pressures with high intermediate reheats. Turbines of 150,000 and 200,000 kW are already in operation at some of the larger thermal power stations of the U.S.S.R.

EXAMPLES OF CONSTRUCTION OF CONDENSING TURBINES

1. Turbines of the Lenin Nevsky Works (L.N.W.)

The Lenin Nevsky Works builds single-cylinder turbines of small capacities to be used as prime movers for alternators, turbo-blowers, etc.

Two of the turbines built by this works are described below. Both these turbines are designed to operate at initial steam conditions of 35 ata and 435°C. Fig. 10-1 shows the sectional elevation of turbine K-6-35 (AK-6). This turbine is built to operate as a prime mover for an alternator of 6,000 kW capacity at a shaft speed of 3,000 r. p. m.; steam is supplied through a set of nozzles. The regulating stage followed by 15 pressure stages receives steam supply to its nozzles through four regulating valves. The regulating stage consists of a two-row disc. The turbine has three extractions for feedwater heating. The first extraction is after the 5th stage; the second—after the 8th and the third extraction after the 11th stage. The front- and rear-end labyrinth gland seals are of the fir-cone type. The discs are forced fit on the turbine shaft by mounting them when hot, which provides the necessary rigidity under normal conditions of operation. The front end bearing is of a combined journal-and-thrust type; the journal bearing sleeve is made spherical in shape so as to obtain an even distribution of thrust on the thrust bearing pads. The turbine and alternator journal bearings (Nos 2 and 3) are located in special casings, and the flexible coupling between the turbine and the alternator. The flexible coupling is lubricated through journal bearing No. 2.

Fig. 10-2 shows the sectional elevation of turbine AK_v-14[1] designed for driving turbo-blowers. The turbine is designed for a nominal capacity of 14,000 kW, at varying shaft speeds within the range of 2,400 to 3,400 r. p. m.

The steam supply to this turbine is through a nozzle system. Fig 10-3 shows the sectional view of the nozzle segment of the turbine. Four regulating valves situated on the turbine stater supply steam to four groups of nozzles, two of which are located in the upper half and the remaining two in the lower half of the casing. The turbine consists of a two-row regulating stage and sixteen pressure stages. The front- and rear-end labyrinth seals, combined journal and thrust bearing and the journal bearings, the flexible coupling and its lubrication are all similar to those of turbine K-6-35 (AK-6). The turbine discs are shrunk onto the shaft by mounting them

[1] A non-standard turbine.

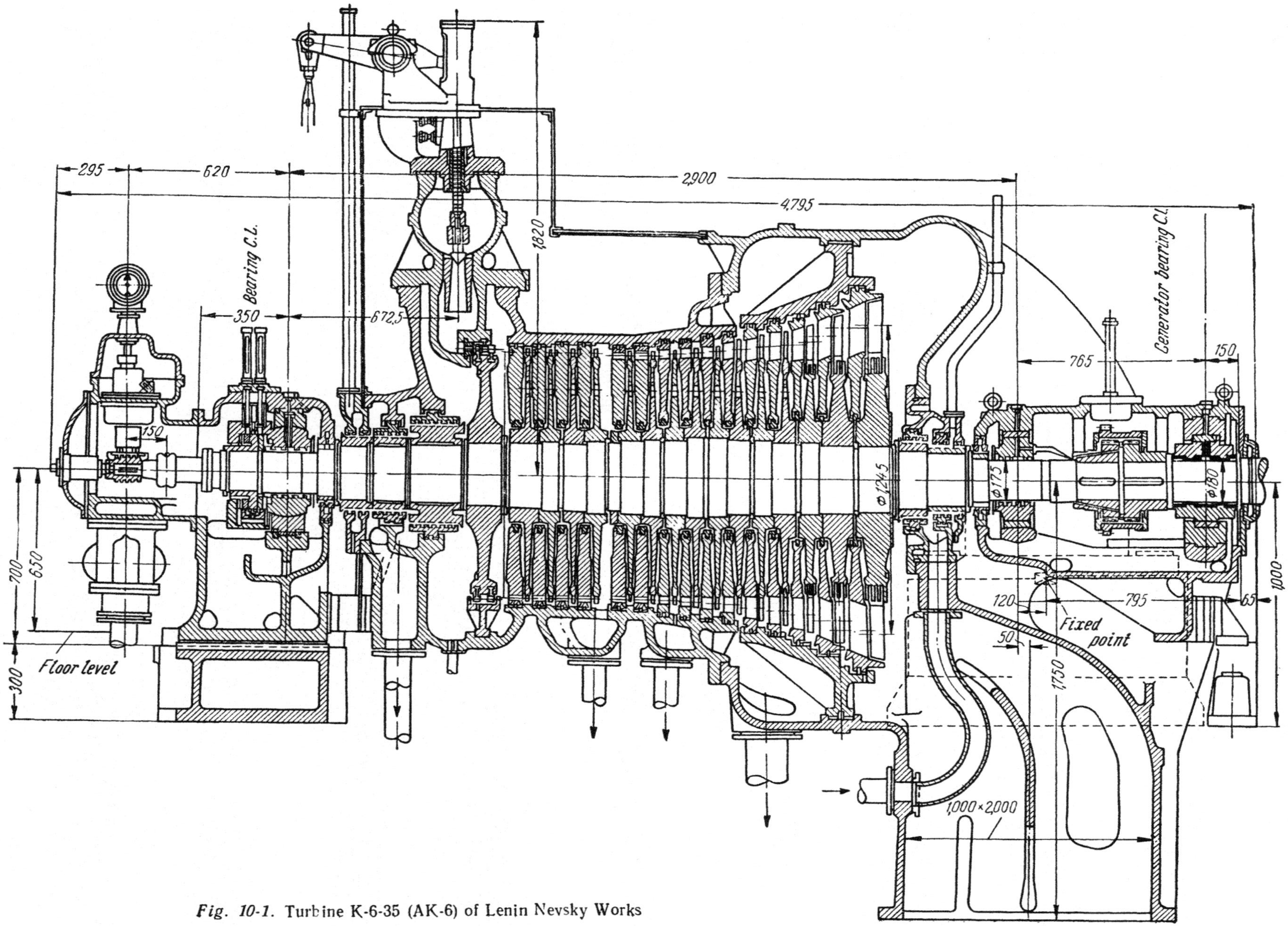

Fig. 10-1. Turbine K-6-35 (AK-6) of Lenin Nevsky Works

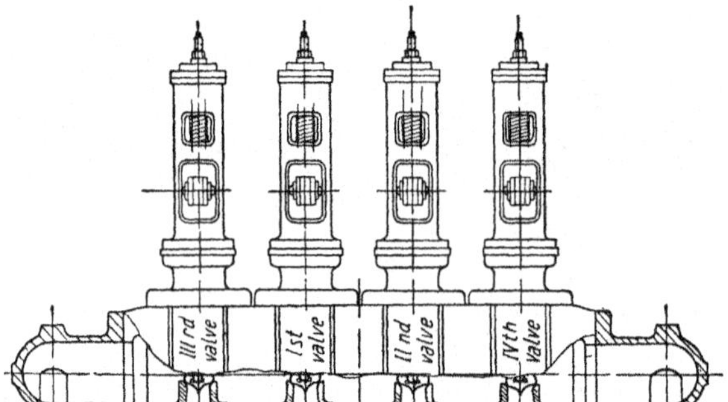

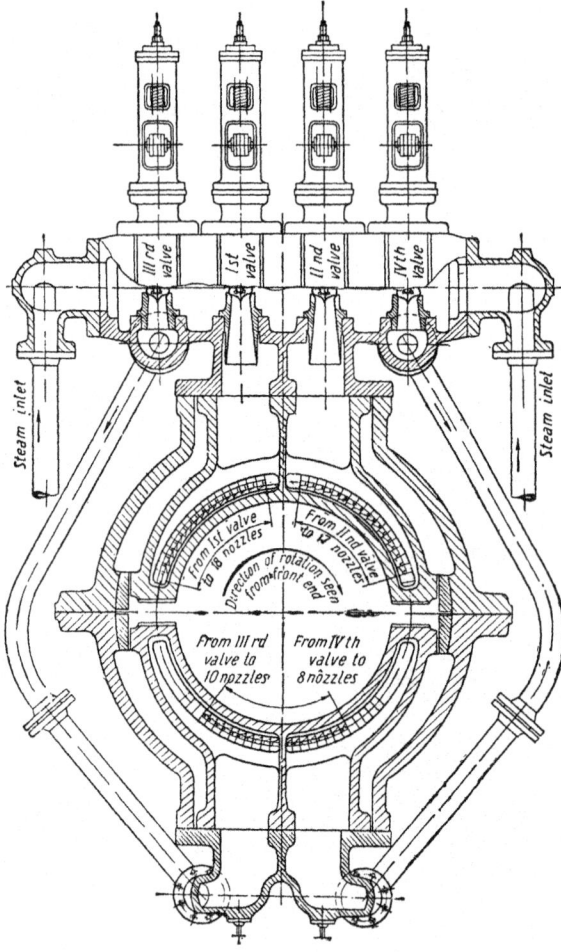

Fig. 10-3. Section of nozzle box, turbine AKv-14

when hot. The centrifugal speed regulator spindle and the main oil pump are driven by the turbine shaft through a worm gearing. The worm rests on independent bearings, its spindle being connected to the turbine shaft through a spline shaft. The overspeed trip lever is also located at the front end of the worm-gear spindle. The turbine is fixed to the base at the exhaust end. Expansion of the turbine, when under operation, is provided for in the high-pressure direction. Axial displacement of the bearing casing is provided for by placing a sliding key between the casing and the bearing.

2. Turbines of the Ural Turbomotor Works (U.T.W.)

The U. T. W. builds condensing turbines with extractions. The capacity of these turbines varies between 12,000 and 50,000 kW operating at medium and high pressures for the inlet steam: 35 ata and 435°C; 90 ata and 500°C; 130 ata and 565°C.

Fig. 10-4 shows the sectional view of turbine K-12-35 (AK-12) which has a capacity of 12,000 kW at the normal speed of rotation of 3,000 r. p. m. It is designed to operate an alternator; the initial steam parameters are 35 ata and 435°C.

The turbine consists of a two-row regulating stage and seventeen pressure stages. Steam supply to the turbine is through a set of nozzles. The quantity of steam flowing to the nozzles of the first stage is controlled by four regulating valves, operated through a system of levers by the main servomotor. The unit has four extractions for feedwater heating. The front- and rear-end gland seals are of the labyrinth type. The front-end labyrinth seal is provided with a bleed into the piping of the second extraction. All the discs are mounted on the shaft when hot and are secured to the shaft by means of keys. The front end of the turbine is supported by a combined journal and thrust bearing, the journal bearing sleeve having a spherical surface. The journal bearing at the rear end is rigidly fixed to the stator.

The turbine and generator rotors are coupled together by means of a semi-flexible coupling. An indicator is provided, between the bearing sleeve and the coupling flanges at the rear end, to show the linear expansion of the turbine when operating. A hand lever is provided to enable the rotation of the rotor manually. The overspeed trip pin is located between the thrust bearing and the spline transmission to the main oil pump. A relay is provided in the front end of the stator to automatically cut off the turbine in case of any serious axial displacement. The turbine unit is provided with hydraulic regulating system developed by the works in conjunction with V. T. I.

3. Turbines of the Kharkov Turbine Building Works (Kh.T.W.)

The Kh.T.W. builds turbines of large capacities at medium, high and supercritical steam pressures. Constructional types and their details are described below.

a) *Turbine AK-50-X*[1]

Fig. 10-5 shows the section through the cylinder of a 50,000 kW turbine at a shaft speed of 1,500 r. p. m. (older type).

The design pressures and temperatures are: $p_0=29$ ata, $t_0=400$°C and $p_2=0.04$ ata; cooling water temperature $t_1^w=15$°C. Fresh steam is sup-

[1] A non-standard turbine.

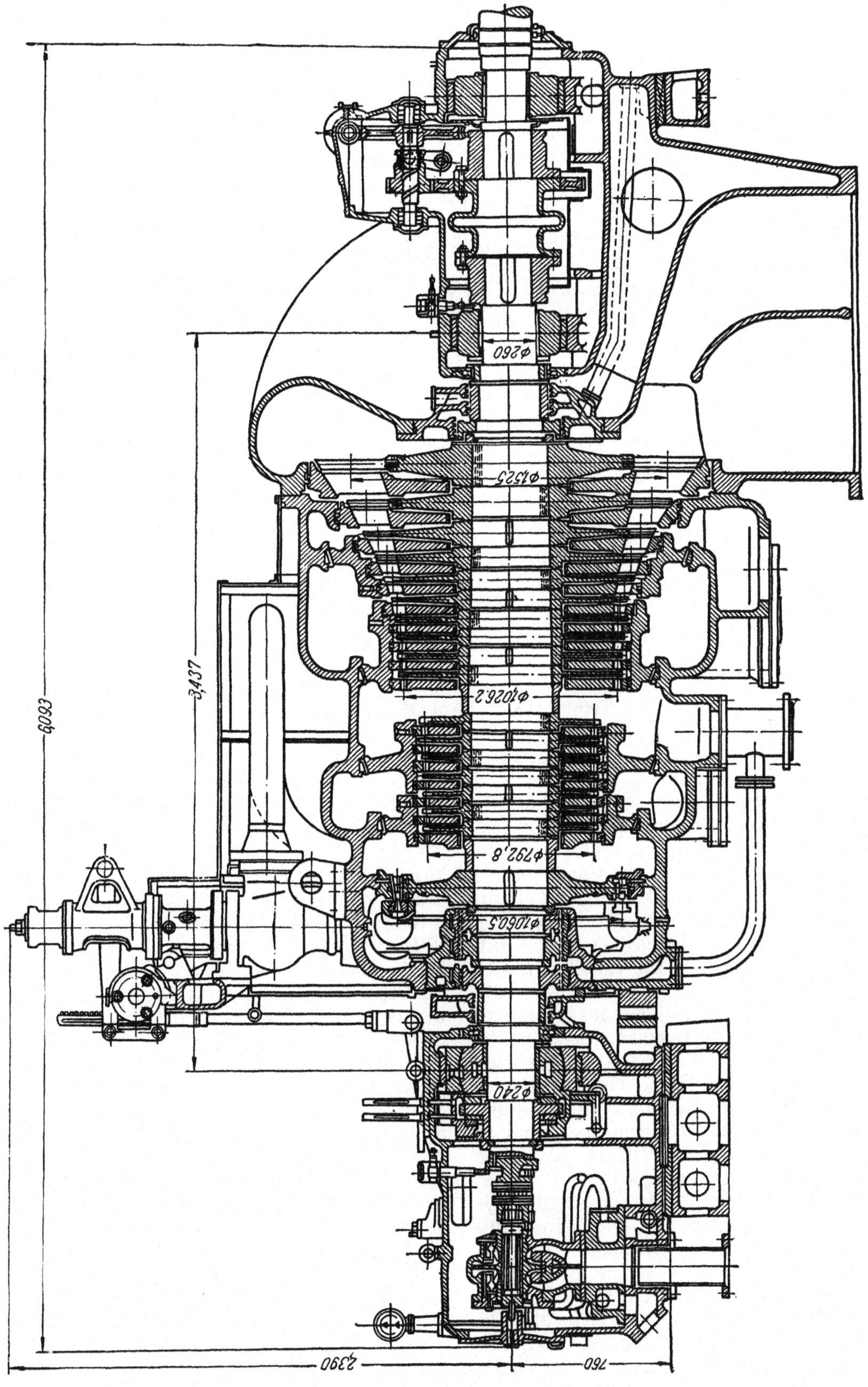

Fig. 10-4. Turbine K-12-35 (AK-12) of Ural Turbomotor Works

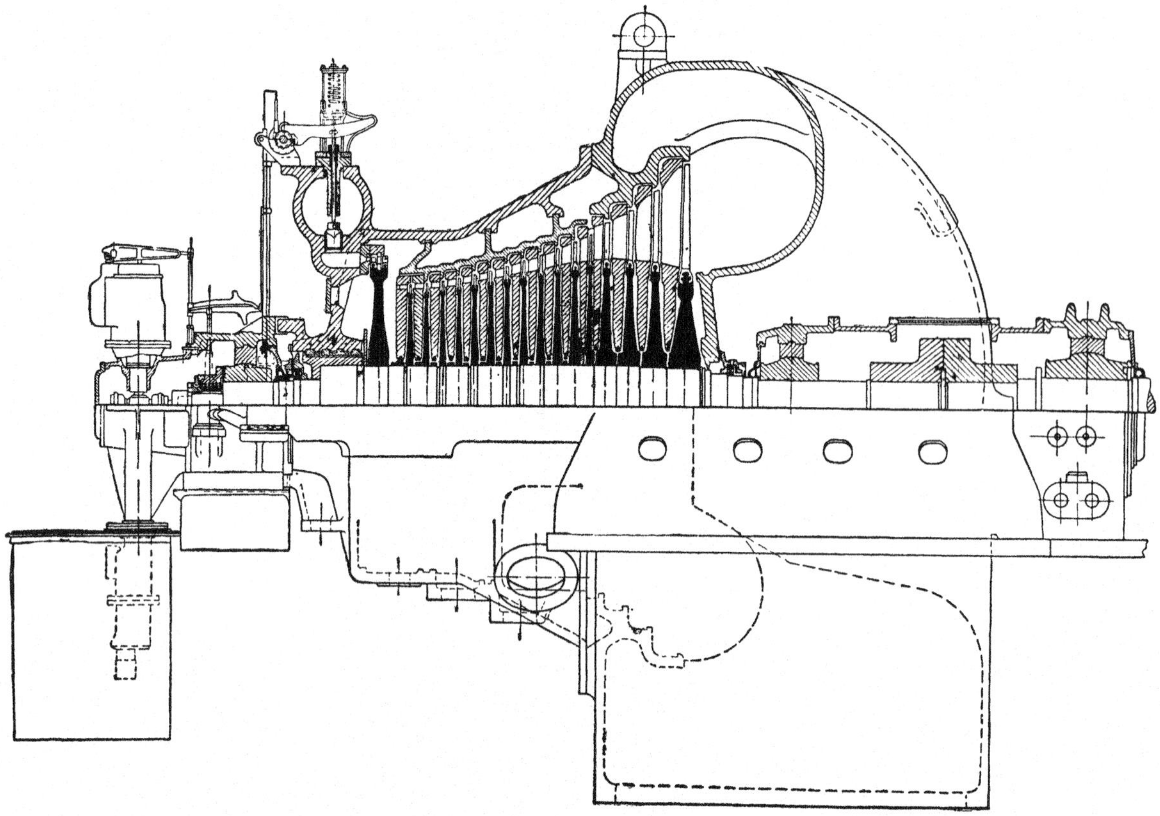

Fig. 10 5. Turbine AK-50-X of Kharkov Turbine Works

plied to the first two-row regulating stage through a set of nozzles. Nine regulating valves located on the front end of the stator control the supply of steam. The steam distribution chest is cast in a single unit with the turbine stator. All the steam is led to the nozzles in the upper half of the stator. The degree of partial admission at full load conditions (50,000 kW) is 40%. The stop valve is fixed to the turbine casing. Fig. 10-6 shows the front elevation of the turbine in section.

Three extractions are provided for regenerative feedheating from the steam spaces after the sixth, tenth and the thirteenth stages. The feed temperature at full load is 152°C. The quantity of steam flowing through the turbine at the designed load conditions is about 225 tons per hour at 50,000 kW and 176.5 tons at 40,000 kW. The efficiencies are $\eta_{oi}=0.82$ and $\eta_m=0.994$.

The front-end gland sealing is combined labyrinth and water sealed, the rear end being only water sealed. Steam leakage through the labyrinth seals at full load is 0.5% of the total steam supplied. The leakage steam from the labyrinth seals is utilised for heating the feedwater in heaters. Pressure after the first stage at maximum load is 16 ata, the heat drop across the regulating stage being about 39 kcal/kg. The mean diameter of the last (low-pressure) few stages is 2,800 mm. Height of blades is 762 mm. Circumferential velocity at the mean diameter $u_{av}=220$ m/sec, and at the rim $u_{max}=280$ m/sec. Large exit cross-sections for the moving blades of the low-pressure stages ensure relatively low carry-over losses. At a load of 50,000 kW at condenser pressure of $p_2=$ 0,04 ata the carry-over loss $h_e=8$ kcal/kg, i. e., 2.9% of H_0.

Diaphragms of the first twelve stages are made of steel and are fastened onto steel liners. The use of steel liners simplifies the casting of the steel cylinder and helps in improving the conditions of warming up. The remaining four diaphragms of the later (low-pressure) stages are of cast iron and are directly mounted on the exhaust housing. The turbine rotor is supported by two journal bearings. The bearing sleeves are made spherical in shape. The turbine rotor is coupled to the generator by a solid coupling. Axial thrust is taken up by a serrated-type thrust bearing placed at the front end of the turbine. The turbine shaft maintains a good degree of stiffness at the critical speed of $n_{cr}=1,910$ r. p. m. The distance between the two journal bearings supporting the turbine shaft is 4,175 mm. The total weight of

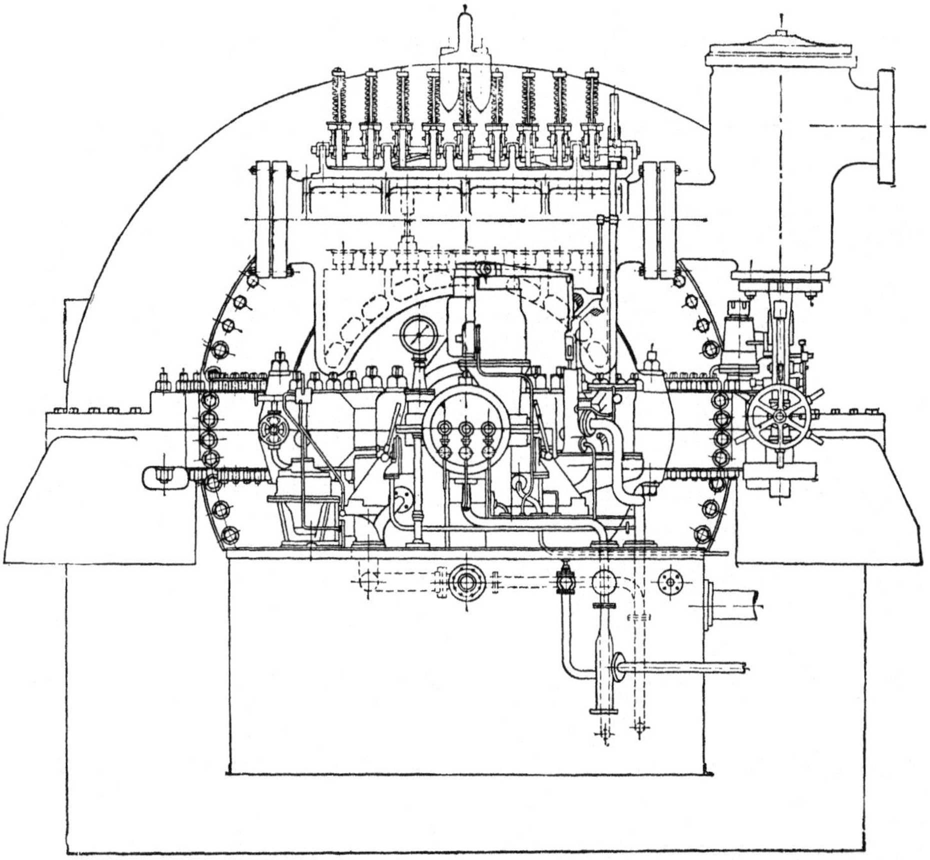

Fig. 10-6. Front view of turbine AK-50-X

the turbine is 240 tons; weight of the rotor being 34 tons.

The turbine is of comparatively simple construction. This turbine has found a wide field of usage because of its reliability and sufficiently high efficiency.

The centrifugal governor and the main oil pump are mounted on the same shaft which is driven by the turbine shaft through a worm pair gearing. The speed of rotation of the centrifugal governor and oil pump is 339 r. p. m. Oil is delivered by the oil pump at a pressure of 7 atm. guage. Fig. 10-7 shows diagrammatically the regulating system employed in the turbine.

Centrifugal governor *1* through levers *2* and *4* displaces servomotor piston *5*. The servomotor piston is connected to a rack and pinion which brings into rotation a spindle carrying a cam *6*. Utilising cams of different shapes mounted on this shaft it is possible to arrange for a consecutive opening and closing of regulating valves *7*. Levers *8* and *4* provide the return connection for the servomotor. In case of parallel operation the variation of load on the turbine is brought about by the synchroniser mechanism by varying the spring tensions.

Overspeed trip gear *15* (ring type) simultaneously operates the two automatic interlocks when the speed increases by more than 10% of the normal speed. When the interlocks are tripped the regulating oil from pipes *16* and *17* flows to the drain. Piston of oil cut-off *9* moves downwards shutting off stop valve *10*. Simultaneously with this oil pressure in chamber *12* falls. The spring *19* pressure now forces down the piston of the cut-off mechanism pulling down with it the automatic lever *4* through coupler *13*. Lever *4* consists of two parts. At the time of normal operation of the turbine both parts of lever *4* are held together by the tension of spring *20*. When coupler *13* moves down the lower portion of lever *4* revolves about hinge *18* and pushes down servomotor piston *5* causing the shutting off of regulating valves *7*. Thus the overspeed trip simultaneously closes both the stop valve and the regulating valves, ensuring tripping of the turbine in case of overspeeding. A drop of pressure in the oil system of the regulators due to any other reason also duplicates the same operation as has been described above.

Oil for lubrication of the turbine and alternator bearings is supplied from the pressure main

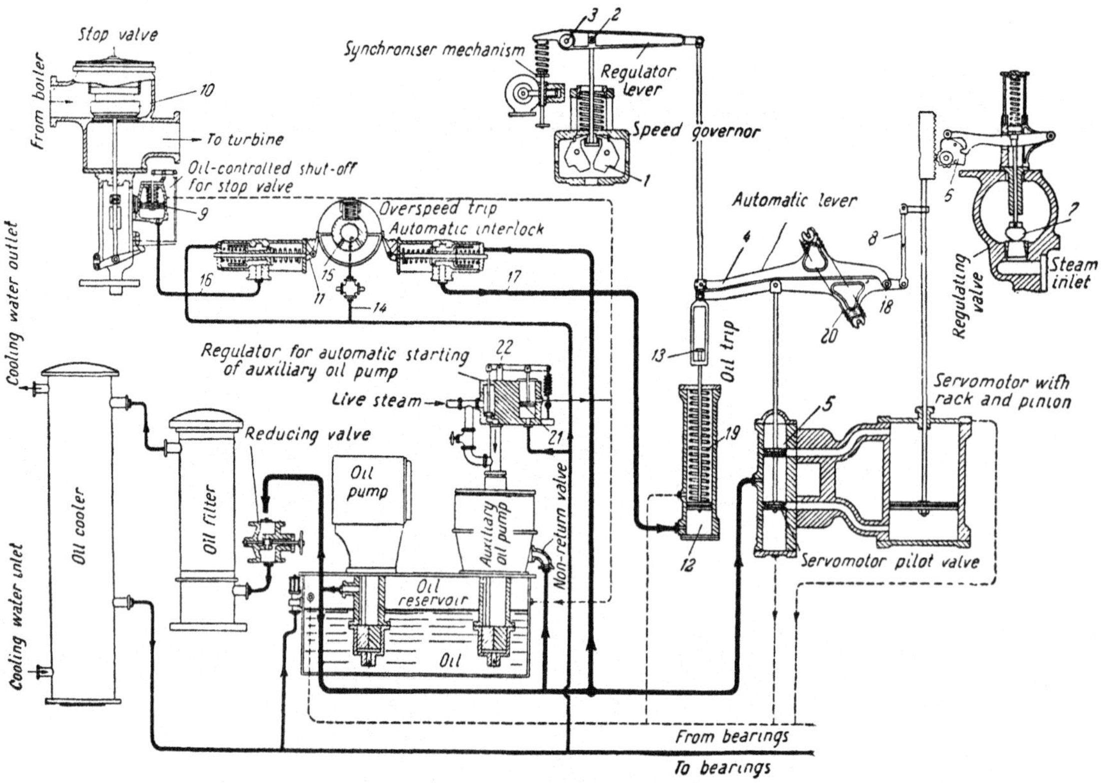

Fig. 10-7. Governing system of turbine AK-50-X

through a reducing valve, filter and an oil-cooler. The turbo-pump and the gear pump for pumping oil are placed on the oil reservoir. The oil system of the turbine is provided with an arrangement for automatically switching on the auxiliary oil pump through a steam-oil regulator. With a fall in the oil pressure of the lubricating system the piston of steam-oil regulator *21* moves downwards about fulcrum *22*, opening the stop valve supplying fresh steam to the turbo-pump.

b) *Turbine K-100-90 (VKT-100)*

Fig. 10-8 shows in section a two-cylinder condensing turbine of 100,000 kW capacity at normal speed of 3,000 r. p. m. The turbine is meant to operate as a prime mover for an alternator (this turbine was built by the factory in 1957). The turbine is designed for the following initial steam conditions: pressure $p_1=90$ ata; temperature $t_1=500$ to 535°C; temperature of cooling water for the condenser at inlet 10°C. The permissible variation of pressure and temperature at rated loads are: pressure 85 to 95 ata and temperature 490 to 540°C.

The turbine has 21 pressure stages, seventeen of which are in the high-pressure region. The low-pressure cylinder is of the double-entry type with four stages for each of the entrances.

Steam supply is through throttle valves with partial bypass to the fifth stage. There is provision for seven extractions for regenerative feed-heating, up to a temperature of 215°C. The nozzles of the first stage and the diaphragms of the following six stages are fixed to the inner casing of the turbine. Diaphragms of the remaining high-pressure stages are fixed to liners. The vertical joints of the high-pressure cylinder are welded. The ends of the turbine are sealed by the usual labyrinth gland seals (without bushing). A regulator is provided for regulating the sealing steam pressure. Steam from the high-pressure cylinder is led to the central portion of the low-pressure cylinder through cast-iron pipes 900 mm dia. The low-pressure stage diaphragms have stiffening ribs. Narrow guide blade segments are secured in diaphragms by sliding them into the slots machined in the discs. The ratio of dia. to blade height for low-pressure stages is 2.82. The front end bearing is a combined journal and thrust bearing, with spherical surface for the bearing sleeve. Two condensers are welded to the exhaust piping of the turbine. These condensers are mounted on a set of springs placed on the condenser supports. The high-pressure rotor is made of forged steel. The low-pressure rotor is made up of discs keyed onto the shaft. The turbine and alternator rotors are joined by a solid coupling. The turbine is provided with an arrangement for slowly revolving the rotor. Load and speed regulation of the turbine is of the hydraulic type. Connected to the turbine shaft are two oil pumps one of which is an impulse pump. The other pump delivers oil to the oil system of the turbine at the normal working pressure.

The turbine is provided with the following protective devices:

1) a vacuum regulator, which gradually reduces the load on the turbine for a vacuum drop to 500 mm Hg and trips the turbine completely when the vacuum drops to 500 mm Hg;

2) axial displacement relay, which trips the load on the turbine if the axial displacement is considerable;

3) automatic condensate level-and-recirculation regulator.

The oil system of a turbine is provided with an auxiliary turbo-pump and an electrical standby pump. The standby pump automatically comes into operation as soon as the pressure of oil in the lubricating system falls below 1 atm. gauge. The electrical oil pump also comes into operation when the turbine rotor is revolved slowly.

c) *Turbine K-150-130 (PVK-150)*

This turbine was manufactured in the year 1958. The maximum continuous rating of the set is 150,000 kW at 3,000 r. p. m. The turbine is made up of two cylinders (Fig. 10-9). Steam is supplied to the turbine at 130 ata and 565°C. There is an arrangement for intermediate reheating at 30 ata up to a temperature of 565°C. At nominal loads the condenser pressure is maintained at 0.035 ata; 20,812 m^3/hr of cooling water at 12°C is required to maintain this condenser pressure at the rated load. Steam is supplied to the moving blades through nozzles. The high-pressure cylinder has a single-row regulating stage and fourteen impulse stages. The low-pressure cylinder is of the double-flow type with six stages on either side. The high-pressure cylinder stator is made up of a double-walled shell. The turbine has eight extractions for feedheating. At the nominal load the feedwater is heated to a temperature of 226°C, the pressures at which extractions are effected being: first—32.5 ata; second—21.15 ata; third—12.5 ata; fourth—7.6 ata; fifth—4.6 ata; sixth—1.45 ata; seventh—0.73 ata; eighth—0.343 ata.

The deaerator is fed with steam from the fourth extraction and operates at a pressure of 6 ata.

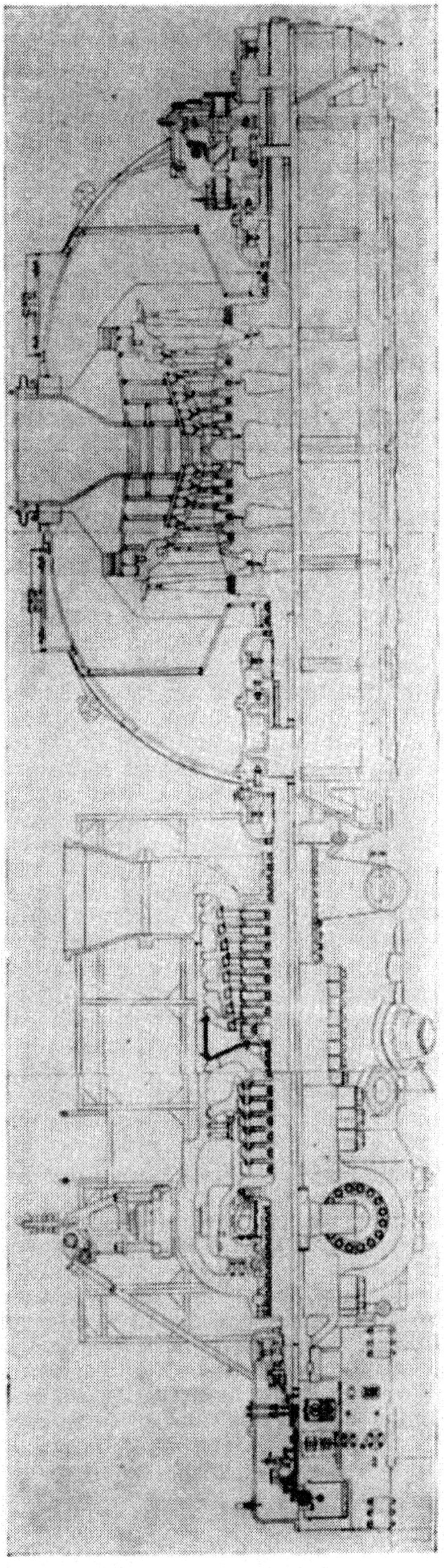

Fig. 10-9. Turbine K-150-130 (PVK-150) of Kharkov Turbine Works

In the event of the turbine operating at lower loads the deaerator can be supplied with steam from the third extraction. Besides being used for feedheating the available heat from the fifth and sixth extractions supply 12 Mcal/hr for heating the recirculating water system of the feedheaters.

At nominal loads the total quantity of steam flowing through the turbine is 472 tons/hr. Carry-over losses account for 10 kcal/kg. The quantity of steam received by the condenser is 309 tons/hr. The guaranteed specific heat consumption is 2,010 kcal/kWh.

Steam is led away for intermediate reheating after the seventh stage of the high-pressure cylinder. The steam chamber after the seventh stage has a separating diaphragm. To avoid large temperature differentials across this diaphragm it is screened and washed on both its sides by steam of the same temperature.

The steam pipes delivering steam after reheat to the high-pressure stages are also of the double-walled type and are shielded and continuously rinsed with "cold" steam in order to reduce the turbine stator temperature. There are special arrangements for heating the bolts and flanges so that the time taken for heating up the turbine before starting may be reduced. The high-pressure cylinder rotor is made of flexible forged steel and weighs 12,800 kg. The distance between the supporting bearing centres is 5,109.5 mm. The low-pressure cylinder is made up of welded forged steel discs without the central orifice. It weighs 36 tons. The low-pressure cylinder bearings are 4,076 mm apart. Mean diameter and height of blades for the last stage are: $d_z=2{,}125$ mm and $l_z=780$ mm. The d_z to l_z ratio is 2.73. At the time of construction of this turbine its last row blades, made of stainless steel, were the longest for a turbine operating at 3,000 r. p. m. The cross-section of exhaust for one stream is 5.21 m^2 and for both 10.42 m^2.

The rotors of high-pressure cylinder, low-pressure cylinder and the alternator are all coupled to one another by means of semi-flexible couplings.

The labyrinth gland seals for the turbine are of the conical tooth type, without a separate sleeve. Steam-air mixture is removed from the labyrinth seals at a pressure of 0.97 ata by a steam-air ejector. The ejector system operates automatically at any load. The journal bearings have spherical bases. The thrust bearing is made in one piece with the front end journal bearing. The turbine is provided with an arrangement for manually revolving the turbine rotor. This device is located in the bearing housing of the low-pressure cylinder at the alternator end. The turbine is provided with hydrodynamic system of regulation. An impeller acts as the source of impulse for changes in speed. The degree of non-uniformity of regulator is 5%. The impeller and the centrifugal-type main oil pump are connected to the shaft of the high-pressure rotor at the front end. The axial displacement relay and the overspeed trip are also situated here. The turbine is also provided with a vacuum regulator which continuously reduces the load with fall of vacuum and at some predetermined vacuum trips the turbine.

4. Turbines of the Leningrad Metal Works (L.M.W.)

Turbines operating at high and supercritical initial steam parameters are built at this works In the very early days of turbo-construction this factory was building turbines at medium initial steam parameters.

The following are some of the types of steam turbines built by the L. M. W.:

a) Turbine Type AK-50-2[1]

This is a single-cylinder condensing turbine of 50,000 kW capacity at normal speed of 3,000 r. p. m. The steam pressure and temperature at inlet are $p_0=29$ ata, $t_0=400$°C and exhaust pressure $p_2=0{,}04$ ata. The cross-sectional view of the turbine is shown in Fig. 10-10.

The turbine has a single-row regulating stage and eleven pressure stages. Mean diameter of the regulating stage is 1,233 mm. The last stage diameter is 1,756 mm; height of blades 576 mm. The circumferential velocity at mean dia. of the last stage is 276 m/sec. The eleventh stage is of the partition type (Bauman stage) with direct delivery of steam to the condenser from the upper half portion. The use of Bauman stage for the eleventh stage enabled the construction of the turbine at 3,000 r. p. m. for the capacity of 50,000 kW. There are four extractions for regenerative feedheating. The front end bearing is a combined journal and thrust bearing. The thrust bearing is of the segment type. The journal bearing sleeve is of the spherical type. Weight of the turbine rotor is 16.5 tons. The distance between the two supporting journal bearings is 3,845 mm. Critical speed of the rotor is 1,760 r. p. m. The turbine and alternator rotors are joined by a flexible coupling. Slow rotation of the turbine rotor is effected by means of an electric drive. The main oil pump and the centrifugal governor spindle are driven through a worm and worm-gearing. Steam is delivered to the blades through nozzles. After passing through the main stop valve steam enters a distribution box *4* provided with

[1] A non-standard turbine.

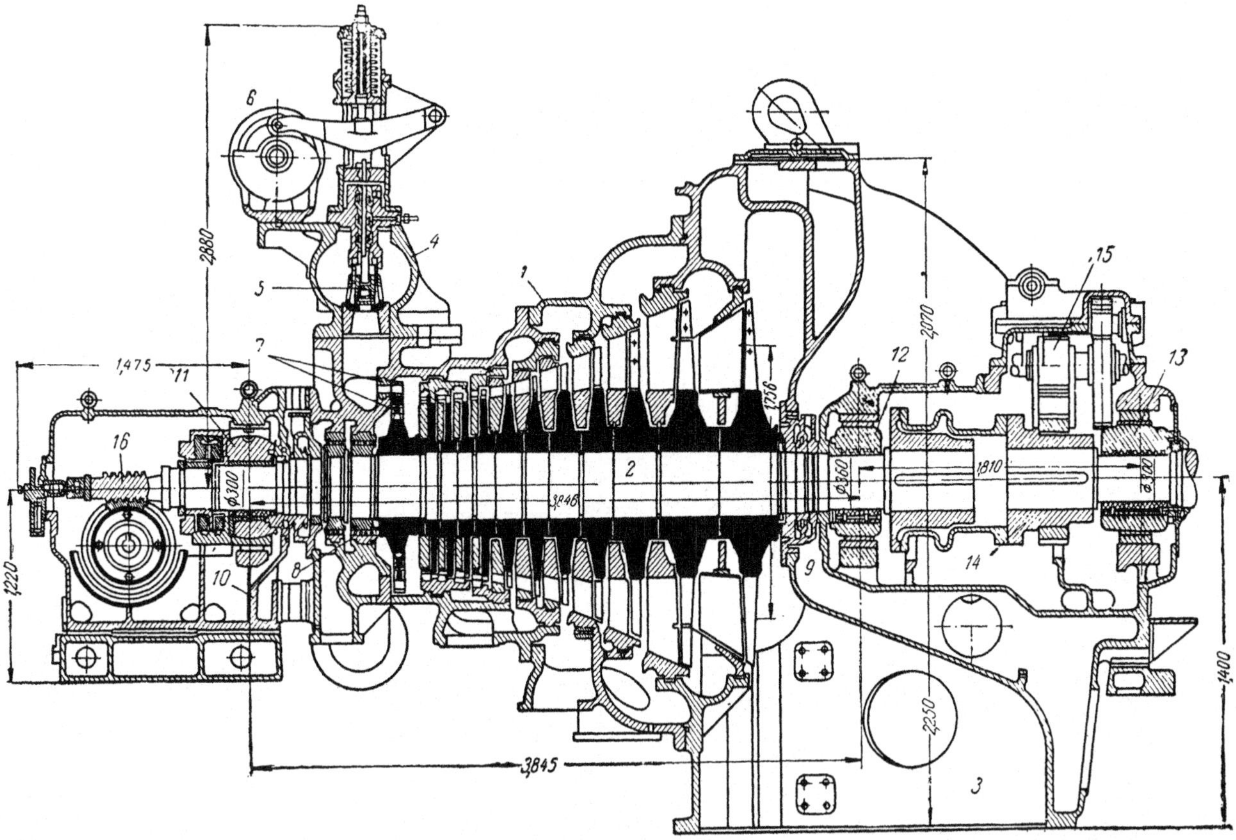

Fig. 10-10. Turbine AK-50-2 (L.M.W.)

1 — casing; *2* — shaft; *3* — exhaust pipe; *4* — valve chest; *5* — regulating valve; *6* — valve actuator; *7* — regulating stage; *8* — H.P. labyrinth packing; *9* — L.P. labyrinth packing; *10* — front bearing casing; *11* — front combined journal and thrust bearing; *12* — rear journal bearing; *13* — generator bearing; *14* — coupling; *15* — shaft turning gear; *16* — worm reduction gearing for speed governor

three regulating valves. The steel steam distribution box is a casting separate from the main turbine housing. Besides there is a fourth regulating valve placed in a separate box at the side of the turbine and attached to the lower casing of the turbine. This regulating valve is operated by an independent servomotor. The three regulating valves at the turbine head are operated by a servomotor of the revolving type through cams fixed on its spindle. The regulating stage has partial admission of steam. The gross weight of the turbine is 161 tons and is considerably smaller than the weight of turbine type AK-50-1 of the same capacity operating at a speed of 1,500 r. p. m.

b) Turbine Type AK-100-1[1]

The use of double flow along with Bauman stages for the low-pressure cylinder considerably increases the capacity of the turbine as compared to that of a unidirectional flow turbine. An example of such a construction is turbine type AK-100-1 built b ythe L.M.W. It has a capacity of 100,000 kW at 3,000 r. p. m. The sectional view of the turbine is shown in Fig. 10-11. Initial steam temperature and pressure are $t_0=400°C$ and $p_0=29$ ata. Exhaust pressure $p_2=0.04$ ata. Blade sizes for the last stage are the same as those of turbine AK-50-2 (Fig. 10-10) whereas the capacity is doubled here.

Turbine AK-100-1 has nozzle governing. Three regulating valves are placed on the turbine head and two at the side. With full opening of all the nozzles the degree of partial admission to the regulating stage is $\varepsilon=0.85$. The regulating stage consists of a single-row Curtis disc of 1,250 mm dia. ensuring a high degree of efficiency for the stage. At the economic loading the internal efficiency of the regulating stage is 0.65 and at the nominal load it reaches a value of 0.75. The turbine is made up of two cylinders. The regulating stage and seven pressure stages make up the high-pressure cylinder. The low-pressure cylinder consists of four stages for each flow direction. Steam from the high-pressure cylinder is deliv-

[1] A non-standard turbine.

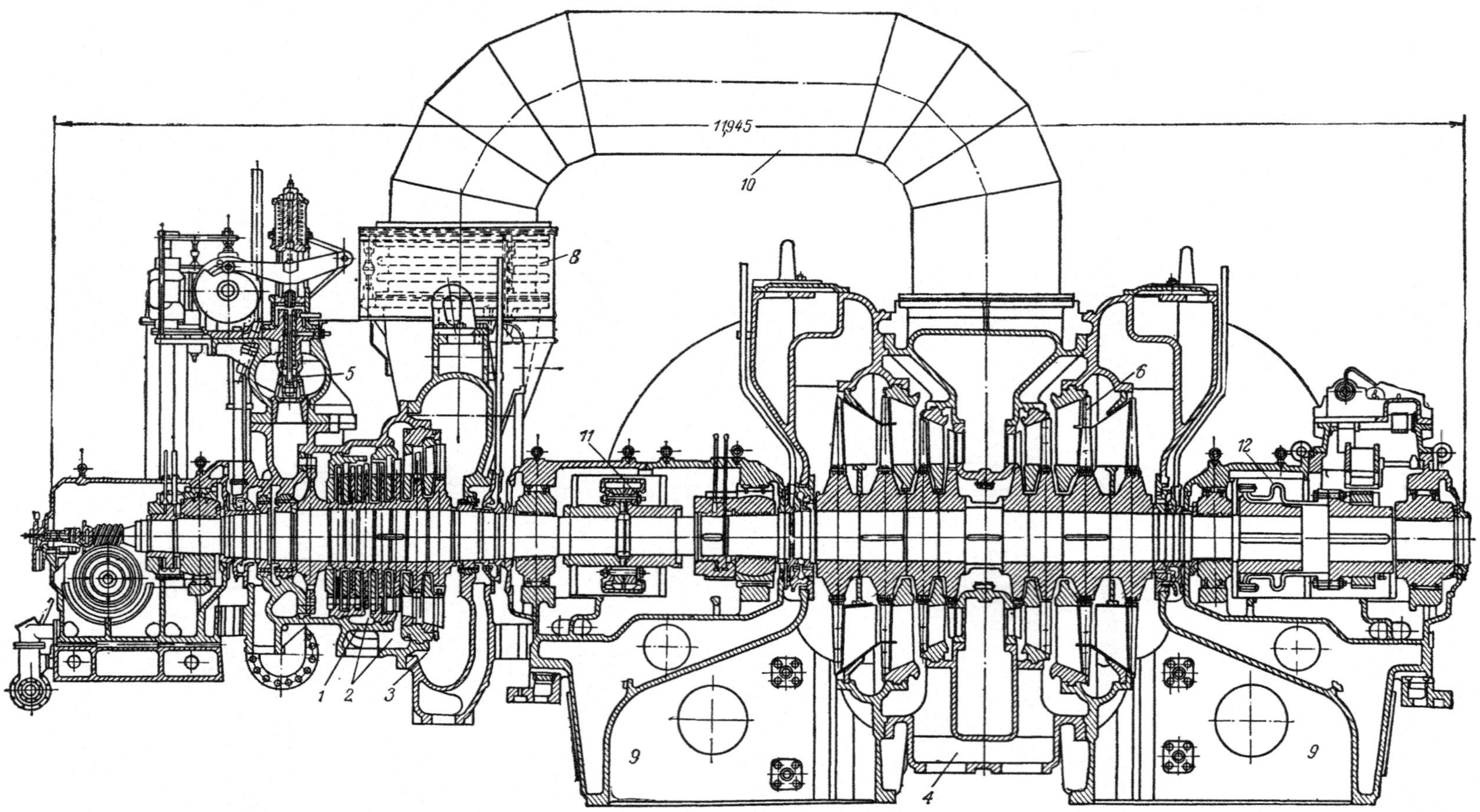

Fig. 10-11. Turbine AK-100-1 (L.M.W.)

1, *2*, *3* and *4*—extraction chambers; *5*—regulating valve; *6*—double-gallery stage; *7*—shaft-turning gear; *8*—compensator; *9*—exhaust pipes; *10*—connecting pipes; *11*—flexible coupling; *12*—semi-flexible coupling

ered through two steam pipes *10* to the central portion of the low-pressure cylinder. Exhaust steam from both flow directions is delivered to the two condensers through two exhaust pipings *9*.

The mean dia. of the last stage is 1,756 mm; height of moving blades l_z=576 mm.

The turbine has four extractions from chambers *1*, *2*, *3* and *4*, the pressures in these chambers at nominal loads being: 8.6, 4.25, 1.5 and 0.3 ata respectively. The total quantity of extracted steam is about 20% of the steam supplied to the turbine. At nominal load the feedwater is heated to a temperature of 167°C and at economic load the temperature of feedwater reaches 158°C. At nominal load with a back pressure of 0.05 ata 36% of the flow is diverted through the Bauman stage directly to the condenser. At the back pressure cited above the exit velocity of steam from the upper half of the Bauman stage is 281 m/sec and the exit velocity from the last stage is about 258 m/sec. The carry-over losses under such conditions of operation are about 8.5 kcal/kg which is about 3% of the total heat drop occurring in the turbine H_0=272 kcal/kg. The internal efficiency η_{0i} of the turbine at economic loads is 0.78. The total quantity of steam flowing through the turbine is, at nominal load, 460 tons/hr and at most economic load D_0=360 tons/hr.

The high- and low-pressure cylinder rotors are joined together by a flexible coupling *11*. Each of the rotors has a separate thrust bearing. The low-pressure cylinder rotor is connected to the alternator rotor with a semi-flexible toothed coupling *12*. The gross weight of the turbine is 289 tons which is about 40% less than that of a turbine of equal capacity operating at 1,500 r. p. m. built by the Kh.T.W.

c) *Turbine Type K-50-90 (VK-50-1)*

The L. M. W. has developed a design for the mass production of single-cylinder uniflow high-pressure turbine of type K-50-90 with maximum capacity of 50,000 kW at 3,000 r. p. m. (Fig. 1-8). The initial and final steam conditions for this turbine are p_0=90 ata, t_0=500°C and p_2=0.036 ata. Five extractions are provided for the regenerative feedheating system with the aim of reducing the quantity of steam flow through the last stage. Circumferential velocity and ϑ at the mean diameter of the last stage were assumed at their maximum allowable values at the time of design of the above turbine: u_{av}=314 m/sec, ϑ=3 conforming to d_{mean}=2,000 mm and l_z=665 mm.

The maximum circumferential velocity at the rim u_{max} =421 m/sec. The leading edges of the moving blades of the seventeenth and eighteenth stages are stellit-coated to protect the blade edges from erosion. The final stage moving blades are given a certain amount of twist in order to obtain a uniform resistance blade section.

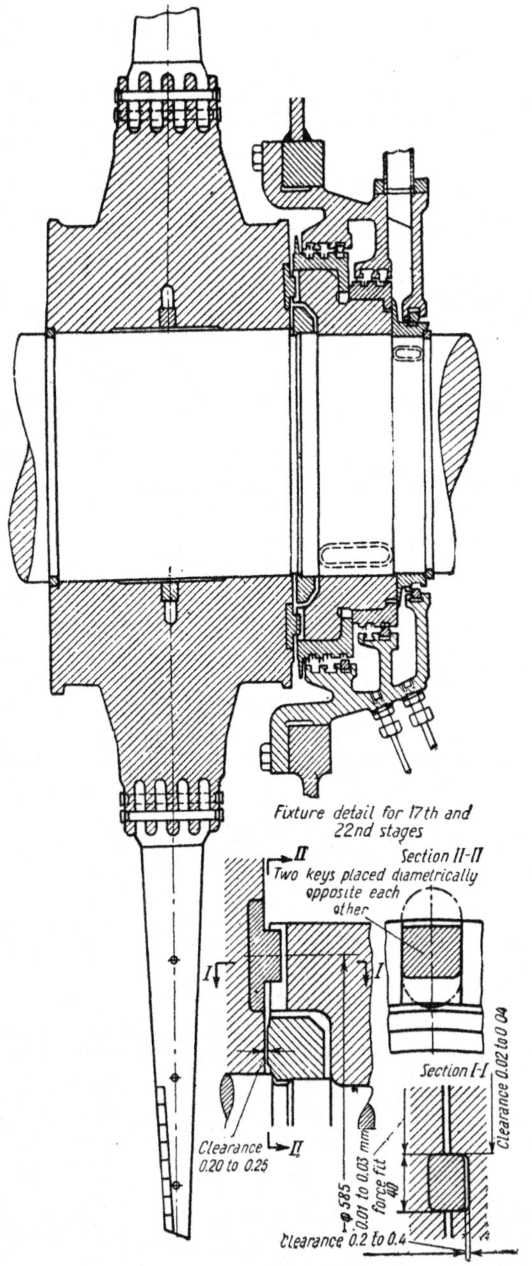

Fig. 10-12. Attachment of last stage blades on shaft of L.M.W. turbines types K-50-90 (VK-50-1), K-100-90 (VK-100-2), etc.

Steam supply to the blade system is through nozzles. The turbine is provided with four regulating valves. The regulating stage is a double-row Curtis stage, and is followed by seventeen pressure stages. The regulating stage disc and the discs of the following ten stages are forged in one piece with the rotor. The discs of the following stages are shrunk on the turbine shaft and are further secured by means of longitudinal keys. In view of the large tangential forces present ($\sigma_{t\,max}=2{,}700$ kg/cm^2) the disc of the last stage is keyed on its side to the rear labyrinth seal bush (Fig. 10-12). The maximum steam flow through the turbine ($N_e=50{,}000$ kW) is 191 tons per hour. At this load the pressure in the regulating stage is about 50 ata and temperature 410°C. The turbine rotor is supported by a pair of journal bearings. The front end bearing *15* is a combined journal and thrust bearing (Fig. 1-8). Distance between centres of the two journal bearings is 4,350 mm. The turbine and alternator rotors are connected by a semi-flexible toothed coupling. There is a provision for the manual rotation of the turbine rotor. The turbine rotor weighs 17 tons. The turbine shaft is a flexible one with critical speed of 1,790 r. p. m. All the diaphragms are fixed in special liners which considerably simplifies the stator design. The high-pressure cylinder stator is made of cast molybdenum steel with 0.5% of Mo. The medium- and low-pressure sections are made up of fabricated steel welded together which considerably reduce the gross weight of the turbine. The gross weight of the turbine is 148 tons. At the nominal load of 50,000 kW the feedwater is heated to a temperature of 212°C.

The theoretical specific heats are:

at 50,000 kW	. .2,280	kcal/kWh
" 40,000 "	. .2,295	" "
" 30,000 "	. .2,335	" "

The turbine has a completely original design for a single-cylinder uniflow machine operating at 3,000 r. p. m. which for that period of turbo-building was an epoch making achievement for the L. M. W.

d) Turbine Type K-100-90 (VK-100-2)

This turbine is a two-cylinder type with double flow in its L. P. cylinder. Steam supply is at $p_0=90$ ata at a temperature of $t_0=500$°C, $p_2=0.033$ ata. The sectional view is shown in Fig. 10-13 (at the end of the book).

Steam is supplied through two stop valves. From the stop valves the steam is led to four regulating valves which are placed in four separate steel steam chests welded to the high-pressure cylinder stator. Two of the regulating valves are placed at the head of the high-pressure cylinder and the remaining two at the side. The front end of the turbine is exactly similar to that of turbine type VK-50. The working part of the turbine consists of one two-row regulating disc and eleven pressure stages. The regulating stage disc and following nine pressure stage discs are made in one piece from forged chrome-molybdenum steel. The last two discs are shrunk on the turbine shaft when hot. The rotor is non-flexible with critical speed of 3,620 r. p. m. Weight of rotor with flanges is 10,330 kg.

From the last stage of the high-pressure cylinder steam is delivered to the central portion of the low-pressure cylinder through two steam pipes *7*. The low-pressure cylinder has ten pressure stages, five for each direction of flow. The L. P. rotor is flexible with critical speed of 1,670 r. p.m. The L. P. cylinder shaft is made of carbon steel. Weight of rotor with half the flange on either side is 21,960 kg. The discs are made from strong chrome-nickel-molybdenum steel. Dimensions of discs and blades of the last stage are the same as were for turbine VK-50-1. The weight of the moving blades of the last stage is 5.27 kg. The discs are force fitted on the turbine shaft and are further secured by longitudinal keys. The low-pressure stage (17 to 22) discs for both flow directions are fixed to the rotor through a radial key attachment to the labyrinth seal bush.

Regenerative feed heating is effected through five extractions from chambers *1*, *2*, *3*, *4* and *5*. The wetness of exhaust steam does not exceed 13%. There is an arrangement for moisture removal from the low-pressure stages. Exhaust from the turbine is condensed in two separate condensers. The front and rear ends of the turbine shaft are sealed by labyrinth packings of the fir-cone type. Steam is bled from the first chamber of the front end labyrinth seal into the third extraction chamber. Bleeding of steam from the second chamber of the front end sealing and the rear end gland seal are effected by connecting them to the steam pipe of the fifth extraction which operates under vacuum. Hence the outer ends of the labyrinth packings are supplied with steam throttled to a pressure of 1.3 ata and cooled by condensate to a temperature of 110 to 120°C. Supply of throttled steam at low temperature to the gland packings helps in cooling the shaft ends. The front end journal pedestal is of cast iron and houses the combined journal and thrust bearing, reducing gear transmission to the main oil pump and speed governor spindle, overspeed trip device and low vacuum trip, main servomotor and high-pressure oil piping. The front end bearing pedestal is similar to that of turbine type VK-50-1.

The turbine is provided with an arrangement for slow rotation of the turbine rotor (about 2 to 4 r. p. m.) in the rear end bearing block. The slow rotation of the turbine shaft is carried out by a small electric motor of 8. 5 kW capacity operating at 750 r. p. m.

The high-pressure cylinder, except for the exhaust end, is made of 0.5% molybdenum steel. The exhaust section is made of gray cast iron (grade СЧ-40). The maximum thickness of flanges for the H. P. cylinder is 275 mm. The midportion of L. P. cylinder is also made of gray cast iron. Exhaust section of the L. P. cylinder is made of welded steel sheetings. The upper half of the H. P. cylinder stator is positioned by four anchors formed by projections on the flanges of the lower half of the cylindrical stator and is supported in front by the bearing pedestal and in the rear by special bearing pads provided on the L. P. cylinder stator. The relative disposition of the H. P. cylinder stator and the bearing pedestal is fixed by the vertical key, permitting expansion in the vertical direction. The front end bearing pedestal rests on the bedplate and is positioned by a longitudinal key grooved into the bedplate. The key is so placed as to coincide with the longitudinal axis of the turbine. Provision of these keys at right angles to each other, one in the vertical direction and one in the horizontal direction, facilitates exact centring of the cylinder and the bearing, without, in any way, obstructing the thermal expansion of the cylinder or the bearing in both the horizontal and the vertical directions. The M. P. cylinder upper casing is secured at its rear end on the special bearing pads by vertical keys.

Table 10-1

Power developed at alternator terminals, kW	Steam flow through stop valves, tons/hr	Final feedwater temp., °C	Guaranteed alternator efficiency, %	Specific heat, kcal/kWh
80,000	301	205	98.8	2,275
90,000	338	207	98.9	2,265
100,000	377	212	99.0	2,250

The fixed point of the turbine is centred by two transverse and one longitudinal keys. The transverse keys are situated at the right and left of the cylinder recessed into the bedplate of the L. P. cylinder. At steady loading conditions linear expansion of the stator reaches a value of 11 mm. Some of the basic dimensions of the turbine are shown in Fig. 10-13.

The total weight of the turbine is 264 tons which is less than that of turbine AK-100-1 of the same capacity.

The guaranteed steam consumption and specific heat in kcal per kg for turbine VK-100-2 are given in Table 10-1; these figures are guaranteed within a 5% variation.

Three extraction pumps, each of capacity 175 m^3/hr, are provided for removal of condensate from the condensers. Two three-stage steam air ejectors are provided for removal of air. The quantity of circulating water at normal operating conditions is 20,000 m^3/hr.

The regulation system of the turbine is shown in Fig. 10-14. The regulating valves are opened by the rotation of camshaft *1* which is operated by the main servomotor *3* through a rack and pinion *2* and connecting levers. The servomotor is in its turn operated by the pilot valve *4*. Displacement of the pilot valve is achieved by the variation of oil pressure in the oil system *A*.

At steady loads the pressure of oil in the oil system *A* is constant. Any variation in the speed of the turbine causes the pilot valve *7* to be displaced. This displacement of pilot valve *7* alters the pressure below the lower piston of pilot valve *5*, causing it to be displaced in its own turn which immediately leads to a change of pressure in the oil system *A*. The pilot valve *4* is held in its neutral position by the oil pressure on one side and spring tension on the other. Thus the variation of oil pressure in the oil pipe *A* causes the pilot valve *4* to be displaced from its midposition. The displacement of pilot valve *4* permits flow of pressure oil to either the upper or lower side of the servomotor piston *3* which as described before operates the regulating valves through a system of levers.

The pressure of oil in the regulating system is 12 atm. gauge, bearing lubrication is at 0.4 atm. gauge. The capacity of the main oil pump is 2,250 l/min. The main oil pump is driven by the turbine shaft through a speed reduction gearing. The turbine is provided with two pin type trip devices designed to trip the turbine in the event of the speed exceeding the normal r. p. m. by 10.5 to 12%. The turbine is automatically tripped by an electromagnetic axial displacement relay in the event of an axial displacement of the turbine by 1 mm towards the alternator side. The relay operates through an oil lever system and closes the main stop valves. These turbines are extensively being used in the various thermal power stations of the U.S.S.R.

e) Turbine SVK-150 [1]

The nominal capacity of this turbine is 150,000 kW at operating speed of 3,000 r. p. m. Steam is supplied to the turbine at $p_0=170$ ata; $t_0=$ 550°C and is exhausted at $p_2=0.03$ ata. The unit consists of three cylinders with intermediate reheating and a double-flow system for the L. P. cylinder. The turbine is shown in Fig. 10-15 (see Appendix).

[1] A non-standard turbine.

Steam is supplied through two stop valves mounted on each of the two supply mains. From the stop valves steam is led to the four regulating valves through a U-shaped piping. The H. P. cylinder casing is of a double shell construction. The nozzle box mouths are welded to the inner casing. The regulating valves are placed in steam chests mounted on the turbine at the head of the H. P. cylinder.

The H. P. cylinder consists of a two-row regulating stage and seven pressure stages. At nominal loads the heat drop occurring in the regulating stage is 40 kcal/kg. The H. P. cylinder rotor is a flexible forged steel construction with critical speed of 2,045 r. p. m. Steam is exhausted from the H. P. cylinder at a pressure of 34 ata. It then enters a reheater and thereafter returns to the front end of the medium-pressure cylinder at a pressure of 29 ata and a temperature of 520°C. The medium-pressure cylinder has 12 pressure stages. The first eight discs are forged in one piece with the shaft and the remaining four are mounted on the shaft and are keyed in place. The first three of these discs have the usual longitudinal keys whereas the fourth disc is secured by radial keys as in the case of the last stage fixing of the L. M. W. turbine K-50-90 and K-100-90 (see Fig. 10-12). The low-pressure cylinder is of the double flow type with three pressure stages for each direction of flow. The last but one stage of the L. P. cylinder is of the Bauman gallery type permitting the sectional area of exhaust to be limited to the value of 12. 6 m^2. All the low-pressure stage discs are mounted on the turbine shaft by hydraulic pressure and are keyed to their hubs by radial keys. The disc hubs are also a force fit on the shaft and are secured in position by longitudinal keys. The turbine is provided with seven extractions for heating feedwater to a temperature of 225°C. The first extraction is from the exit chamber of the H. P. cylinder. The second, third, fourth, fifth and sixth extractions are from the medium-pressure cylinder from the steam spaces after the third, sixth, ninth and eleventh stages and the exit end of the medium-pressure cylinder respectively. The seventh extraction is from the first stage of the L. P. cylinder. The gland packings are of the fir-cone type with carrier bushings. Steam is bled from the first compartment of the front end gland packing into the steam pipe of the first extraction. Also from the second compartment of the front end seal of the H. P. cylinder, first compartment of the rear end seal of the H. P. cylinder and the first compartment of the front end seal of the medium-pressure cylinder steam is bled into the steam pipe of the third extraction. From the remaining compartments of the gland packings of the H. P. cylinder as well as the medium-pressure cylinder the outgoing steam is utilised for condensate heating. Steam is supplied to the gland sealings from a steam cooler at a throttled pressure of 1.03 ata and a temperature of 110°C. This steam, further, ensures lower temperatures for the shaft at the high- and medium-pressure cylinder labyrinth sealings. The turbine is provided with a device for the slow rotation of the rotor. This device is situated in the rear end bearing housing of the L. P. cylinder with an electric drive to the midportion of coupling between the turbine and the alternator. The H. P. and M. P. cylinder rotors are connected together by a flexible coupling and have a combined journal and thrust bearings. The M. P. and L. P. cylinder rotors are connected together by semi-flexible coupling. The main dimensions of the turbine are shown in Fig. 10-15. The turbine is designed for a steam flow of 445 tons/hr. The quantity of steam entering the condenser at nominal load and normal operation of the regenerative feedheating system is 303 tons/hr. For these conditions of operation the specific steam loading on the exit section of the L. P. cylinder is 24.1 tons/m^2 hr. At condenser pressures of 0.03 ata and load of 150,000 kW the carry-over losses are 9.5 kcal/kg, i. e., 2.2% of the adiabatic heat drop occurring in the turbine as a whole. The calculated value of specific heat consumption with operation at nominal load, normal initial steam conditions and a condenser pressure of 0.03 ata is 2,008 kcal/kWh.

f) Turbine K-200-130 (PVK-200-1)

Turbine K-200-130 has a capacity of 200,000 kW at 3,000 r. p. m. Initial pressure and temperature of the supply steam are 130 ata and 565°C, condenser pressure $p_2=0.035$ ata. There is intermediate reheat up to a temperature of 565°C. The turbine forms a compact unit with a steam boiler of 640 tons/hr capacity. The turbine is shown in Fig. 10-16. The machine consists of three cylinders with double flow in the L. P. cylinder, and direct exhaust from the upper gallery of the Bauman stage into the condenser, which is the last but one stage in the L. P. cylinder.

Steam is supplied to the turbine through two automatic stop valves placed at the front end of the H. P. cylinder. From the stop valves steam flows to four regulating valves placed at the front of the H. P. cylinder. High-pressure cylinder housing is of welded cast steel. The nozzle boxes are welded to the H. P. cylinder and the steam chests. The first stage nozzles are made up of four segments and are placed in the nozzle boxes. The H. P. cylinder consists of one regulating stage and eleven pressure stages. The high-pressure cylinder diaphragms are held by three liners. The H. P. cylinder rotor is of forged steel construction (grade P2) and has a critical speed of 1,750 r. p. m.

H. P. cylinder gland packings are of the sleeveless type. The turbine shaft has grooves turned on it and the packing foils are held in liners and pressed into the grooves by springs. The exhaust from the H. P. cylinder at a pressure of 25 ata and a temperature of 340°C is taken to an intermediate reheater. The reheated steam at a pressure of 21.2 ata and a temperature of 565°C flows to the regulating valves of the M. P. cylinder through two safety valves mounted on each of the four steam pipes.

The medium-pressure cylinder consists of eleven pressure stages. The first three diaphragms are recessed into the cylinder casing and the following eight stages are secured by two liners. The first seven discs of the M. P. cylinder rotor are formed by turning the steel forging, the remaining four stages are shrunk on the shaft. The critical speed of the M. P. cylinder rotor is 1,780 r. p. m. The front end gland packing is sleeveless and the rear end one is mounted on bushings.

From the exhaust of the M. P. cylinder steam is led to the central portion of the L. P. cylinder through a 1,520 mm dia. pipe. The pressure and temperature of steam at entry to the L. P. cylinder are 1.6 ata and 235°C. The steam flow is divided into two streams flowing in opposite directions. Each of the two flows has four identical stages. Exhaust steam from each of these streams is condensed in two separate condensers welded to the two respective exhaust pipes.

The L. P. cylinder stator is made up of three parts. The central portion is made from cast iron СЧ-21-40, and the two exhaust ends are made of welded steel fabrication. All the eight discs of the L. P. cylinder rotor are shrunk on the shaft. The discs are further secured by longitudinal keys. The critical speed of the L. P. rotor is 1,610 r. p. m. The end gland seals are of the bush type. These bushes are mounted on the turbine shaft while hot, i. e., shrunk fit.

The H. P., M. P. and L. P. cylinder rotors are supported by five bearings; the L. P. rotor by two bearings and the H. P. and M. P. rotors by three bearings. The H. P. and M. P. rotors are connected by a rigid coupling. Steam supply to the H. P. and M. P. cylinders is from the side of the middle combined journal and thrust bearing. Such an arrangement has led to the decrease in the length of the unit by 1.5 metres as well as reduced the loading on the thrust bearing as a result of the opposite flow directions in the H. P. and M. P. cylinders. This is deemed to be very necessary when there is a large degree of reaction in the moving blades. The M. P. and L. P. cylinder rotors and the turbine and alternator rotors are connected by semi-flexible couplings. The bearing pedestal between the L. P. cylinder and the alternator houses an arrangement for the slow rotation of the turbine rotor to facilitate uniform heating and cooling of the rotor at the time of starting and stopping the turbine. The blades of the high-pressure cylinder are made according to the profiles of high-efficiency blades. Blades with a ratio of $d_{mean}/l_b \leqslant 10$ are made with a twist. Thus the efficiency of the H. P. and M. P. cylinders for this unit is much greater than that of turbines of the earlier periods. The mean diameter of the last stage disc is 2,100 mm with blade heights of 765 mm, the ratio

$$d_{mean}/l_z = \frac{2,100}{765} = 2.75$$

and the circumferential velocity at the mean diameter $u_{mean}=330$ m/sec. The maximum disc diameter for the last stage moving blades from the exhaust side is 2,870 mm, and the maximum circumferential velocity $u_{max}=450$ m/sec. The weight of the L. P. cylinder rotor when assembled is 36 tons. The main dimensions of the turbine are given in Fig. 10-16. The generation capacity at the shaft of each of the cylinders is: H. P. cylinder —62,000 kW, M. P. cylinder — 91,000 kW, L. P. cylinder — 51,000 kW. All turbine parts exposed to high temperatures are made of alloyed steels (pearlite class). The H. P. cylinder casing, nozzle boxes and steam chests, regulating valves and the M. P. cylinder up to its vertical joint are made of chrome-molybdenum-vanadium steel 15Х1М1Ф. The H. P. and M. P. cylinder rotors and the L. P. cylinder shaft are made from P2 grade steel. All the shrunk-fit discs are made from 34ХН3М steels. Manufacture of turbine parts from pearlite class steels considerably reduced their costs. The cost of PVK-200-1 turbine was found to be 30% lower than that of turbine SVK-150-1 (turbine parts in the high temperature zone for this turbine were made out of austenite steels which are about six to seven times costlier than the pearlite steels).

The regulation system of turbine K-200-130 in contrast to the governing system of turbines without intermediate reheat includes additional protection against increase of turbine r. p. m. by the reheat steam. For this purpose four regulating valves are mounted on the medium-pressure cylinder. These regulating valves are also operated by the main servomotor provided for the H. P. cylinder regulating valves. Besides, two safety valves are provided on the reheat sup-

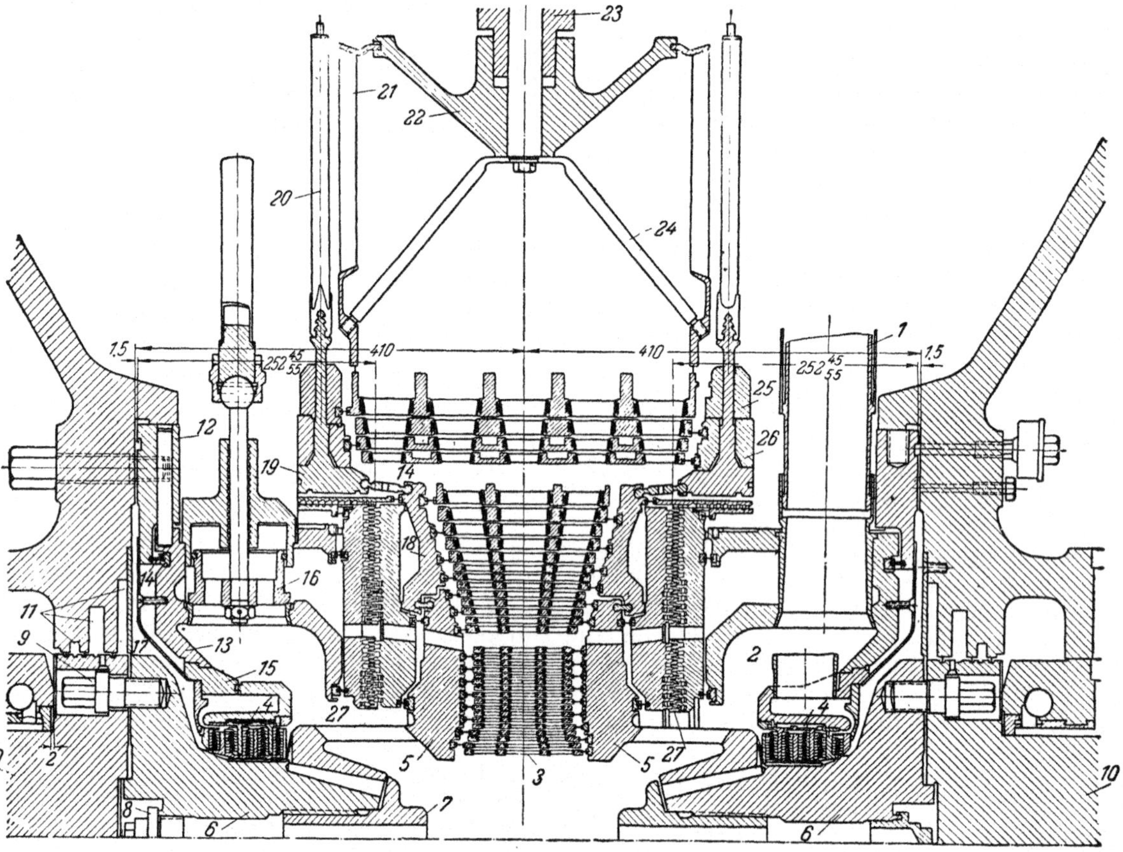

Fig. 10-17. Ljungström turbine 1,400 kW at 3,000 r.p.m.

1 — steam inlet main; *2* — annular chamber; *3* — first row of blades; *4* — labyrinth packing; *5* — blade holders; *6* — turbine shaft; *7* — set screw; *8* — plug; *9* — coupling bolts; *10* — alternator shaft; *11* — oil collector grooves; *12* — ring; *13* — steam chest; *14* — flexible ring (dumb-bell); *15* — labyrinth seal steam chest; *16* — bypass valve; *17* — isolator strips; *18* — central portion of rotor; *19* — outer portion of rotor; *20* — L.P.-stage moving blades; *21* — L.P.-stage guide blades· *22* — casing; *23* — bolt; *24* — stiffeners; *25* and *26* — rings; *27* — radial labyrinth seals

ply mains which exhaust the steam directly into the condenser in case of load tripping. These valves operate in a manner similar to that of the automatic stop valves. A main oil pump of the centrifugal type with a capacity of 7,000 litres per minute is provided for the supply of oil to the lubrication and governing system of the turbine. The main oil pump is situated in the front bearing housing and its spindle is connected to the turbine rotor by a flange coupling. Oil for governing is supplied at a pressure of 20 kg/cm², and for lubrication of bearings a special two-stage injector is provided in the oil sump. Thus the oil system dispenses with the use of reduction gearing and reducing valve considerably simplifying the working of the oil system as well as enhancing its reliability of operation.

For the starting and stopping periods the turbine is provided with a centrifugal auxiliary oil pump with electric motor drive (a. c.). Another motor-driven centrifugal pump is provided for oil supply for lubrication at low turbine speeds. This motor is automatically switched on when the oil pressure in the lubricating mains decreases due to any cause. Further the turbine is provided with a standby pump (driven by a d. c. motor) to supply oil when the turbine is operated only to supply the needs of domestic feeders. This pump is operated by a set of accumulators and is automatically switched on when the lubricating oil pressure drops down to 0.45 kg/cm².

5. Ljungström Turbine

Fig. 1-7 diagrammatically shows the arrangement of a radial Ljungström turbine. Each concentric blade ring operates simultaneously as a guide and a moving blade. Thus the presence of fixed blades with their attendant losses is completely dispensed with. The expansion of steam occurs only in the working blades since every one of the blades is a working blade. Since the two discs carrying the blades rotate with the

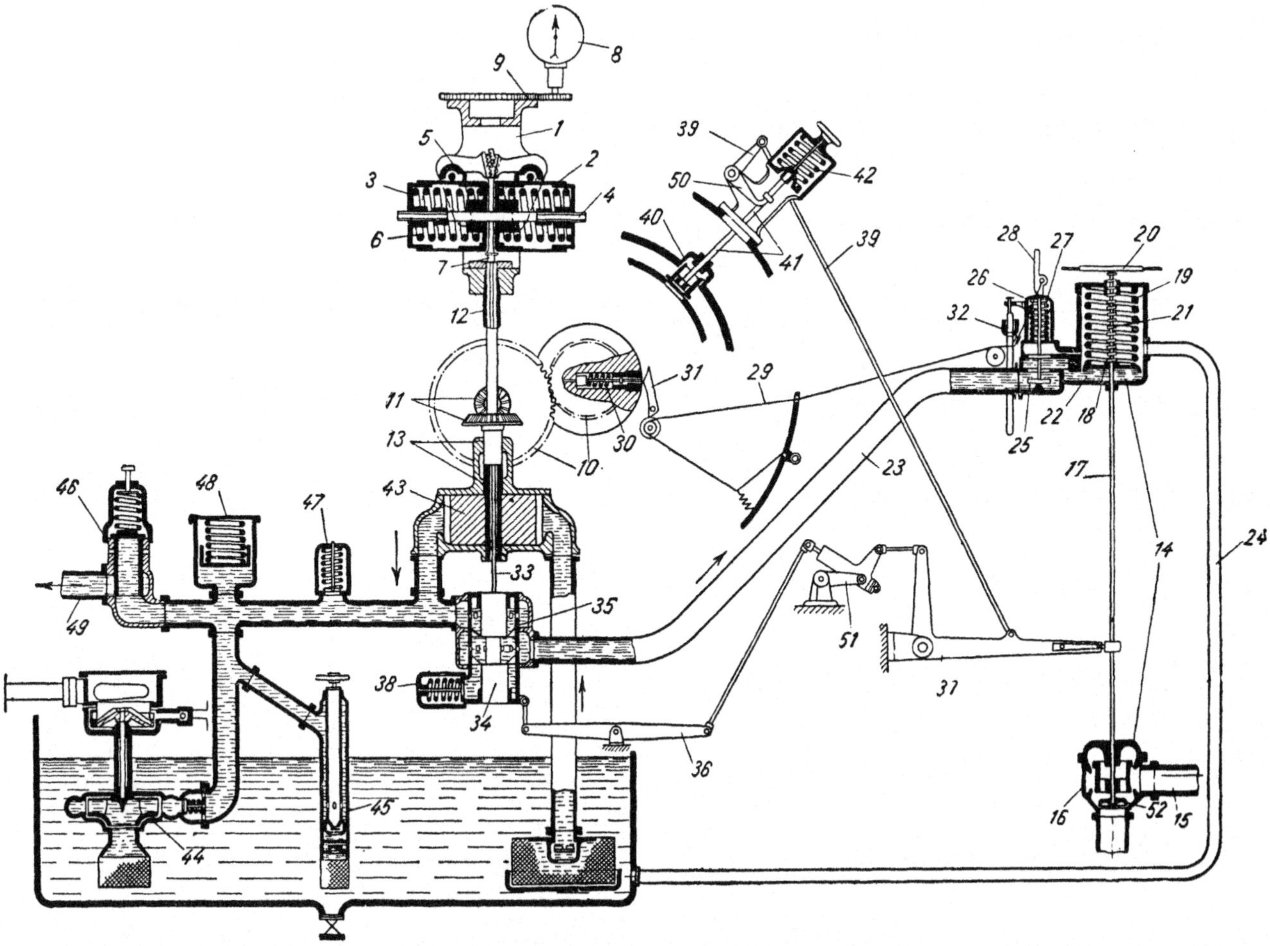

Fig. 10-18. Regulation system of Ljungström turbine

1 — centrifugal regulator; *2* — weights; *3* — washer; *4*—spring spindle; *5* — pins; *6* — spring; *7* — connecting rod to servomotor pilot valve; *8* — tachometer; *9* — transmission to tachometer; *10* and *11* — transmission to regulator; *12* — regulator spindle; *13* — oil pump spindle; *14* — steam supply system; *15* — steam to turbine; *16* — throttle valve; *17* — throttle valve and stop valve spindle; *18* — servomotor piston; *19* — servomotor spring; *20* — handwheel; *21* — handwheel spindle; *22* — oil under pressure; *23* — oil supply to servomotor; *24* — drain oil pipe; *25* — overspeed trip regulator valve; *26* — cover; *27* — spring to valve *25*; *28* — overspeed trip lever; *29* — steel cable to overspeed trip; *30* — trip pin; *31* — overspeed trip catch; *32* — indicating device; *33* — pilot valve spindle; *34* — pilot valve; *35* — pilot valve cylinder; *36* — pilot valve stabiliser lever; *37* — transmission lever; *38* — relief valve for oil; *39* — bypass valve operating lever; *40* — bypass valve; *41* — bypass valve spindle; *42* — spring; *43* — main oil pump; *44* — auxiliary oil pump; *45* — standby hand pump; *46* — reducing valve; *47* — safety valve; *48* — alarm; *49* — oil pipe to bearings; *50* — lever; *51* — rocking lever; *52* — stop valve

same speed but in opposite directions the relative circumferential velocity becomes twice the speed in a reaction turbine, with fixed guide blades, say of the axial type. For a given u/w_2 ratio (w_2 — relative exit velocity of steam from the blades) the velocity w_2 is also doubled. Since the kinetic energy developed in a stage is proportional to the square of the exit velocity (w_2), the energy developed in this turbine is four times that in an axial reaction turbine. Consequently the number of stages required in a radial turbine, all other quantities remaining the same, is considerably smaller than in an axial reaction turbine. Thus the radial Ljungström reaction turbine is a very compact turbine. The first blade ring diameter is made usually small, with a view to obtain blades of sufficient heights. The length of low-pressure blades is smaller than in the case of axial turbines; this is necessitated from the limitations of material strength, etc. The radial turbine has two generators revolving in opposite directions. The turbine rotors are attached to the cantilever ends of the generator shafts. The maximum rating of a radial turbine is limited by the blade size which is comparatively small. Medium- and high-capacity radial turbines are now usually made with one or two of their low-pressure stages as axial to provide greater flow areas for the high specific volume steam.

An example of such a construction is the 14,000 kW; 3,000 r. p. m. double-motion turbine shown in Fig. 10-17. Steam from main pipe *1* enters the blade system *3* through an annular chamber *2*. From the exit of the last row of radial blades the steam is divided into two streams each flowing in the two respective opposite axial directions and flowing through the two axial stages enters the condenser through the exhaust piping. Since the length of the moving blades, even for the first stage, has been found to be sufficiently large, these blades have been divided into three sections for the initial stages and into five sections for the L. P. stages by circular discs, thus decreasing the cantilever ends for the blades and the resulting bending moments.

Each of the discs consists of three parts: internal *5*, middle *18* and external part *19*. The last named carries the axial stages *20* as well. These disc sections are joined together by flexible dumb-bell expansion rings. The inner part of disc section *5* is joined to the hollow shaft *6* by keys and setscrews *7*. The turbine shaft is coupled to the alternator shaft *10* with the help of coupling bolts *9*. The turbine has both axial *4* and radial *27* labyrinth packings for the shafts and discs respectively. Steam supply is through throttle valves, as is the usual practice for all turbines of this type. At nominal loads there is an intermediate steam supply to the eighteenth stage through stop valve *16*. The valve *16* is relieved by steam pressure. The upper piston portion of the stop valve is made of a cast iron ring. The operational data gathered for the operation of radial turbines in the U.S.S.R. show that these turbines are in no way less reliable than their axial counterparts.

In spite of the large leakage losses from the gland packings these turbines are found to be quite economical. The test data of a 10,000 kW radial turbine with steam supply at 29 ata and a temperature of 400°C show that the efficiency of the turbine is as much as 82%. At the present time these turbines are also being built for operation at high steam pressures and temperatures, as well as with back pressures with high efficiencies in the region of design conditions of operation.

The governing system of the turbine is shown in Fig. 10-18. The drive for the centrifugal regulator is obtained from the turbine shaft through two pairs of toothed gearings *10* and *11*. In the event of a variation of speed the centrifugal regulator displaces the pilot valve *34*, either increasing or decreasing the area of oil flow to the oil pipe *23*. Thus the opening and closing of the throttle valve *16* is controlled by the oil pressure under servomotor piston *18*. The pilot valve is brought back to its central position by the system of levers *36* and *37*. The turbine reaches its nominal capacity when valve *40* is opened completely by the levers *39* and *50*. Under normal conditions of operation valve *25* of the overspeed trip gear is kept closed and only the lower oil space of the servomotor *18* is connected to the pressure oil main *23*. If the shaft speed increases by 10% above the normal speed of operation the overspeed trip comes into operation and through the steel cable *29* opens the valve *25*. The two oil spaces of the servomotor piston are now connected to each other and at the same time the pressure main is cut off from the servomotor. Since now there is an absence of oil pressure acting on the servomotor piston, the piston is pushed down by the spring force thus shutting off the throttle valve *16*.

Chapter Eleven

BACK-PRESSURE AND MIXED-PRESSURE TURBINES

11-1. BACK-PRESSURE TURBINES

Back-pressure turbines are used in industries where both electrical energy and process steam are required at the same time. The exhaust steam from these turbines is utilised for process and heating purposes. The capacity of a back-pressure turbine depends to a very large extent on the quantity of steam passing through the turbine to the consumer of process steam. Thus a back-pressure turbine would not be able to simultaneously satisfy both the electrical power and process steam requirement of an industry in its entirety. Hence back-pressure turbines are mostly run in parallel with condensing turbines, the back-pressure turbine supplying that amount of the electric power which is generated by it as a consequence of the quantity of steam passing through it to the process steam consumer. The power requirement in excess of this is supplied by the condensing turbine. Back-pressure turbines supplying steam for heating purposes only have many major drawbacks since their operation is mainly a function of the seasonal changes and, consequently, they have a very limited field of application. The construction of back-pressure turbines is much simpler than that of a condensing turbine. The last stage dimensions of a back-pressure turbine are considerably smaller than those of a condensing turbine of similar capacity. Back-

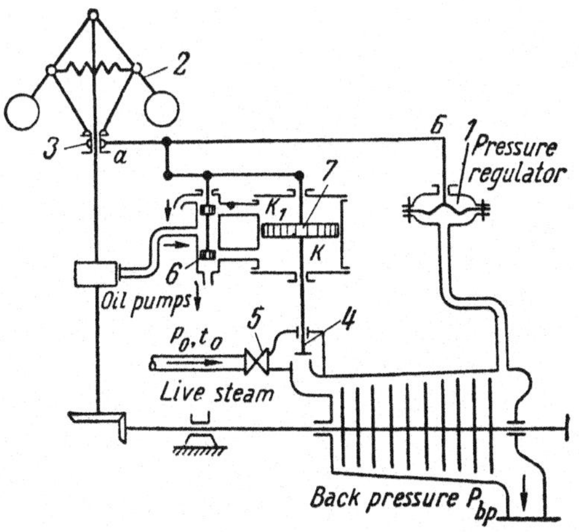

Fig. 11-1. Schematic diagram of speed regulation for a back-pressure turbine

pressure turbine rotors are made of constant mean diameter, or with a very slight increase in diameter towards the exhaust end. The present practice of construction is to keep the disc diameters constant instead of a constant mean diameter for the rotor. The rotor construction of the back-pressure turbine does not in any way differ from the construction of the high-pressure stages of a condensing turbine. The general arrangement of regulation used for a back-pressure turbine is shown in Fig. 11-1.

Regulator *1* is cut off at the time of initial heating and starting the turbine. Sleeve *3* of the centrifugal regulator is lowered down, the servomotor piston is brought to its upper dead centre and valve *4* is opened.

Oil for governing and lubrication is supplied by a subsidiary oil pump. Regulating valve *4* supplying fresh steam to the turbine is kept completely open by the regulating oil pressure on the lower face of the servomotor piston. The turbine is filled with steam by gradually opening the valve connecting the exhaust end of the turbine to the process steam consumer, being supplied by process steam from another back-pressure turbine or from the steam mains through a reducing valve. The turbine is uniformly heated up by slightly opening stop valve *5* and allowing it to run at very slow speed. After initial heating valve *5* is gradually opened to bring the turbine speed to its normal speed of operation. The centrifugal regulator now comes into operation and the turbine is run on no load. After this valve *5* is opened to its maximum and the turbine is synchronised in parallel with the system, some of the system load being transferred onto it. Immediately after loading the turbine back-pressure regulator *1* is put into commission and the turbine now operates according to the process steam requirements of the consumer.

The speed governing of the turbine according to the process steam requirement is effected by the centrifugal regulator in conjunction with the speeder gear. When the turbine speed exceeds the normal r. p. m. the speeder gear springs exert such a force as to exclude the centrifugal regulator from load fluctuations on the system. Thus with variations of load the sleeve *3* of the centrifugal regulator remains still at its former position. The amount of steam flowing through the turbine is now no more a function of load (controlled by the speed regulator) but of the process steam demand made by the consumer, and the control of steam flow is transferred to the back-pressure regulator. The frequency of the system is controlled by the condensing turbines running in parallel.

The principle of operation can be more easily understood by the following example. Let us suppose that the demand for process steam increases. The back pressure P_{bp} will now decrease. Lever *aB* is lowered, relative to point *a*, and pilot valve piston *6* is pushed downwards. As a consequence of this oil under pressure is permitted to flow into chamber *K* and regulating valve *4* begins to open more increasing the quantity of steam flow through the turbine. The upward movement of servomotor piston *7* displaces the pilot valve piston *6* upwards again, brings it to its original position, cutting off the pressure oil supply to the servomotor chamber *K* and drain oil from K_1. In case of increased back pressures an exactly opposite operation takes place, increasing oil pressure in chamber K_1 and draining oil from chamber *K*. The topping turbines may also be considered as one of the types of back-pressure turbines. The exhaust steam from a topping turbine enters a medium-pressure turbine where it is expanded to its final value.

The topping turbines are a product of reconstruction and renovation of old medium-pressure turbines operating at some of the power stations to enable them to operate at higher pressures and temperatures. Topping turbines are built for high initial pressures and temperatures since they exhaust at relatively high back pressures (15 to 29 ata) to medium-pressure turbines. According to the current practice these are $p_0=90$ ata and $t_0=535°C$. The topping turbines may be built for even higher initial steam parameters such as 300 ata and 650°C. The exhaust from a topping turbine may be fed to a medium-pressure turbine either directly or after reheat. Both back-pressure turbines and topping turbines are, as a rule, built with nozzle control governing, and with a very developed system of gland sealings.

11-2. EXAMPLES OF CONSTRUCTION: BACK-PRESSURE TURBINES

a) Turbine R-4-35/3 (AR-4-3) of the Kaluga Turbine Works (K.T.W.)

Fig. 11-2 shows the constructional details of a 4,000 kW turbine operating at 3,000 r. p. m. built by the K. T. W. The steam is supplied at a pressure of 35 ata and a temperature of 435°C. The back pressure is 3 ata. The permissible initial pressure fluctuations are from 32 to 37 ata with temperatures ranging from 420 to 445°C, and back-pressure variation from 2 to 4 ata. Besides, the turbine can be operated continuously at 29 ata and 400°C with back-pressure fluctuations in the range of 2 to 4 ata. This is a single-cylinder impulse-type turbine with a two-row regulating stage and nine pressure stages. The steam supply is through a set of nozzles. Six regulating valves mounted on the casing control the steam flow to the turbine. The turbine rotor is made up of ten discs mounted on the shaft as a shrunk fit. The end seals are of the labyrinth type with foils caulked into liners. The front end bearing is a combined journal and thrust bearing. The turbine and alternator rotors are coupled together with a flexible coupling. The fixed point of the turbine is at its exhaust end so that linear expansion is towards the front end journal and thrust bearing. The turbine is provided with a manually operated slow rotation device.

The regulating system is supplied with oil under a pressure of 3.5 to 6.5 kg/cm^2, and the lubrication system with oil at a pressure of 0.5 kg/cm^2. The main oil pump is of the centrifugal type and is connected to the turbine shaft. Besides, there are two standby oil pumps one run by a turbine and the other by an electric motor operating on a. c. supply. This motor comes into operation as soon as the lubricating oil pressure drops below 0.2 kg/cm^2.

b) Turbine R-6-10 (AR-6-10) of the Lenin Nevsky Works (L.N.W.)

This is an impulse turbine with a capacity of 6,000 kW at 3,000 r. p. m. (Fig. 11-3) designed to operate with steam supply at a pressure of 35 ata and a temperature of 435°C. The back pressure for this turbine is 10 ata. The turbine can be operated for long periods at its nominal load with the following allowable fluctuations of temperature and pressure: initial pressure from 32 to 37 ata; initial temperature from 420 to 445°C and back pressure from 8 to 13 ata.

The rotor consists of one single-row regulating stage followed by five pressure stages. Steam distribution is through a nozzle system. The quantity of steam flow through the turbine is controlled by six regulating valves mounted on the turbine casing. The end labyrinth gland sealing is of the fir-cone type. The labyrinth seals are mounted on liners. The front end is supported by a combined journal and thrust bearing. The turbine and alternator rotors are interconnected by a flexible coupling. There is a device for manually rotating the turbine shaft. The fixed point of the turbine is shown in Fig. 11-3. The regulation system of the turbine is shown in Fig. 11-4. The regulating valves are operated by a servomotor of the piston type. The servomotor receives its impulse from a back-pressure regulator. If the turbine operates with load regulation the impulse to the servomotor piston is given by the speed regulator. The speed regulator and the gear oil pump are driven by the turbine shaft through a worm gearing. For the starting and stopping periods the turbine is provided with an auxiliary oil pump driven by a small impulse steam turbine. An electric motor driven standby pump is also provided to supply oil to the bearings in the event of the lubricating oil pressure falling below 0.3 atm. gauge. An axial displacement relay is provided to trip the turbine by shutting off the stop valve in case of axial displacement.

c) Radial Turbine VR-15 of Siemens-Schuckert Werke

The Siemens-Schuckert Werke have been building radial turbines with fixed guide blades. For such turbines in contrast to the Ljungström turbines, only one alternator is required, so that the turbine becomes more compact. Fig. 11-5 shows the sectional view of such a turbine of 15,000 kW capacity. The turbine has two discs carrying the blades and operates at a speed of 3,000 r. p. m. The turbine has nozzle control governing instead of the usual throttle governing used in Ljungström turbines. The nozzle control governing is decidedly an advantage. Steam passing through the first regulating stage and the following reaction stages goes to the centre of the turbine. From here it flows along the turbine shaft and then flowing outwards through the radial stages on the second disc goes out through the exhaust. Such a construction enables the axial thrusts of the two discs to be mutually balanced. The thrust bearing helps in fixing the rotor in the proper alignment and bears any surplus thrust left over from the mutual thrust cancelling. The circumferential velocity for this turbine is half that of a similar capacity Ljungström turbine. Hence single-disc radial turbines with fixed guide blades have comparatively more number of stages than in the Ljungström turbine which is a serious drawback. Another drawback for such tur-

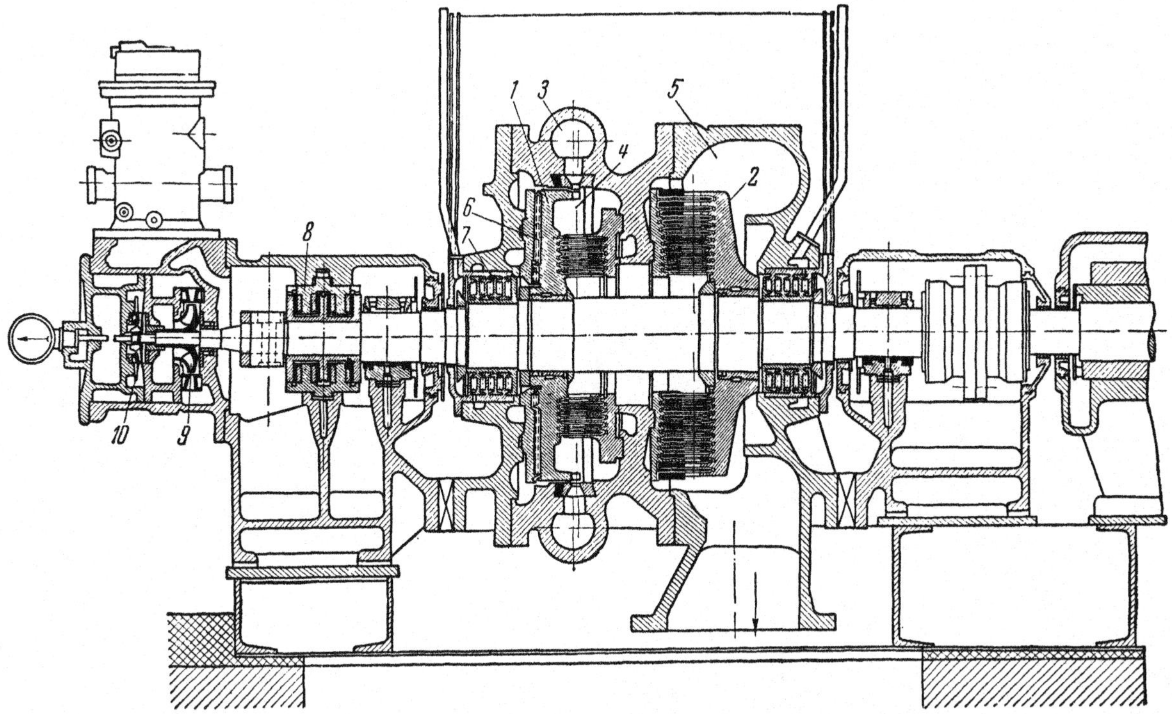

Fig. 11-5. Turbine VR-15 (Siemens-Schuckert Werke)

1 – steam chest; *2* – regulating stage; *3* – first disc; *4* – second disc; *5* – guide blade diaphragms; *6* – radial labyrinth packing; *7* – shaft labyrinth packing; *8* – thrust bearing; *9* – lubricating oil pump; *10* – regulating oil pump

bines is the difficulty in ascertaining the proper alignment of the rotor blades and the labyrinth seals at the time of assembly. For small pressure drops these turbines are built with a single disc mounted over the overhanging end of the alternator shaft, so that the high-pressure labyrinth sealing is dispensed with. This leads to a decrease in leakage losses and consequently to an increase in the efficiency of the turbine. The use of radial stages in the high-pressure regions, especially for very high steam pressures and temperatures, favourably influences the efficiency of a turbine.

11-3. EXAMPLES OF CONSTRUCTION: TOPPING TURBINES

a) Topping Turbine of the Kh.T.W.; 25,000 kW Capacity at 3,000 r.p.m., Type VR-25-1[1]

Initial steam pressure and temperature $p_0=$ 90 ata and $t_0=500°C$. Back pressure 31 ata. The turbine is shown in Fig. 11-6.

The turbine consists of a two-row regulating stage and six pressure stages. The blade discs are machined on the forged rotor which reduces the length of the rotor and the turbine as a whole. The critical speed for the turbine rotor is 4,230 r. p. m. The rotor is made of chrome-molybdenum-nickel steel.

The internal relative efficiency (theoretical) is $\eta_{0i}=0.787$. Steam temperature at the turbine exit is 361°C. The steam can, therefore, be used directly in the medium-pressure cylinder without reheat. At nominal loads the mass flow of steam through the turbine is 379 tons/hr. The turbine is provided with nozzle control governing. Steam is supplied to the nozzles and to the first regulating stage through six regulating valves, placed in two steam chests at the head of the turbine; one steam chest is mounted on the upper casing and the other on the lower casing diametrically opposite. The first two regulating valves open simultaneously and the remaining four open in a sequence. The regulating valves are opened and closed by the main servomotor through a system of levers. The back pressure of the turbine is controlled by a pressure regulator operating independent of the speed regulator. Under normal conditions of operation the centrifugal speed regulator is driven at higher r. p. m. The regulation system has an oil pressure of 9 atm. gauge and the lubrication system 0.5 atm.gauge. The turbine casing is made of a double-walled shell. The outer wall of the shell is made of carbon steel, and the inner of molybdenum steel. The external shell weighs 15 tons and the internal one 6.5 tons. The de-

[1] A non-standard turbine.

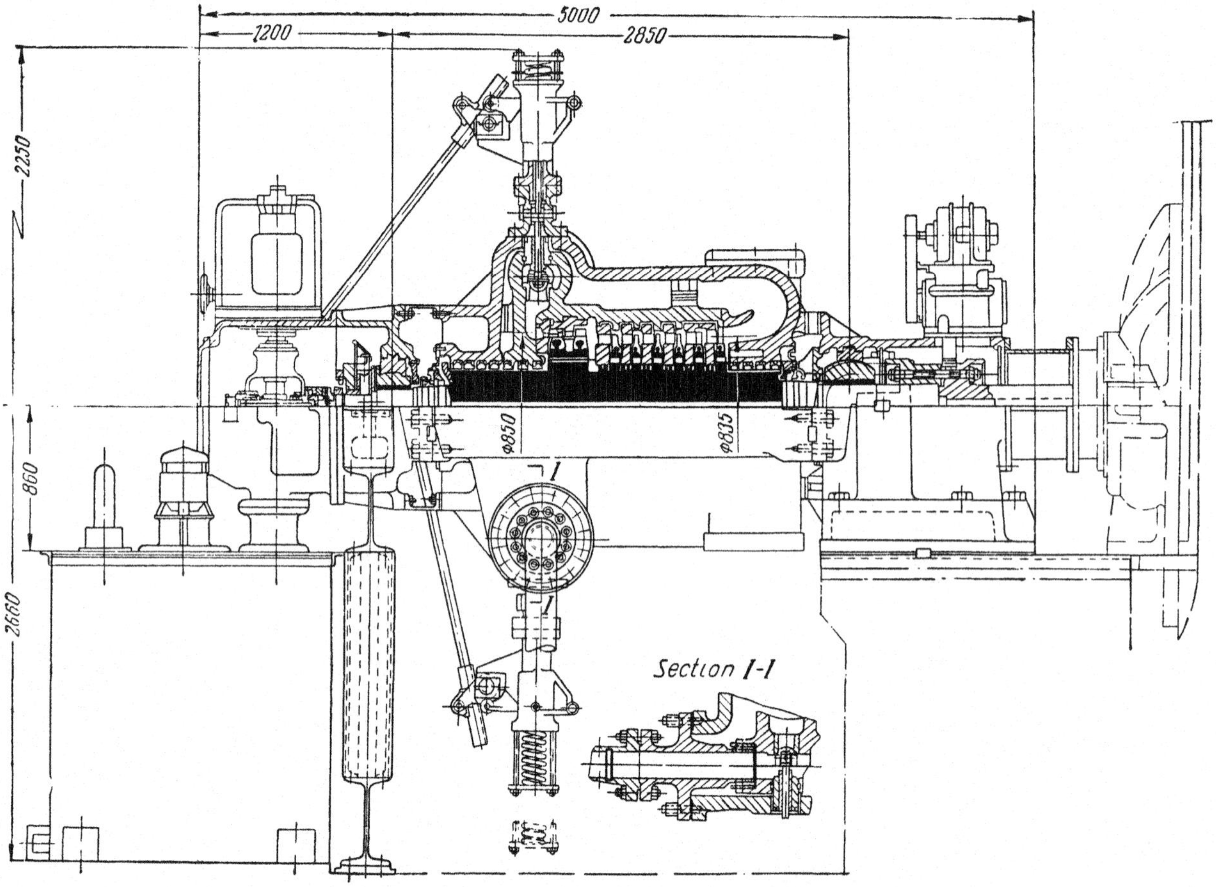

Fig. 11-6. Turbine VR-25-1 of Kharkov Turbine Works

tailed drawing (inset) section I-I shows the method of steam supply to the regulating valves. The use of a double-shell construction for the turbine casing helped in reducing the dimensions of the inner casing made of costly alloy steel. The welded steel diaphragms are directly fixed to the inner casing. Both the blades and the diaphragms are made of chrome-molybdenum-nickel steel with 0.5% of Mo.

At the high-pressure end the turbine has labyrinth packing as shown in Fig. 11-7. There is a hydraulic seal at the end of the labyrinth seals in order to cool the turbine shaft. The liners carrying the labyrinth seals have freedom of motion in the radial direction since they are held on to the turbine shaft by springs. Special grooves have been cut in the shaft at the labyrinths to compensate shaft expansion thus avoiding the use of the usual liner. Such a construction enabled the factory to decrease the labyrinth seal diameter and consequent leakage losses. The front end is supported by a combined journal and thrust bearing (segment type). Both the journal bearings have spherical supporting bushes. The distance between centres of the two bearings is 2,850 mm. Total length of the turbine is 5,000 mm, and its gross weight 56.5 tons.

The fixed point for the turbine is situated at the rear end journal bearing and lies on the axis of the lateral keys placed between the pedestal sole and the bedplate frame (Fig. 11-6). The front end bearing is fixed in such a way as to be flexible in the axial direction and rigid in the lateral direction. Front bearing pedestal and the oil reservoir are fixed to the front bedplate. The turbine is provided with a slow rotor revolving device, driven by an electric motor.

The Kh. T. W. topping turbine type VR-25-2 is also designed on the same initial steam parameters as for turbine type VR-25-1, but with a back pressure of 18 ata. The rotor has a two-row regulating stage and eight pressure stages. The total quantity of steam flowing through turbine VR-25-2 at rated loads is 267.5 tons/hr. The exhaust temperature is 302°C. Most of the VR-25-2 turbine components are the same as those of the VR-25-1 turbine.

Both these turbines had a relatively low efficiency. The Kh. T. W. has now come out with a new design for topping turbines increasing both the exhaust steam temperature and the efficiency. These turbines operate with higher initial steam temperatures.

b) The Kh.T.W. 25,000 kW Topping Turbines R-25-90/31 (VRT-25-1) and R-25-90/18 (VRT-25-2)

These turbines are designed to operate with fresh steam at 90 ata and 535°C, and a back pressure of 31 ata for turbine R-25-90/31 and 18 ata for turbine R-25-90/18. Both these turbines have a single-row regulating stage. The heat drop occurring in these stages being small, the number of pressure stages is more than in the case of turbine VR-25. The VRT-25 turbines are very similar in design (stators, steam distribution and other components) to VR-25 turbines. Turbine R-25-90/31 has in all eight stages. High efficiency is ensured by the use of good blade profiles for both the moving and the fixed blades in both these turbines.

Type R-25-90 turbines can, if required, develop nominal power with 85 ata initial pressure, 500°C initial temperature at a back pressure 3 ata more than the design value of 31 ata. The guaranteed specific steam consumption at nominal capacity is $d_e = 10.32$ kg/kWh for turbine R-25-90/18 and $d_e = 14.84$ kg/kWh for turbine R-25-90/31. Both these turbines have hydraulic system

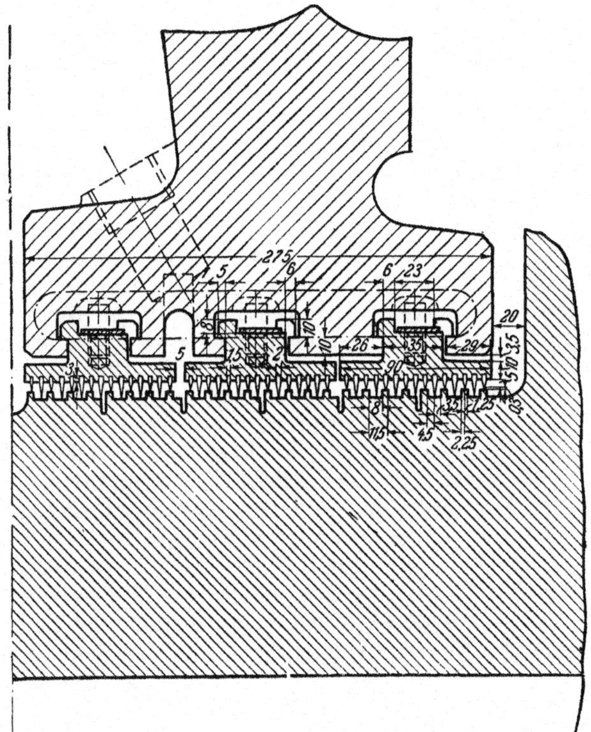

Fig. 11-7. Construction of front labyrinth packing of Kharkov Works turbines types VR-25, R-25-90 (VRT-25), K-100-90 (VKT-100), K-150-130 (PVK-150), etc.

of regulation. The regulation impulse is signalled by an impeller built in a single unit with the main centrifugal oil pump. The main oil pump at the normal shaft speed of 3,000 r. p. m. supplies the oil system with 100 m^3/hr of oil at a pressure of 15 kg/cm^2. The regulation system and the lubrication mains are supplied with oil through special injectors. The pressure at the suction piping of the main oil pump and impeller is 0.3 to 0.4 atm. gauge. This pressure is built up by the main injector. The nozzle of this injector receives its oil supply from the pressure mains of the main oil pump. The impeller transmits the impulse to a special piston-type meter. This meter is in itself a first-stage amplifier transmitting the impulse to the pilot valve of the second-stage amplifier; from the last named oil flows to the space below the servomotor piston. The servomotor has a single operating side with a spring on the other side to balance the oil pressure. Through a lever system the servomotor operates the regulating spindles which control the opening and closing of the regulating valves. When the turbine operation is to be governed by the process steam requirement the back-pressure regulator is put into commission. This regulator has a three-stage amplification for transmission of the pressure variation impulse. The initial impulse is transmitted to a spring-membrane element. From here the impulse is further transmitted to the piston of the pilot valve and thence to the servomotor. The displacement of the servomotor piston through a rack and pinion arrangement causes the regulation valve camshaft to revolve thus controlling the regulation valves.

The oil system of the turbine is provided with auxiliary oil pumps: an a. c. motor-driven oil pump, delivering 116 m^3/hr of oil, is provided for oil supply to the bearings and the governing system at the time of starting the turbine; two motor-driven low-pressure oil pumps, one run on a. c. and the other on d.c., are also provided, each delivering 90 m^3/hr at a static head of 25 metres of water, for supply of oil to the lubrication system.

The turbines are also provided with axial displacement relays. In case of an axial displacement of the turbine unit the axial displacement relay energises an electromagnet which automatically shuts off the main stop valve and the regulating valves.

11-4. CONDENSING TURBINES WITH CONTROLLED EXTRACTIONS (PASS-OUT TURBINES)

1. Basic Concepts

A condensing turbine with intermediate extractions is a machine which can supply simultaneous-

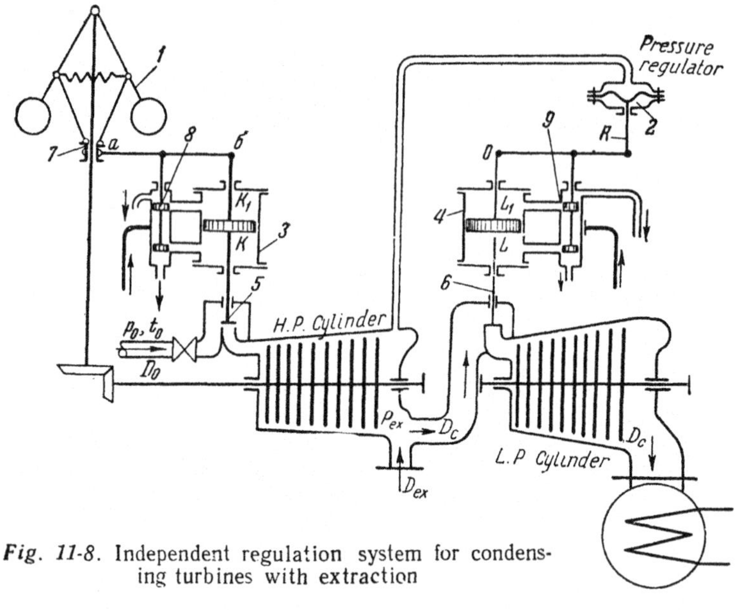

Fig. 11-8. Independent regulation system for condensing turbines with extraction

ly various combinations of heat and electrical loads. Steam from the controlled extractions at constant pressure is used for heating and process purposes in industries. In the absence of heat load demand this turbine can operate as a normal condensing turbine, its operation now being completely dependent on the electrical load requirements. Consequently the regulation system of the turbine must be so designed as to independently govern the turbine for heat and electrical loads.

Fig. 11-8 shows the flow diagram for the regulation system of a condensing turbine with intermediate extractions. The signal from the centrifugal regulator is transmitted only to the fresh steam regulating valve, and the impulse from the pressure regulator is transmitted to the regulating valves of the low-pressure cylinder. D_0 kg of fresh steam enters through regulating valve *5*. At the exit of the high-pressure cylinder steam is bifurcated into two streams: D_{ex} flows out to the consumer for heating and process purposes and D_c through the second regulating valve *6* enters the low-pressure cylinder from where it is directed to the condenser. Under normal conditions of operation both the centrifugal regulator *1* and the pressure regulator *2* are adjusted to be in their normal working positions.

With load variations the r. p. m. of the turbine varies, sleeve *7* of the centrifugal regulator is displaced either upwards or downwards and along with this pilot valve piston *8* is also displaced. Depending upon the flow of pressure oil to either the upper or lower chamber K or K_1 of servomotor *3*, regulating valve *5* opens or closes increasing or decreasing the quantity of steam flow D_0 through the turbine.

The extraction pressure p_{ex} which is a function of the quantity of bled steam D_{ex}, governs the flow of steam to the low-pressure cylinder. With an increase in D_{ex} p_{ex} falls, lever R is pulled upwards displacing the pilot valve piston *9* relative to the fulcrum point *O* on the servomotor piston spindle. Pressure oil enters space L_1 of servomotor *4* and the regulating valve *6* begins to close. The closing of the regulating valve will continue until the pilot valve piston does not regain its mid-position closing the oil ports.

The disadvantage of this type of regulation is the protracted period for the complete operation. For example, a variation in the load on the turbine would in the first instance cause only the governing of fresh steam flowing to the H. P. cylinder. This in its turn causes the extraction pressure p_{ex} to vary, and only after this variation the back-pressure regulator starts to function, controlling the flow of steam through the L. P. cylinder. This type of governing is now very rarely found for extraction turbines.

A more developed regulating system for extraction turbines is the compound regulation system.

Fig. 11-9 shows the principles of operation of such a system. In it the displacement of centrifugal regulator sleeve *2* causes a simultaneous operation of the regulating valves of both the high- and the low-pressure cylinders. Similarly the variation of back pressure also causes both the regulating valves to operate in unison. For example, a decrease in the load on the turbine causes the regulator sleeve *2* to move upwards. The rigid lever *bcd* connected with the sleeve is displaced upwards revolving with point *b* as its fulcrum and in the process pilot valves *3* and *4* are displaced upwards. Regulating valves *5* and *6* begin to close reducing the quantity of steam flow both through the high- and the low-pressure cylinders. The two regulating valves continue to close until the pilot valve pistons are not returned back to their initial positions. For any required operating conditions the lever arms can be so adjusted as to ensure constant extraction pressure as well as constant quantity of extraction steam. This type of regulating system is designated as an independent regulation system.

If the pressure of the extraction steam varies, sleeve *2* remains steady at its position and the

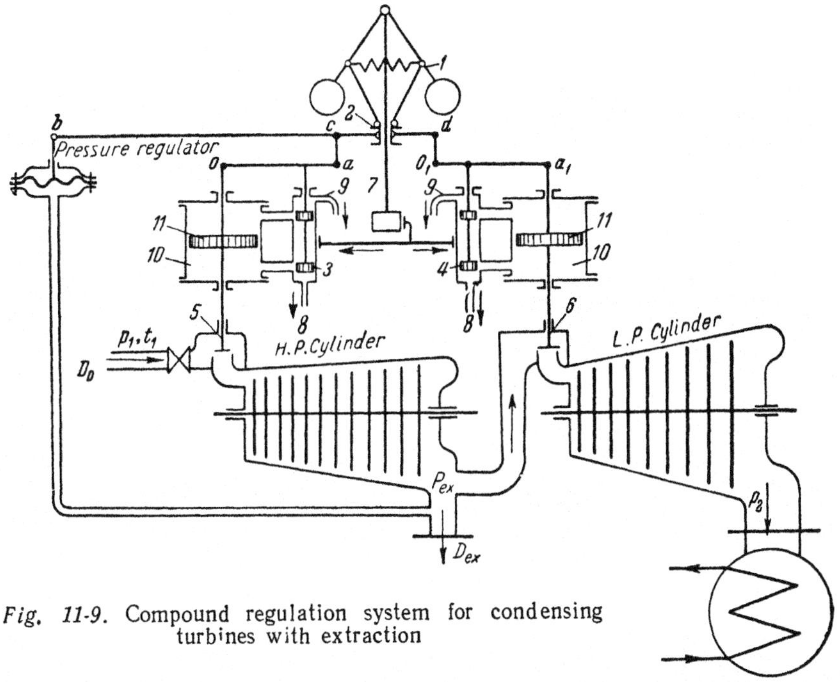

Fig. 11-9. Compound regulation system for condensing turbines with extraction

lever *bcd* rotates with the regulator sleeve as its fulcrum. If, for example, the extraction pressure p_{ex} increases lever *ac* and the extreme end of lever *b* are moved upwards and lever O_1d is pushed down. Pilot valve *3* moves upwards and pilot valve *4* down causing regulating valve *5* to close and valve *6* to open. In case of independent regulation a variation in the extraction pressure p_{ex} causes the regulating valves *5* and *6* to operate in such a way as to maintain the power developed constant with varying extractions to suit the consumer's demand.

2. Consumption Curves for Turbines with Single Pass-Out

The condition curves for a turbine permit the determination of fresh steam requirements for given loads and extraction quantities. The heat drop process for a turbine with extraction is shown in Fig. 11-10. H_i' and H_i'' are the heat drops usefully utilised in the H. P. and the L. P. cylinders. $\Delta p = p_{ex} - p_{1ex}$ is pressure loss in the connecting piping and the regulating valves of the L. P. cylinder. The magnitude of Δp depends upon the steam distribution in the L. P. cylinder, D_c—the quantity of steam flow to the condenser, and may vary in a wide range from 3-5 to 30%, or even more, of the extraction pressure p_{ex}.

The equation for power developed by an extraction turbine may be written as

$$N_e = \frac{D_c H_i + D_{ex} H_i'}{860} \eta_m \eta_g, \qquad (11\text{-}1)$$

where D_c—steam flowing into the condenser;

D_{ex}—quantity of steam taken away from the pass-out;

$H_i = H_i' + H_i''$ —the useful heat drop for the steam flowing to the condenser through the H. P. and the L. P. cylinders.

Equation (11-1) enables us to calculate the power developed by the turbine for any given mass flow of steam D_c and D_{ex}. By alternately varying D_c and D_{ex} in this equation we can obtain a curve for N_e as a function of D_c with D_{ex} = const, or D_{ex} with D_c remaining constant and thus obtain a diagram of condition curves for the turbine.

As an approximation the condition curves may be drawn with sufficient accuracy, assuming N_e to vary linearly for variations of D_c and D_{ex}.

a) Consumption Line for Condensing Operation ($D_{ex}=0$).

In this case equation 11-1 becomes

$$N_e = \frac{D_c H_i}{860} \eta_m \eta_g. \qquad (11\text{-}1a)$$

The maximum quantity of steam flowing to the condenser $D_{c\,max}$ at nominal load N_e^{nom} is

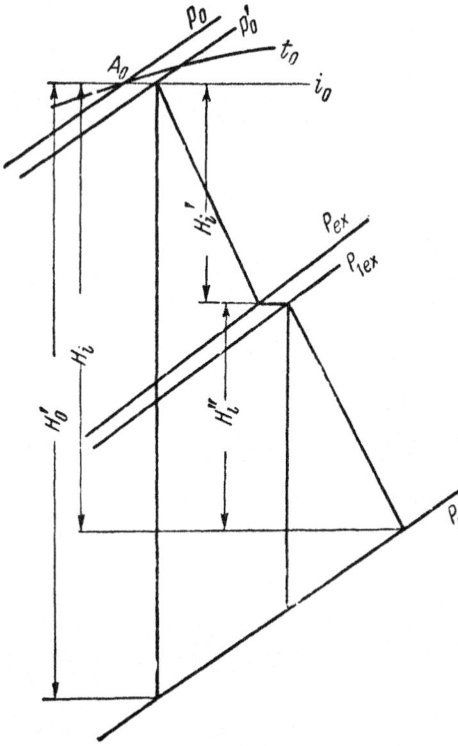

Fig. 11-10. *I-s* diagram for a pass-out turbine

a known quantity from theoretical calculations. Plotting N_e^{nom} along the x axis and $D_{c\ max}$ along the y axis we obtain point k (Fig. 11-11).

The maximum quantity of steam flowing into the turbine is also known. The turbine develops nominal power at D_{0max} kg of steam flowing through it (point a), and at some fraction of its maximum load the turbine delivers the maximum quantity of extraction steam $D_{ex\ max}$ (point r_0). Line ma gives the maximum steam flow through the turbine. On the diagram point a determines the relation between N_e^{nom}, D_{omax} and $D_{ex\ th} = D_{ex}^{III}$.

The energy losses occurring in the turbine when running light may be approximately determined by the equation

$$\Delta N_n = N_i - N_e^{nom} = \left(\frac{1}{\eta_m \eta_g} - 1\right) N_e^{nom}, \quad (11\text{-}2)$$

where η_m, η_g — mechanical efficiency of the turbine and alternator efficiency.

Drawing ΔN_n along the x axis we obtain point O_1 giving the turbine condition at the moment of starting. If it is assumed that the variation of power developed is a linear function of the steam flow then line O_1k will satisfy these conditions. However, the actual curve of variation of power developed with steam flow is a complicated one. N_e as a function of D_c is basically determined by the steam distribution and the variation of η_{oi} as a function of D_c.

Intercept D_{xx} approximately gives the steam flow at no load.

b) Pass-Out Pressure Line ($D_c = 0$).

When D_c is assumed to be constant equation (11-1) becomes

$$N_e = \frac{D_{ex} H_i'}{860} \eta_m \eta_g. \quad (11\text{-}1b)$$

When $D_{ex} = D_{omax}$ the power developed is

$$N_e' = \frac{D_{0\ max} H_i'}{860} \eta_m \eta_g,$$

where H_i' is obtained from theoretical calculations and η_m and η_g from experimental data. We shall plot the value obtained for N_e' on the diagram to obtain point r. The straight line joining point O_1 and r approximately gives the relation between steam flow and power developed when all the steam entering the H.P. cylinder is exhausted from the pass-out.

The intercept D_{xx}' gives the steam flow for no load operation of the turbine with full back pressure in the pass-out pipe. Line O_1r is only of theoretical interest since such an operating condition is inadmissible: in any case some minimum quantity of steam has to be allowed into the L. P. cylinder in order to remove the heat developed in the L. P. stages due to windage and friction. The minimum quantity of steam to be allowed into the L. P. cylinder should not be less than 10 to 20% of $D_{c.\ max}$ and in some particular cases not less than 5 to 10% of $D_{c\ max}$.

c) Consumption Curve for $D_{c\ min} = $ $=$const.

We shall assume a value $0.2D_{c\ max}$ for $D_{c\ min}$ for plotting the condition curves in Fig. 11-11. With this condition D_{ex} will now vary between the limits of zero and $D_{0\ max} - D_{c\ min}$.

Equation (11-1) for this condition becomes

$$N_e = \frac{D_{c\ min} H_i + D_{ex} H_i'}{860} \eta_m \eta_g. \quad (11\text{-}1c)$$

If now in this equation D_{ex} is assumed to be zero then the value of N_e depends only on $D_{c\,min}$, i. e., N_e will now lie along the line O_1k. Point k_0 on the line O_1k is obtained from the conditions:

$$D_{c\,min}=0.2D_{c\,max} \text{ and } D_{ex}=0.$$

Since the variations of N_e with respect to D_{ex} is determined by the inclination of line O_1r to the abscissa, line k_0r_0, parallel to O_1r, is obtained from equation (11-1c) when $D_{c\,min}$ = const.

d) Constant Pass-Out Line (D_{ex}=const).

For plotting this line values are obtained by keeping D_{ex}=constant in equation (11-1) and varying the quantity of steam flow D_c. When the steam flow is $D_{c\,min}$ the power developed N_e depends only on D_{ex}. Hence for any values of D_{ex}, and $D_{c\,min}$ equal to $0.2D_{c\,max}$, N_e lies on the line k_0r_0; viz., point k_1 is given by the equation

$$N_{e1}=\frac{D_{c\,min}H_i+D'_{ex}H'_i}{860}\eta_m\eta_g=N'_{ec}+N'_{eo}$$

when the quantity of fresh steam flowing through the turbine is $D_{01}=D_{c\,min}+D'_{ex}$.

Point k_2 is given by the condition

$$N_{e2}=\frac{D_{c\,min}H_i+D^{II}_{ex}H'_i}{860}\eta_m\eta_g=N'_{ec}+N''_e;$$

$$D_{02}=D_{cmin}+D''_{ex},$$

and point k_3 by

$$N_{e3}=\frac{D_{c\,min}H_i+D^{III}_{ex}H'_i}{860}\eta_m\eta_g=$$

$$=N'_{ec}+N'''_e;\ D_{03}=D_{c\,min}+D'''_{ex}$$

If in equation (11-1) D_{ex} is assumed to be constant then N_e varies only as a function of D_c. Since the variation of N_e with respect to D_c is given by the inclination of line O_1k to the abscissa On, lines $k_1k'_1$, $k_2k'_2$ and k_3a, which are parallel to k_0k, give the variation in the power developed with respect to steam consumption D_0 with constant steam extraction from the pass-out. For example, it is required to determine the quantity of steam flowing through the turbine when the power developed at the alternator terminals is equal to N_{ex} at an extraction of D''_{ex}. The power developed N_{ex} is measured along the x axis (point x) and a perpendicular is drawn at x to intersect the constant extraction line D''_{ex} at point y. From point y a line parallel to the x axis is drawn to intersect the ordinate Om in z. The length Oz gives the quantity of steam flow through the turbine as D_{0x}.

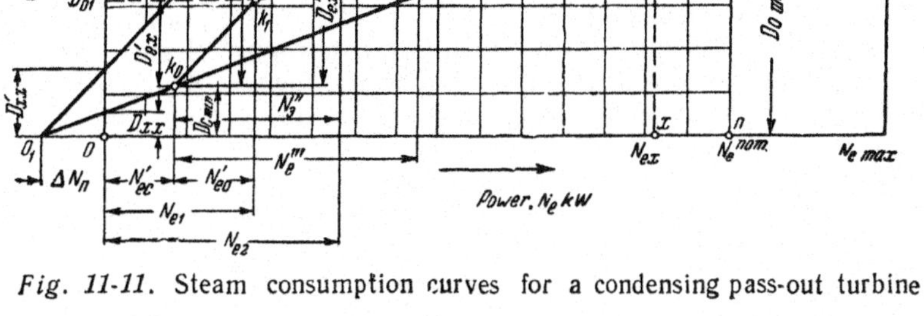

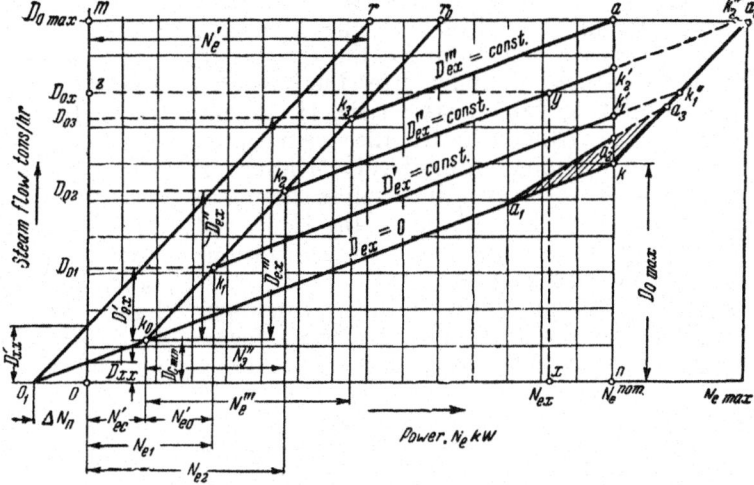

Fig. 11-11. Steam consumption curves for a condensing pass-out turbine

11-5. EXAMPLES OF CONSTRUCTION: PASS-OUT TURBINES

a) Turbine P-1.5-35/5 (AP-1.5) of Kh.T.W.

This is a single-cylinder impulse turbine with controlled extraction. The capacity of the turbine is 1,500 kW at an r.p.m. of 8,000 (Fig. 11-12). The turbine drives an alternator at a speed of 3,000 r.p.m. through a reduction gearing. The initial steam pressure and temperature at the stop valve are 35 ata and 435°C. The turbine has a regulated pass-out for industrial purposes of about 12 tons/hr at a pressure of 5 ata and two extractions for feed heating up to a temperature of 150°C. The allowable pressure variations for the supply steam is from 32 to 37 ata with temperature variation in the limits of 420 to 445°C. The turbine continues to develop its nominal capacity even when the cooling water temperature reaches 33°C, if the initial steam pressure and temperature are not less than 35 atm. gauge and 435°C respectively. The turbine can be operated continuously at its nominal capacity for long periods with constant pass-out of 12 tons/hr, at initial pressure 29 ata and temperature 400°C. The pass-out pressure may vary between 4 to 6 ata.

Steam supply to the turbine blades is through a nozzle system, the quantity of steam being regulated by six regulating valves provided at the front end of the turbine mounted on the turbine stator. Steam supply to the low-pressure cylinder is controlled by a revolving diaphragm. The rotor consists of a two-row regulating stage and eight pressure stages. The shaft gland packings at its ends are of the labyrinth type. The sealing foils are caulked into liners. The front end bearing is a combined journal and thrust bearing with self-aligning bearing pads. The rotor is of forged alloy steel and is joined to the alternator rotor by a flexible coupling. The H.P. end portion of the turbine stator is made of steel with the steam chests welded onto it. Exhaust end of the turbine is made of cast iron. The main oil pump is of the centrifugal type. It is directly coupled to the turbine shaft and has its own journal bearing. Also provided are a turbine-driven auxiliary oil pump and a motor-driven standby oil pump which automatically comes into operation when the lubricating oil pressure drops below 0.2 atm. gauge. The fixed point for the turbine is situated on the turbine axis.

b) Turbine P-6-35/5 (AP-6) of the Lenin Nevsky Works

This turbine is also a single-cylinder impulse turbine with regulated steam pass-out. The turbine has a capacity of 6,000 kW at shaft speed of 3,000 r.p.m. and runs an alternator (Fig. 11-13). Steam is supplied to the turbine at a pressure of 35 ata and a temperature of 435°C. The permissible variations in the initial steam pressure and temperature are from 32 to 37 ata and 420 to 445°C. The pass-out from the turbine is at a pressure of 5 ata, the quantity of steam extracted for industrial purposes being 40 tons/hr. The turbine also has two simple extractions for feedheating purposes. Steam distribution to the blades is through a nozzle system both for the H. P. cylinder and the L. P. cylinder after the pass-out. Twelve regulating valves, six on the H. P. and six on the L. P. cylinder, control the steam supply to each of the cylinders respectively. The use of a large number of regulating valves helps in reducing throttling losses both in the H. P. and the L. P. cylinders in the event of large variations in the quantity of steam flow.

The H. P. cylinder rotor has a two-row regulating stage followed by five pressure stages, and the L. P. cylinder rotor has one two-row regulating stage and seven pressure stages. The labyrinth seals at the shaft ends are of the fir-cone type mounted on sleeves. The front end of the shaft is supported by a combined journal and thrust bearing. The turbine and the alternator shafts are joined together by a flexible coupling. The turbine is provided with an axial displacement relay which automatically shuts off the stop valve and the regulating valves in case of axial displacement of the rotor. The fixed point of the turbine very nearly passes through the condenser axis (Fig. 11-13). The main oil pump is of the gear type and is driven by the turbine shaft through a worm and worm gearing. The turbine is also provided with an auxiliary oil pump for starting and stopping purposes and a standby pump for emergencies. The standby pump is driven by an electric motor and is automatically switched on when the lubricating oil pressure drops below 0.3 atm. gauge.

c) Turbine AT-12-1[1] of Kirov Works

This is a single-cylinder "semi-basement" type turbine with a capacity of 12,000 kW at 3,000 r. p. m.; the turbine is meant to supply the motive power to an alternator (Fig. 11-14).

Steam supply is at 29 ata and 400°C, $p_{ex}=1.2$ to 2.5 ata and $p_2=0.05$ ata. The low-pressure cylinder is designed for a steam flow of 30.5 tons/hr and in the absence of pass-out does not develop more than 60% of the rated capacity of the turbine. The H. P. cylinder consists of a two-row velocity wheel and nine pressure stages; the L. P. cylinder has six pressure stages. The pass-out is from chamber *3* after the tenth stage, the quantity of steam extracted being 60 tons/hr. Besides this pass-out the turbine has two simple bleeds for feedwater heating. The first of these two bleeds is from chamber *2* after the fourth stage going to the high-pressure feed heater. The second bleed is from the steam space after the twelfth stage supplying heating steam to the L. P. heater.

Steam distribution is of the mixed type: the H. P. cylinder is provided with nozzle control, whereas the L. P. cylinder has throttle governing. Steam to the H. P. cylinder nozzles is supplied through three regulating valves provided at the front end of the turbine; and to the L. P. cylinder through a revolving diaphragm *4*. The blade discs are shrunk on the shaft. The front end gland sealing is of the fir-cone type. The low-pressure end is water sealed. The front end bearing is a combined journal and thrust bearing with spherical base. The rear end journal bearing is held rigid in its position. The pedestal of the rear end journal bearing is cast in a single piece with the exhaust piping. The alternator bearing also forms an integral part of this casting. The turbine and the alternator rotors are joined together by a

[1] A non-standard turbine.

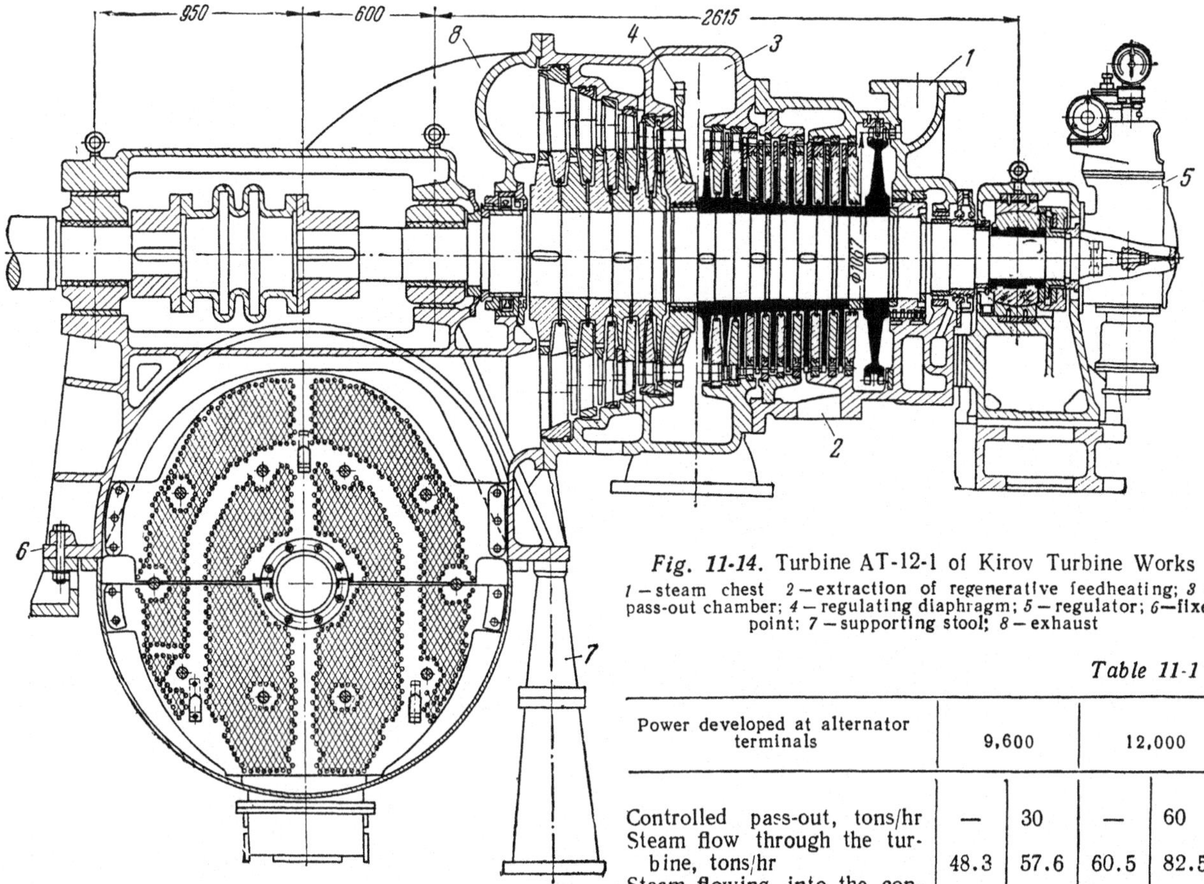

Fig. 11-14. Turbine AT-12-1 of Kirov Turbine Works

1 — steam chest *2* — extraction of regenerative feedheating; *3* — pass-out chamber; *4* — regulating diaphragm; *5* — regulator; *6* — fixed point; *7* — supporting stool; *8* — exhaust

Table 11-1

Power developed at alternator terminals	9,600		12,000	
Controlled pass-out, tons/hr	—	30	—	60
Steam flow through the turbine, tons/hr	48.3	57.6	60.5	82.5
Steam flowing into the condenser, tons/hr	39.5	20.3	48.6	13.0
Cooling water temperature at inlet to the condenser, °C	20	15	20	15
Vacuum in the condenser, %	93.7	96.4	92.1	96.7
Internal efficiency of the H. P. cylinder η_{oi}, %	77	82	81	82
Internal efficiency of the L. P. cylinder η_{oi}, %	75.9	75.9	75.7	77.5
Mechanical efficiency of the turbine, %	97.3	97.3	97.3	97.3
Alternator efficiency, %	95	95	95.7	95.7
Specific steam consumption, kg/kWh	5.03	6.0	5.04	6.87
Final temperature after feedheating, °C	144	150	153	160

semi-flexible coupling. The turbine shaft is rigid. The fixed point for the set is located under the alternator bearing. The gross weight of the turbine along with the condenser is 59 tons. The turbine performance at loads of 9,600 and 12,000 kW is given in Table 11-1.

Fig 11-15 shows the main features of the regulating system. The H. P. regulating valves are operated by a revolving servomotor *11*. The throttling diaphragm *19* which controls steam flow to the L. P. cylinder is operated by a piston-type servomotor *18*. Besides the two main servomotors *11* and *18* there are two auxiliary servomotors *5* and *24*. These last two operate the two pilot valves *9* and *16* in the H. P. and L. P. cylinder circuits respectively.

The impulse for regulation is given by a centrifugal regulator *1* and a pressure regulator *22*. From gear oil pump *33* oil is supplied to the main pilot valves *9* and *16*. From pilot valve *9* oil goes to the vacuum relay pilot valve *27* and the trip mechanism *28*. The oil space of pilot valve *16* is connected to the pressure regulator pilot valve *23*. The centrifugal regulator pilot valve *2* receives oil from the trip mechanism *34*.

A displacement of pilot valve *2* causes the displacement of servomotor *5* and the pistons of pilot valves *9* and *16*, causing the simultaneous opening or closing of the regulating valves and the diaphragm distributor. The pilot valves *9* and *16* are brought back to their initial positions through a lever arrangement connecting them to the main servomotors. If the pressure of steam in the pass-out varies piston of pilot valve *23* is displaced causing the auxiliary servomotor *24* and the main pilot valves *9* and *16* to be displaced. The pistons of pilot valves *9* and *16*

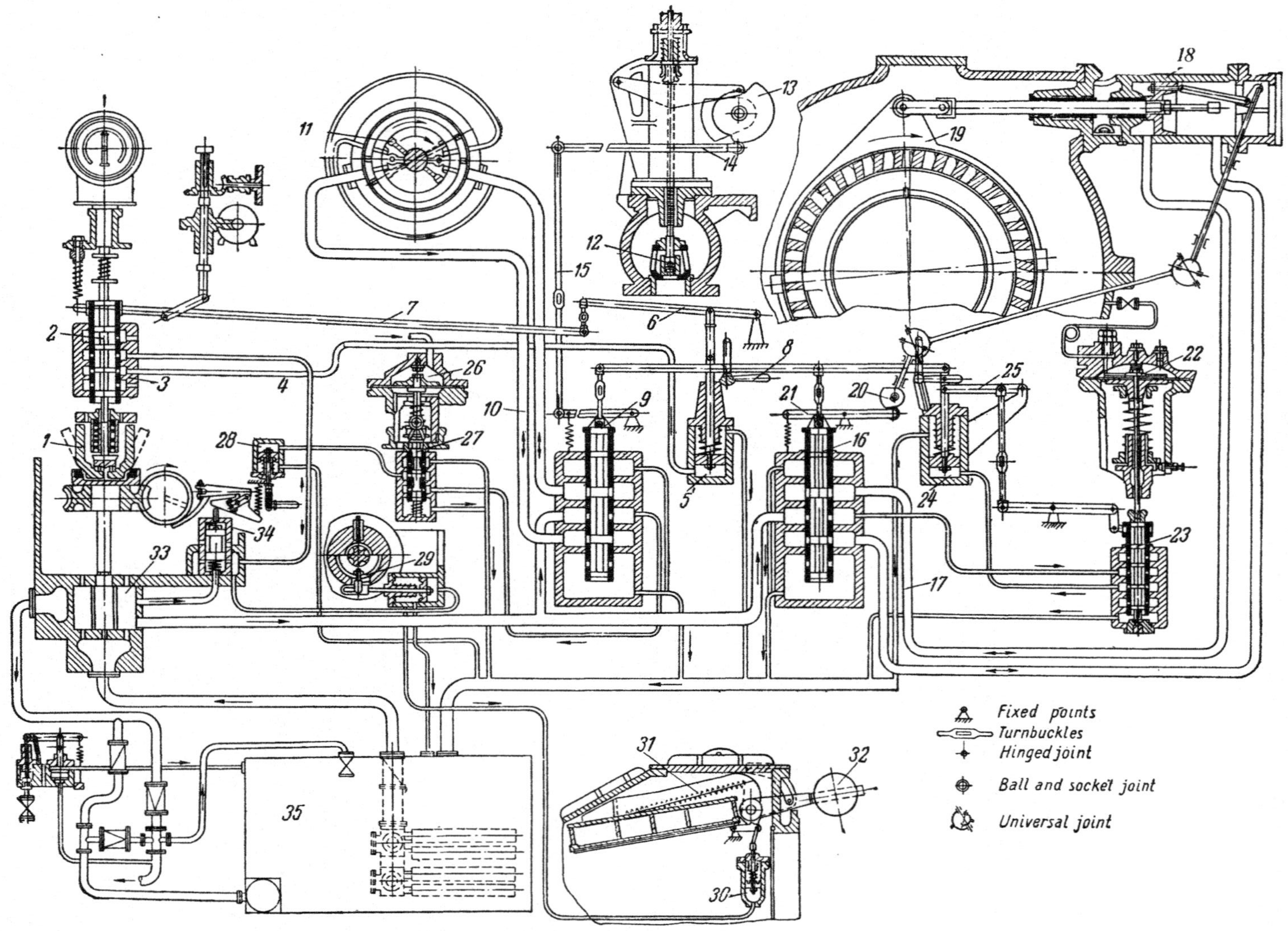

Fig. 11-15 Regulating system for turbine AT-12-1 of Kirov Turbine Works

are displaced in opposite directions. The pilot valve *23* is brought back to its original position by the lever *25*. Let us suppose that the pressure in the pass-out chamber increases. Piston of pilot valve *23* moves down. Pressure oil is supplied into the lower oil space of auxiliary servomotor *24* displacing it upwards. Piston of pilot valve *16* moves up and pilot valve *9* moves down isolating pilot valve *23*. The regulating valves of the H. P. cylinder start closing reducing the steam flow through the H. P. cylinder. On the other hand the regulating diaphragm of the L. P. cylinder begins to open increasing the flow through the L. P. cylinder to the condenser.

When the turbine reaches its maximum permissible speed, pilot valve *34* connects the pressure oil from the trip mechanisms *29* and *30* to the drain which causes the main stop valve and the non-return valve on the pass-out line to be quickly shut off.

d) 25,000 kW Turbine of the Leningrad Metal Works Type AP-25-2[1]

Initial steam pressure $p_0=29$ ata, initial steam temperature $t^\circ=400°C$, $p_{ex}=10$ ata. The turbine is designed to operate continuously for long periods with the following allowable pressure and temperature variations: p_0 from 27.5 to 31 ata and t_0 from 385 to 410°C. The turbine is shown in section in Fig. 11-16 (see Appendix). The turbine is a single-cylinder construction with the H. P. section consisting of a single-row velocity wheel followed by four pressure stages. The low-pressure part of the turbine has a single-row regulating stage and nine pressure stages.

The turbine has nozzle control governing, steam being supplied to the H. P. regulating stage through four regulating valves. The four nozzle groups of the L. P. cylinder are supplied with steam through a regulating diaphragm. The turbine rotor is flexible with a critical speed of 1,940 r.p.m. The shaft seals at both ends are of the labyrinth type with fir-cone type of development. The labyrinth elements are mounted on sleeves. The front end is supported by a combined journal and thrust bearing. The journal has a spherical base to allow for adjustments. The turbine is provided with rotor turning device at a shaft speed of 4 r.p.m. Steam from the controlled pass-out is utilised for process purposes. The pass-out pressure can be varied between 8 and 13 ata.

The turbine is further provided with two simple bleeds for feedwater heating. The turbine can develop a power of 30,000 kW with steam extraction from the pass-out. Details of turbine performance at various loads are given in Table 11-2.

Table 11·2

Power developed at alternator terminals, kW	Quantity of pass-out steam at 13 ata, tons/ hr	Alternator efficiency, %	Specific steam consumption, kg/kWh	Final feed temperature at exit from the last heater, °C
25,000	150	98.2	10.72	178
25,000	100	98.2	8.88	178
20,000	100	98.1	10.08	178
15,000	100	98.1	11.82	178
25,000	0	98.2	5.46	178

Note: The figures of specific steam consumption do not include steam used in ejectors.

e) High-Pressure Steam Turbine of L.M.W.: Capacity 25,000 kW at 3,000 r.p.m., Type T-25-90 (VT-25-4)

Steam pressure supply $p_0=90$ ata; temperature $t_0=500°C$; $p_{ex}=1.2$ ata. Permissible steam pressures and temperatures at turbine inlet for continuous operation: pressure from 85 to 95 ata and temperature from 495 to 505°C, p_{ex} from 1.2 to 2.5 ata.

Fig. 11-17 shows the sectional elevation of the turbine (see Appendix). The turbine is made in a single cylinder consisting in all of twenty stages. The H. P. portion consists of a two-row regulating stage and fifteen pressure stages. The L. P. section consists of a single-row regulating stage followed by three pressure stages. Steam supply is through a nozzle system. The first regulating stage is supplied with steam through four regulating valves. The two nozzle groups of the L. P. section are supplied with steam through a regulating diaphragm. The first eleven stage discs are machined out from a single forging. The rotor is flexible with a critical speed of 1,850 r.p.m. The labyrinth packings are of the fir-cone type. The front end is supported by a combined journal and thrust bearing. The journal bearing has spherical outer sleeve to enable it to adjust itself on the seat. The turbine is provided with a shaft rotation device. The turbine and alternator rotors are joined together by a toothed semi-flexible coupling. Steam from the pass-out is utilised for heating purposes The pass-out pressure can be varied between 1.2 to 2.5 ata. Besides the regulated pass-out, the turbine is further provided with four simple extractions for regenerative feedheating. The turbine can develop 30,000 kW, both with or without the pass-out being effected. The minimum steam flow through the low-pressure portion at pass-out

[1] A non-standard turbine.

pressure of 1.2 ata is about 8 tons/hr. The operating characteristics of the unit are given in Table 11-3.

Table 11-3

Power developed at alternator terminals, kW	Quantity of pass-out steam at a pressure of 1.2 ata, tons/hr	Alternator efficiency, %	Specific steam consumption, kg/kWh	Final feedwater temperature, °C
25,000	100	98.2	5.42	221
25,000	40	98.2	4.63	202
20,000	40	98.1	4.70	193
15,000	40	98.0	5.06	183
25,000	0*	98.2	4.12	197

* Pass-out regulator is switched off.

Note: The figures of specific steam consumption do not include steam used in ejectors.

Governing System of Turbines T-25-90 (VT-25-4) and AP-25-2

The hydraulic governing system of these turbines is shown in Fig. 11-18.

When the turbine operates with extraction from the pass-out variation in the electrical load on the turbine causes both the H. P. regulating valves and the regulating diaphragm for the L. P. section to open increasing the steam flow through the turbine. On the other hand if the quantity of pass-out steam varies the regulating mechanisms for the H. P. and the L. P. sections move in opposite directions.

The opening of the H. P. regulating valves is effected by servomotor *14* through pilot valve *15*. The regulating diaphragm is controlled by servomotor *10* through pilot valve *11*. The displacement of pilot valves *15* and *11* is caused by the variation of oil pressure in the two oil mains *A* and *B*. This variation is again a result of the operation of the integrating pilot valves N_1 and N_2. The pilot valves N_1 and N_2 are displaced because of the change in oil pressures in the two chambers K_1 and K_2, caused by these chambers purging their pressure oil to the drain through the speed regulator or pressure regulator pilot valves.

The construction of the integrating pilot valves for turbines VT-25-4 and AP-25-2 is similar differing only in some of their dimensions.

11-6. BACK-PRESSURE TURBINE WITH PASS-OUT, TYPE APR[1]

Back-pressure turbines with pass-out extraction are mostly used in industry. Both the exhaust steam and the extraction steam from the pass-out can be utilised for process purposes as well as heating, etc. Thus these turbines can supply both electrical power and process steam at various pressures and temperatures. The capacity of such turbines depends on the quantity of steam flowing through them which again is directly dependent on the requirements of the consumer. Hence these turbines are run in parallel with condensing turbines so that the electrical loads can be taken care of by the condensing turbines. Usually these turbines are provided with nozzle control governing. The general arrangement of a back-pressure turbine with pass-out is shown in Fig. 11-19. Under normal conditions of operation the centrifugal speed governor is set at a higher speed so that it does not come into operation for changes in the electrical loads on the network. The turbine operation thus solely depends upon the process steam demand of the consumers, the power developed by the set remaining constant irrespective of the changes in the system load as long as the process steam demand remains constant.

The regulating valves of the H. P. stages are controlled by the back-pressure regulator *2* increasing or decreasing the quantity of steam flowing through the H. P. cylinder. When the back pressure increases the back-pressure regulator membrane is displaced upwards relative to the fulcrum point *a*. The regulating valve *4* partially closes decreasing the steam flow through the H. P. cylinder and the pass-out; the quantity of steam flow through the L. P. cylinder, however, remains unchanged. The back-pressure regulator *3* controls both the fresh steam and the pass-out regulating valves *4* and *5* respectively. If the back pressure decreases the membrane of regulator *3* deflects downwards, lever *bc* rotates about point *b* as its fulcrum, the regulating valves *4* and *5* open more and the quantity of steam flowing through the turbine increases with the pass-out steam pressure p_{ex} and quantity D_{ex} remaining constant.

Turbines of this type, however, have not found much industrial usage since they can be efficiently replaced by condensing turbines with two pass-outs (see 11-8 and 11-9).

11-7. EXHAUST-PRESSURE AND MIXED-PRESSURE TURBINES

The steam supply to these turbines is obtained from the exhaust of non-condensing reciprocating engines such as steam hammers, presses, etc. Since the quantity of steam available from these machines sharply varies, their exhaust main is sometimes connected with the extraction main

[1] A non-standard turbine.

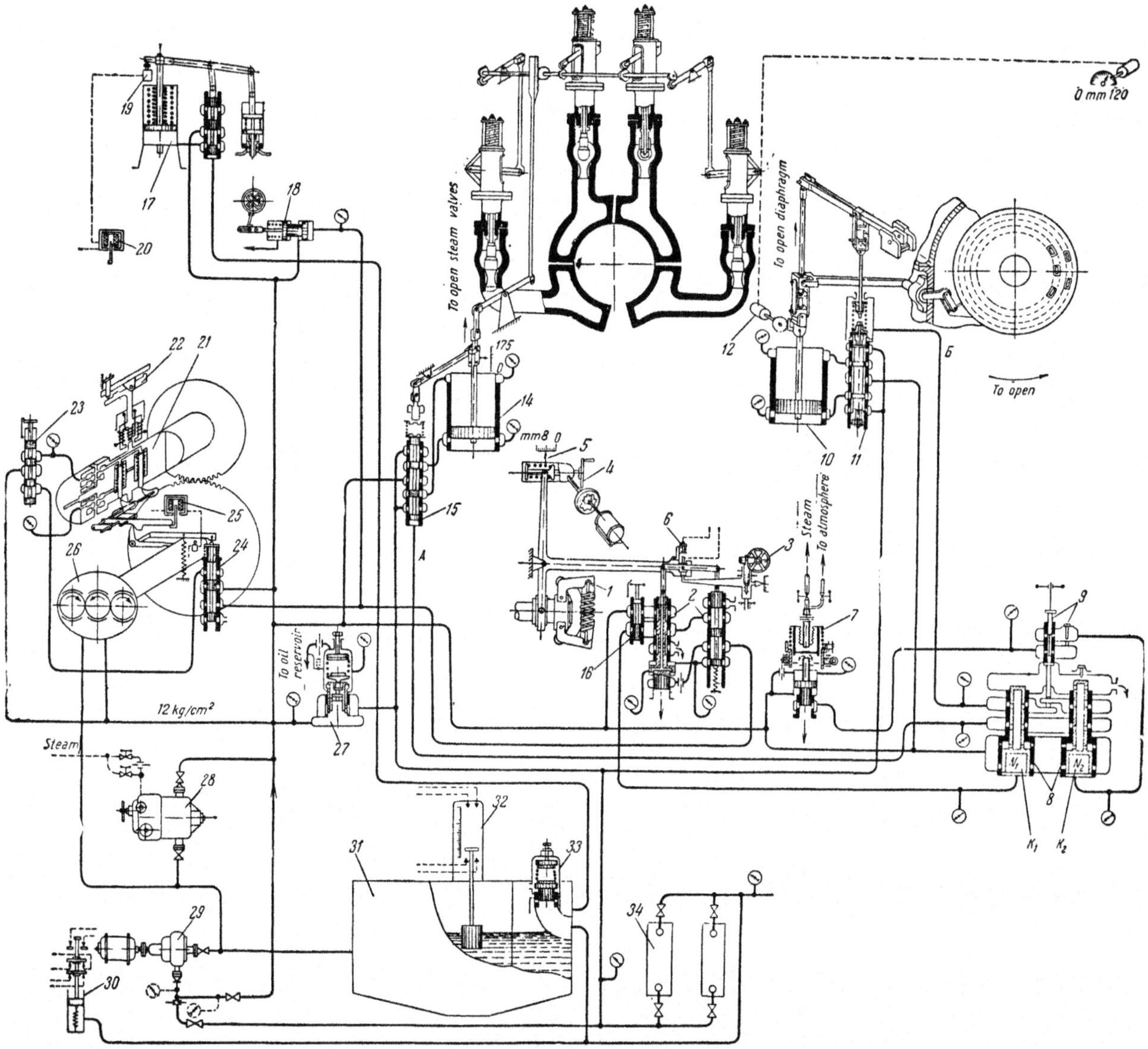

Fig. 11-18. Regulating system of turbine T-25-90 (VT-25-4) and AP-25-2.

1 – speed regulator; *2* – speed regulator pilot valve block; *3* – power limit switch; *4* – speed changer (synchroniser); *5* – sleeve displacement indicator; *6* – electric contact for remote control indicator "decrease"; *7* – pressure regulator 1.2 to 2.5 ata; *8* – integrating pilot valve block; *9* – switch for changing over from throttle to pressure regulator; *10* – servomotor for diaphragm regulator from 1.2 to 2.5 ata; *11* – pilot valve for servomotor *10*; *12* and *13* – remote selsyn indicators for operation of pass-out servomotor; *14* – H.P. regulating valve servomotor; *15* – pilot valve for servomotor *14*; *16* – pilot valve for overspeed trip testing; *17* – automatic stop valve; *18* – oil cut-off for automatic stop valve; *19* – final cut-off switch; *20* – pass-out main non-return valve solenoid switch-off; *21* – overspeed regulator; *22* – axial displacement indicator; *23* – pilot valve for overspeed trip testing by simulation; *24* – overspeed trip pilot valve; *25* – electrical switch-off for axial displacement trip; *26* – main oil pump; *27* – reducing valve; *28* – auxiliary oil pump; *29* – motor-driven oil pump; *30* – relay for switching on motor-driven pump; *31* – oil reservoir; *32* – oil level indicator with electrical contact for remote indication; *33* – non-return valve; *34* – oil cooler

of a back-pressure turbine or that of a pass-out turbine. Thus the exhaust-pressure turbine utilising some of the exit steam from a back-pressure turbine, along with the latter forms a sort of "two shaft" condensing unit with regulated pass-out for heating purposes, etc. The above combination of two turbines operating under different steam conditions makes it possible to utilise the back-pressure turbine more and more, the exhaust from which is utilised for heating purposes. In the winter season when the amount of steam used for room-heating purposes is the maximum the exhaust-pressure turbine works at very light loads since the power developed by this turbine entirely depends on the quantity of exhaust steam made available to it. Hence these tur-

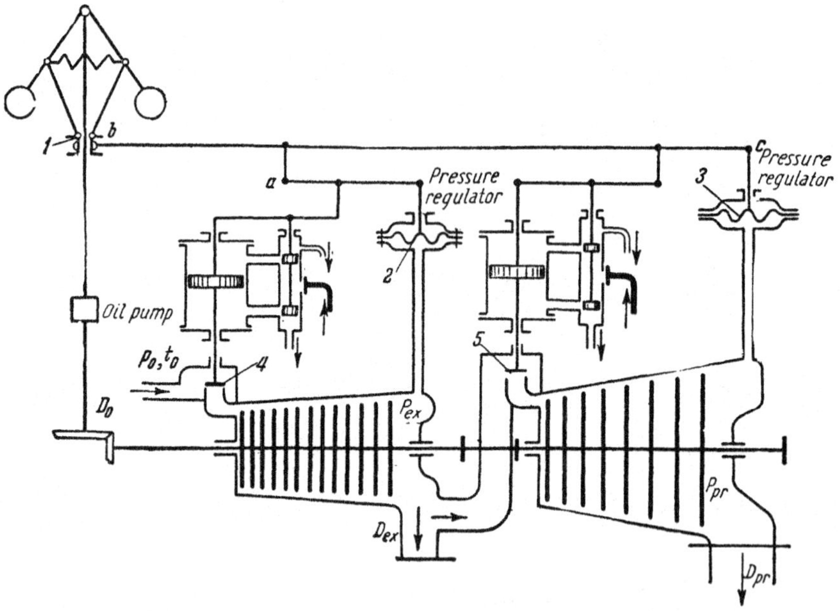

Fig. 11-19. Regulation system of back-pressure turbine with a pass-out

bines are operated in parallel with base load condensing turbines.

The 6,000 kW L.M.W. turbine type MK-6-1[1] is one such turbine. This turbine operates at the usual speed of 3,000 r.p.m., at a pressure of 1.2 ata and 110°C. The sectional view of the turbine is shown in Fig. 11-20.

The rotor consists of a single-row regulating stage and three pressure stages. Steam supply to the nozzles of the regulating stage is through a rotating diaphragm. The diaphragm opens the steam gates fully for a load of 4,800 kW (economic load). The nominal capacity of the turbine (6,000 kW) is developed when the intermediate steam supply valve *3* opens, supplying steam to the second stage of the turbine. The axial thrust exerted on the blades due to reaction and difference in hub diameters is balanced by a dummy piston provided before the regulating stage. Further, to fix the rotor in the proper axial direction and to bear the residual axial thrust a segment-type thrust bearing is provided, combined with a journal bearing, at the steam supply end. The shaft ends are water sealed.

Table 11-4

Power developed at alternator terminals, kW	Alternator efficiency	Specific steam consumption, kg/kWh
2,400	0.914	15.9
3,600	0.933	14.2
4,800	0.944	13.5
6,000	0.954	15.6

[1] A non standard turbine.

The guaranteed steam consumption rates for turbine MK-6-1 at initial steam pressure and temperature 1.2 ata and 110°C and 4,900 m^3/hr of cooling water at an inlet temperature of 30°C are given in Table 11-4.

Turbine MK-6-1 can be operated in combination with turbine type APR, the latter operating continuously.

In some cases mixed pressure turbines find industrial application. Steam at different pressures and temperatures is supplied to either different groups of nozzles of the regulating stage or different stage nozzles.

11-8. TURBINES WITH TWO PASS-OUTS

Turbines with two extractions (pass-outs) have found wider application in the past few years in the U.S.S.R. Steam extracted from the first pass-out is utilised for process purposes and from the second for domestic heating. These turbines function in accordance with both the electrical and the heating steam loads.

The schematic diagram of this type of turbine is shown in Fig. 11-21. The load variation signal is transmitted to the regulating valves from the speed governor and the two pressure regulators. The three regulators independently regulate the control valves of all the three stages. Thus the signal from any of the three regulators operates all the three regulating valves simultaneously. If the load on the network varies the turbine speed undergoes a change. This impulse is transmitted to the regulating valves through the centrifugal speed governor. During this operation points *b* and *a* remain fixed as the pressures and extraction quantity through the two pass-outs remain constant. Let us suppose, for example, that the load on the alternator terminals increases. This causes the centrifugal governor sleeve to move downwards. The lever arrangement *dgmk* is also displaced accordingly, revolving the lever *gde* about *e* as its fulcrum and lever *nkpa* about point *a* as its fulcrum resulting in the opening of the regulating valves of the medium-pressure cylinder. Simultaneously

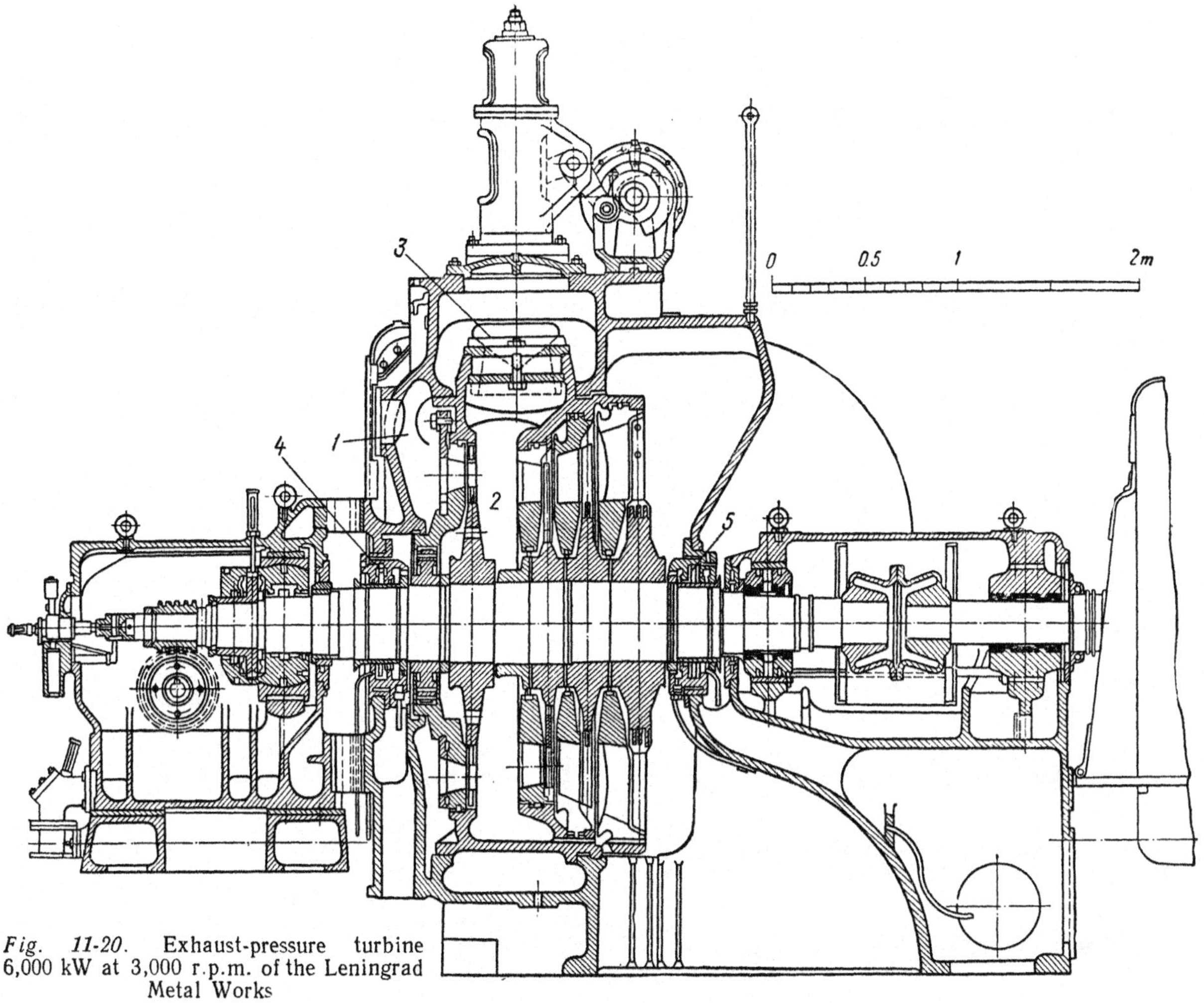

Fig. 11-20. Exhaust-pressure turbine 6,000 kW at 3,000 r.p.m. of the Leningrad Metal Works

with this the lever *bcf* also drops lower revolving about point *b* as its fulcrum so that the L. P. cylinder regulating valves also open. The mass flow through the turbine increases, the pass-out pressures, however, remaining unchanged, since the back-pressure regulation is independent of the speed regulation. The pilot valves are returned back to their central positions through the system of lever arrangements shown in Fig. 11-21.

If the process steam pressure p_p changes the system would operate with points *e*, *d*, *c*, *m* and *k* as fixed. For example, if the pressure p_p decreases lever *nkpa* would revolve about point *k* as its fulcrum and *bcf* about point *c* as its fulcrum so that the regulating valves of the H. P. cylinder open and those of the M. P. and L. P. cylinder close resulting in larger steam flow for process purposes with the pass-out for domestic heating as well as the power developed by the turbine remaining constant. If the pressure in the second pass-out alters points *a*, *b* and *g* would be fixed for the lever system operation. Increase in pressure p_h causes the regulating valves of the H. P. and M. P. cylinders to close and those of the L. P. cylinder to open more. In this case the power developed and the quantity of steam extracted for process purposes must remain unaltered.

The heat-drop process of a turbine with two pass-outs is shown in Fig. 11-22 on the *i-s* diagram. The equation for power developed by a turbine with pass-outs will be

$$N_e = \frac{D_c H_i + D_{pr} H_i' + D_h (H_i' + H_i'')}{860} \eta_m \eta_g, \quad (11\text{-}3)$$

where D_c —steam flowing to the condenser;
H_i —heat drop utilised in the turbine;
D_{pr} —steam pass-out for process purposes;
D_h —steam for domestic heating;
H_i' —heat drop utilised in the H. P. cylinder;

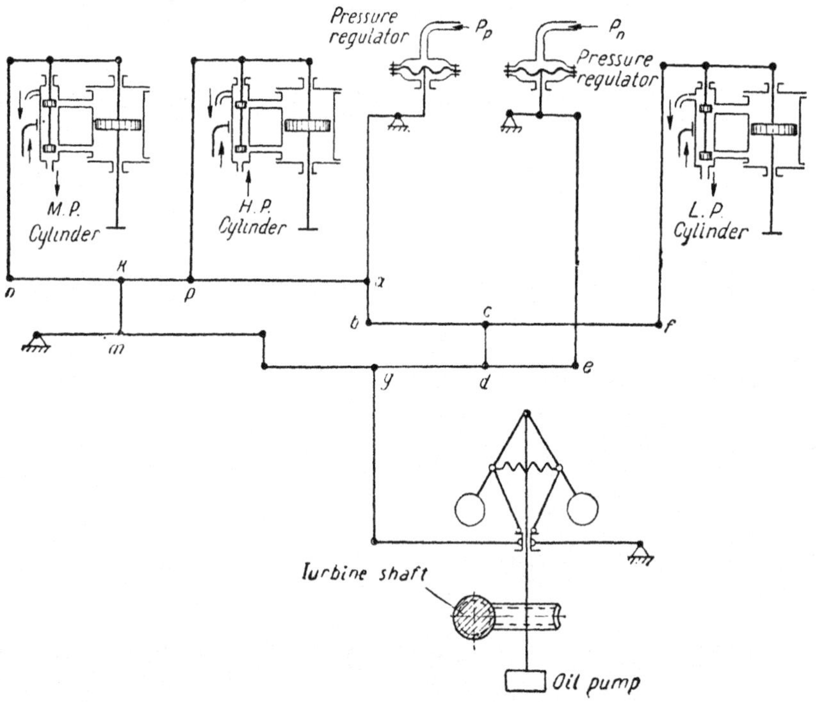

Fig. 11-21. Regulation system of a condensing turbine with two pass-outs

H_i'' —heat drop utilised in the M. P. cylinder.

If now we add and subtract $D_h H_i'''$ from the numerator of equation (11-3) we obtain after a simple rearrangement

$$N_e = \frac{(D_c + D_h) H_i + D_{pr} H_i' - D_h H_i'''}{860} \eta_m \eta_g, \quad (11\text{-}3a)$$

where H_i'''—heat drop utilised in the L. P. cylinder.

This formula may be further expressed as

$$N_e = \frac{D_c^I H_i + D_{pr} H_i^I}{860} \eta_m \eta_g - \frac{D_h H_i^{III}}{860} \eta_m \eta_g, \quad (11\text{-}3b)$$

where $D_c' = D_c + D_h$—steam flow to the condenser without steam pass-out for domestic heating;

or

$$N_e = N_e^I + \Delta N_e^{II}, \quad (11\text{-}3c)$$

where N_e —power developed at the alternator terminals with $D_h = 0$;

ΔN_e^{II} —correction for the steam extracted for domestic heating.

The operation curves for a turbine with two pass-outs can be plotted from formula (11-3b). Assuming $D_h = 0$ equation (11-3b) becomes

$$N_e = N_e' = \frac{D_c' H_i + D_{pr} H_i'}{860} \eta_m \eta_g,$$

which is the formula for power developed by a single-extraction turbine. With the help of this equation as was shown in 11-4 the diagram of steam consumption curves is plotted as though for a single pass-out turbine (the upper portion of Fig. 11-23 above the axis OA). Point A in Fig. 11-23 conforms to the nominal capacity of the turbine. From the diagram it can be inferred that with $D_h = 0$ the power developed by the turbine at the normal steam consumption would be greater than the nominal value; this is shown in Fig. 11-23 by the intercept AA_1.

When the turbine supplies steam for domestic heating the power developed must reduce by an amount

$$\Delta N_e'' = \frac{D_h H_i'''}{860} \eta_m \eta_g.$$

The correction for power consumed for heating purposes is given in the lower portion of Fig. 11-23. When $D_h = 0$ and $\Delta N_e'' = 0$ (point O in the figure).

When the extraction of heating steam is a maximum we have

$$\Delta N_e'' = \frac{D_{h\,max} H_i'''}{860} \eta_m \eta_g,$$

(point a_0); here η_m and η_g may be given the same values as for nominal loads. If it is assumed that $\Delta N_e''$ is a linear function of D_h line Oa_0 will give the correction for power developed depending upon the steam extraction from the pass-out for domestic heating. Lines a_1a_1', b_1b_1', c_1c_1' and d_1d_1' show the maximum steam extraction for domestic heating that can be effected when the pass-outs for process purposes remain constant.

These lines are given by the condition

$$D_{h\,max} = D_0 - D_{c\,min} - D_{pr}, \quad (11\text{-}3d)$$

where D_0—mass flow of steam through the H. P. cylinder.

For the various operating conditions characterised by the points a, b, c and d, the value of D_h is equal to zero. This is given by points a_1, b_1, c_1 and d_1. Points a_1', b_1', c_1' and d_1' are plotted according to the values obtained from the equation (11-3d) for $D_{h\,max}$ with various D_{01}, D_{02},

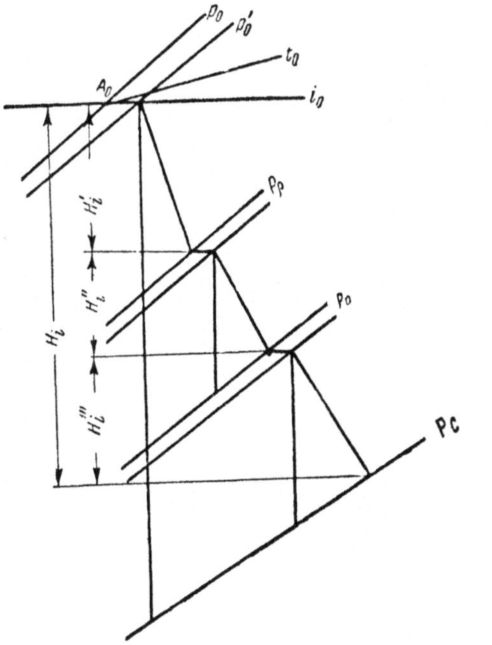

Fig. 11-22. I-s diagram for a turbine with two pass-outs

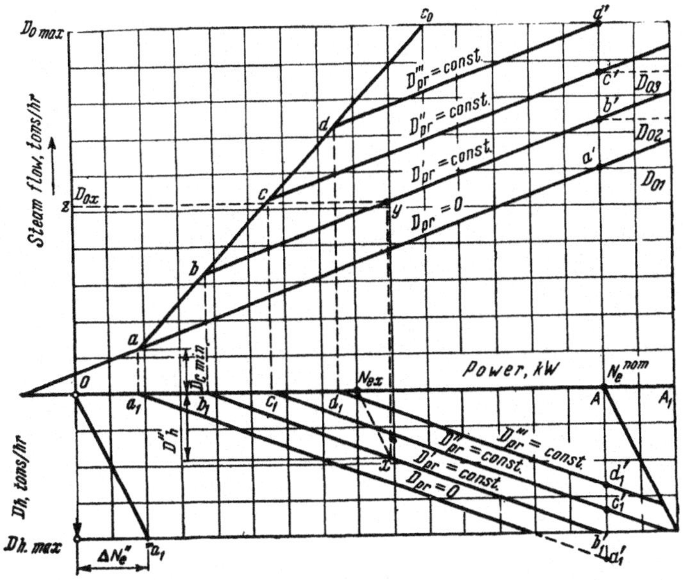

Fig. 11-23. Diagram of steam consumption curves for a turbine with two pass-outs

D_{03}, $D_{0\,max}$ and extractions $D_{pr}=0$, D'_{pr}, D''_{pr}, and D'''_{pr} for process steam. The lines joining points a_1 and a_1', b_1 and b_1', c_1 and c_1', d_1 and d_1' give the maximum possible $D_{h\,max}$ for given D_0 and D_{pr}.

Example: The power developed at the alternator terminals is N_{ex}, steam for domestic heating — D_h'' and for process — D'_{pr}. Find the quantity of steam supplied to the turbine.

Measuring N_{ex} along the x axis and D_h'' along the y axis in the lower portion of diagram 11-23 for the given process steam pass-out of D'_{pr} we obtain point x. Drawing a vertical line from here to intersect the constant process steam pass-out point y is obtained. From point y a line parallel to the x axis is drawn giving point z at the $OD_{0\,max}$ line. The length Oz gives the quantity of steam flowing through the H.P. cylinder, i.e., the total quantity of steam supplied to the turbine.

11-9. EXAMPLES OF CONSTRUCTION: TURBINES WITH TWO PASS-OUTS

a) Turbine PT-12-35/10 (APT-12-1) of the Bryansk Locomotive Works (B.L.W.)

Fig. 11-24 shows the sectional view of a single-cylinder condensing turbine with two pass-outs. The capacity of the turbine is 12,000 kW at 3,000 r.p.m. The turbine is designed to operate at an initial pressure of 35 ata and steam temperature of 435°C. The permissible fluctuations for initial pressure and temperature are 32 to 37 ata and temperature from 420 to 445°C. The design pressure in the first pass-out is 10 ata with extraction of 50 tons/hr. The permissible pressure fluctuations for the first pass-out are from 8 to 13 ata, the pressure in the second pass-out is designed to be 1.2 ata at an extraction of 40 tons/hr, the pressure variations allowable being from 1.2 to 2.5 ata for the second pass-out. The turbine has also, besides the two pass-outs, extractions for feedheating. The turbine is nozzle governed. Steam supply to the turbine is through four regulating valves for the H. P. stages and for the medium- and low-pressure stages steam supply is through regulating diaphragms controlling the quantity of steam flowing to the nozzles of these stages. The turbine has, in all, fifteen stages. The H. P. section has a two-row regulating stage followed by three pressure stages. The M. P. section has a single-row regulating stage and six pressure stages and the L. P. section has a single-row regulating stage with three pressure stages following it. The labyrinth seals at the turbine ends are of the fir-cone type mounted on sleeves. The blade discs and the labyrinth sleeves are shrunk on the turbine shaft. The turbine shaft is of flexible construction. The front end is supported by a combined journal and thrust bearing with the journal of the self-aligning type. The turbine is provided with a linear expansion indicator and a shaft revolving device. The governing system is hydraulic. The main oil pump is of the centrifugal type and is directly driven by the turbine shaft. The regulating oil pressure is 8 atm. gauge; lubricating oil supply is at 0.5 atm. gauge. There is an axial displacement relay to trip the turbine in the event of any untoward axial displacement. The turbine is also provided with an auxiliary oil pump run by a turbine and a motor-driven oil pump which automatically comes into operation when the lubricating oil pressure falls below 0.2 atm. gauge.

The fixed point of the turbine is situated on the condenser axis and the turbine is permitted to expand, when heated, in the direction of the front end bearing.

b) 25,000 kW Turbine of the L.M.W., Type PT-25-90/10 (VPT-25-3)

Steam supply at $p_0=90$ ata and $t_0=500°C$; $p_{pr}=10$ ata and $p=1.2$ ata. Allowable fluctuations in steam pressures and temperatures: p_0—from 85 to 95 ata; t_0—from 495 to 505° C; p_{pr}—8 to 13 ata; p—from 1.2 to 2.5 ata. The turbine is shown in Fig. 11-25.

The turbine has nineteen stages in a single cylinder. The H.P. portion consists of a two-row regulating disc followed by eight pressure stages. The blade discs for these stages are machined out directly on the forged rotor. The M. P. section consists of a single-row regulating stage and five pressure stages. Finally, the L.P. section consists of a single-row regulating stage followed by three pressure stages. The blade discs of the M. P. and L. P. sections are made separately and mounted on the shaft. The turbine is supplied with steam through nozzles. The first stage nozzles are supplied with steam through four regulating valves. The four nozzle groups of the M. P. section are supplied with steam through a regulating diaphragm replacing four regulating valves and the L. P. section through a regulating diaphragm in place of two regulating valves. The turbine shaft is of flexible construction with a critical speed of 1,800 r.p.m. The front end bearing is a combined journal and thrust bearing with spherical seating for self-alignment. The end seals for the turbine shaft are of the fir-cone labyrinth type mounted on sleeves. The turbine is provided with a shaft rotating device.

The first steam pass-out is utilised for industrial purposes and the second one for domestic heating. The turbine is further provided with three bleeds for regenerative feedheating. The turbine can develop up to 30,000 kW with the two pass-outs as mentioned above[1]. The minimum mass flow of steam through the L. P. section is 8 tons/hr at a pressure of 1.2 ata in the second pass-out. Some of the operation characteristics of turbine VPT-25-3 are given in Table 11-5.

Table 11-5

Power developed at alternator terminals, kW	Pass-out quantity, tons/hr		Alternator efficiency, %	Specific steam consumption, kg/kWh	Final feed water temp., °C
	10 ata	1.2 ata			
25,000	72	54	98.2	6.66	203
25,000	130	0	98.2	7.38	206
25,000	0	100	98.2	5.60	199
25,000	0*	0*	98.2	4.15	185
20,000	50	40	98.2	6.49	194
17,000	40	40	98.1	6.78	188

* The pass-out regulators are put out of operation.

Note: The temperature of condensate returned from industrial usage as well as from the domestic heating circuit or of treated water make-up is assumed to be 100° C.

Governing of Turbine PT-25-90/10 (VPT-25-3)

The operating pressure of oil in the regulating system is 12 kg/cm². The governing system is designed to ensure: a constant power generation with variations in the quantity of steam extracted from the pass-outs, constant pass-out pressures in case of change of electrical load on the turbine, and in the event of pressure variations in one of the pass-outs, a constant pressure for the second pass-out and constant power developed.

This complex system of governing is obtained by connecting the speed regulator and the two pressure regulators to an intermediate hydraulic device (integrating pilot valves) which ensures the necessary interrelation between the three regulating devices, and the three regulating valve servomotors.

Let us examine the system of regulation from the schematic diagram presented in Fig. 11-26 and the full diagram in Fig. 11-27 (see Appendix).

A change in the electrical load on the turbine or the frequency of the network causes the centrifugal regulator to displace the valve *1* (Fig. 11-26). Let us suppose that the load has increased. Centrifugal regulator sleeve moves downwards, the pressure of oil in pipe *D* increases

[1] L. I. Tubyansky and L. D. Frenkel, *High-Pressure Steam Turbines of the L.M.W.*, Gosenergoizdat, 1953.

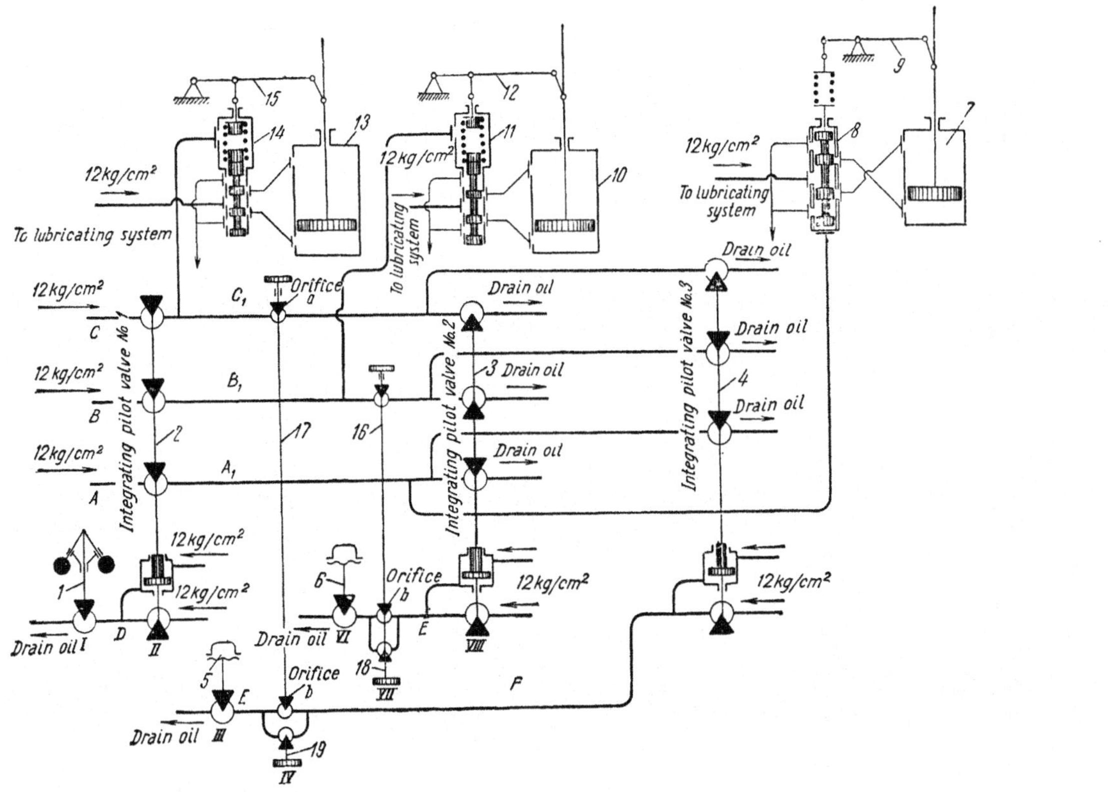

Fig. 11-26. Regulating system of turbine PT-25-90/10 (VPT-25-3) (L.M.W.)

1 – speed regulator; *2* – integrating pilot valve No. 1; *3* – integrating pilot valve No. 2; *4* – integrating pilot valve No. 3; *5* – pressure regulator 1.2 to 2.5 ata; *6* – pressure regulator 8 to 13 ata; *7* – H.P. regulating valve servomotor; *8* – pilot valve for servomotor 7; *9* – differential lever; *10* – servomotor for pass-out 8 to 13 ata; *11* – pilot valve for servomotor *10*; *12* – differential lever; *13* – servomotor for pass-out 1.2 to 2.5 ata; *14* – pilot valve for servomotor *13*; *15* – differential lever; *16* – changeover switch for 8 to 13 ata pass-out; *17* – changeover switch for 1.2 to 2.5 ata pass-out; *18* – throttle for regulator *6*; *19* – throttle for regulator *5*

displacing the integrating pilot valves upwards causing increased oil pressure in the pipes A_1, B_1 and C_1. These increased pressures cause pilot valve *8* to move upwards and pilot valves *11* and *14* to move downwards displacing servomotors *7*, *10* and *13* to move upwards opening the regulating valves of the H. P. section and revolving the regulating diaphragms of the M. P. and L. P. sections so that more steam flows through them. In the event of a decrease in the load on the turbine the system operates in reverse, i. e., closes the regulating valves and reduces steam flow through the regulating diaphragms of the L. P. and M. P. sections.

If there is an increase of pressure in the first pass-out throttle VI moves down, oil pressure in pipe *E* increases causing the integrating pilot valve No. 2 to move up which causes a fall of oil pressure in pipe A_1 and a rise of oil pressure in pipes B_1 and C_1. Pilot valves *8*, *11* and *14* move down causing the servomotor piston *7* to move down and *10* and *13* to move up. The regulating valves of the H. P. section reduce the steam flow and the regulating diaphragms of the M. P. and L. P. sections open allowing more steam.

The above operation leads to the decrease in steam flowing out from the first pass-out with the power developed and the extraction quantity from the second pass-out remaining unchanged. If the pressure in the second pass-out decreases the operation would be in a reverse order. With variation in the steam pressure in the second pass-out throttle III moves down (increase in pressure) or moves up (pressure decrease) causing an increase or decrease in oil pressure in pipe *F*. In accordance with this integrating pilot valve No. 3 either moves up or down. If pilot valve No. 3 moves up the oil passing to drain from pipes A_1 and B_1 increases causing a fall in the oil pressure and flow of oil from pipe *C* to drain decreases causing a rise in pressure. This causes the regulating valves of the H. P. section and the first regulating diaphragm to reduce the steam flow through them whereas the quantity of steam flow through the second regulating diaphragm increases (since it is opened more). If the pressure of oil in pipe E decreases the operation in the regulating system is just the opposite.

c) 50,000 kW Turbine Type PT-50-130/7 (VPT-50-4) of the Ural Turbomotor Works (U.T.W.)

Initial steam conditions: $p_0=130$ ata, $t_0=565°C$. Cooling water temperature $t_{cw}=20°C$. Permissible variations in the steam pressures and temperatures: p_0—from 125 to 135 ata; t_0—from 555 to 570°C. The turbine is made in two cylinders having two pass-outs, one for process purposes and the other for domestic heating (see Fig. 11-28). The process steam has a pressure of 7 ata with permissible variation between 5 and 9 ata. The pressure of the second pass-out for domestic heating purposes may vary between the limits of 0.5 and 2 ata. Besides these two, the turbine has one more extraction from a higher point of the 0.6 to 2.5 ata pass-out which enables the heating of the system water within a very wide range. The rated quantity of steam extracted from the pass-outs is 118 tons/hr for industrial purposes and 76 tons/hr for domestic heating (two pass-outs). With the normal operating pressures and temperatures for steam at inlet and at the two pass-outs the turbine develops nominal power. The nominal power may be developed with various combinations of the parameters:

a) initial pressures and temperatures between 125 and 135 ata, and 555 to 570°C;

b) cooling water temperature up to 33°C;

c) pressure of steam in the first pass-out between 5 and 9 ata;

d) reduction of pass-out steam from both pass-outs up to zero.

The maximum quantity of steam available for process purposes at nominal capacity and zero extraction from the second pass-out is 160 tons/hr at a pressure of 7 ata.

The maximum quantity of steam that can be extracted for domestic heating is 120 tons/hr. However, this can be done only for a particular process steam pass-out or particular quantity of steam passing through to the condenser at the nominal turbine capacity. The turbine is further provided with five simple bleeds for feedheating purposes. The first of these is after the H. P. chamber at a pressure of 34 ata. The second, third, fourth and the fifth bleeds are after the second, fourth, ninth and the eleventh stages of the L. P. chamber. There is one more bleed in the chamber after the eleventh stage (higher point of the second pass-out for heating). At nominal loads the feedwater is heated up to a temperature of 230°C.

The H. P. cylinder consists of a two-row regulating stage and eight pressure stages. The L. P. cylinder consists of fifteen stages. The medium- and the low-pressure portions of the turbine have each a single-row regulating stage. The H. P. and L. P. cylinders are in opposition with respect to steam supply, i. e., the direction of steam flow in the two cylinders is in the opposite directions reducing the thrust on the bearing. The two rotors are joined by a solid coupling so that a single thrust bearing could be used for both the rotors. The placement of the thrust bearing between the two rotors facilitates the linear expansion of the two rotors in the respective directions. The shaft ends are sealed by labyrinth packings without sleeves but with the labyrinth foils extending into grooves turned directly on the turbine shaft. These grooves are meant for temperature compensation to safeguard the labyrinth foils from fouling. The fixed point of the turbine is situated on the axis of the rear bearing of the L. P. cylinder. Hence the expansion of the M. P. and H. P. cylinder rotors and stators takes place in a direction away from the generator. The turbine as usual is provided with a shaft revolving device. Steam supply to the H. P. and M. P. cylinders is through nozzles. From the main stop valve the steam flows to four regulating valves two of which are placed on the upper housing and the remaining two on the lower housing diametrically opposite to the other two. These regulating valves are operated through a distributor camshaft and a toothed sector actuated by a double acting servomotor.

The M. P. and the L. P. cylinders are supplied with steam through regulating diaphragms operated by servomotors. The regulating diaphragm of the M. P. cylinder supplies steam to three groups of nozzles, i. e., it replaces three regulating valves. The L. P. cylinder regulating diaphragm supplies and controls the steam flow to all the nozzles of the regulating stage at one and the same time (i. e., throttle governing). The regulation system is of the compound type. It consists of three regulators, a speed regulator and two pressure regulators. The three regulators and the main servomotors are connected with each other through a hydraulic system with two amplifiers. The three regulators are of exactly the same type and are of the membrane type. The pressure regulator membranes are acted upon by the steam pressures, the speed regulator membrane by the pressure of regulating oil from the impeller. The main oil pump and the governor impeller are both directly mounted on the turbine shaft. The pump supplies the regulating system with oil at 14 atm. gauge; the lubricating system is supplied with oil by two injectors at an oil pressure of 1 atm. gauge measured before the oil coolers. Since there is a provision in this turbine for the heating of feedwater through various ranges the regulating system is equipped with an arrangement for changing over the domestic heating pass-out pressure regulator from the steam chamber after the 22nd stage to the chamber after the 20th stage. With a single boiler operation (base load) the pressure regulator is connected to the chamber after the 22nd stage which is connected with the first boiler water circuit. In the case of feedheating by parts the second boiler (peak load) is also brought into operation and the pressure regulator is switched over to the chamber after the 20th stage. An ingenious device is provided in this turbine for protection against rotor speeding in the event of load shedding. The servomotor of the first amplifier and the pilot valve of the second amplifier are designed in such a way that with a drop in the load on the turbine the non-uniformity of regulation drops sharply decreasing the tendency on the part of the turbine to speed up. The turbine is provided with two overspeed trip devices of the ring type. The auxiliary oil pump for starting the turbine is motor driven. The turbine is further provided with two motor-driven oil pumps as standby pumps to be used in case of main oil pump failure. One of the pumps is driven by an a. c. motor and the other by a d. c. motor. The turbine is provided with the following protective devices:

1) axial displacement relay;
2) low vacuum trip;
3) relay for switching on standby motor-driven oil pumps;
4) impeller oil pressure drop indicator relay;
5) relay for switching on condensate pumps.

The works has incorporated the following check devices for observing the working of some of the important components:

1) temperature indicators for the thrust bearing pads and journal bearings;
2) temperature indicators for studs, walls and flanges of the stator;
3) thermal expansion indicators for the rotor (placed in the front and rear bearing pedestals);
4) stator thermal expansion indicator;
5) vibration indicator, facilitating the control of vibration of all turbine and alternator bearings.

d) L.M.W. Turbine Type PT-50-130/13 (VPT-50-3)

The Leningrad Metal Works is now producing a series of condensing turbines of 50,000 kW capacity at 3,000 r.p.m. with two pass-outs. These

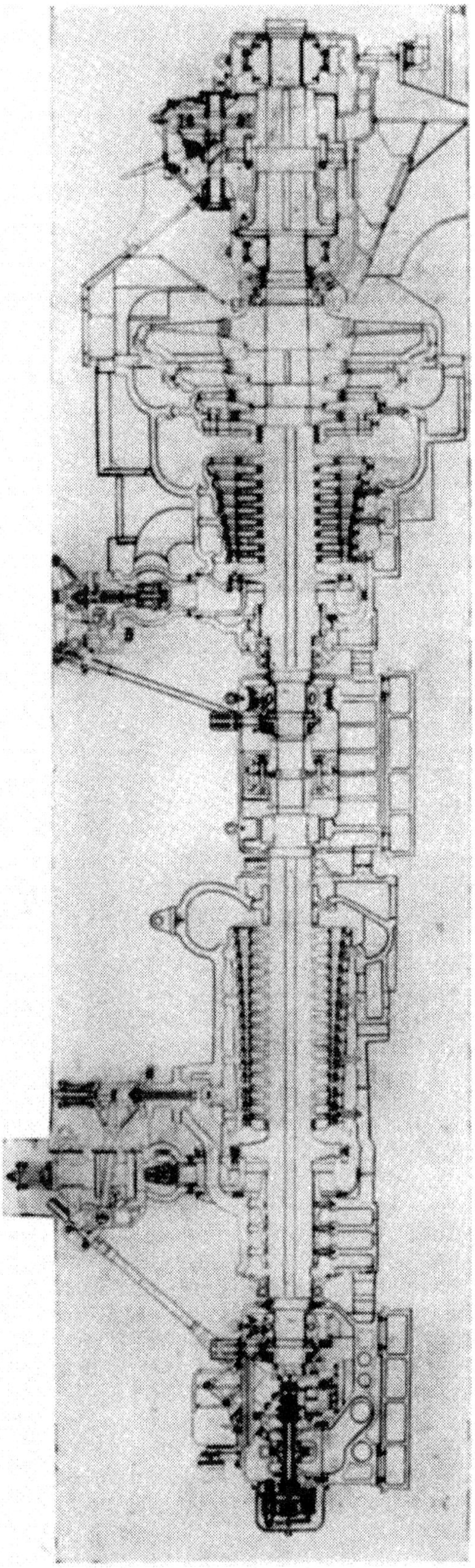

Fig. 11-29. **Turbine PT-50-130/13 (VPT-50-3) (L.M.W.)**

turbines are designed on a unit system basis (boiler-turbine unit) at initial steam parameters $p_0=130$ ata and $t_0=565°C$.

The sectional view of the turbine is shown in Fig. 11-29. The turbine consists of two cylinders with a total of thirty stages. The H. P. cylinder has a single-row regulating disc followed by sixteen pressure stages. The first pass-out (process steam) at a pressure of 13 ± 3 ata is provided at the exit from the H. P. cylinder. The low-pressure cylinder has thirteen stages; the M. P. section has a single-row regulating stage and eight pressure stages and the L. P. section of the L. P. cylinder has one single-row regulating stage followed by three pressure stages. The second pass-out (domestic heating) at a pressure of 1.2 ata to 2.5 ata is provided in the steam chamber after the ninth stage of the L. P. cylinder. Besides these two, the turbine has five simple bleeds for feedwater heating. The use of single-row regulating stages made with high grade blade profiles enabled the manufacturer to obtain high efficiencies for these turbines.

These turbines are provided with nozzle control governing; steam supply to the H. P. and the M. P. portions of the turbine is through four regulating valves each and to the L. P. section through a regulating diaphragm with two rows of steam ports effectively replacing two regulating valves. The regulating valves are placed directly on the turbine, two on top and two at the side. The H. P. section has in addition a fifth regulating valve controlling the flow of steam from the regulating stage to the fifth stage. This valve opens simultaneously with the fourth regulating valve. The maximum permissible pressure and temperature in the H. P. regulating stage are 104 ata and 540°C. The quantity of steam extracted from the pass-outs at nominal turbine capacity is 115 tons/hr for process purposes and 90 tons/hr for domestic heating.

The H. P. rotor is of forged steel. The end seals for the shaft are of the labyrinth type, without liners but with the labyrinth strips extending into grooves cut directly on the turbine shaft. The L. P. cylinder is a combined forged and assembly construction. The medium-pressure section is of forged steel whereas the L. P. portion is made up of discs mounted onto the shaft. The front end labyrinth seal for the L. P. cylinder is of the same type as for the H. P. cylinder. The rear end gland packing, however, is with a sleeve mounted on the shaft. The two rotors are joined by a flexible coupling and each has its own thrust bearing. The thrust bearings form a single unit with the journal bearings. The turbine is provided with a shaft-revolving device driven by an electric motor.

Oil supply. Governing and protection devices for the turbine.

Oil at a pressure of 20 atm. gauge is supplied to the governing system by a centrifugal pump mounted directly on the turbine shaft. From the pump the oil flows to the regulating system and to special oil injectors provided in the oil reservoir. The oil injectors are meant for the supply of lubricating oil to the bearings and to the suction side of the centrifugal oil pump.

For starting, the turbine is provided with a motor-driven auxiliary oil pump. It is also provided with an a. c. motor-driven oil pump for bearing lubrication when the turbine is stopped or when it is being slowly revolved by the shaft rotation device. A motor-driven oil pump, operated by a storage battery is provided for emergencies. For obtaining a constancy of extraction pressures from the two pass-outs the turbine is provided with high sensitivity cellophane-type pressure regulators. The turbine is equipped with the following protective devices: two overspeed trip gears (centrifugal type), axial displacement relay, low vacuum trip relay and a remote turbine tripping gear.

11-10. THERMAL EXPANSION OF TURBINES

When finally lined up on the bedplate a turbine has a single fixed point relative to which it expands when heated up during its operation. Usually the fixed point is situated below the L. P. cylinder and lies on the vertical plane of the turbine. Thermal expansion of the turbine in the radial and the axial directions is provided for by special guide keys, placed on the bedplate and the stator of the turbine. The thermal expansion takes place during starting and loading of the turbine and during unloading and stopping the rotor and stator of the turbine contract. Fig. 11-30 shows the various guide keys, etc., for the alignment of turbine AT-25-1 of the L.M.W. The L. P. cylinder stator is supported on bedplates *1* and *2*. The fixed point of the turbine lies at the intersection of the axes of two guide keys *3* and *4*. Key *4* prevents expansion towards the rear of the turbine. The oblique faced keys *5* permit both longitudinal and transverse expansion of the L. P. cylinder. Displacement of the L. P. cylinder causes the central bearing to move ahead with keys *15* and *16* guiding it. The front end of the L. P. cylinder and the central supporting bracket are centred with the help of vertical key *8*. The H. P. cylinder rests on the central and the forward bearing pedestals. The axial alignment of the H. P. cylinder is fixed by two vertical keys *11* and *12*. The correct cen-

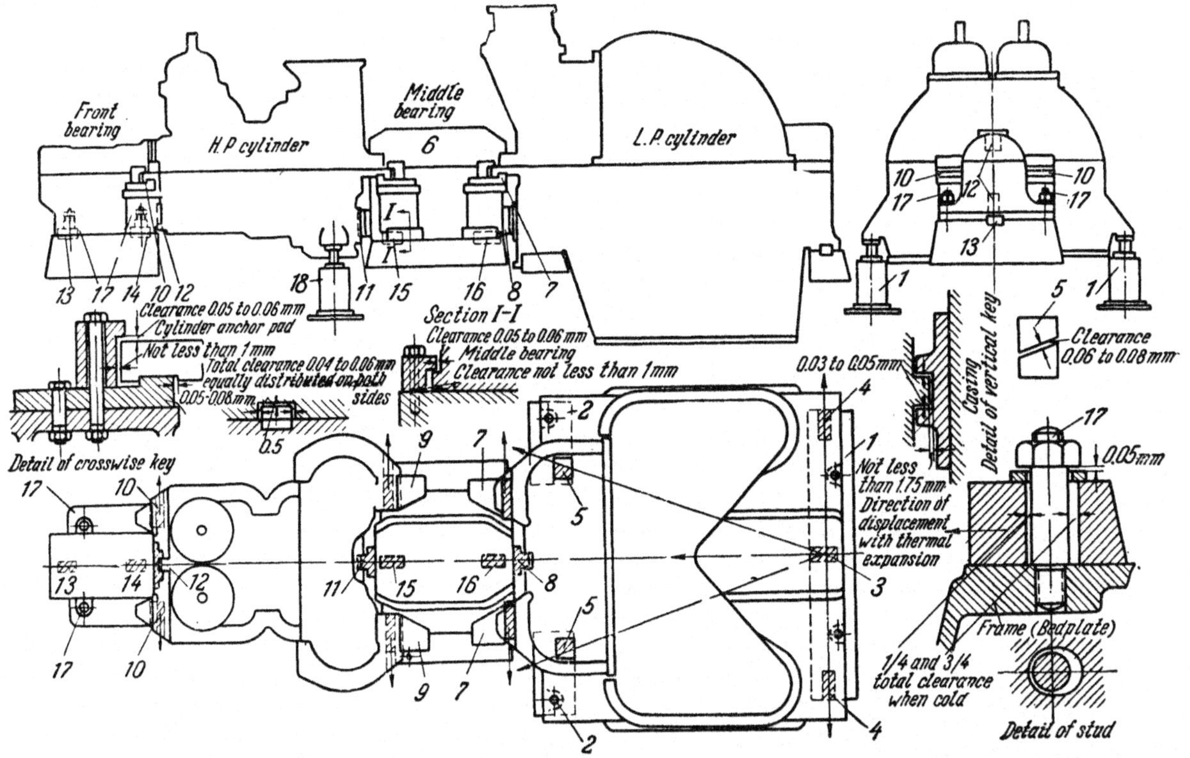

Fig. 11-30. Method of supporting and fixing 25,000 kW L.M.W. turbine on the foundation

1-2 – bedplates (L.P. cylinder); *3* – longitudinal key (L.P. cylinder); *4* – crosswise keys (L.P. cylinder); *5* – wedge keys; *6* – middle bearing; *7* – crosswise keys (L.P. cylinder); *8* – vertical key (H.P. cylinder); *9* and *10* – crosswise keys (H.P. cylinder); *11* and *12* – vertical keys (H.P. cylinder); *13* and *14* – longitudinal keys (front bearing); *15* and *16* – longitudinal keys (middle bearing), *17* – front bearing bolts; *18* – supporting stools

tering of the H. P. cylinder, while thermal expansion takes place, is ensured by longitudinal keys *13* and *14* situated under the front bearing bracket and transverse keys *9* and *10* as well as vertical keys *11* and *12*. To ensure free movement of the front bearing bracket the latter is fixed to the bedplate with special bolts *17*. These bolts are slid into oval holes which permit free movement of the block. A clearance of 0.05 mm is provided between the nut and the washer of these bolts for this purpose. The sliding surfaces of the bracket are ground finished to freely slide over the bedplate bearing surface.

The longitudinal displacement of the front bearing bracket is about 9 mm for turbine AT-25-1.

The maximum increase in the length of the turbine depends on its construction and steam conditions. The maximum linear expansion for a two-cylinder high-pressure turbine sometimes reaches a value of about 11 to 12 mm.

PART TWO

DESIGN AND CONSTRUCTION OF STEAM TURBINE COMPONENTS

Chapter Twelve

CONSTRUCTION OF CYLINDERS AND THEIR DETAILS

12-1. FORMS OF CYLINDERS, MATERIAL AND DESIGN

The turbine stator has a complicated shape often varying in diameter along its length to incorporate the steam chambers for supply of steam to its various stages, to accommodate the rotor with growing diameters at its L. P. end and the exhaust pipings as well as pass-out chambers or extraction points if any. The exhaust pipings of large condensing turbines are especially noted to have large dimensions. Various forms of stator construction can be seen from the Figs 10-1, 10-4, 10-5, 10-8, 11-13, 11-17, 11-24 and 11-25. The turbine cylinder usually has a horizontal flange, indispensable for assembling the turbine, and one or two vertical flanges which help in considerably simplifying their moulding as well as machining.

The material used in general for the manufacture of turbine cylinders is steel and cast iron, welded fabrication is also often used. The L.M.W. in its latest designs (VK-50, VK-100, SVK-150, PVK-200, SKK-300, etc.) has been using welded fabrication for the L. P. end of these turbines. In these turbine constructions only the bearing brackets are made of cast material.

The cylinders for turbines of small and medium capacities with initial steam pressures of 12 to 16 ata and a temperature of 250°C are usually made of cast iron. Some of the factories, however, have been using special pearlite cast iron for turbines operating in the range of 350°C or even up to 400°C if the turbines are of sufficiently small capacity. The most widely used material for turbine cylinder construction is carbon steel. It is used for turbines operating with initial steam temperatures of 400 to 425°C at pressures of 35 to 40 ata.

The high-pressure portion of turbines operating at high, superhigh and supercritical initial steam conditions is made from special costly alloy steels with additions such as chromium, nickel, molybdenum, tungsten, vanadium and other rare elements.

Along with the widely acknowledged single-shell construction, for high-temperature high-pressure turbines the double-shell construction has come into prominence at the present time. An example of such a construction is shown in Fig. 11-6. The space between the two walls of the shell is filled with medium-pressure steam. Therefore the wall thickness for the shell and the flange dimensions is found to be considerably smaller than in the usual single-shell construction. The complexity of double-walled construction more than makes up for by the saving of costly alloy steels used, since in a double-wall construction only the inner wall is made from alloy steel, the outer shell being made from cheaper steel or sometimes from ordinary carbon steel. The use of a double-shell construction simplifies flange design and reduces the temperature differentials for the stator flanges, etc.

A stator must be sufficiently strong satisfying the strength and stiffness requirements under normal conditions of operation. The use of lugs, cylindrical ribs and the end covers increase the stiffness of the cylinder. The exhaust pipes for large-capacity turbines are provided with special ribs inside them in order to increase their stiffness. These ribs also act as guides for the steam flowing to the condenser thus reducing heat losses in the exhaust piping.

Calculation of Wall Thickness

Because of the very complicated shape of the turbine cylinder the exact calculation of the wall thickness becomes very difficult. Neglecting the effect of side walls, stiffening ribs, flanges, the pressure and temperature variation along the length, etc., we may consider the cylinder to be drum shaped. In this case the tensile forces acting on the stator walls may be expressed by the formula

$$\sigma = \frac{Dp}{2\delta}, \qquad (12\text{-}1)$$

where D —internal diameter of the cylinder, cm;
p —gauge pressure acting on the walls, kg/cm²;
δ —thickness of the cylinder wall, cm.

The pressures allowable for cast iron are about 200 kg/cm² and for steels about 500 kg/cm².

From considerations of facility of casting, etc., the walls of the cylinders are often made much thicker than what is obtained from the above equation (12-1).

Flanges and Bolts

The flanges of a turbine cylinder operate under conditions of compression and bending. Their design is, however, based only on the bending forces present.

Fig. 12-1 shows the section of a flange with the basic dimensions and the forces acting on the flange and the bolts. The flange thickness is h, distance between bolt hole centres is t. The distance b between the edges of the bolt holes is chosen according to the strength of the material used for the flange and the bolts. The remaining dimensions shown in Fig. 12-1 are arbitrarily chosen to suit the design under consideration.

The flange design is based on the gauge pressure $\Delta p = p_i - p_0$ (p_i—pressure of steam inside

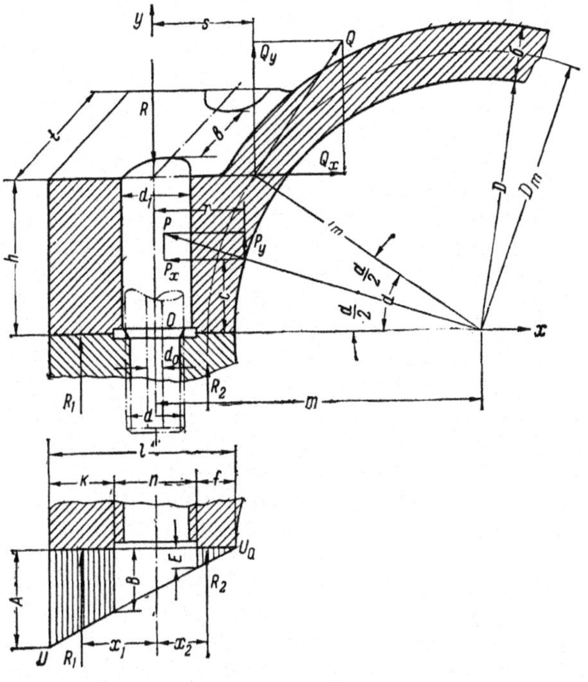

Fig. 12-1. Basic dimensions of a turbine flange

the cylinder and p_0—pressure outside). For cylinders of double-wall construction p_0—outside pressure for the inner wall and inner pressure for the outer wall.

The main forces acting on the flanges due to the pressure difference are obtained from:

$$\left.\begin{aligned}
&Q = \frac{Dt}{2}\Delta p;\\
&Q_x = Q\sin\alpha = \frac{Dt}{2}\Delta p\sin\alpha;\\
&Q_y = Q\cos\alpha = \frac{Dt}{2}\Delta p\cos\alpha;\\
&P = Dt\,\Delta p\sin\frac{\alpha}{2};\\
&P_x = P\cos\frac{\alpha}{2} = Dt\,\Delta p\sin\frac{\alpha}{2}\cos\frac{\alpha}{2} =\\
&\quad = \frac{Dt}{2}\Delta p\sin\alpha;\\
&P_y = P\sin\frac{\alpha}{2} = Dt\,\Delta p\sin^2\frac{\alpha}{2}.
\end{aligned}\right\} \qquad (12\text{-}2)$$

The moment acting on the flange due to the pressure difference is given by

$$M_p = Q_y s - Q_x h + P_y r + P_x c, \qquad (12\text{-}3)$$

where

$$s = m - \frac{D_{av}}{2}\cos\alpha;$$

$$r = m - \frac{D}{2}\cos\frac{\alpha}{2};$$

$$c = \frac{D}{2}\sin\frac{\alpha}{2}.$$

The value of α is obtained from the constructional details and dimensions of the turbine stator and flanges.

To ensure a tight joint between the flanges we shall assume that for some tensile force R exerted on the bolts, and at the pressure under consideration (reaction force) the pressure exerted on the frange face along its length l varies in a linear fashion and is given by the line u_0u. This condition shows that the flanges are closely pressed together at the inner edge (point u_0 in Fig. 12-1). The moment of the two opposing reaction forces R_1 and R_2 acting on the flange face is given by the equation

$$M_r = R_2x_2 - R_1x_1. \qquad (12\text{-}4)$$

From elementary mechanics we know that for any system in equilibrium the sum of the moments of all the forces acting in it is equal to zero. Since the flange is to be designed for sufficient strength at its minimum section bh we shall take the sum of moments of all the forces acting with respect to this section

$$M_p + M_r = Q_y s - Q_x h + P_y r + P_x c + R_2x_2 - R_1x_1 = 0, \qquad (12\text{-}5)$$

(here the plus sign indicates the moment for an anticlockwise direction and the minus sign for a clockwise direction). The forces acting on the section bh of the flange can be obtained from the equation

$$R_1x_1 = M_p - R_2x_2. \qquad (12\text{-}6)$$

In equation (12-6) the unknowns are x_1, x_2, R_1 and R_2 which may be determined from the following relations.

From Fig. 12-1 we find that,

$$\left.\begin{aligned} x_2 &= \frac{n}{2} + \frac{f}{3}; \\ x_1 &= \frac{n}{2} + k_1 = \frac{n}{2} + \frac{k}{3} \times \frac{2A+B}{A+B}, \end{aligned}\right\} \qquad (12\text{-}7)$$

where k_1—distance from the side B of the trapezoid to its centre of gravity.

In the above equation the values of A and B are not known. To determine the value of x we shall write the relation

$$B = \frac{n+f}{l} \times A. \qquad (12\text{-}7a)$$

Substituting the value of B obtained from equation (12-7a) in equation (12-7) we have

$$x_1 = \frac{n}{2} + \frac{k}{3} \times \frac{2l+n+f}{l+n+f}. \qquad (12\text{-}7b)$$

The values of R_1 and R_2 can be determined from the following equations

$$\left.\begin{aligned} R_1 &= \frac{M_p}{x_1 + x_2N}; \\ R_2 &= NR_1. \end{aligned}\right\} \qquad (12\text{-}8)$$

We find from Fig. 12-1 that the value of $N = R_2/R_1$ is given by

$$N = \frac{fE}{k(A+B)}; \qquad (12\text{-}9)$$

and again

$$E = \frac{f}{l} \times A. \qquad (12\text{-}10)$$

Substituting the value of B obtained from equation (12-7a) and the value of E from equation (12-10) in equation (12-9) we have

$$N = \frac{f^2}{k(l+n+f)}. \qquad (12\text{-}11)$$

Substituting the value of N from (12-11) in equation (12-8) the values of R_1 and R_2 can be determined.

The bending force on section bh of the flange is determined from the relation.

$$\sigma_f = \frac{R_1x_1}{W_f} = \frac{6R_1x_1}{bh^2}. \qquad (12\text{-}12)$$

Forces of 300 kg/cm² may be allowed for the higher grades of cast iron. For steels in the temperature range of up to 350°C, the value of σ_f may be taken up to 50% of the flow limit. In case of flanges operating under high-temperature conditions the allowable stresses are chosen taking into account the creep factors for the material under consideration.

The force exerted on the bolts will be

$$R = R_1 + R_2 + P_y + Q_y. \qquad (12\text{-}13)$$

The tensile stress at the minimum bolt section, i. e., at the root section of the threads will be

$$\sigma_b = \frac{R}{F_{min}} = \frac{4R}{\pi d_{min}^2}, \qquad (12\text{-}14)$$

where d_{min}— bolt diameter at the thread root.

Assuming the stress exerted the diameter is obtained as

$$d_{min} = \sqrt{\frac{4R}{\pi\sigma_b}}. \qquad (12\text{-}14a)$$

Taking into consideration the shortcomings of the design method, the creep effects on bolts at high temperatures, the unknown initial bolt tension at assembly, it is usually recommended by the various turbine builders, on the basis of previous experience, to take an initial tension equal to about 20 to 40% of R so that the initial tension experienced by the bolts when tightened will be

$$R_0 = (1.2 \text{ to } 1.4)R. \qquad (12\text{-}15)$$

The tightness of the flange joints for the period of operation between general overhauls (8,000 to 20,000 hrs) depends largely on the initial bolt tension as well as the method of tightening.

The flange bolts of medium- and low-pressure cylinders are usually tightened cold with the help of a long-handed spanner. Bolts with diameters of 70 to 80 mm are tightened after heating them red either by a gas torch or with high-frequency electric current. The flange bolts and nuts are provided with through holes to facilitate their heating as shown in Fig. 12-2. After the initial tightening the bolts are then heated in pairs one on each side of the cylinder to a predetermined temperature. The bolts expand permitting their further tightening through the required angle. The above method of tightening the bolts by preheating eliminates addition-

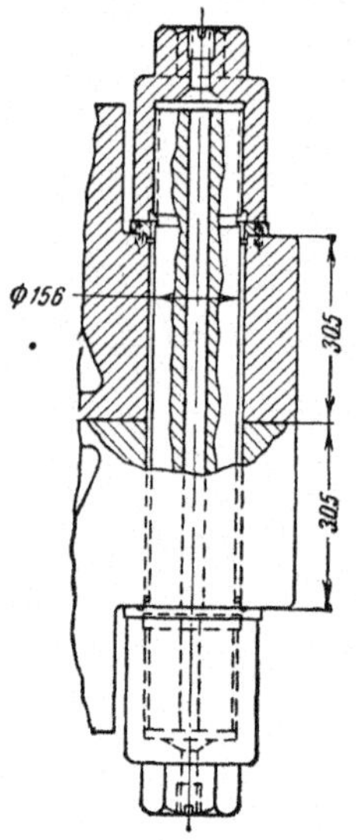

Fig. 12-2. Flange bolt for H.P. cylinder

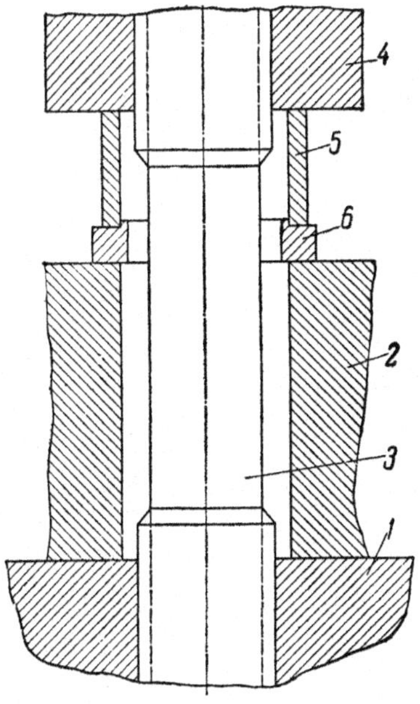

Fig. 12-3. Flange bolts for B.B.C. high-pressure turbine with back pressure

1 – lower flange; *2* – upper flange; *3* – bolt; *4* – nut; *5* – cylindrical ring; *6* – special jointing ring

al twisting stresses that inevitably occur in cold-tightening of bolts.

Some of the leading turbine builders use special rings between the joints of massive flanges in order to obtain the necessary elasticity of jointing (Fig. 12-3).

The material used for the manufacture of bolts depends on the stress requirement and the temperature to which they are heated under normal conditions of operation. Flange bolts for turbines operating in the range of temperatures not higher than 300 to 320°C may be made from carbon steel No. 35 or 45. For higher temperatures use is made of chrome-molybdenum steels 30XM (C—0.25 to 0.35%; Si—0.17 to 0.37%; Mn—0.4 to 0.7%; Cr—0.8 to 1.1%; Ni—0.4%; Mo—0.15 to 0.25%; S⩽0.04% and P⩽0.04%) and 35XM (C—0.3 to 0.4%; Si—0.17 to 0.37%; Mn—0.4 to 0.7%;

Table 12-1

Permissible Stresses for Flange Bolts (kg/cm²).

Type of steel \ Temperature, °C	315	315-400	400-455	455-510
Carbon steel (0.35%)	1,270	—	—	—
Chrome-molybdenum steel. .	1,750	1,550	—	—
Tungsten-molybdenum steel (1% W, 0.6% Cr, 0.5% Mo)	1,750	1,550	1,270	900

Cr—0.8 to 1.1%; Ni⩽0.4%; Mo—0.15 to 0.25%; S⩽0.04% and P⩽0.04%) as well as of tungsten-molybdenum steels (1% W, 0.6% Cr, 0.5% Mo).

The stresses permissible in bolts are indicated in Table 12-1.

12-2. CONSTRUCTION OF NOZZLES AND GUIDE BLADES

a) N o z z l e s. The nozzles of the first stage are made in various shapes. In the past nozzles were made in cast segments; for high- and medium-pressure turbines from steel and for low-pressure turbines from cast iron (Fig. 12-4 *a*). The surfaces of the nozzles were hand finished which involved great labour. Inspite of the careful hand finish the surface finish was far from satisfactory which led to large losses in nozzle friction.

A more satisfactory method of making nozzles is of the built-up type with milled blades as shown in Fig. 12-4,*b*. Another method of making milled nozzles is shown in Fig. 12-4,*c*.

This method has also found a very wide field of usage. Nozzles cast in one piece with the first diaphragm and machined all over are now being very widely used. This type of nozzle is

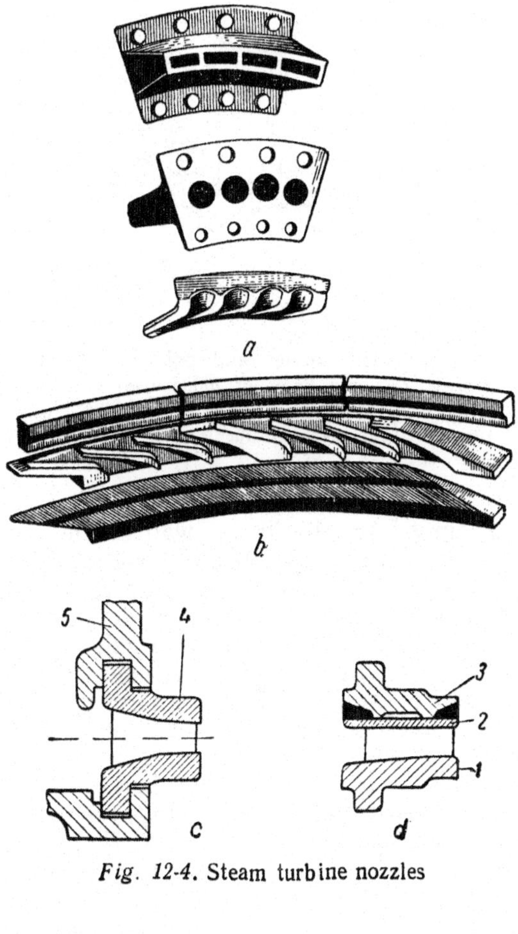

Fig. 12-4. Steam turbine nozzles

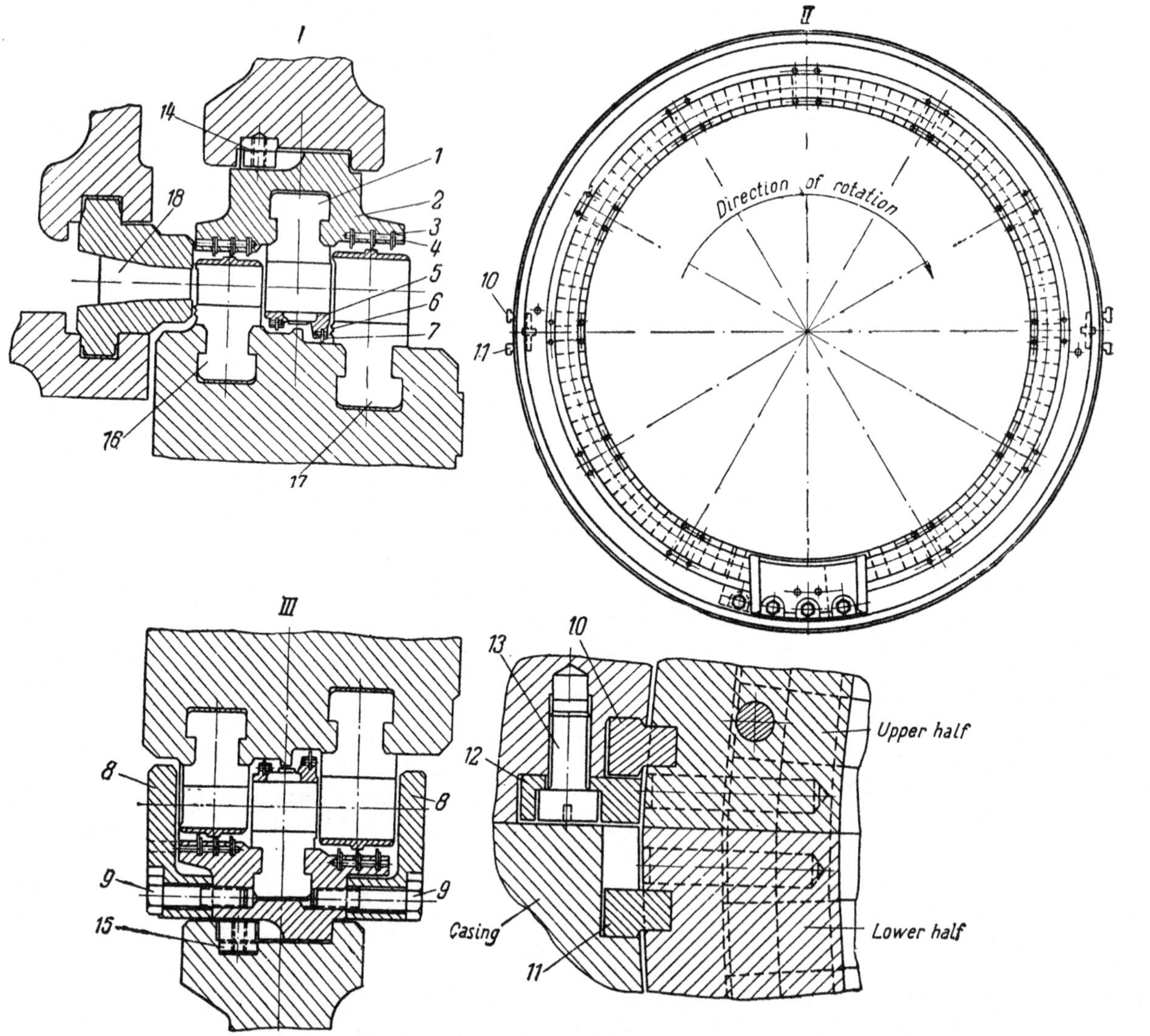

Fig. 12-5. Nozzles and guide blades for the regulating stage of L.M.W. high-pressure turbine

1 – guide blade; *2* – upper linear for quide blades; *3* – labyrinth seal strips (for moving blades); *4* – labyrinth strip holder; *5* – shroud ring for guide blades; *6* – labyrinth strips for guide blades; *7* – strip holder; *8* – side shields; *9* – bolts for fixing the shields; *10* – upper liner lugs; *11* – lower liner lugs; *12* – upper liner lug key; *13* – screws for fixing the keys to the liners; *14* – upper liner alignment key; *15* – lower liner alignment key; *16* and *17* – first-and second-row moving blades of the regulating stage

shown in Fig. 12-4,*d*. The inner annular ring of diaphragm *1* forms the inside wall of the nozzle and the outer wall is obtained by mounting strips *2* at the crown of the nozzle orifices. These strips are welded to the crown of the blade profiles to give a smooth surface for the nozzle passages. A special strengthening ring *3* is now welded to the outside of these strips as shown in Fig. 12-4,*d*. Built-up nozzle segments of the type shown in Figs 12-4,*c* and 12-4,*d* are being used by the L.M.W. for the high-pressure turbines built there. The L.M.W. makes the nozzle rings for its H.P. turbines from stainless steel. The nozzle segments of welded or built-up blade sections are fixed to the front end of the turbine cylinder or directly to the nozzle box as it is done for the high-pressure turbines of the L.M.W.

Depending upon the theoretical heat drop occurring in the regulating nozzles their passages are made either convergent or convergent-divergent. The heat drops occurring in the nozzles of high-pressure turbines are relatively small and since convergent nozzles are known to be more stable in their efficiencies at loads other than design ones they are preferred to convergent-divergent nozzles for these turbines.

b) Regulating Stage Guide Blades. Two-row regulating stages are widely

Fig. 12-6. Nozzles built up on diaphragm

used in the first stages of large high- and medium-capacity turbines. Steam flowing out of the first row of moving blades is directed into a set of guide blades which are meant to change the direction of steam flow to conform to the entry angle of the second row of moving blades. If the turbine is provided with partial admission the regulating stage guide blades also conform to this partial admission. However, in case of turbines with a large degree of partial admission the guide blades after the first stage are provided all along the circumference of the disc. The guide blades are fixed in special liners which are then attached to the turbine casing. A typical construction of nozzles and guide blades for a two-row regulating stage of a high-pressure turbine is that of the L.M.W. turbine shown in Fig. 12-5. The guide blade system of turbine VK-100-2 and other high-pressure turbines built by the L.M.W. consists of two liners: one upper and one lower. These liners are placed in slots cut in the upper and lower cylinder halves. The liners are properly aligned with the help of keys *14* and *15* and lugs *10* and *11*. These lugs are inserted into slots milled in the casing. The lugs in the upper half of the casing are held by keys *12* fixed to the casing by screws *13*. Thus when the upper casing is unbolted and removed the guide liners of the upper half also go up with it. Labyrinth seals *3* and *6* are provided for the moving and the guide blades to reduce steam leakage. These labyrinth seals are especially of great importance for high-pressure turbines, having some dergee of reaction in the guide blades of the regulating stage. The labyrinth strips are made sharp at their ends, about 0.5 mm thick, and have a radial clearance of about 1.5 to 2.5 mm.

c) B u i l t - U p N o z z l e s. The nozzle passages and diaphragms of the pressure stages are usually of built-up construction (for the first few stages) as shown in Fig. 12-6 (outdated con-

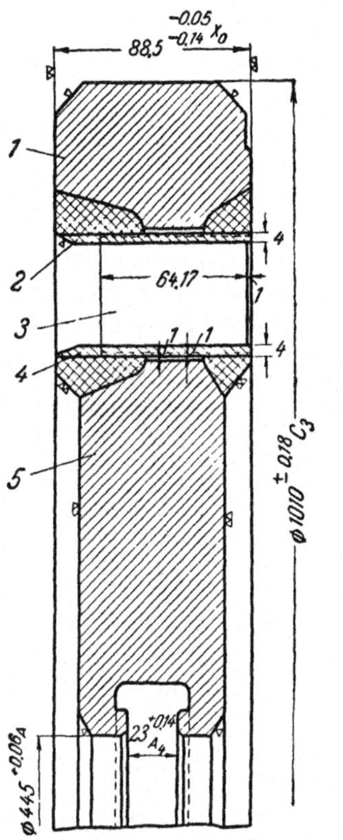

Fig. 12-7. Welded nozzles of Kharkov Turbine Works
1 – liner; *2* – upper strip; *3* – guide blade; *4* – inner strip; *5* – body of the diaphragm

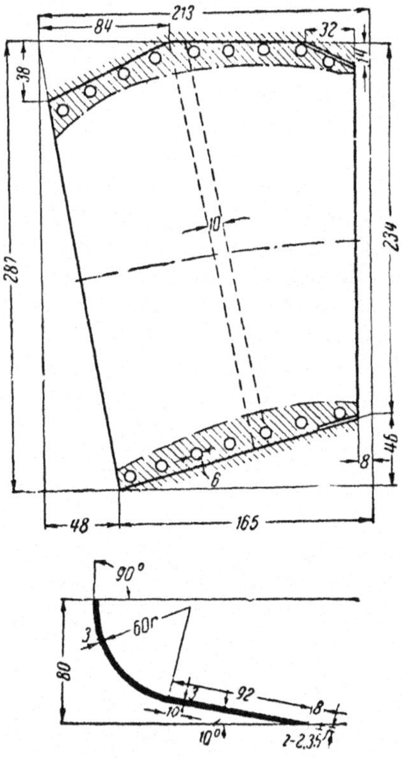

Fig. 12-8 Guide blades from steel stampings imbedded in cast iron diaphragm

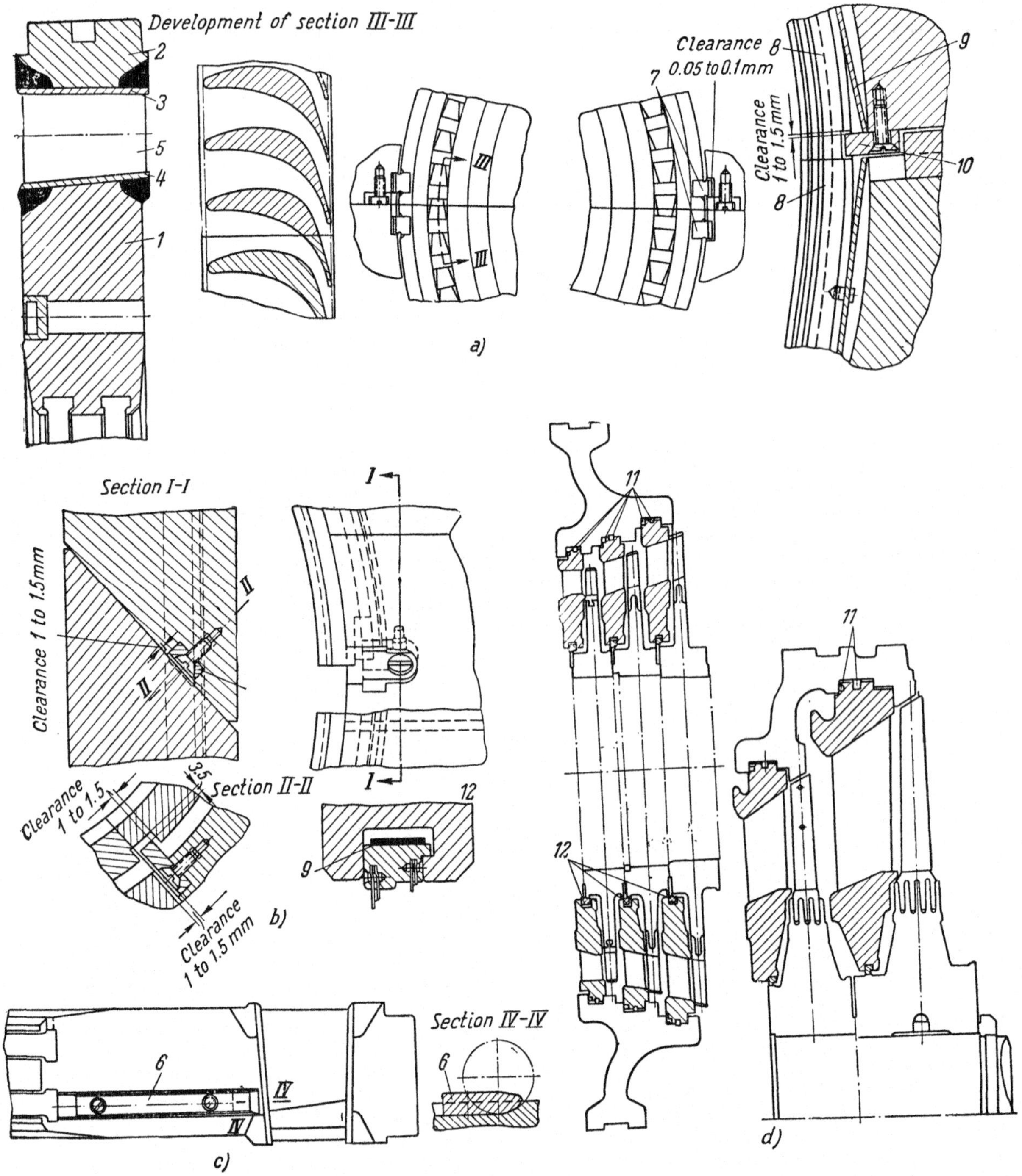

Fig. 12-9. Diaphragms of Leningrad Metal Works

a — welded diaphragm; *b* — L.P. cast iron diaphragm jointing; *c* — method of steam seal fixing to the diaphragm; *d* — cast iron diaphragms;

1 — body of the diaphragm; *2* — rim; *3* — outer shrouding strip; *4* — inner shrouding strip; *5* — guide blade profiles; *6* — locking key; *7* — lugs for fixing diaphragms in H.P. stages; *8* — diaphragm labyrinth seals; *9* — flat spring; *10* — labyrinth seal fixing lug; *11* — radial and axial lugs for centring cast iron diaphragms; *12* — cast iron diaphragm labyrinth seals

struction) or welded construction as shown in Fig. 12-7 (extremely cheap as well as strong construction) or made by casting the diaphragm with the steel blades imbedded in it as shown in Fig. 12-8 (usually for the low-pressure stages of a condensing turbine). For obtaining a better and stronger jointing between the steel blades and the cast iron diaphragm it is usual to provide either holes or slots in the blade profiles at their sides. The imbedded portion of the blades is coated with tin to safeguard the blades from the acidic corrosive action of the molten cast iron.

12-3. CONSTRUCTION OF DIAPHRAGMS

The diaphragms are made up of two parts, one half of which is fixed in the lower half of the cylinder and the other in the upper cylinder half. The diaphragms are either placed directly in slots milled in the cylinder or fixed to special liners provided in the turbine casing. A certain amount of gap is provided between the diaphragms and the cylinder slot or the liners to allow for an unhindered expansion of the diaphragms. The radial clearance usually is of the order of 0.003 to 0.004 times the diaphragm diameter and the axial clearance from 0.1 to 0.3 mm.

The diaphragms are coated with a thin layer of graphite before inserting them into the slot or before fixing them in the liners to prevent the diaphragm from sticking to either the liner or the turbine casing.

The first few stages of modern high-pressure turbines are of forged construction with the blade profiles welded to them; the L. P. diaphragms are usually of cast iron with the blades imbedded

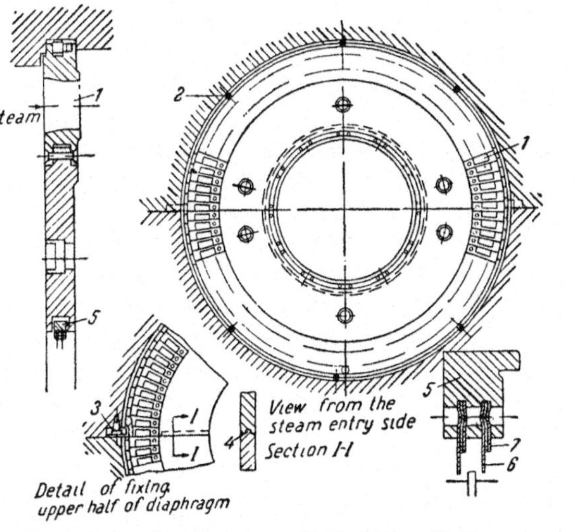

Fig. 12-10. Diaphragm with built-up blades

1 – nozzle profile guide blades, *2* – centring lugs; *3* – lug for fixing diaphragm; *4* – locking key; *5* – labyrinth seal ring; *6* – seal strips; *7* – stiffeners

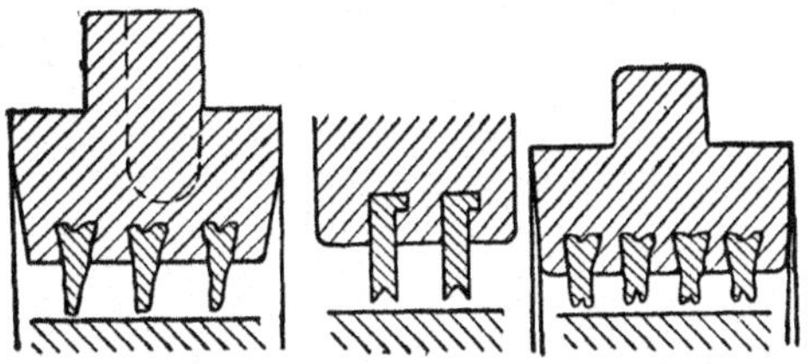

Fig. 12-11. Labyrinth packing for diaphragms

in them (Fig. 12-9). Forged diaphragms with milled blades welded to them were used widely in the older constructions for steam temperatures of 200-250 to 400° C. The upper half of the diaphragms are fixed to the cylinder casing in such a way as to form an integral part of the upper casing. All the diaphragms are invariably provided with steam seals of the labyrinth type at their inner circumferences, i. e., at their hub seatings. There are various methods in practice for the construction and fixing of the sealings to the diaphragms; these may be rigid or flexible. The flexible segment type of construction is shown in Fig. 12-9 where the flat spring *9* permits radial movement of the sealing elements. Fig. 12-10 shows the method adopted for fixing a steel diaphragm with built-up milled blade sections *1* to the turbine casing. The correct radial alignment of these blade segments is ensured by six lugs *2*. These lugs are made of steel in the high-pressure section and of copper in the L. P. stages.

The upper and the lower halves of the diaphragm are fixed tightly with the help of locking key *4*. The upper half of the diaphragm is fixed to the cylinder or the liner by means of lug *3*. The diaphragms are provided with labyrinth seals *5* at their inner diameters in order to reduce steam leakage through these radial clearances. These seals usually consist of six segments. These segments are kept pressed onto the turbine hub by a flat spring so that the radial clearance is always a minimum.

The radial clearance δ between the labyrinth seals and the disc hub is, for high-pressure stages, from 0.25 to 0.4 mm and for low-pressure stages from 0.3 to 0.5 mm. If the strips are fixed rig-

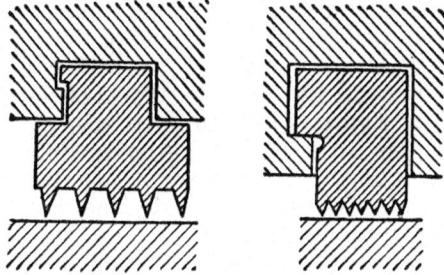

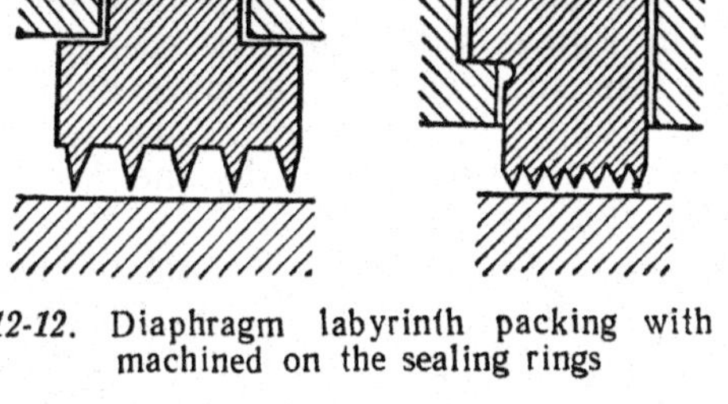

Fig. 12-12. Diaphragm labyrinth packing with strips machined on the sealing rings

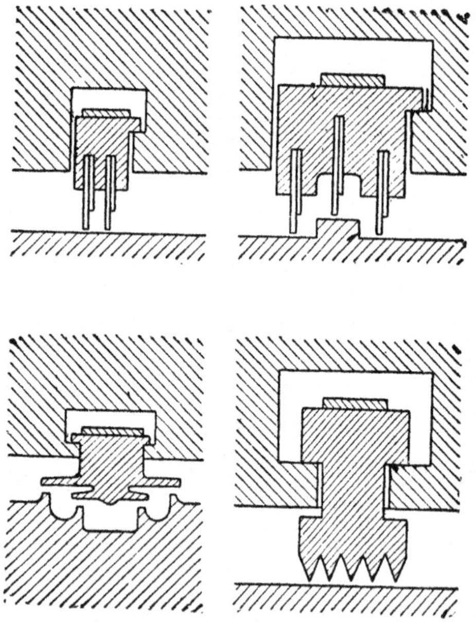

Fig. 12-13. Types of labyrinth seals for diaphragms

idly to the diaphragm these clearances are of the order of 0.4 to 0.6 mm.

Fig. 12-7 shows a diaphragm made by the Kharkov Turbine Works. In this diaphragm the blades are welded directly to its rim. The guide blades *3* are placed in the slots provided in the upper and the lower shrouding strips *2* and *4* and are welded to them. This ring of blades and upper and lower shrouds is then placed between the body of the diaphragm *5* and its outer rim *1* and finally welded to give the necessary diaphragm construction. Since diaphragms experience large pressure loads, especially in the high-pressure regions, great care has to be taken while choosing the material for their construction as well as the method of fastening them to the cylinders. The diaphragms are made of special chrome, molybdenum and nickel steels for the high-temperature regions of high-pressure turbines. The first few stages of medium-pressure turbines are also sometimes made of 0.5% molybdenum steels. The labyrinth seal strips are fixed in either a carrier ring or are directly caulked into grooves in the diaphragm body (Fig. 12-11). Labyrinth seals of the machined segment type are shown in Fig. 12-12. Fig. 12-13 shows some of the flexible types of labyrinth seals. The segments in which the seal strips are caulked are inserted into grooves cut in the diaphragm body. These carrier segments are kept pressed onto the turbine hub by the flat springs provided at their back.

Fig. 13-14 shows the fastening of liners to the cylinder casing for the L.M.W. turbine type VK-100-2. A clearance of 1.5 mm is given between the liner rim and the groove in the casing to allow for radial expansion. The upper and the lower halves of the liners are fastened together at their horizontal flanges by special pins after inserting the diaphragms into their grooves. Two bolts *7* are provided for correct centring of the two liners in position. The diaphragms must be placed in the liners in such a way as to maintain a uniform radial clearance between the diaphragm and the liner for unhindered expansion.

12-4. DIAPHRAGM CALCULATIONS

The diaphragms and the blades mounted on them are subjected to bending loads because of the pressure difference existing between their two sides. If the pressures acting on the two sides of the diaphragm are p_1 and p_1' (p_1—steam pressure before the diaphragm and p_1' after it) the load on the diaphragm is $\Delta p = p_1 - p_1'$. There are various practical methods of calculating the diaphragm dimensions for strength and stiffness based mostly on the experience of the various builders as also experimental data available. Theoretical methods of calculations are found to be very cumbersome. The simplified methods of calculations assuming the diaphragm to be a circular disc do not give satisfactory results.

One method of diaphragm calculation recommended by A. A. Moiseyev [Ref. 6] and systematised by G. S. Zhiritsky [Ref. 1] is given below.

1. W e l d e d d i a p h r a g m s a n d c a s t i r o n d i a p h r a g m s w i t h c a s t-i n b l a d e s.

The maximum stress for a diaphragm of variable thickness (Fig. 12-15) is given by the equation

$$\sigma = k_\sigma \cdot \frac{\Delta p \left(\frac{D}{10}\right)^3 t_{max}}{I} \ [\text{kg/cm}^2], \qquad (12\text{-}16)$$

where k_σ— a coefficient depending on the ratio d/D (the values are taken from graph given in Fig. 12-16);

D — the internal diameter of the turbine casing, cm;

t_{max}— the maximum thickness of the diaphragm, cm;

I — moment of inertia which is assumed to be

$$I = I_t + I_{rim}, \quad \text{if} \quad I_t > I_{rim};$$
$$I = 2I_t, \quad \text{if} \quad I_t < I_{rim};$$

I_t— moment of inertia of the diaphragm relative to the x-x axis, cm^4;

I_{rim}— moment of inertia of the diaphragm rim relative to the x-x axis, cm^4.

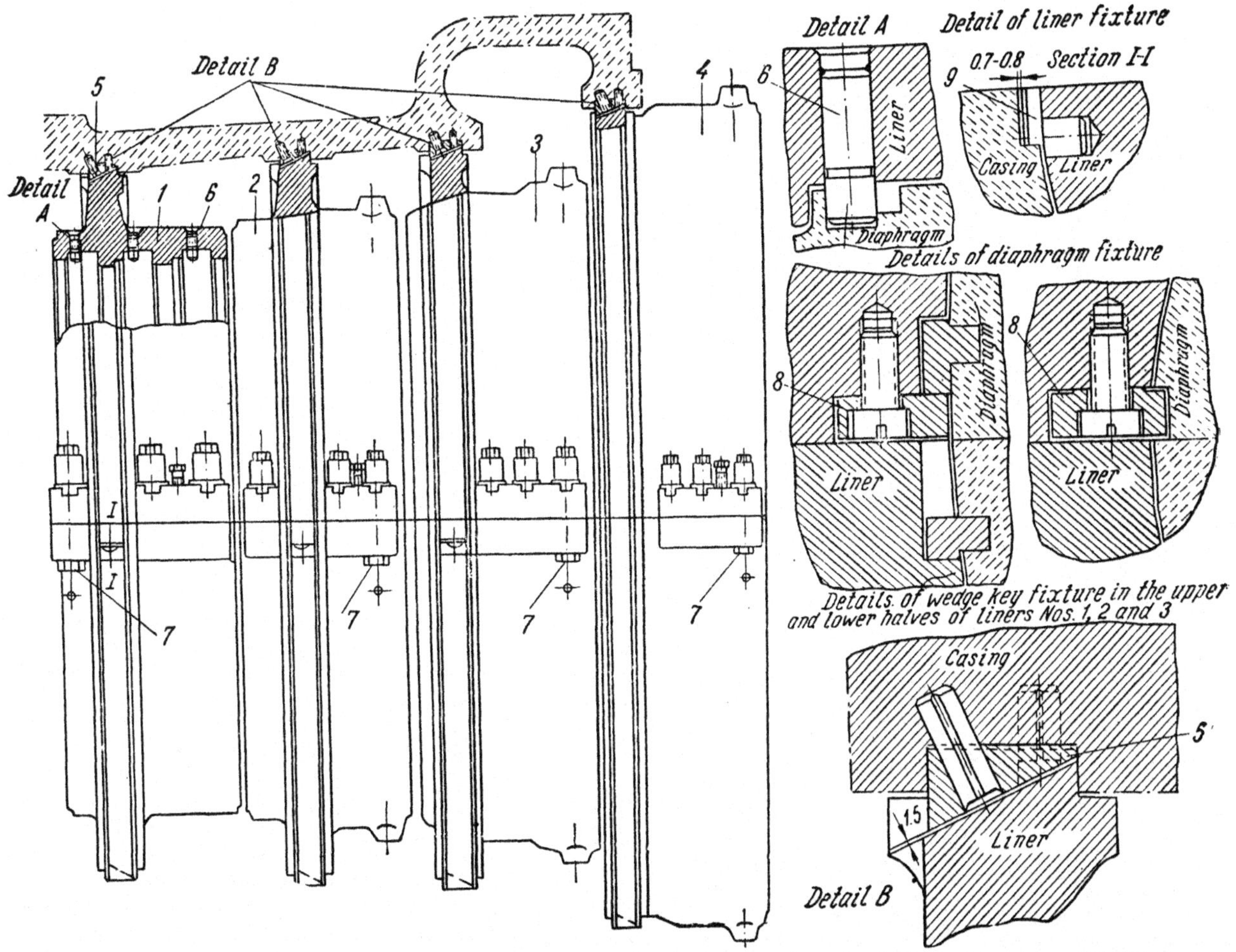

Fig. 12-14. Method of fixing diaphragm liners to the turbine casing (L.M.W. turbine type VK-100-2)
1, 2, 3 and *4* – liners; *5* – wedge keys; *6* – lugs for centring diaphragms; *7* – centring bolts for liners; *8* – keys and lugs holding the upper half of the diaphragms in position when the upper casing of the turbine is removed; *9* – liner lugs

The maximum bending force on blades welded to, or imbedded in, diaphragms due to the pressure difference is (stress due to gas-dynamic forces is neglected):

$$\sigma_u = \frac{1.2\Delta p D_m (D_m - d) h}{z_1 W_x}, \qquad (12\text{-}17)$$

where z_1— number of blades on a diaphragm;
W_x— moment of resistance of the blade section about the x-x axis (Fig. 12-17), cm^3;

$$W_x = \frac{I_x}{y},$$

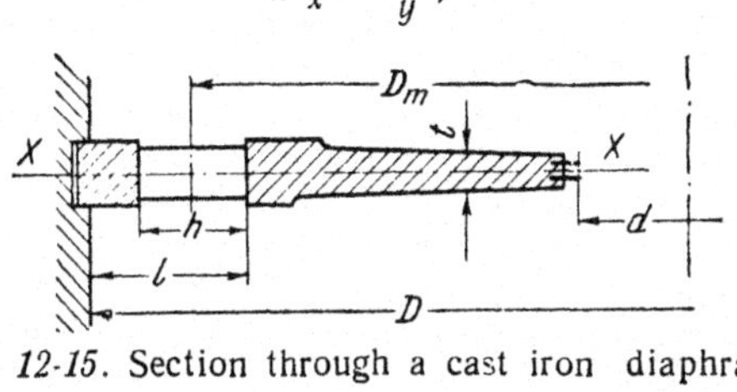

Fig. 12-15. Section through a cast iron diaphragm

I_x —moment of inertia of the blade section about the x-x axis, cm^4;
y—the distance between the x-x axis and the outer edge of the blade section, cm.

The maximum deflection of the diaphragm on its diameter d is obtained by adding up the deflection of the blades and the diaphragm body

$$\Delta = \Delta_b + \Delta_d,$$

where Δ_b—deflection of blades and Δ_d—deflection of the diaphragm body.

The deflection of blades is obtained from the equation

$$\Delta_b = \frac{0.2\Delta p D_m (D_m - d) h^3}{E I_x z_1} \text{ [cm]}, \qquad (12\text{-}18)$$

where E—modulus of elasticity of the material used for the blades.

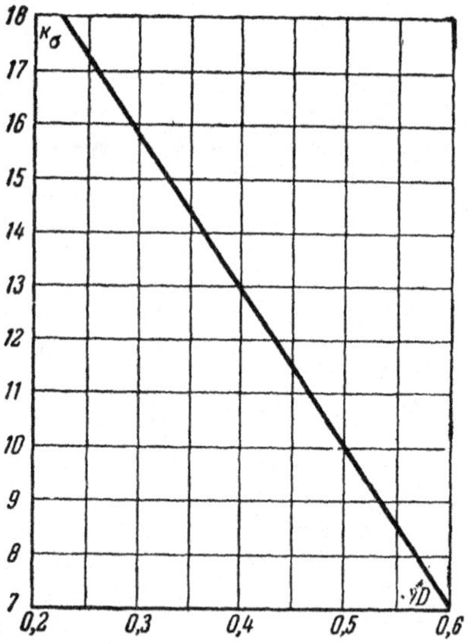

Fig. 12-16. Coefficient k_σ for equation (12-16)

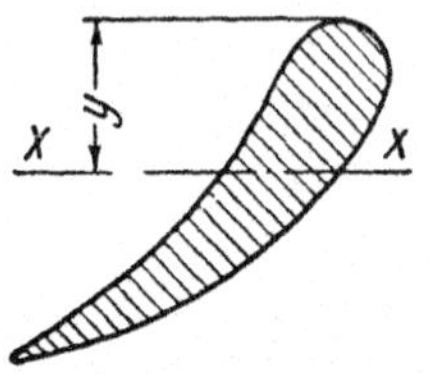

Fig. 12-17. Cross-section of a guide blade

The deflection of the diaphragm body is

$$\Delta_b = k_\Delta \frac{\Delta p \left(\frac{D}{10}\right)^6}{EI} \text{[cm]}, \qquad (12\text{-}19)$$

where k_Δ—coefficient chosen from Fig. 12-18. This coefficient is a function of the ratio d/D for determining the deflection on the disc diameter (at the labyrinth seals at the inner diameter) and a function of d/D and l/D for the determination of deflection at the root of the blades;

E—modulus of elasticity of the material used for the diaphragm body (usually for cast iron, $E = 0.985 \times 10^6$ kg/cm²);

I—moment of inertia as in equation (12-16).

The deflection on diameters d and $D-2l$ is equated with the axial clearance between the diaphragm and the blade disc of the stage under consideration.

2. Built-up diaphragms.

The maximum stress on the diaphragm body (Fig. 12-19) is obtained from the equation

$$\sigma = k'_\sigma \frac{\Delta p D^3}{l_1 t^2} \text{[kg/cm}^2\text{]}, \qquad (12\text{-}20)$$

where the coefficient k'_σ is obtained from Fig.

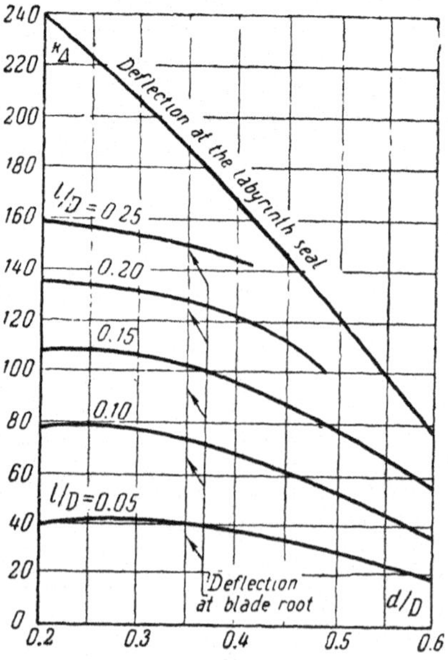

Fig 12-18. Coefficient k_Δ for equation (12-19)

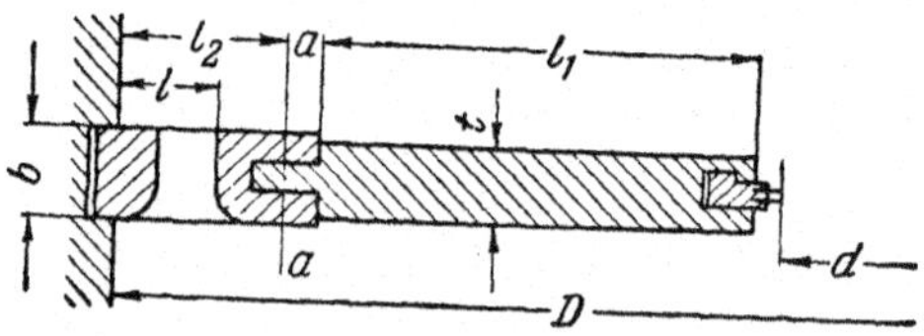

Fig. 12-19. Section through a built-up diaphragm

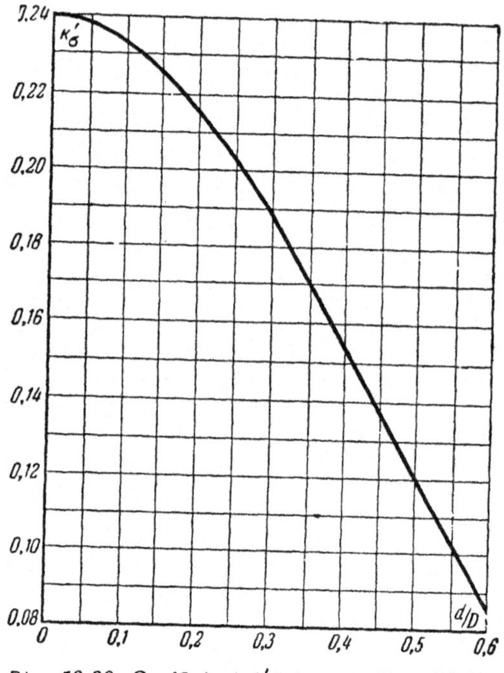

Fig. 12-20. Coefficient k'_σ for equation (12-20)

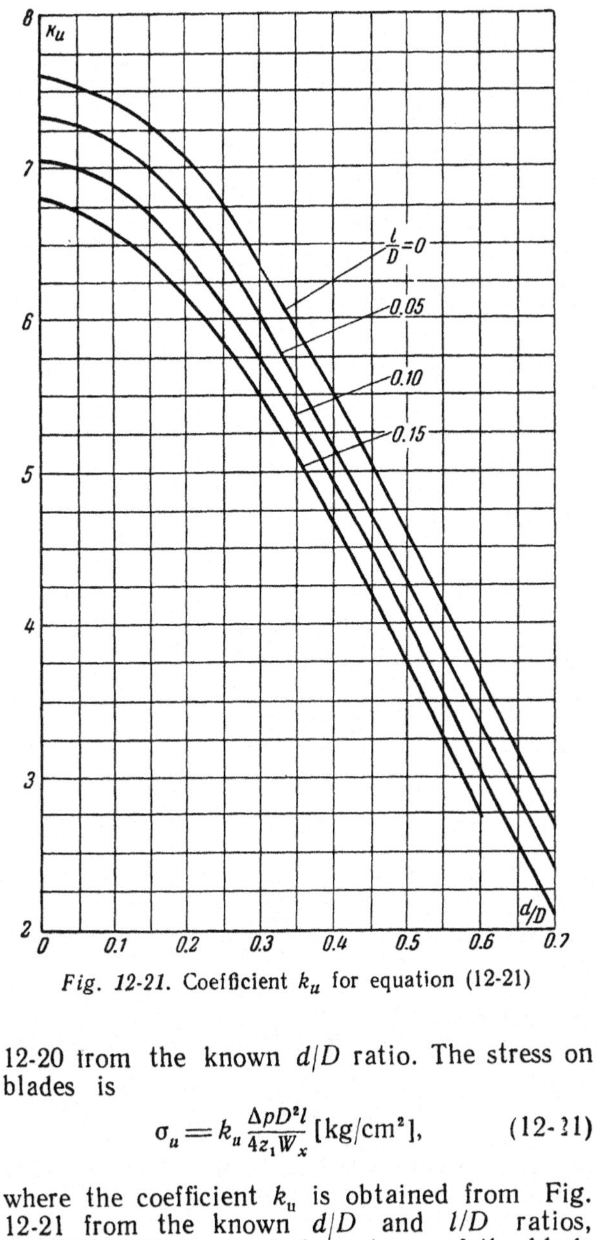

Fig. 12-21. Coefficient k_u for equation (12-21)

12-20 from the known d/D ratio. The stress on blades is

$$\sigma_u = k_u \frac{\Delta p D^2 l}{4 z_1 W_x} \text{ [kg/cm}^2\text{]}, \qquad (12\text{-}21)$$

where the coefficient k_u is obtained from Fig. 12-21 from the known d/D and l/D ratios, and W_x is the moment of resistance of the blade section about the x-x axis (Fig. 12-17). The deflection of a built-up diaphragm is obtained from the equation

$$\Delta = k'_\Delta \frac{D^4 \Delta p}{E t^3} \text{ [cm]}, \qquad (12\text{-}22)$$

where k_Δ—a coefficient obtained from Fig. 12-22 from the d/D and l_2/D ratios.

For built-up diaphragms the rigidity of blade fixture on the diaphragm is also checked. The bending moment at section a-a along the axis of blade jointing (Fig. 12-19) is determined from,

$$M_u = k_m D l_2 \Delta p \text{ [kg cm/cm]} \qquad (12\text{-}23)$$

where k_m is taken from Fig. 12-23 knowing the ratios d/D and l_2/D. Table 12-2 gives the permissible bending moments at the blade joint according to the thickness of the blade ring as well as the steam temperatures. The moment M_u obtained from equation (12-23) should in no case exceed $M_{per.}$ given in Table 12-2.

Table 12-2

The Maximum Allowable Bending Moments at the Blade Joint in kg cm/cm

b	Steam temperature, °C			
	370	400	425	450
23	270	225	180	135
30	500	410	320	250
38	770	635	500	385
46	1,090	905	680	545
53	1,450	1,180	905	725
63	2,070	1,685	1,290	1,035
75	2,880	2,340	1,790	1,440

The maximum permissible stress in the diaphragms is chosen in conformity with the material used and the prevailing temperatures.

For diaphragms made out of pearlite cast iron the maximum stress may be allowed up to 600 kg/cm². If the diaphragms are made of carbon steel rolled sheets stresses of 700 to 800 kg/cm² may be allowed for a temperature of 350° C. At 400° C the maximum stress allowable for the same material is 500 to 600 kg/cm². For diaphragms made of alloy steels with 0.5% of Mo the maximum stress at a temperature of 400° C can be allowed up to 600-800 kg/cm². For very high-temperature regions diaphragms are made of chromemolybdenum steels 15XM, 20XM and 35XM. For these steels the allowable stress at a temperature of 350° C is about 1,200 kg per cm², and at a temperature of 400° C up to 800 kg/cm². In order to avoid fouling of diaphragms by the rotating discs the axial clearance between them is taken 3 times the value obtained from design calculations.

The moment of resistance of the blades can be fairly accurately determined by graphic methods. The blade section is generally drawn to about 5 to 10 times its original size. The blade section area is divided into 12 to 30 parts, depending upon its size, by lines parallel to the axis about which the moment is to be determined. For convenience of calculations it is recommended that these lines be drawn at equal intervals. For bending moment calculations the reference axis used is the $x_0 x_0$ axis shown in Fig. 12-24.

Each elementary area between two parallel lines (ΔF_i) may be obtained by treating these

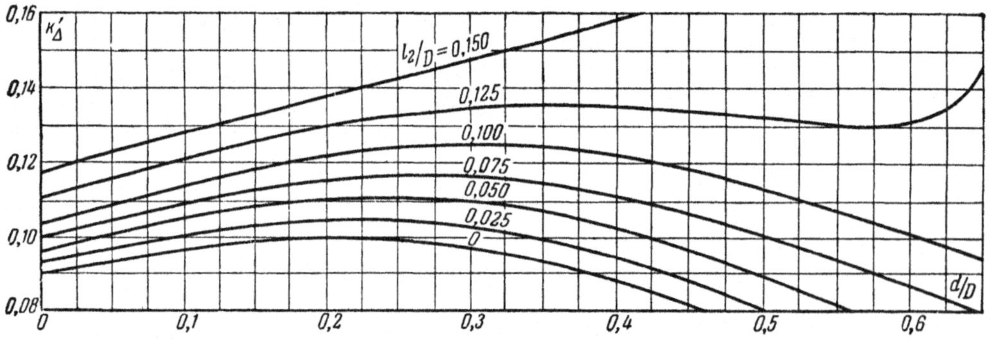

Fig. 12-22. Coefficient k'_Δ for equation (12-22)

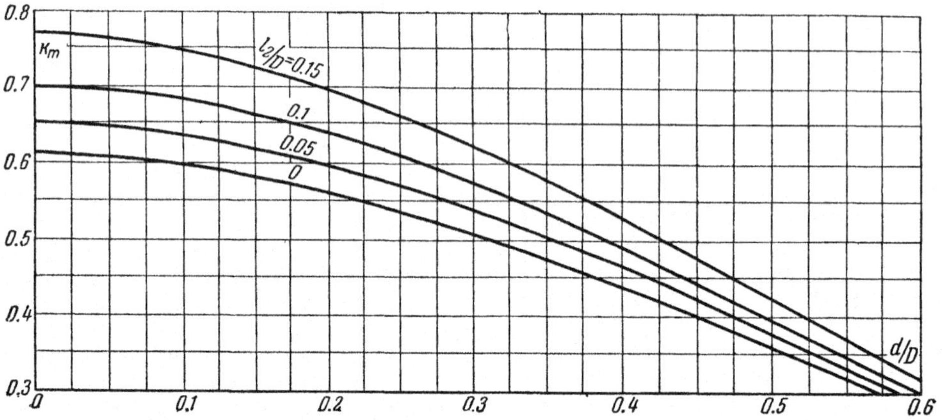

Fig. 12-23. Coefficient k_m for equation (12-23)

areas as triangles, trapezium or segments, or by planimetering.

From Fig. 12-24 we can obtain the distance of centre of gravity from the x_0x_0 axis as

$$l_0 = \frac{\Delta F_1 l_1 + \Delta F_2 l_2 + \Delta F_3 l_3 + \ldots + \Delta F_n l_n}{\Delta F_1 + \Delta F_2 + \Delta F_3 + \ldots + \Delta F_n}, \quad (12\text{-}24)$$

where n—the number of elementary sections into which the blade area is divided.

The value l_0 is the distance of the centre of gravity of the area under consideration from the x_0x_0 axis and at the same time expresses the

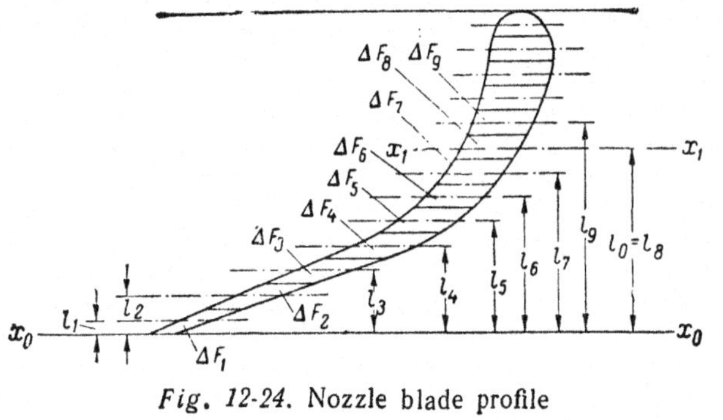

Fig. 12-24. Nozzle blade profile

distance from the x_1x_1 axis to the outermost edge of the section area.

The moment of inertia of the blade section may be determined approximately as the sum of the moments of inertia of the elementary sections,

$$I_x = \sum_1^n \Delta F_i (l_0 - l_i)^2. \quad (12\text{-}25)$$

Moment of resistance W_{1x} can now be determined as

$$W_{1x} = \frac{I_x}{l_0}. \quad (12\text{-}26)$$

The numerical values of moment of inertia and the moment of resistance are calculated for this profile in example in 13-2.

The stresses permissible for guide blades depend on the material used and the temperatures at which the stage operates. For cast-in blades the stresses usually assumed are about 400 to 500 kg/cm².

12-5. LABYRINTH PACKINGS FOR SHAFT ENDS

The labyrinth packings especially at the high-pressure end are of multifin type making it possible to reduce the amount of steam leakage and

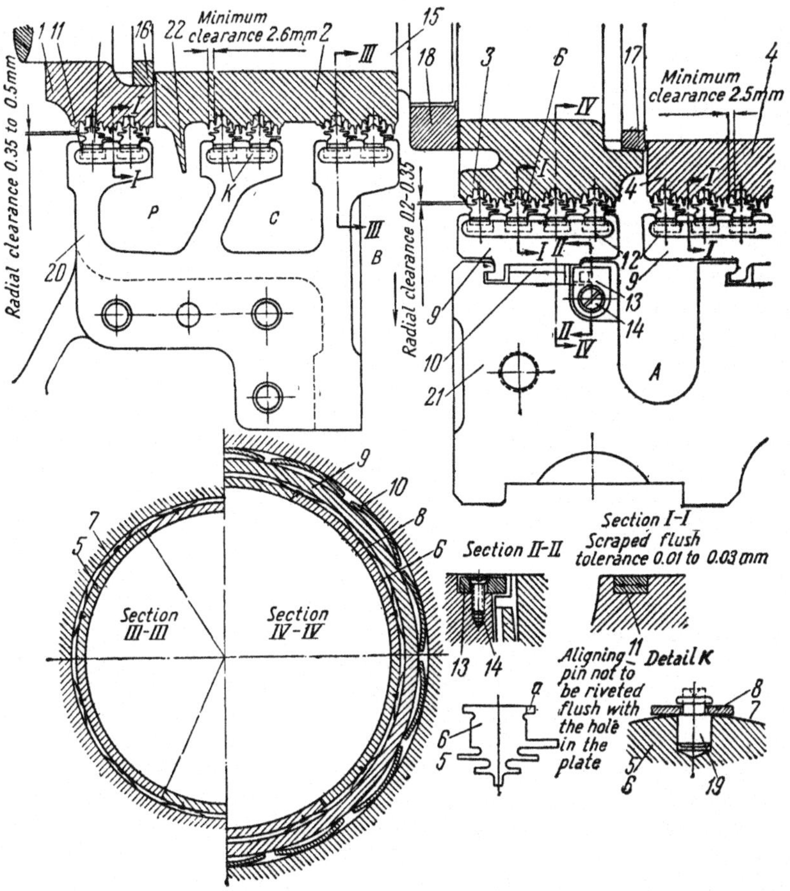

Fig. 12-25. Details of the H.P. labyrinth packing for rotor shaft of L.M.W. turbine

1, 2, 3 and *4* — labyrinth packing sleeves with sealing strips, *5* and *6* — sealing rings made up of six segments each; *7* and *8* — flat springs; *9* — liners in two halves; *10* — springs; *11* and *12* — strips for fixing labyrinth elements; *13* — stop lug; *14* — screw; *15* — turbine shaft; *16* and *17* — thrust rings in two halves; *18* — circular nut; *19* — aligning pin; *20* — exit-end liner; *21* — front-end liner; 22 — protective ring; *a* — two slots each 3 mm depth in the upper half of each ring; detail *k* — fixture of alignment pin with flat springs 7 (or *8*); *A* — bleed chamber connected to heater No. 3; *B* — bleed to heater No. 1; *C* — chamber for sealing with "cooled" steam; *D* — last chamber

its influence on the efficiency of the turbine (see Chapter Five, "Steam Leakage Through End Labyrinth Packings"). Different turbine builders use different methods of labyrinth sealings which may be in the form of metallic rings or water seals, or carbon rings.

Construction of Labyrinth Packings

Fig. 11-7 shows the details of the labyrinth packings made by the Kharkov Turbine Works. The sealing strips are caulked or rolled into grooves cut in the liners fixed to the turbine casing. The turbine shaft has grooves cut on its surface. The labyrinth strips protrude into these grooves while the stator and the rotor are in position. The resulting labyrinth changes the direction of steam flow and offers an effective resistance to it thus discouraging leakage from the shaft ends. Besides these, the turbine shaft has additional deeper grooves which are primarily meant for the compensation of local deformation due to thermal expansion caused by the rubbing of the turbine shaft against the labyrinth strips in the event of untoward axial displacement of the rotor.

Fig. 12-25 shows the details of a labyrinth seal used by the L. M. W. for high-pressure end packing. Four sleeves *1*, *2*, *3* and *4* with machined fins are shrunk on the turbine shaft. Fir-cone type of labyrinth sealing elements each made up of six segments are placed in such a way as to mate with the fins on the sleeves. The sealing elements are located in liners *9* and *20* and are pressed onto the shaft by springs *7*. Besides these, the labyrinth elements are pressed on the turbine shaft by steam pressure. The details of the shapes of sealing elements used in the turbines of the L. M. W. are shown in Fig. 12-26 where the upper two figures present sealing rings of the fir-cone type. Each of the sealing elements is made up of rings with fins on either both sides (left figure) or on one side only (right figure). The double-fin type of labyrinth seals is also used in the high-pressure diaphragms of L.M.W. turbines. For medium- and low-pressure diaphragms the type of seals used is shown in Fig. 12-26, *c*, *d* and *e*.

The high-pressure turbines of the L.M.W. have vacuum bleeds from the H. P. labyrinth packings. Steam seeping in from the regulating stage into chamber *A* is bled to H. P. heater No. 3. The remainder of the steam is sucked into vacuum bleed chamber *B* through the next labyrinth rings. Steam bled from the vacuum bleed chamber is taken to L. P. heater No. 1. Such an arrangement enabled the builders to achieve shaft cooling to a temperature of 125 to 140° C at a pressure not higher than 1.5 ata by steam supply to chamber *C*. Steam supplied to chamber *C* divides up and flows partly into vacuum bleed chamber *B* and partly into the last chamber *D*. In the last chamber *D* the sleeve *2* has a protective projec-

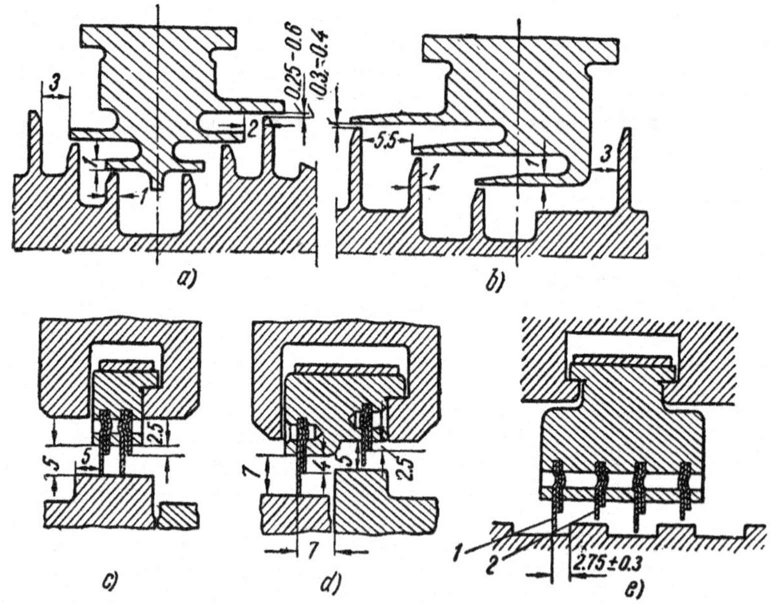

Fig. 12-26 Forms of labyrinth packing (L.M.W.)
a – fir-cone type of seal for H.P. end; *b* – L.P.end seal; *c, d* and *e* – diaphragm labyrinth packing

tion *22* which deflects the condensate (present in large quantities at starting) radially towards the bottom of chamber *D* where a few drain holes are provided. Thus any possibility of this condensate entering the close-by bearing is avoided.

The radial clearances for L.M.W. labyrinth seals is usually between 0.25 to 0.4 mm. These radial clearances are increased to about 0.5 to 0.6 mm in the case of sealing adjacent to the regulating stage.

Fig. 12-27 shows the labyrinth packing used in AEG turbines operating at high pressures. The labyrinth sleeves *2* are rigidly fixed to the turbine casing. The details of clearances and types of construction are shown in the inset in Fig. 12-27. The labyrinth fins are made sharp at their ends (0.2 mm) so that local heating would be reduced to a minimum in case of accidental fouling.

Water Sealing

Water seals are used for the low-pressure ends of turbine shafts. These consist of a water buffer between the atmosphere and the vacuum end of condensing turbines. Fig. 12-28 shows a water seal consisting of a flanged liner. When the turbine shaft rotates the water held in the flange chamber is thrown onto the flanges at the periphery forming a water lock. Since the pressure on the two sides of the ring is different there is a difference in the water levels of the two sides; the pressure difference ($\Delta p = p_1 - p_2$[kg/cm²]) between the two sides being balanced by the difference in water levels. This pressure balance may be obtained from the expression

$$\Delta p = \frac{\gamma}{g} h r_x \omega^2 = \frac{\gamma}{g}(r_2 - r_1)\frac{r_2 + r_1}{2}\omega^2 =$$

$$= \frac{\gamma}{2g}(r_2^2 - r_1^2)\omega^2, \qquad (12\text{-}27)$$

where γ —specific gravity of water, kg/cm³;
g—acceleration due to gravity, 981 cm/sec²;
h—difference in the heights of water columns, cm;

$$r_x = \frac{r_2 + r_1}{2}.$$

The remaining notations are self-explanatory from Fig. 12-28. The seal is supplied with water from a special tank at a certain water head, about 4.5 to 6 metres. The radius r_2 depends on

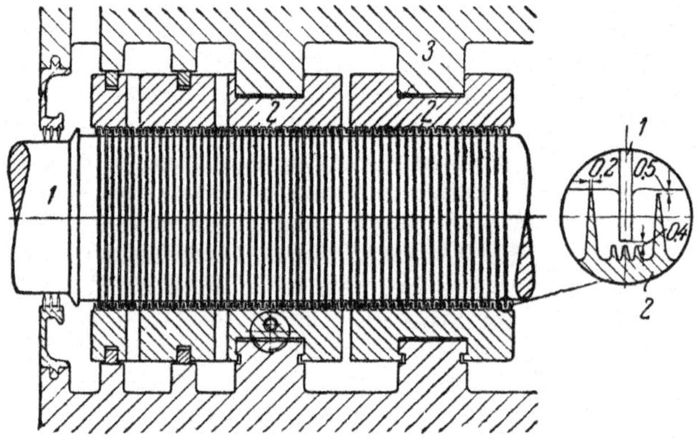

Fig. 12-27. Labyrinth packing of AEG turbine

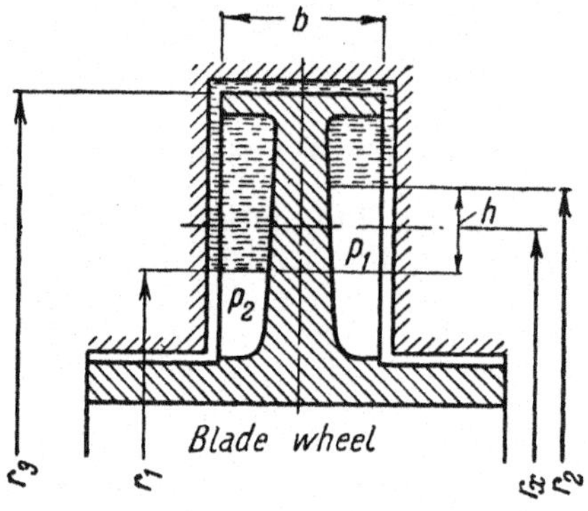

Fig. 12-28. Water-sealed gland packing

the height at which the water supply tank is placed. It may be obtained from the equation

$$0.1H = \frac{\gamma}{g} h_1 \frac{r_3+r_2}{2}\omega^2 = \frac{\gamma}{2g}(r_3-r_2)(r_3+r_2)\omega^2 = \frac{\gamma}{2g}(r_3^2-r_2^2)\omega^2. \qquad (12\text{-}28)$$

The radius r_1 is chosen from design considerations, r_2 — from equation (12-27) and r_3 — from equation (12-28). The main disadvantage of water sealing is the large amount of energy losses especially at higher circumferential velocities. The value of these losses may be determined from Flugel's equation:

$$N \approx 3{,}500\left(\frac{n}{1{,}000}\right)^3 r_3^5\left[2-\left(\frac{r_1}{r_3}\right)^5-\left(\frac{r_2}{r_3}\right)^5\right] + 7{,}050\left(\frac{n}{1{,}000}\right)^3 r_3^4 b\ [\mathrm{kW}], \qquad (12\text{-}29)$$

where b—width of wheel;
n—r. p. m.;
r and b—the dimensions as shown in Fig. 12-28, in metres.

This power loss is used up in heating and partially evaporating the water in the sealing ring. The quantity of water thus used is replenished constantly from the water reservoir. When starting or stopping the turbine the water seal is supplied with fresh steam from the vacuum side. Both the sides of the water seal are provided with a few labyrinth seals. The strips on the bearing side prevent water or moisture from leaking into the latter. The H. P. ends are also sometimes provided with water sealing. The vapour in the seal is exhausted to the atmosphere, which increases the possibility of condensate entering the bearing.

The greatest advantage of a water-sealed gland packing is a very low loss due to steam leakage to atmosphere while at the same time the shaft can be effectively cooled by the sealing water. However, the design complications involved are such that these seals are now used very rarely.

Carbon Ring Seals

Carbon ring packings are used by some factories for shaft end seals. These are used mostly for turbines of medium pressures and small circumferential velocities at the hub, not more than 30 m/sec and in very rare cases up to 50 m/sec. Fig. 12-29 shows the sectional view of a carbon ring packing. It consists of six carbon rings placed between L-shaped cast iron liners. Each of the rings is made up of three sections held together and pressed onto the turbine shaft by a binding wire and helical spring. The carbon rings are held in the required position by flat springs *4*.

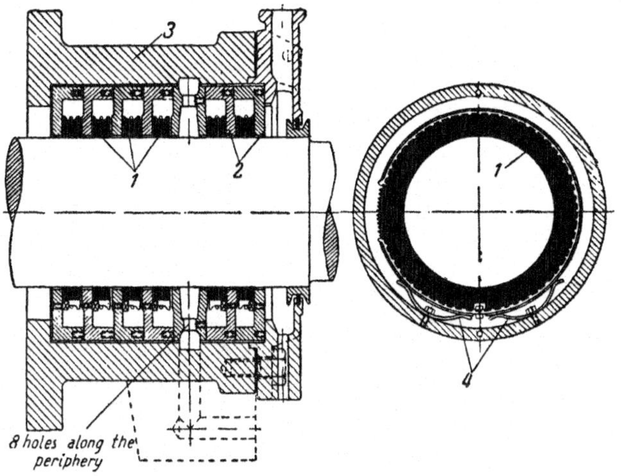

Fig. 12-29. Carbon-ring gland packing

The clearances provided between the carbon rings and the shaft are larger at the H. P. end than the L. P. side; larger from the steam side and smaller from the bearing side for H. P. glands, and vice versa for the L. P. glands. The clearances vary from 0.07 to 1 mm depending upon the diameters of sleeves and the relative position of the rings. For large diameter shafts the clearances provided may be much larger than 1 mm. The carbon rings have graphite content and hence are self-lubricating.

12-6. JOURNAL BEARINGS

The rotor of a steam turbine is supported by two or more journal bearings. The bearings are supplied with lubricating oil usually at a pressure of 0.4 to 0.7 atm. gauge. A certain clearance is provided between the shaft and the bearing surface for the accommodation of the lubricating oil film.

Prof. Petrov was the first who showed that the satisfactory operation of a journal bearing is based on the molecular friction of the lubricating oil particles adhering to the shaft and the bearing surfaces. On the basis of hydraulic principles Prof. Petrov deduced that with liquid friction the actual frictional resistance depends solely on the type of lubricating oil used and not on the bearing surfaces. Fig. 12-30,*a* shows the shaft position at rest. When the shaft begins to revolve oil is entrained into the clearance space between the shaft and the bearing surface. At some shaft speed the oil in the clearance space attains sufficient pressure to support the shaft and keep it floating (Fig. 12-30,*b* and *c*). The shaft centre is situated eccentrically with respect to the bearing centre. According to the theory of Humbell with increasing shaft speed the shaft centre exe-

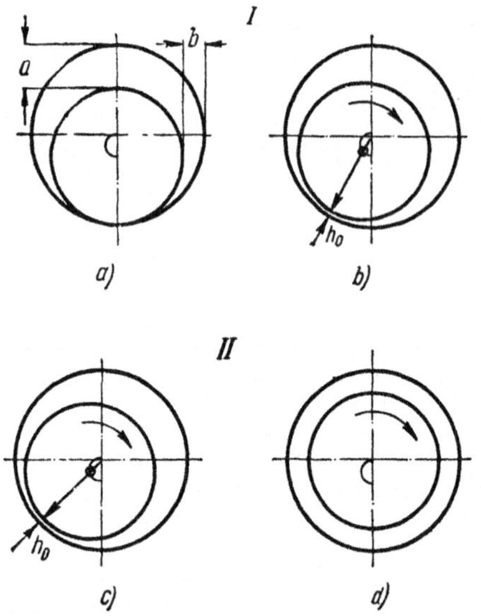

Fig. 12-30. Shaft position in the bearing at various speeds

a — at $n=0$; *b* — at slow speeds; *c* — at high speeds; *d* — at $n=\infty$

cutes a semi-circular path trying to coincide with the bearing centre. The shaft and the bearing centres coincide at an infinite speed of rotation (Fig. 12-30,*d*). Thus h_0—the minimum clearance—goes on increasing from zero when the shaft is at rest to some value depending upon the r.p.m. A sustained oil film between the shaft and the bearing surfaces is developed only when the oil pressure is able to balance the weight of the shaft.

Fig. 12-31 shows the pressure distribution of lubricating oil in the clearance space for a correctly designed bearing. The variation of local pressure p is shown in the diagram, where it can be seen that the maximum pressures attained are some tens of atmospheres. On the exit side of the clearance the pressure of oil suddenly drops which may even cause some vaporisation of the oil. The oil pressure similarly varies along the bearing axis as well (Fig. 12-32).

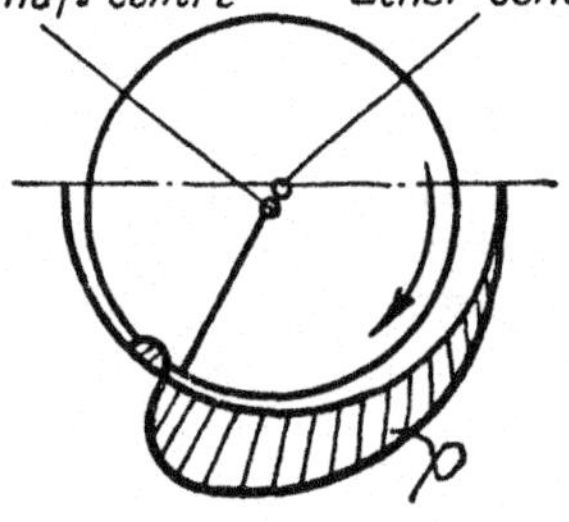

Fig. 12-31. Distribution in the oil film between the shaft and bearing

The oil is supplied to the bearings so that the oil film is not disturbed. To avoid dry friction the shafts are ground and the bearing surfaces are scraped. A high degree of surface finish is necessary since any roughness would lead to the breakage of the lubricating film causing dry friction. The total of the shaft and bearing non-uniformities should not exceed 0.01 mm. Clearances a and b are chosen according to the shaft diameter (Fig. 12-30). The permissible clearances for journal bearings based on operation data of steam turbines are given in Table 12-3.

The supports for the journal bearings are of many types such as the rigid cylindrical and spherical self-aligning type, etc. The spherical self-aligning type of bearings are to be preferred since they give satisfactory operation even with large shaft deflections. Fig. 12-33 shows the constructional details of a bearing support made by the L.M.W. This is a rigidly fixed support and is stationed in the bearing pedestal with the help of four supporting pads *1*. The correct radial alignment of the bearing support is achieved by inserting a few packing strips between these pads and the supports.

The bearing support consists of two parts held together by bolts. Strip *4* prevents the support from rotating with the shaft. Oil is supplied through a hole *2* provided at the side of the bearing. A thin metal diaphragm limits the oil flow in such a way as to allow only the required quantity of oil between the shaft and bearing. The upper part of the bearing support is provided with a wide oil passage *3* for better oil circulation and cooling of the shaft surface. Oil after lubrication is removed from the bearing from the side as shown in the figure.

In order to provide a better hold for the oil film the bearing surface covered by angle θ is scraped by a mandrel having a diameter larger than that of the shaft by the required amount of clearance a as shown in Table 12-3. The table also gives the lateral clearances between the shaft surface and the bearing support.

Besides circular lining, oval white metal lining is also used to obtain larger side clearances; the brass is bored circular with temporary liners which are removed while assembling the bearing so that the horizontal clearance is greater to give an oval shape. The white metal surface where the oil film forms must have a smooth surface finish since irregularities of surface finish can cause the film to break which may lead to dry friction.

The bearing supports are usually made of cast iron. Turbine bearing brasses are usually made with a high tin-content babbit metal. In the U.S.S.R. the most commonly used babbit

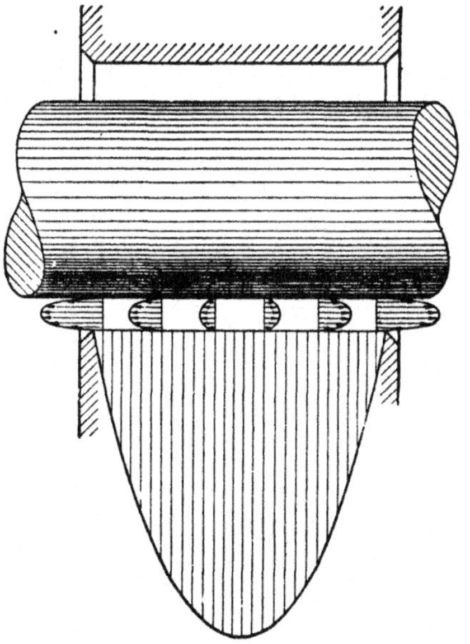

Fig. 12-32. Oil pressure along the bearing axis

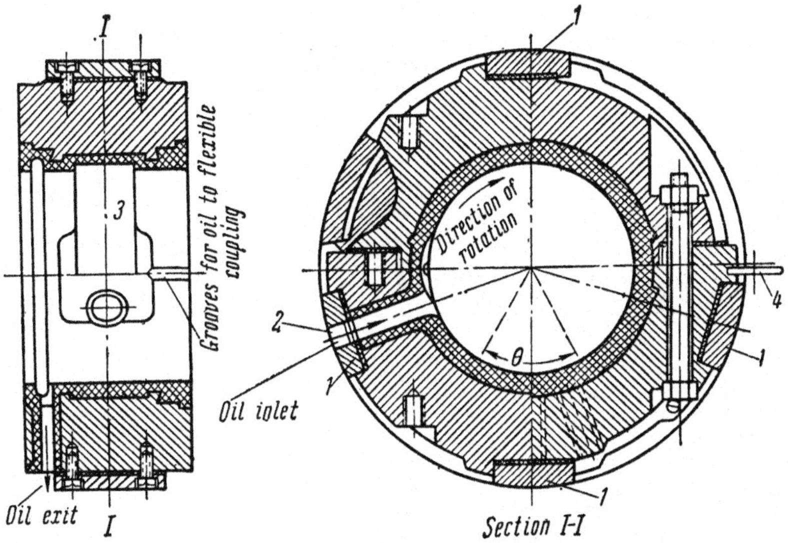

Fig. 12-33. L.M.W. journal bearing

Table 12-3

Shaft dia., mm	Bearings without white metal lining				Bearings with white metal lining			
	Top clearance, mm		Side clearance, mm		Top clearance, mm		Side clearance, mm	
	min	max	min	max	min	max	min	max
50	0.15	0.25	0.10	0.15	0.10	0.12	0.15	0.20
100	0.20	0.30	0.10	0.20	0.10	0.15	0.20	0.25
150	0.30	0.40	0.15	0.25	0.20	0.25	0.30	0.40
200	0.40	0.55	0.20	0.30	0.20	0.30	0.35	0.45
250	0.50	0.65	0.25	0.35	0.25	0.35	0.45	0.55
300	0.60	0.75	0.30	0.40	0.30	0.45	0.55	0.62
350	0.70	0.85	0.35	0.45	0.35	0.50	0.62	0.70

metal is Б-83 (GOST 1320-41) of the following composition: tin 83%, antimony 11% and copper 6%.

12-7. DESIGN OF JOURNAL BEARINGS

We shall introduce the following nomenclature for design purposes:

P—force exerted by the shaft on the bearing, kg;
L—length of the bearing surface, cm;
d—shaft diameter, cm;
a—clearance between the shaft and the bearing surface (from Table 12-3), cm;
e—eccentricity between the shaft centre and the bearing centre, cm;
y_s—shaft sag at the bearing end, mm;
h_0—minimum oil film thickness between the shaft and the bearing surface, mm;
$u = \pi dn/60$—circumferential velocity of the shaft surface at the turbine speed of n r.p.m.

We shall carry out the design according to the method suggested by M. I. Yanovsky for a 180° journal.

The coefficient (load criterion) of bearing load is obtained from the expression

$$\Phi_v = \frac{P\left(\frac{a}{d}\right)^2}{Lu\mu}, \tag{12-30}$$

where μ—average viscosity of oil which depends upon the type of oil and the temperature of operation; for oil type TZOUT (GOST 32-53) μ may be taken as 0.30×10^{-6} kg sec/cm².

Knowing the load criterion (coefficient Φ_v) and the ratio $\varepsilon = d/L$ we can obtain the value of relative eccentricity $\chi = 2e/a$ with the help of the two graphs shown in Fig. 12-34. With the value obtained for χ and ε the coefficient of resistance Φ_s can be determined from the curves given in Fig. 12-35,*a* and *b*. The friction coefficient for the bearing is obtained as

$$f = \frac{a}{d}\frac{\Phi_s}{\Phi_v}. \tag{12-31}$$

Then the work done against friction will be

$$A_r = \frac{fPu}{100}\,[\text{kg m/sec}] \tag{12-32}$$

or

$$N_r = \frac{A_r}{102}\,[\text{kW}]. \tag{12-33}$$

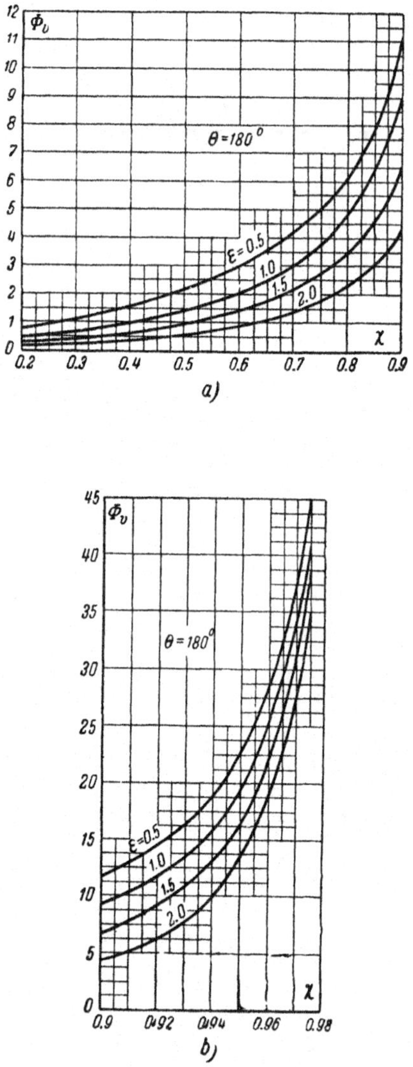

Fig 12-34. Coefficient Φ_v (load criterion)

The heat equivalent of this work will be

$$Q_r = \frac{A_r}{427}\,[\text{kcal/sec}]. \qquad (12\text{-}34)$$

If radiation losses are neglected then the quantity of oil required to carry away the heat generated by friction will be

$$q_o = \frac{Q_r}{\gamma c\,(t_2 - t_1)} = \frac{A_r}{427\gamma c\,(t_2 - t_1)} = \frac{fPu}{42{,}700\gamma c\,(t_2 - t_1)}\,[\text{l/sec}],$$

where t_1—temperature of oil at inlet assumed between 35 to 45° C conforming to the conditions of operation;

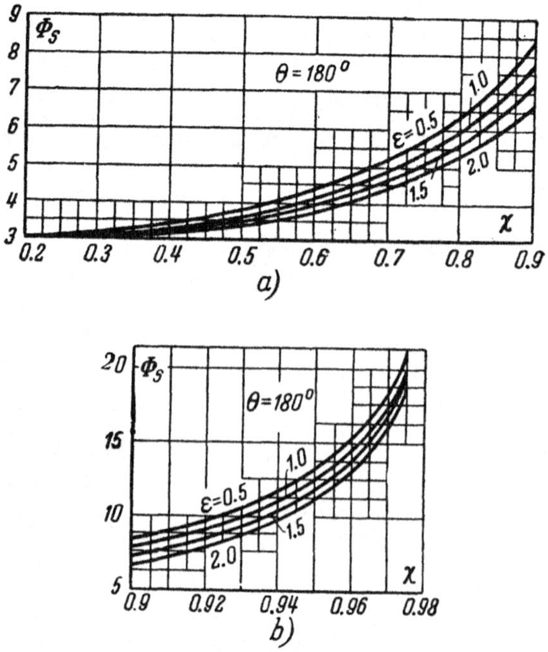

Fig. 12-35. Coefficient Φ_s

t_2—oil temperature at exit from the bearing; may be taken as $t_1 + (10 \text{ to } 15)°$ C;

γ—specific gravity of the lubricating oil kg/litre;

c—average thermal capacity of the oil kcal/kg° C.

The oil film thickness is obtained from the relation

$$h_0 = \frac{a}{2}(1 - \chi). \qquad (12\text{-}36)$$

For a satisfactory operation of the journal bearing the oil film thickness h_0 must be greater than, or equal to, $y_s + 0.01$ mm (where 0.01 is the total of the non-uniformities of shaft and bearing surfaces). The temperature of oil at the exit from the bearing should not exceed 60°C since at higher temperatures the lubricating oil deteriorates very quickly becoming unserviceable for further usage. The ratio d/L is usually assumed from 1 to 1.2. However, for heavily loaded bearings as well as in the presence of reduction gearings higher values are assumed.

12-8. CONSTRUCTION OF THRUST BEARINGS

Thrust bearings are meant to take the axial thrust present in a turbine as well as to fix its rotor position. In practice two types of thrust bearings are widely used: the collar type and the segment type. At present the use of collar type of thrust bearings is almost discontinued since

Table 12-4

Design quantities	Dimension	Values obtained from calculations		
L	cm	16	18	20
$\varepsilon = d/L$	—	1.25	1.11	1.0
$\Phi_v = \dfrac{P\left(\frac{a}{d}\right)^2}{Luu}$	—	0.830	0.737	0.664
χ	—	0.39	0.35	0.30
Φ_s	—	3.2	3.1	3.0
$f = \left(\dfrac{a}{d}\right)\dfrac{\Phi_s}{\Phi_v}$	—	0.00963	0.0105	0.0113
$A_r = \dfrac{fPu}{100}$	kg m/sec	605	660	710
$N_r = \dfrac{A_r}{102}$	kW	5.92	6.46	6.96
$Q_r = \dfrac{A_r}{427}$	kcal/sec	1.42	1.55	1.66
$q_o = \dfrac{60Q_r}{\gamma c(t_2 - t_1)}$	litres/min	23.1	25.2	27.1
$h_o = \dfrac{a}{2}(1-\chi)$	mm	0.152	0.162	0.175

Example 12-1. Design a journal bearing with the following data: $p = 2{,}000$ kg; $d = 20$ cm; $a = 0.05$ cm; $u = 3{,}140$ cm/sec; $t_1 = 40°$ C; $t_2 = 50°$ C; $\gamma = 0.92$ kg/litres; $c = 0.4$ kcal/kg° C; $\mu = 0.35 \times 10^{-6}$ kg sec/cm².

The bearing is designed for various lengths L. We shall make use of the curves given in Figs 12-34 and 12-35 to obtain the values of χ and Φ_s. The order of design calculations is given in Table 12-4 from where it can be seen that with increasing bearing lengths the friction losses N_r and the oil film thickness h_0 increase. The length of the bearing is so chosen as to keep $h_0 \geqslant y_s + 0.01$ mm.

the segment type is found to be much more advantageous.

Fig. 12-36 shows a collar type thrust bearing made by the Kh.T.W. The thrust bearing is fixed by bolts to the journal bearing support. The rotor is fixed in the required position by putting packing strips between the flanges of the thrust bearing. The thrust bearing is provided with six annular bearing liners made of babbit castings. The shaft collars mesh into the grooves formed by these liners. The bearing is lubricated with oil supplied under pressure through a central hole and radial grooves provided in the turbine shaft. The collar type thrust bearings are found to be unsatisfactory because of the difficulties encountered in providing proper lubrication and an equally distributed axial pressure on its bearing surfaces. However, their one advantage is that they can stand large thrust fluctuations. The pressures usually assumed for collar bearings is from 2 to 8 kg/cm². The axial clearance between the liners and the collars is usually from 0.25 to 0.5 mm. The white metal lining thickness is about 1.5 mm and is so calculated as to avoid large shaft displacement in the event of overheating when the white metal may flow out.

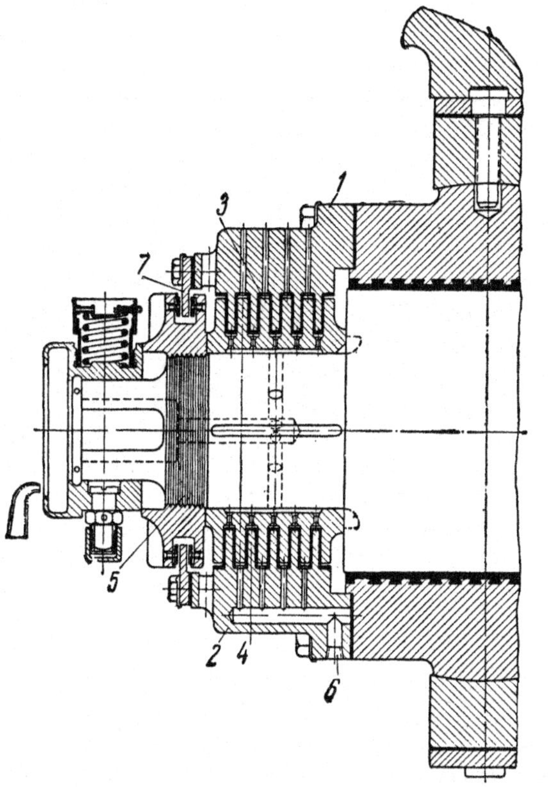

Fig. 12-36. Collar-type thrust bearing of Kharkov Turbine Works

1 – upper half of bearing support; *2* – lower half of bearing support; *3* – collar, *4* – flanges in two halves; *5* – packing strips; *6* – adjustable ring; *7* – lugs holding the flanges in position

Fig. 12-37 shows a segment type thrust bearing of earlier design made by the L.M.W. The bearing consists of a thrust disc *3*, mounted on the rotor shaft, thrust pads *1* found on the disc surface, self-aligning pad *2* with spherical seating held in position by the support *7*. Support *7* is rigidly fixed in the bearing pedestal. The axial displacement necessary for proper adjustment of the rotor position is achieved by varying

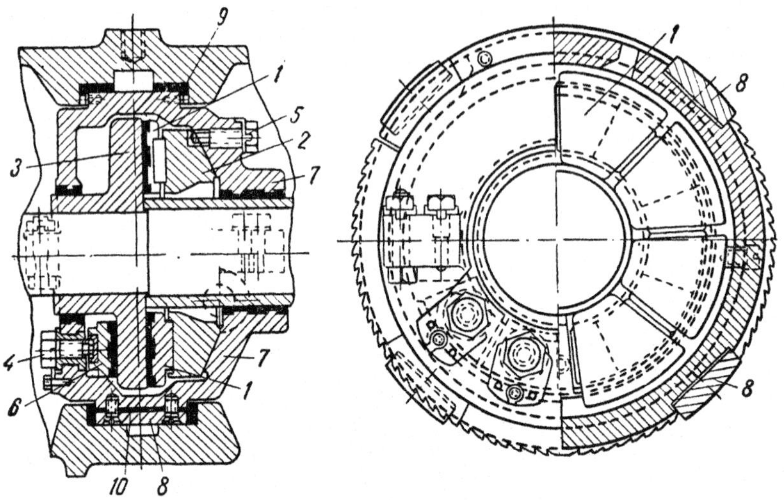

Fig. 12-37. Segment-type thrust bearing of L.M.W.

1 – thrust pads; *2* – aligning pads; *3* – thrust disc; *4* – dummy thrust pad bolts; *5* – locking bolt; *6* – dummy pads; *7* – bearing support; *8* – supporting pads; *9* – packing strips for axial alignment; *10* – packing strips for radial alignment

the number of packing strips *9* placed between the bearing support *7* and the bearing pedestal. The radial alignment of the bearing is achieved with the help of packing elements *10* placed between the support *7* and the pads *8*. The thrust pads *1* are housed in the aligning pads *2*. The aligning pads can adjust themselves to the required shaft sag by finding their proper position on the spherical seating of support *7*.

A locking bolt *5* prevents the aligning pads from rotating with the shaft. A clearance of 1.5 to 2 mm is provided between the stop bolt *5* and the slot in the aligning pads *2*. The second side of the thrust disc has dummy pads *6* held in the required position by bolts *4*. These pads should be positioned in such a way that the distance between the dummy pads and the thrust pads *1* is more than the thrust disc thickness by 0.25 to 0.5 mm. This clearance ensures a satisfactory oil film thickness between the disc and the thrust pads. The lubricating oil is supplied to the bearing under pressure at the shaft surface. The thrust bearing is closed from both ends to prevent the escape of lubricating oil. It has been found that the use of thrust bearings of the type described above resulted in several cases of turbine shafts breaking due to non-uniform loading of the thrust pads. This irregular pressure distribution gave rise to an unforeseen force couple culminating in accidents.

The construction of a combined journal and thrust bearing is a much more satisfactory one where all the shortcomings of the bearing described above have been eliminated. Fig. 12-38 shows a combined journal and thrust bearing manufactured by the L.M.W. The journal bearing portion of the unit has a spherical form which ensures correct alignment conforming to the shaft position. Further the ability to self-alignment ensures proper pressure distribution on the thrust pads of the bearing minimising the possibility of unforeseen force couples. The lubricating oil is supplied to the bearing through an annular passage *1*, passes to chamber *2* along the shaft face and thence is removed through exit holes *3*.

Fig. 12-39 shows a combined journal and thrust bearing of a different construction. The journal bearing portion of the unit is placed between two thrust bearings. The thrust discs which fix the position of the turbine rotor are mounted on the turbine shaft. The correct positioning of the bearing supports in the pedestal is achieved by plac-

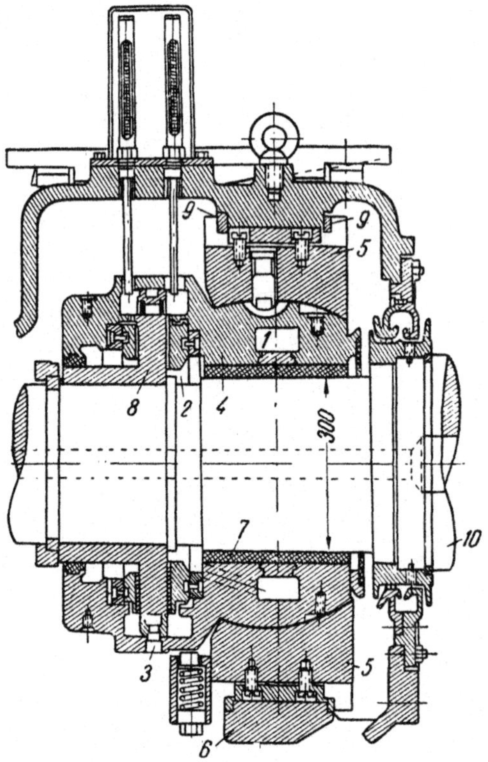

Fig. 12-38. Combined journal and thrust bearing of the L.M.W.

1 – annular oil passage; *2* – chamber for oil supply to thrust bearing; *3* – orifice for removal of oil from bearing; *4* – journal and thrust bearing support; *5* – upper and lower support holders; *6* – journal bearing pads; *7* – white metal lining for journal bearing; *8* – thrust disc; *9* – packing strips for axial alignment; *10* – turbine shaft

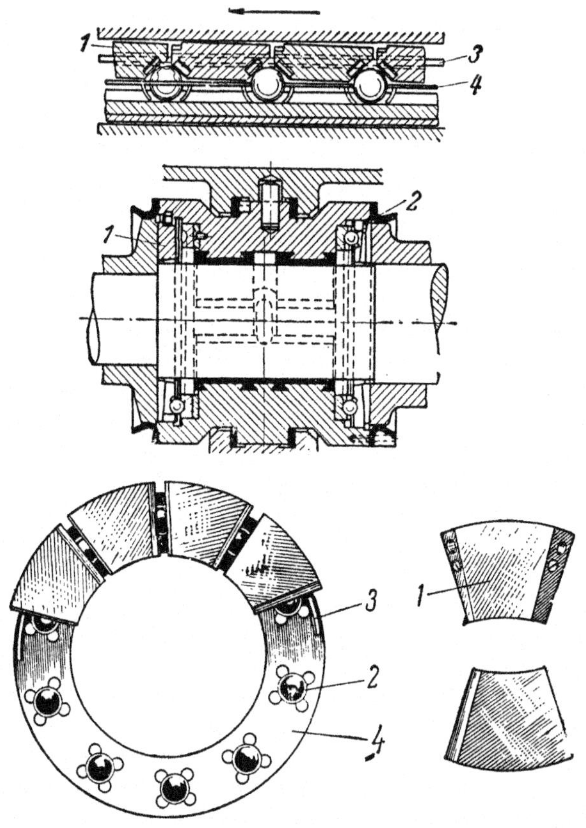

Fig. 12-39. Detailed sectional view of a combined journal and thrust bearing (Brown Bover Corporation)
1 — thrust pad; *2* — steel balls; *3* — binding wire; *4* — cage

ing a number of packing strips between the support and the body of the pedestal.

The thrust pads of the thrust bearing are connected together by a steel wire *3* holding them in their seats. Besides each of these thrust pads can adjust itself on its spherical seat *2* which are free to move in the holes *4*. Steel washers are recessed into the thrust pads to bear the load coming on them. The provision of steel balls enables the thrust pads to assume the required position. The upper sketch in Fig. 12-39 gives the developed view of the thrust disc and pads from which it can be clearly seen that a wedge-shaped clearance is formed between the thrust disc and the bearing pads. The oil pressure in the clearance is not uniform, increasing with decrease of clearance. The journal portion of the combined bearing is supplied with oil at its centre. Some portion of this oil is utilised for the lubrication and cooling of the journal bearing and the remaining portion is led to the thrust bearing for its lubrication. The advantages of the segment type of thrust bearing are its compactness and the uniform pressure distribution on all the thrust pads. The construction of segment type of thrust bearings is such that a thin film of oil between the pads and the disc is always assured and hence large specific pressures of up to 15 to 25 kg/cm^2 may be safely assumed for their design.

Chapter Thirteen

CONSTRUCTION OF TURBINE ROTORS AND THEIR COMPONENTS

13-1. MATERIALS AND CONSTRUCTION OF MOVING BLADES

Moving blades in a turbine are meant for the conversion of the kinetic energy of the flowing steam into the mechanical work on the turbine shaft. The work done by steam is transmitted to the shaft through the disc on which the blades are mounted. Various methods are in use for the attachment of blades to the drum or discs of a turbine. Fig. 13-1 shows some of the more common methods in use. Short blades having small centrifugal forces are generally made with T-shaped tangs and are attached to the disc as shown in Fig. 13-1, *a*, *b* and *c*. If the blade and its tang are of the same width spacers are used to get the proper blade passage. The blades and the spacers are inserted in the groove all around the disc through slots provided at a few points on the disc periphery. After complete assembly of all the blades and spacers these slots are either plugged or blades with tangs conforming to the slot shape are inserted and fixed by riveting them in place. If spacers are used for blade assembly some sort of locking arrangement is used to hold them against the centrifugal forces developed while in operation. Depending on the type of locking arrangement used the blade spacing at this place may be equal or greater than normal. The blade adjacent to the lock, in case of larger blade spacing, experiences forces greater than those of the other blades resulting in increased non-uniformities of flow after the blades and higher losses. The types of blade tangs shown in Fig. 13-1,*a* and *b* have the same advantages and disadvantages. The bearing surfaces of the blades are the areas between dimensions *p* of the surface *M* and the tang collars mating with the surface along the outer diameter of the discs. Clearances as shown in the figure are provided to eliminate the devel-

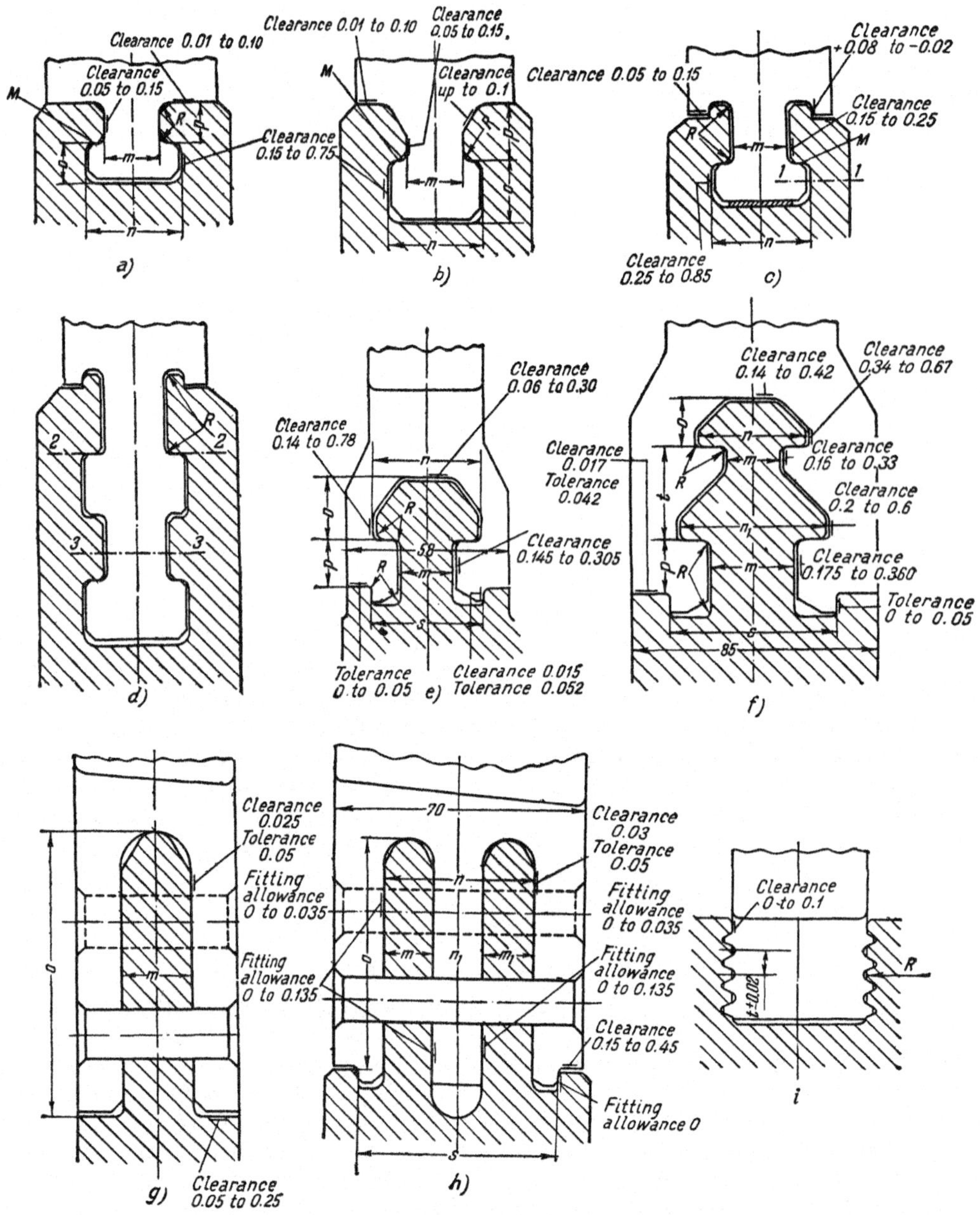

Fig. 13-1. Methods of blade attachments

opment of stresses on the lateral faces of the blade tang.

The blade tangs shown in Fig. 13-1,*c* and *d* do not differ much from the ones shown in *a* and *b* however, their bearing surfaces are developed in a completely different manner. The presence of side grips on the blades and collars on the discs forms such a bearing surface that the opening out of the disc jaws is restricted when bending forces are produced at section *1-1* while the disc is in rotation. The bearing faces along the disc circumference increase the rigidity of fixture in the axial direction. A thin packing foil is sometimes placed between the blade tang and the groove as shown in Fig. 13-1,*c* which gives additional bearing surface and strengthens the blade fixture. The use of such shims is found to be highly desirable.

The double T-shaped blade tang shown in Fig. 13-1,*d* increases the rigidity of blade fixture in the axial direction but is not suitable from the production point of view since it entails heavier manufacturing costs. An alternative method of blade attachment is shown in Fig.

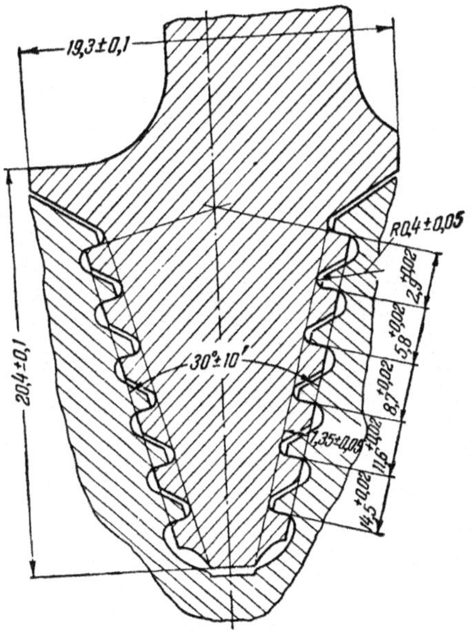

Fig 13-2. Conical serrated-root attachment

13-1,*e* and *f* which is used by the Kh.T.W. for the attachment of both short and long blades to the discs of high-pressure turbines type VR-25 and others; the one shown in *e* is used for short blades and that in *f* for long blades.

Fig. 13-1,*g* and *h* show straddle attachment with forked tangs. The number of straddles depends upon the type and dimensions of the blade. The blades are fixed to the disc with the help of rivets which are placed in tightfitting drilled countersunk holes and caulked from both ends. The serrated root blades shown in Fig. 13-1, *i* are used for short blades.

The method of blade attachment employed for the low-pressure stages of L.M.W. high-pressure turbine types VK-50, VK-100 and others is shown in Fig. 10-12.

Some of the manufacturers make use of the conical serrated root type of attachment for the longer blades as shown in Fig. 13-2. This arrangement helps in keeping the bearing stress within the permissible limits both on the tang and on the disc at their mating surfaces. Figs 13-1 and 13-2 show the margins for filing to facilitate the mounting of the blades on the disc.

The blades are built up in segments consisting of from 5-6 to 10-12 blades threaded on a root wire and assembled on a former, and are held together by the shrouding strip.[1] Further it is usual to braze the whole assembly in a silver bath.

[1] A. V. Levin, *Blades and Discs of Steam Turbines*, Gosenergoizdat, 1953. P. N. Shlyakhin, *Vibration of Turbine Blades*, Gosenergoizdat, 1946.

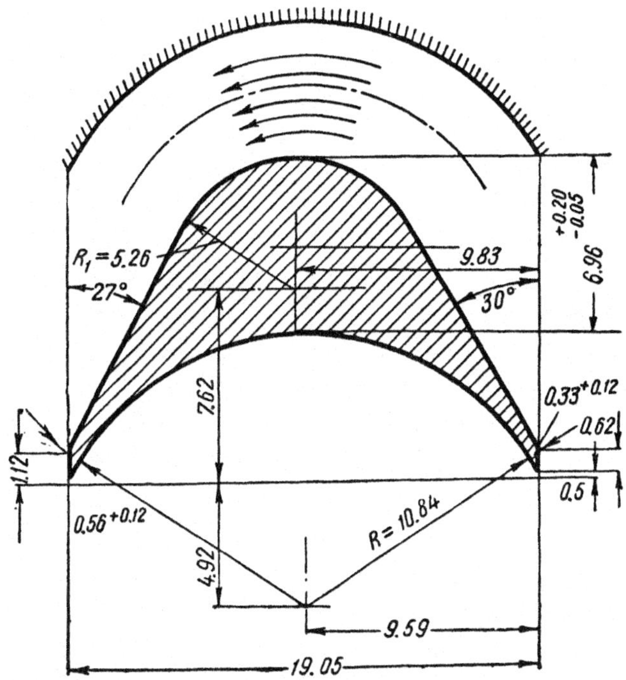

Fig. 13-3. Profile of a moving blade

Fig. 13-3 shows on an enlarged scale the section of a constant-section moving blade. The limits to which the blade has to be manufactured are shown on the sketch. The arrows indicate the direction of steam flow, and the working surface of the adjacent blade shows the blade passage formation. The ends are rounded to a dia. of 0.5 to 0.8 mm. The width of blades is chosen depending upon their length; it is very rare to find blades with widths less than 0.1 time their length and in any case less than 12 to 14 mm. A width of 20 to 25 mm is used in turbines of medium and large capacities (in the low-pressure stages of condensing turbines, considerably more). The regulating stage blades of medium- and large-capacity turbines are usually made with integral roots and shrouding with widths of 30 to 40 mm and sometimes of up to 65 to 100 mm (L.M.W. and Kh.T.W. turbines types VK-100, VR-25, etc.).

Certain standard dimensions based on blade width are now followed to avoid multifarious blade constructions. These standard blade widths also form the basis on which further designs are carried out. The long blades for the low-pressure stages are made with varying section along the length. The root portion of the long blades is quite large reaching from 70 to 100 mm in width and 15 to 20 mm in thickness. The width and thickness of these blades go on diminishing from the root to the tip reducing their weight and the

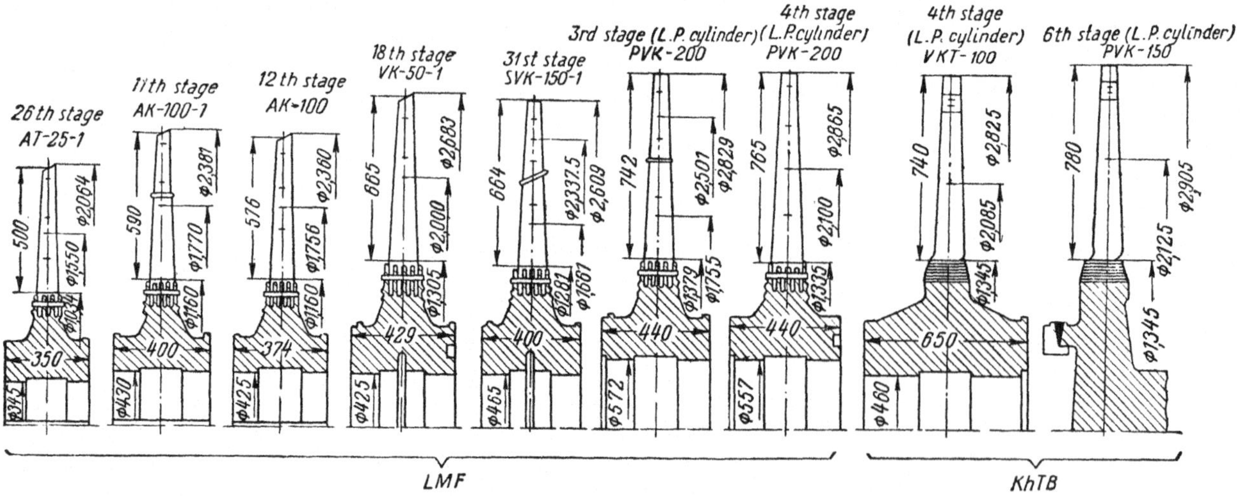

Fig. 13-4. L.P. stage blades of large-capacity turbines manufactured by L.M.W. and Kh.T W.

stresses induced due to centrifugal forces. Shockless entry of steam into the blade passages is ensured by varying the blade entry angles along the height. Fig. 13-4 shows the low-pressure blades of an L.M.W. turbine operating at 3,000 r.p.m. The root and tip dimensions and their sectional view are shown in this figure. The blades are held together only by a binding wire. The shrouding is dispensed with to reduce stresses in them. Since these blades operate in the region of wet steam flow stellit strips are brazed to the back of the blades at the leading blade tips to safeguard against erosion. The moving blades of large-capacity stationary turbines are made of nickel and chromium steels (stainless type). Regulating stage blades operating at high pressures and temperatures are made of high quality alloy steels (e. g., the Kh.T.W. uses ЭИ69[1] grade of steel for the regulating stage blades of turbine VR-25).

Experience shows that nickel steel blades operating in wet steam regions are very much subjected to both corrosion and erosion. Fig. 13-5 shows a segment of nickel steel blades which were rendered unfit for operation after working for 44,782 hours, because of the effects of corrosion and erosion of the blade tips. The surfaces of the blades were found to be heavily pitted all along their lengths. Stainless steel blades, however, were found to be free from any such effects and hence blades of condensing turbines are now made either from stainless steel or high-quality alloy steels.

[1] The chemical composition of this steel is: C—0.4 to 0.5%; Si—0.3 to 0.8%; Mn—0.2 to 0.4%; Cr—13 to 15%; Ni—13 to 15%; W—2 to 2.7%; Mo—0.4 to 0.6%.

13-2. DESIGN OF BLADES

The centrifugal forces exerted on the blades, shrouding, binding wire and the force exerted by the steam on the blades are all summed up to ob-

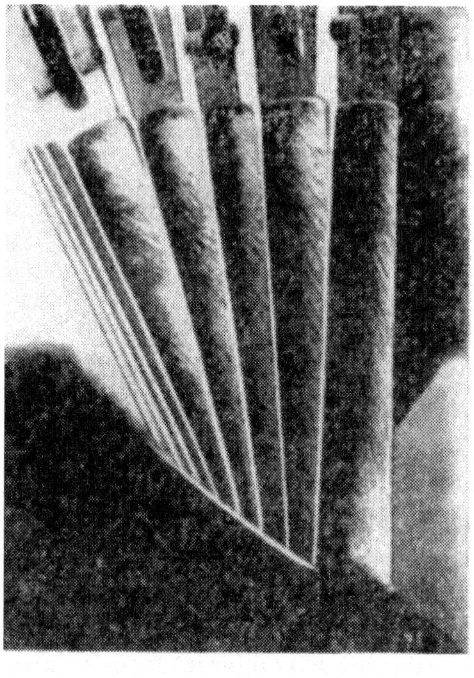

Fig. 13-5. Nickel steel blades removed from disc after long period of operation

tain the total force inducing stresses in the blades. The centrifugal forces cause tensile and bending stresses of constant magnitude whereas the steam pressure causes bending stresses of a varying nature. The constant stresses due to centrifugal forces are known as static stresses and those due to the steam pressure are known as dynamic stresses. The blades are designed for strength on the basis of the total effects of both the static and dynamic stresses since the blades are subjected to these stresses at one and the same time.

1. Tensile Stresses in Turbine Blades Due to Centrifugal Forces (Centroids of the Blade Section on One Radial Line)

The most dangerous section of a constant section blade is the one at the root since it is weakened by the presence of riveting holes, etc. In some cases the blade section is made larger at the root to reduce stress concentration. Twisted blades and blades of varying section also have their weakest section at the root.

For blades with constant blade section along the length the stresses at the weakest section are

$$\sigma = \frac{C_0}{F_0} = \frac{C_b + \Sigma C_s}{F_0}, \qquad (13\text{-}1)$$

where $C_0 = C_b + \Sigma C_s$—centrifugal forces of the blade, shroud, etc., kg;
F_0—area of the weakest blade section (root section) cm².

The centrifugal force of a constant section blade will be

$$C_b = \frac{G_b}{g} r_{av}\omega^2 = \frac{F_0 l\gamma}{g} r_{av}\omega^2, \qquad (13\text{-}2)$$

where G—weight of blade, kg;
l—height of blade, cm;
γ—specific weight of material from which the blades are made, kg/cm³;
r_{av}—mean-diameter radius, cm;
ω—angular velocity.

The centrifugal force of the shrouding is obtained as

$$G_s = \frac{G_s}{g} r_s\omega^2 = F_s t_s \gamma r_s \frac{\omega^2}{g}, \qquad (13\text{-}2a)$$

where G_s—weight of the shrouding strip, kg;
F_s—cross-sectional area of shrouding strip, cm²;
t_s—length of each shrouding strip, cm;
r_s—radius of the shrouding strip centroid, cm.

The centrifugal force of the binding wire will be

$$C_w = \frac{G_w}{g} r_w\omega^2 = F_w t_w \gamma r_w \frac{\omega^2}{g}, \qquad (13\text{-}2b)$$

where G_w—weight of wire for length t_w, kg;
F_w—cross-sectional area of wire, cm²;
t_w—length of each binding wire, cm;
r_w— radius at which the wires are placed measured from their centre, cm.

Substituting expressions (13-2), (13-2a) and (13-2b) in equation (13-1) and writing $\omega = \pi n/30$ we have,

$$\sigma = \frac{\pi^2 n^2 \gamma}{900 g}\left[l r_{av} + \frac{F_s}{F_0} t_s r_s + \frac{1}{F_0}\sum (F_w t_w r_w)\right].$$

With $\gamma = 0.00785$ kg/cm³ and $g = 981$ cm/sec², we obtain

$$\sigma = 0.88 \times 10^{-7} n^2 \left[l r_{av} + \frac{F_s}{F_0} t_s r_s + \frac{1}{F_0}\sum (F_w t_w r_w)\right]. \qquad (13\text{-}3)$$

The centrifugal force of a blade with varying section will be

$$C_b' = \frac{G_b'}{g} r_{cg}\omega^2, \qquad (13\text{-}4)$$

where G_b'—weight of blade from the tip to the section under consideration, kg;
r_{cg}—radius at which the centre of gravity of the section under consideration is situated, cm.

For blades of varying section where the sectional area changes uniformly along the length the weight G_b' may be obtained as

$$G_b' = \frac{F_1 + F_2}{2} l\gamma, \qquad (13\text{-}4a)$$

where F_1— sectional area at the root or at the section under consideration;
F_2—tip section;
l—length of the blade or the length of the portion under consideration.

The centre of gravity of the blade or the portion under consideration is obtained by equating moments of small elementary sections. Knowing the centre of gravity the radius r_{cg} can be easily found.

Substituting the value of G_b' from equation (13-4a) in (13-4) we have

$$C_b' = \frac{F_1 + F_2}{2g} l\gamma r_{cg}\omega^2. \qquad (13\text{-}4b)$$

The centrifugal stresses at the root of a blade with uniformly varying section along its length from equations (13-3) and (13-4b) are

$$\sigma' = 0.88 \times 10^{-7} n^2 \left[\frac{F_1 + F_2}{2} l r_{cg} + \frac{F_s}{F_1} t_s r_s + \frac{1}{F_1}\sum (F_w r_w t_w)\right]. \qquad (13\text{-}5)$$

2. Bending Stresses Due to Centrifugal Forces (Blades with Centroids Not on a Radial Line)

When the centroids of various sections do not fall on one radial line the centrifugal force does not act at the centre of gravity of the blade

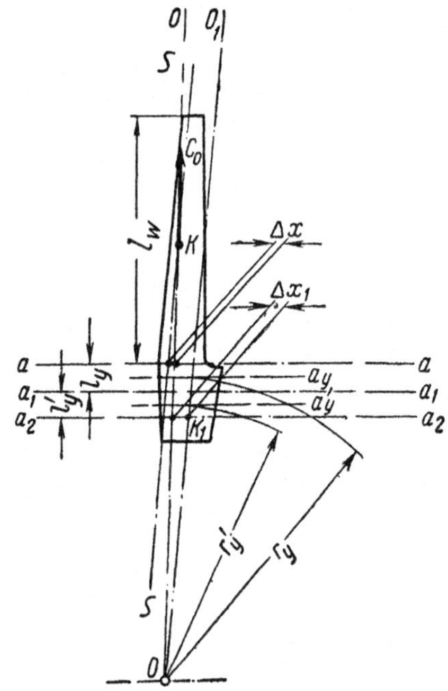

Fig. 13-6. Long blades with centroids not on the radial line

in which case a bending stress is experienced at the section where the centre of gravity lies.

Fig. 13-6 shows the setting of a blade with the radial line passing through its centre of gravity K. Line OO is the radial passing through the turbine axis (point O) and the blade centroid (point K). Line SS passes through the centroids of the transverse sections, along its entire length l. Line OO_1 passes through the turbine axis and the centroid of section a_2a_2 (point K_1).

The moment of centrifugal force C_0 with respect to the root section aa is given by the equation

$$M_b = C_0x = C_b\Delta x_{aa} + + C_s\Delta x_s + \sum (C_w \Delta x_w), \qquad (13\text{-}6)$$

where Δx—distance between the c. g. of the root section aa and the radial line passing through the c. g. of the blade, shrouding and binding wire;

Δx_s and Δx_w—distance between the c. g. of the shrouding and binding wire and the radial line passing through the blade c. g. respectively.

The bending stress at section aa is obtained as

$$\sigma_b = \frac{M_b}{W_x} = \frac{C_0\Delta x}{W_x}, \qquad (13\text{-}7)$$

where W_x—the minimum resisting moment of the section aa.

The expression may be simplified further as

$$\sigma_b = \frac{C_0\Delta x}{W_x} \approx \frac{C_0\Delta x_{aa}}{W_x}.$$

Substituting the value of C_0 from equations (13-2), (13-2a) and (13-2b) in equation (13-7) with the simplification as indicated above we have

$$\sigma_b = \frac{0.88 \times 10^{-7} n^2 \Delta x_{aa}}{W_x} [F_0 l r_{av} + F_s t_s r_s + + \sum (F_w t_w r_w)] \qquad (13\text{-}7a)$$

for constant section blades; and

$$\sigma_b = \frac{0.88 \times 10^{-7} n^2 \Delta x_{aa}}{W_x} \left[\frac{F_1 + F_2}{2} l r_{cg} + F_s t_s r_s + + \sum (F_w t_w r_w)\right] \qquad (13\text{-}7b)$$

for blades with varying section along their length. The centrifugal forces in equation (13-7b) are those which are obtained from equation (13-4b).

The centrifugal forces of the blades and the shrouding cause stresses at the root section of the blades. It is usual to calculate these stresses for the weakest section such as section a_2a_2 in Fig. 13-6.

Example 13-1. Calculate the stress caused by centrifugal force in a blade of constant section at its root if $F_0 = 0.82$ cm²; $l = 140$ mm; $r_{av} = 560$ mm; $\gamma = 0.00785$ kg/cm³; $n = 3{,}000$ r. p. m.; number of blades on the disc $z = 240$. The blades are held only by a shrouding strip; width of shrouding is 26 mm and thickness 2 mm.

The cross-sectional area of the shrouding strip $F_s = 2.6 \times 0.2 = 0.52$ cm².

The radius at which the c. g. of this section is situated is

$$r_s = r_{av} + 0.5l + 0.5s = 560 + 0.5 \times 140 + 0.5 \times 2 = = 631 \text{ mm}.$$

Blade spacing at the shrouding

$$t_b = \frac{2\pi r_s}{z} = \frac{2\pi \times 631}{240} = 16.5 \text{ mm}.$$

The stress in blades from equation (13-3) will be

$$\sigma = 0.88 \times 10^{-7} \times 3{,}000^2 \left(14 \times 56 + \frac{0.52}{0.82} 1.65 \times 63.1\right) = = 672 \text{ kg/cm}^2.$$

Bending and tensile stresses are induced at section a_2a_2 by the centrifugal forces of the blade, shrouding and the tail portion of the blade above section a_2a_2.

The tensile stresses caused by the centrifugal forces at section a_2a_2 may be determined from the relation

$$\sigma'' = 0.88 \times 10^{-7} n^2 \left[\frac{F_0}{F'_y} l r_{av} + \frac{F_s}{F'_y} t_s r_s + + \frac{1}{F'_y} \sum (F_s t_s r_s) + \frac{F_y}{F'_y} l_y r_y + l'_y r'_y\right] \qquad (13\text{-}8)$$

for blades of constant section;
and

$$\sigma''' = 0.88 \times 10^{-7} n^2 \left[\frac{F_1 + F_2}{2F_y'} l r_{cg} + \frac{F_s}{F_y'} t_s r_s + \right.$$

$$\left. + \frac{1}{F_y'} \sum \left(F_w t_w r_w + \frac{F_y}{F_y'} l_y r_y + l_y' r_y' \right) \right] \quad (13\text{-}8a)$$

for blades of varying section,
where F_y and F_y' — cross-sectional areas of the blade at the sections $a_y a_y$ and $a_y' a_y'$, cm^2;
l_y and l_y'—the heights between sections aa and $a_1 a_1$, and $a_1 a_1$ and $a_2 a_2$;
r_y and r_y' —radius at which the c.g.s of these blade portions are situated.

The bending stresses caused by the centrifugal forces at the root section of the blade may be determined from the equation

$$\sigma_b'' = \frac{0.88 \times 10^{-7} n^2}{W_x'} [F_0 l r_{av} \Delta x_b + F_s t_s r_s \Delta x_s +$$

$$+ \sum (F_w t_w r_w \Delta x_w) + F_y l_y r_y \Delta x_y] \quad (13\text{-}9)$$

for constant section blades, and

$$\sigma_b = \frac{0.88 \times 10^{-7} n^2}{W_x'} \left[\frac{F_1 + F_2}{2} l r_{cg} \Delta x_b + F_s t_s r_s \Delta x_s + \right.$$

$$\left. + \sum (F_w t_w r_w \Delta x_w) + F_y l_y r_y \Delta x_y \right] \quad (13\text{-}9a)$$

for varying section blades where Δx_b, Δx_s, Δx_w and Δx_y are the distances between the c.g. of the section $a_2 a_2$ and the radial lines passing through the c.g.s of the blade, shrouding, binding wire and the tail portion of height l_y.

The total tensile and bending stresses from centrifugal forces allowable for nickel and chrome-nickel steels are about 1,000 to 1,200 kg/cm^2 and for stainless steels from 1,500 to 1,600 kg/cm^2, and sometimes even more; for example, the L.P. stage blades of large-capacity high-pressure condensing turbines are designed for tensile and bending stresses of 2,000 to 2,200 kg/cm^2 caused by the centrifugal forces.

3. Bending Stresses Due to Steam Pressure

The steam flowing out of a nozzle does not have a constant velocity all along its periphery. The steam thus impinges on the moving blades with a periodically varying velocity and induces variable dynamic stresses at the blade root. These stresses are difficult to determine since the velocity perturbations of steam are unknown. Hence it is assumed that the velocity of the steam flow is constant all along the periphery of the nozzles exerting a constant static force on the blades.

With the above assumption the following constant forces acting on the moving blades may be determined:

a) Force in the direction of rotation

$$P_u = \frac{102 N_u}{\varepsilon u z} = \frac{102 \times 3{,}600 G_0 h_u}{860\, \varepsilon u z} = \frac{427 G_0 h_u}{\varepsilon u z}, \quad (13\text{-}10)$$

where N_u—the power developed at the rim of the disc, kW;
ε—degree of partial admission;
u—mean circumferential velocity, m/sec;
z—number of blades on the disc;
G_0—quantity of steam flowing through the stage, kg/sec;
$h_u = h_0 + h_e^{pr} - h_n - h_b - h_e$—the heat drop utilised in the stage.

b) Force exerted because of pressure difference between the two faces of the blade acting in the direction of the turbine axis

$$P_a = lt\,(p_1' - p_2), \quad (13\text{-}11)$$

where l—height of blades;
t—blade spacing at the mean dia.;
p_1' and p_2—steam pressure before and after the moving blades (for pure impulse turbines $P_a = 0$).

c) Force exerted in the direction of the turbine axis because of the change in momentum of the flowing steam

$$P_a' = \frac{G_0}{g \varepsilon z} (c_{1a} - c_{2a}), \quad (13\text{-}12)$$

where c_{1a} and c_{2a} —the axial components of the steam velocities at inlet and exit.

Fig. 13-7 shows the resultant direction of the force exerted by the steam on a moving blade.

The resultant force is given by the diagonal of the rectangle, i.e.:

$$P_0 = \sqrt{P_u^2 + (P_a + P_a')^2}. \quad (13\text{-}13)$$

The force exerted in the direction of the axis xx is

$$P = P_0 \cos \varphi. \quad (13\text{-}14)$$

Assuming the force in the xx direction as constant for all the blade length the bending moment about yy is determined as

$$M_x = \frac{Pl}{2}. \quad (13\text{-}15)$$

The bending stresses due to steam pressure is therefore

$$\sigma_b = \frac{M_x}{W_y} = \frac{Pl}{2W_y} = \frac{P_0 l}{2W_y} \cos \varphi, \quad (13\text{-}16)$$

where W_y—the smallest resistance moment of the blade relative to the yy axis, cm^3.

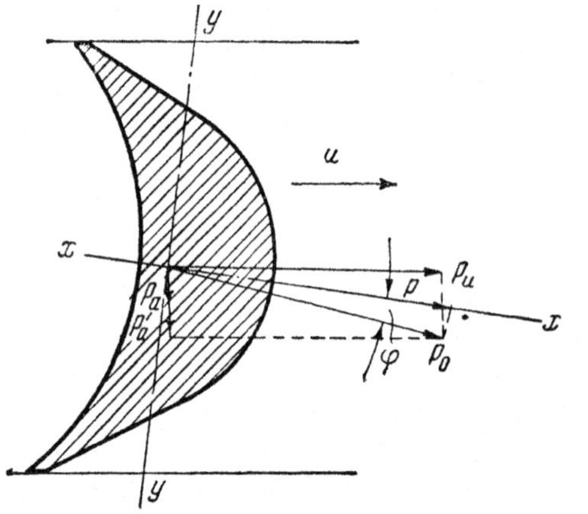

Fig. 13-7. Bending forces on a blade

In the case of impulse turbines with negligible reaction the forces P_a and P'_a are neglected and the blade axis yy is taken parallel to the turbine axis. Equation (13-16) then becomes

$$\sigma_b = \frac{P_u l}{2W_y}. \qquad (13\text{-}17)$$

The bending stress σ_b is assumed:

a) for turbines with partial admission as

$$\sigma_b \leqslant 190 \text{ kg/cm}^2;$$

b) for stages with full admission

$$\sigma_b \leqslant 380 \text{ kg/cm}^2.$$

The bending stresses for the low-pressure stages of condensing turbines are usually taken as 100 kg/cm² or even less because of the larger centrifugal forces present in them. To obtain some stress relief in the relatively long blades of the L.P. stages of large condensing turbines it is usual to set the line of centroids along the blade length on a radial slightly forward of the radial line passing through the blade c.g. so that some of the bending stresses induced by the centrifugal force are cancelled by the bending stresses caused by impulse.

The moment of resistance W_x may be determined graphically as given in 12-4.

The graphical determination of the moment of inertia and the moment of resistance for the blade

Table 13-1

No. of elementary strip	Area of elementary strips, cm²	Distance from axis x_0x_0 to the c. g. of the elementary strip, cm	Moment of area about x_0x_0 axis, cm³	Distance from main axis xx to the c. g. of elementary strip, cm	Square of the distance from main axis xx to the c. g. of the elem. strip, cm²	Moment of inertia of the elem. strip about the main axis xx, cm⁴
1	0.250	0.25	0.0625	6.773	45.874	11.468
2	0.600	0.75	0.4500	6.273	39.391	23.635
3	0.725	1.25	0.9063	5.773	33.328	24.163
4	0.800	1.75	1.4000	5.273	27.805	22.244
5	0.925	2.25	2.0813	4.773	22.782	21.073
6	1.125	2.75	3.0938	4.273	18.259	20.541
7	1.375	3.25	4.4688	3.773	14.236	19.574
8	1.698	3.75	6.3675	3.273	10.713	18.191
9	2.175	4.25	9.2438	2.773	7.690	16.726
10	2.825	4.75	13.4188	2.273	5.167	14.596
11	3.945	5.25	20.7113	1.773	3.144	12.031
12	5.470	5.75	38.4525	1.273	1.621	8.867
13	6.115	6.25	39.2188	0.773	0.598	3.657
14	5.870	6.75	39.6225	0.273	0.075	0.440
15	5.820	7.25	42.1950	0.227	0.052	0.303
16	5.340	7.75	41.3850	0.727	0.529	2.825
17	5.070	8.25	41.8275	1.227	1.506	7.635
18	4.800	8.75	42.0000	1.727	2.983	14.318
19	4.450	9.25	41.1625	2.227	4.960	22.072
20	4.025	9.75	39.2438	2.727	7.437	29.934
21	3.475	10.25	35.6188	3.227	10.414	36.189
22	2.700	10.75	29.0250	3.727	13.891	37.506
23	1.125	11.25	12.6563	4.227	17.867	20.100
	$M^2\Delta F$	Me	$M^3\Delta Fe$	$M(e_0 - e)$	$M^2(e_0 - e)^2$	$M^4\Delta F(e_0 - e)^2$

$M^2\Sigma\Delta F = 70.703$ cm²; $M^3\Sigma\Delta Fe = 496.6118$ cm³;

$M^4 I_x = M^4\Sigma\Delta F(e_0 - e)^2 = 388.088$ cm⁴;

$e_0 = M^3\Sigma\Delta Fe/M^2\Sigma\Delta F = 496.6118/6 \times 70.703 = 1.17$ cm;

$W_x = M^4 I_x/M^4 e_0 = 388.1/64 \times 1.17 = 0.2555$ cm³.

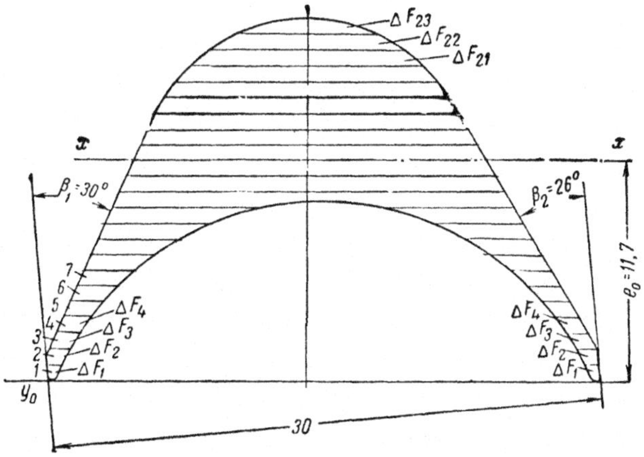

Fig. 13-8. Moving blade profile for determining the moment of resistance about the *xx* axis

profile shown in Fig. 13-8 is as follows (the blade is drawn to a scale of 6 : 1).

The blade section is subdivided into 23 elementary parts. The results of these calculations are given in Table 13-1. The first column gives the No. of each section and the following columns give the necessary quantities for obtaining the moment of inertia and the bending stress of each of these elementary sections about the *xx* axis from which I_x and W_x are determined.

13-3. VIBRATION OF BLADES

If a blade or an assembly of blades is acted upon by an instantaneous force free vibrations are set up because of the inherent resilience of the blade or blade assembly. The frequency of these vibrations depends on the dimensions of the blade or blade assembly as well as the method of their mounting on the disc. The amplitude of these vibrations continuously goes on decreasing because of the resistance of the surrounding medium and the internal friction of the material concerned, in other words, these are damped vibrations.

Curves *I* in Fig. 13-9, *a*, *b* and *c* show cases of free damped vibrations. When a blade or an assembly of blades is acted upon by external forces of periodic nature forced vibrations are set up. Especially important is the case when the frequency of the external force is equal or less than the natural frequency of the blade or blade assembly in which case resonance is set up. While in resonance the amplitude of the vibrations goes on increasing from the minimum to a certain maximum value.

Curves *II* in Fig. 13-9,*a*, *b* and *c* show forced vibrations for various harmonics. A comparison of curves *I* and *II* shows that for a given blade (or a blade assembly) the frequency of the forced

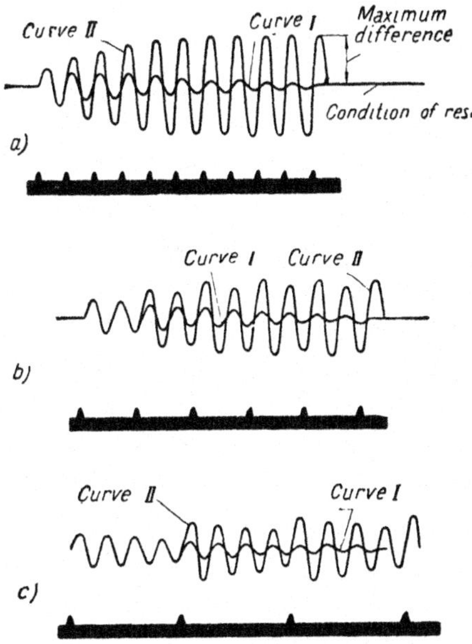

Fig. 13-9. Natural and forced vibrations of blades

and free vibrations is the same (*a*) or less (*b*) and (*c*). Thus the superposition of a periodic external force with resonating frequency brings about only an increase in the amplitude of the vibrations without affecting their frequency.

Under normal operating conditions only forced vibrations are possible in steam turbines. The frequency of these vibrations may or may not be in resonance with the frequency of the periodic external force. The steam flowing out from the nozzles causes the blades to vibrate in various forms such as:

a) vibrations in the plane of the disc—tangential vibrations;

b) vibrations in a plane perpendicular to the direction of rotation—axial vibrations; and

c) twisting around the blade axis itself or the axis of a blade assembly—torsional vibrations.

1. Tangential Vibrations of Blades

The vibrations of blades mounted on the disc as separate elements differ from those of the blades mounted on the disc in the form of segments.

Blades mounted directly onto the disc may have the following types of vibrations:

a) intense vibrations of the blade tip without any nodes (Fig. 13-10,*a*) known as the first tone vibrations with 1/4 wave;

b) vibrations with a single node (Fig. 13-10,*b*) known as the second tone with 3/4 wave;

c) vibrations with twin nodes (Fig. 13-10,*c*) known as the third tone with 5/4 wave, etc.

The blades in an assembly may undergo the following types of vibrations:

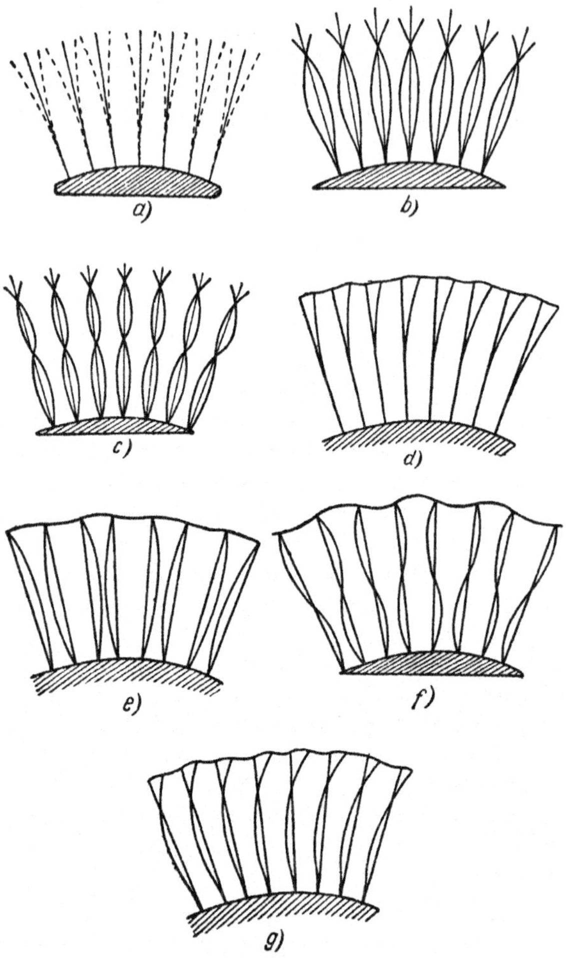

Fig. 13-10. Various forms of vibrations

a) intensive vibrations of the blade tips in phase with each other known also as the first tone with 1/4 wave (Fig. 13-10,*d*).

b) vibrations of blades with anchored tips without nodes but with phase differences (Fig. 13-10,*e*) also of the first tone with 2/4 wave. Such vibrations are possible only if the blades are not secured by additional fasteners such as binding wire, etc.

These types of vibrations are known as internal or auto-vibrations. Auto-vibrations can take place with a single node at the centre (Fig. 13-10,*f*). Such vibrations are known as the second tone but with a complete wave. It is the characteristic property of auto-vibrations to vibrate in contraphase.

c) the blades may vibrate in phase with a single node as shown in Fig. 13-10,*g*; these are of the second tone with 3/4 wave.

In addition there are various other types of secondary vibrations. These have a very high frequency and in general are very difficult to determine and hence are not considered here. Both from theory and practical experience it is found that the most dangerous vibrations are those with the first tones having 1/4 and 2/4 wave (Fig. 13-10*a*, *d* and *e*).

2. Axial Vibrations

In axial vibrations the blades vibrate in a plane perpendicular to the plane of the disc (Fig. 13-11). In actual turbine practice these vibrations are very rarely to be met with. These are directly connected with the disc vibrations. Tuning of the blades out of the dangerous region depends on the tuning of the discs themselves.

3. Torsional Vibrations of Blades

In this form of vibrations the blades perform an angular twist of a certain magnitude to and fro repeating the operation a certain number of times in a given time interval (Fig. 13-12). The blade tips, in such types of vibrations, oscillate at a greater rate at the ends of the blade assembly than at the centre. These vibrations are usually found in blades of varying section especially of the twisted type which are mainly used in the low-pressure stages of a turbine.

13-4. CAUSES OF BLADE VIBRATIONS

The vibrations of turbine blades are caused by the irregularities of steam flowing out from nozzles or guide blades, i. e., as a result of the periodic nature of the steam flow. The perturbation forces are classified into two categories to facilitate tuning of the blades out of the dangerous regions of vibrations:

a) forces having a frequency equal to nz (where n—revolutions per second of the rotor and z—number of nozzles or guide blades along the periphery of the disc);

b) forces having frequencies equal to the rotor r.p.s. or its multiple, i. e., equal to kn where k is an integer 1, 2, 3, etc.

In the first case the perturbations are caused by the partitions which constitute the guide passages. When the steam flows through these canals the frictional forces appearing at these wall surfaces cause the flow to slow down and hence the velocity of steam flow is not uniform all across the nozzle or guide blade section, but varies from a certain minimum at the walls to the maximum value at the centre of the blade passage. The moving blades come into contact with these non-uniform periodic forces and begin to vibrate. If now the natural frequency of these blades coin-

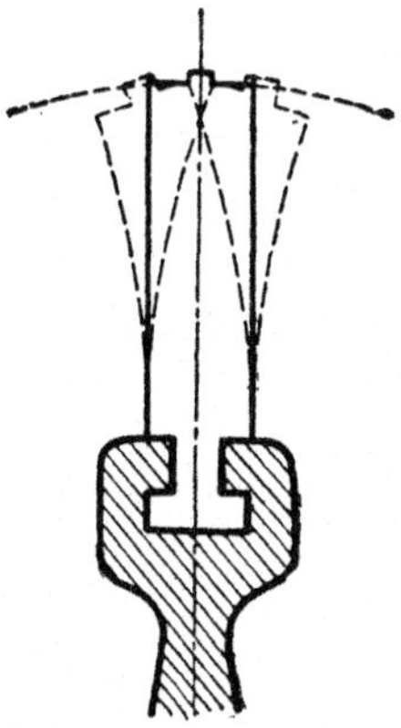

Fig. 13-11. Axial vibrations

cides with the frequency of the perturbation forces, resonance is set up. The conditions at which resonance may be set up is expressed by the following relation

$$f = nz, \qquad (13\text{-}18)$$

where f—the natural frequency of vibration of a blade, c.p.s.

In the second case the perturbation forces are caused by the following:

1) disturbance of flow at the diaphragm joints;

2) local disturbances caused by the differences in the guide passage dimensions (resulting from inaccuracies in manufacture);

3) rotor vibrations caused by improper balancing. If the frequency of natural vibrations of the blades coincides with that of the perturbation forces or its harmonics resonance is set up. The conditions under which a resonating vibration is set up is expressed by the relation

$$f_s = kn, \qquad (13\text{-}19)$$

where f_s—the natural frequency of the blade segment, c.p.s.

Both these forms of resonating vibrations are dangerous from the point of view of safe turbine operation. The stresses induced in a blade while in resonance depend upon the magnitude of the perturbing force or its harmonics as well as the duration of its action. If, say, the natural frequency of the blades coincides with the frequency of the perturbing force $f_s/n=1$ (Fig. 13-9,*a*, the first harmonic $k=1$) then the resonating impulse occurs at the beginning of the second cycle. Here the blades having some initial amplitude are subjected to an additional impulse and deflect from their central position to a marked extent. It is easily seen that with each succeeding resonating impulse this deflection would be on the increase with a consequent increase in the stresses induced since the latter are directly proportional to the amplitude of vibrations. However, the increase in amplitude continues only to a certain limit, since the surrounding medium, the internal friction of the material of the blade, the shrouding and binding wire, etc., and the fixture of the blade at its root all combine to counteract the tendency of the amplitude to increase indefinitely. The maximum resonance of the blades when in resonance occurs when the energy of the impulse of steam flow is completely absorbed by the internal friction of the blades and their fastenings, friction at the blade root, and the resistance of the surrounding medium. Resonating impulses, even of small magnitudes, can cause deflections inducing stresses which may easily overstep the elastic limits of the material concerned resulting in damaged blades.

Curve *II* in Fig. 13-9,*a*, shows the vibrations under the influence of a resonating impulse reaching a certain maximum value.

The vibration of the same blades when acted upon by the perturbation forces of a different frequency would be altogether different.

If, for example, the frequency of the perturbing force is half the natural frequency of the blades $f_s/n=2$ (the second harmonic $k=2$) then the resonating impulse begins to act at the beginning of the third period of vibrations. In this case the blades are acted upon by consecutive resonating impulses at smaller deflections than in the first case. The deflection of the blades goes on increasing with each succeeding impulse until a stable amplitude is reached. However, the maximum amplitude of vibrations remains smaller than in the first case. In Fig. 13-9,*b* curve *II* shows the force vibrations under the action of the second harmonic resonating impulse.

In the case of the third harmonic ($k=3$) the expression $f_s/n=3$ and each succeeding resonating impulse is transmitted to the blades at the beginning of the fourth cycle of vibrations. Fig. 13-9,*c* curve *II* shows the forced vibrations for a fourth order harmonic ($k=4$). It is easily seen that with the increase in the order of the harmonic ($k=4.5$, etc.) each of the resonating impulses is impressed on the blades at the beginning of the fourth, fifth, etc., cycle of natural vibrations. In view of the damping capabilities of the blades the maximum deflection of the blades decreases with the increase in the order of the harmonics. This, however, is true only if the periodic time of the perturbing force changes without a change in its magnitude.

Comparing these above-mentioned cases of vibrations we may conclude that in resonance the maximum deflection of the blades goes on decreasing with increase in the order of the harmonics.

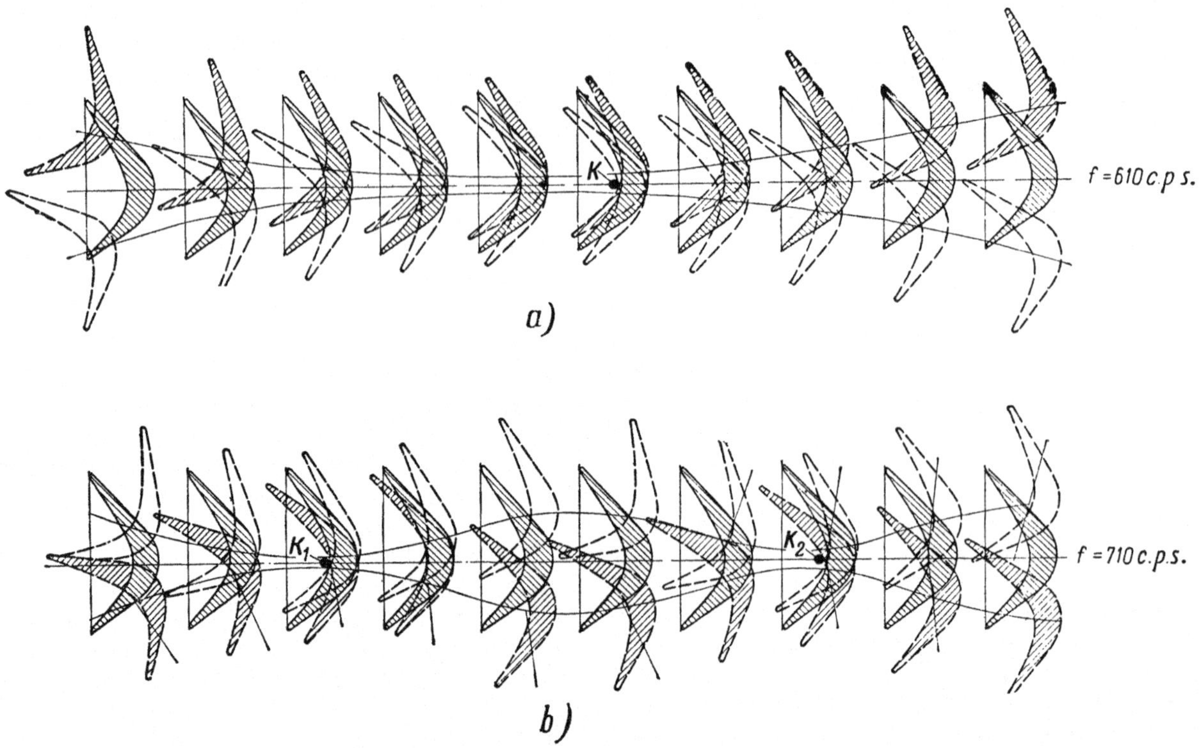

Fig. 13-12. Torsional oscillations

a — torsional oscillations of blade assembly with single node *K*; *b* — torsional oscillations of blade assembly with two nodes K_1 and K_2

Consequently the higher the order of harmonic, the smaller are the stresses induced in the blades due to vibrations.

All the types of vibrations considered above are characteristic of the low-pressure stages of a condensing turbine. At the same time segments of blade assemblies may vibrate under the influence of perturbing forces having a frequency smaller than the revolutions per second of the rotor. These vibrations may very well be in the first tone with 1/4 wave (Fig. 13-10,*d*). With harmonics of the lower order, resonance or the region close to it spells danger for the operation of the blades since the amplitude and consequently the stresses in the blades are liable to a rapid growth which may result in the blades being damaged permanently.

13-5. EXPERIMENTAL METHODS OF TUNING THE BLADES OUT OF THE REGIONS OF DANGEROUS VIBRATIONS

The natural frequencies of vibrations of blades lie within very wide limits starting from 40 to 5,000-10,000 cycles and even higher. Depending upon the frequency of vibrations the blades are subdivided into two groups: low-frequency blades and high-frequency blades. Blades with a frequency of up to 400 cycles per second are grouped under the first category and those with frequencies more than 400 are grouped under high-frequency blades. Such a grouping of the blades according to their frequencies is purely conventional, the division being made on the basis of the marked difference between the methods of tuning of low- and high-frequency blades.

Various means are used while designing the blades to avoid dangerous blade vibrations. Usually the blades are assembled in separate units and are held together by the shrouding strip and binding wire, etc. The following types of vibrations must be avoided for blades assembled to form segments:

a) in phase vibrations of blades with the first tone of 1/4 wave (Fig. 13-10,*d*);

b) contraphase vibrations of blades with the first tone and 2/4 wave (Fig. 13-10,*e*) as well as vibrations of the type shown in Fig. 13-10,*g*;

c) axial vibrations of blades as shown in Fig. 13-11.

1. Tuning of Blades Against Tangential Vibrations of the First Tone with 1/4 Wave

Only low-frequency blades of the low-pressure stages of large-capacity condensing turbines are tuned for these vibrations.

In the Soviet Union it is the widespread practice to tune blade assemblies which have a vibration frequency lying within the limits of

$$f_d \leqslant 7n, \tag{13-20}$$

where f_d—the dynamic frequency of natural vibrations of the blade assembly.

The relation between the dynamic frequency and the static frequency is given by the expression

$$f_d = \sqrt{f_{st}^2 + Bn^2}, \tag{13-21}$$

where f_{st}—static frequency of natural vibrations of the blade assembly;

B—proportionality coefficient accounting for the influence of rotation;

n—revolutions per second of the rotor.

For blades of constant section the value of coefficient B may be determined from the equation

$$B = 0.8\frac{D_{mean}}{l} - 0.85, \tag{13-22}$$

where D_{mean} —mean diameter of the disc at the mean height of the blades;

l—length of the blade.

While elaborating the initial design of blades the effects of vibration are taken into consideration. Initially a few blades are made and assembled to form a segment which is then experimentally tested to obtain the value of f_{st}.

The blade assembly is tuned by either varying the blade profile or altering the method of blade fixture in the segment. In actual practice the various manufacturing concerns very rarely make use of completely new blade profiles as it involves the preparation of new milling cutters, attachments and checking devices. In most of the cases tuning of the blades can be satisfactorily carried out by a slight alteration of the diameter or the disposition of the binding wire. The blade segment so tuned becomes the model on the basis of which all other remaining blades of the disc are manufactured.

Usually the frequencies of different blade segments mounted on a disc differ from each other which may be due to various reasons such as: inaccuracies in the manufacture of blades, differences in the jointings at the shroud and the root of the blades, differences in the rigidity of blade assembly both by the shroud strip and binding wire, quality of soldering, etc.

The difference between the maximum static frequency $f_{st\ max}$ and the minimum $f_{st\ min}$ is known as the frequency spread. The percentage frequency spread may be expressed by the equation

$$\Delta f = \frac{f_{st\ max} - f_{st\ min}}{f_{st\ min}} \times 100\%. \tag{13-23}$$

The quantity Δf indicates the quality of blading on a disc and should not be higher than 8%.

The dynamic frequency of vibrations of a blade assembly from equations (13-19) and (13-21) can be expressed in terms of the resonance r.p.s. of the rotor by the relation

$$f_d = kn_{cr} = \sqrt{f_{st}^2 + Bn_{cr}^2}, \tag{13-24}$$

where n_{cr}—critical speed of the rotor when the blades are in resonance.

Solving this equation for n_{cr} we have

$$n_{cr} = \frac{f_d}{k} = \frac{f_{st}}{\sqrt{k^2 - B}}. \tag{13-25}$$

This equation contains two unknowns: n_{cr} and k(k—always an integer 1, 2, 3, 4, etc.) which are interdependent, and can be solved by a proper selection of the value of k. Two different values should be selected for k so that at one value, say k_1, the resonating r.p.s. would be higher than the normal operating speed and with the second value k_2 the resonating r.p.s. would be lower. Since the vibrations of blade segments on a disc differ from each other two frequencies $f_{st\ max}$ and $f_{st\ min}$ are chosen to determine the vibration characteristics of the blades. Thus from equation (13-25) we may write

$$\left.\begin{aligned} n_{cr}^{I} &= \frac{f_{st\ min}}{\sqrt{k_1^2 - B}}; \quad n_{cr}^{III} = \frac{f_{st\ min}}{\sqrt{k_2^2 - B}}; \\ n_{cr}^{II} &= \frac{f_{st\ max}}{\sqrt{k_1^2 - B}}; \quad n_{cr}^{IV} = \frac{f_{st\ max}}{\sqrt{k_2^2 - B}}. \end{aligned}\right\} \tag{13-26}$$

Thus there are four resonating speeds for the blades of each of the disc, two speeds higher than normal and two lower than normal. In the event of a single value for k we may have only two resonance speeds one higher and one lower than the normal speed of rotation, i. e., $n'_{cr} < n_0 < n''_{cr}$. This conforms to the resonance of separate blade assemblies having vibration frequencies of f_{stx} lying between

$$f_{st\ min} < f_{stx} < f_{st\ max}.$$

The margin in r.p.s. against resonance is expressed by the equation

$$\Delta n = \frac{n_0 - n_{cr}}{n_0} \times 100\%, \tag{13-27}$$

where n_0—normal speed of operation, r.p.s.

The value of n is obtained for all the n_{cr} calculated from equation (13-26) with the help of the above relation.

The vibration characteristics are now valued on the basis of the minimum margin against resonance on either side of the normal operating speed. The current standards of valuation of vibration characteristics of 1/4 wave of the first tone is given in Table 13-2.

Table 13-2

Harmonics	Vibration characteristics of blades	
	Satisfactory if the margin against resonance is not less than, %	Unsatisfactory if the margin against resonance is less than, %
2nd	15	15
3rd	8	8
4th	6	6
5th	5	5
6th	4	4

At $k=7$ the blades may be operated in resonance.

2. Tuning of Blades Against Tangential Vibrations of the First Tone with 2/4 Wave

It is found both from theory and practical experience that the absence of strengthening wires or some such arrangement results in dangerous vibrations of the first tone with 2/4 wave. These vibrations are caused by the action of perturbation forces having a frequency of nz. The equation $f_x=nz$ for the case under consideration becomes the resonance condition for the internal vibrations in a blade assembly (f_x—natural frequency of vibrations of the blades; first tone with 2/4 wave). These vibrations make their appearance when the perturbation forces having a frequency of nz are unable to induce vibrations in the blade assembly as a whole and in phase. The number of nozzles or guide blade passages along the periphery of a diaphragm are always less than the number of moving blades on a disc. The impulses of the steam flowing out from the nozzles asynchronously act on the moving blades causing them to vibrate with phase differences. These vibrations are found to be dangerous and must be avoided by tuning the blades.

The frequency of vibrations in the first tone with 2/4 wave depends upon the rigidity of blade fixture and varies between very wide limits. Theoretically the frequency of vibrations f_x in the first tone with 2/4 wave depends upon the value of the static frequency of vibration of each individual blade and is given by the relation

$$4f<f_x<8f, \qquad (13\text{-}28)$$

where $f_x=4f$ conforms to the case when the blades are fixed in the disc and the shroud by the slotted hole arrangement and $f_x=8f$—when the blade fixture is completely rigid both at the root and the shroud. The static frequency of each of the individual blades, except at the two fixture ends, may be obtained experimentally or according to the equation

$$f=\frac{0.56\psi}{l^2}\sqrt{\frac{EIg}{\gamma F}}, \qquad (13\text{-}29)$$

where

ψ—coefficient taking into consideration the rigidity of blade fixture at the root and tip;

l—the length of the blade, cm;

E—modulus of elasticity, kg/cm^2;

I—lowest moment of inertia of the blade section, cm^4;

g—acceleration due to gravity, cm/sec^2;

γ—specific gravity of the material of blades, kg/cm^3;

F—the area of cross-section of the blade, cm^2.

The equation (13-29) is valid only for blades of constant section. From equations (13-18) and (13-28) we can determine the zone of dangerous vibrations occurring inside a blade assembly

$$4f\leqslant nz\leqslant 8f, \qquad (13\text{-}30)$$

from which we may further write

$$4\leqslant\frac{nz}{f}\leqslant 8. \qquad (13\text{-}31)$$

From the above it follows that if the blades are secured together only by the shroud strip then the relation nz/f should be either less than 4 or more than 8. If the ratio nz/f lies within the limits of 4 and 8 the blade assembly must be further secured by a binding wire or else the number of guide blades z must be altered. Sometimes the blade profile is altered in such a way as to obtain nz/f either less than 4 or more than 8.

For blades of low-frequency natural vibrations the types of vibrations considered in paragraphs 1 and 2 are eliminated simultaneously. For the high-frequency blades, however, the methods outlined above form the basis which has to be further elaborated to obtain satisfactory tuning. For individual blades, i.e., blades which do not form an integral part of a blade assembly, it is found that resonance would occur when $f=nz$ (where f—the natural frequency of vibrations in the first tone with 1/4 wave of each individual blade). This would be a dangerous zone of operation for the blades. Their profiles, therefore, must be altered or alternatively the number of guide blades must be increased or decreased so that the value of f would be either less than 0.75 nz or more than 1.25 nz.

13-6. DESIGN AND CONSTRUCTION OF ROTORS

Various methods are employed for the construction of rotors for the reaction and impulse-reaction turbines. If the rotor diameters are small then they are usually made as a solid forging in one piece. The regulating stage disc which is usually of a larger diameter is either machined directly on the solid forging or made separately and shrunk on the rotor shaft.

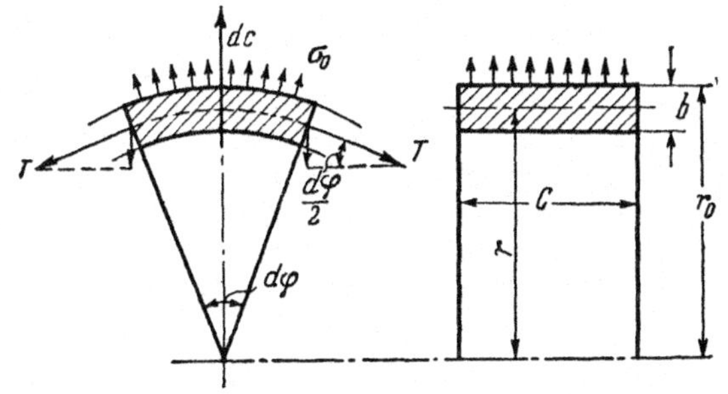

Fig. 13-13. Stresses in a hollow cylinder element

Large diameter rotors are made by welding a number of discs (as made by Brown Bover Co.) at the periphery. The wheels may be either continuous discs without central holes or with central alignment holes. The transverse rigidity of such rotors is found to be exceedingly high. The complete rotor is thus made up of the regulating stage, the welded discs and the balance piston which is also usually welded onto the rotor shaft. The stress calculation of such rotors is therefore directly dependent on the calculations of the discs. The theory of the freely rotating ring forms the basis on which hollow turbine rotors are designed.

Let us consider an elementary length c of the rotor subtending an elementary angle $d\varphi$ at the rotor axis (Fig. 13-13). The centrifugal force of the element under consideration while the rotor is revolving will be

$$dc = dmr\omega^2 = \left(\frac{\gamma}{g} r d\varphi bc\right) r\omega^2, \qquad (13\text{-}32)$$

where $dm = \frac{\gamma}{g} r d\varphi bc$ — mass of the element;

γ —specific gravity of the material, kg/ cm^3;

ω —angular velocity of the rotor, radians per sec.

The centrifugal force induces a stress σ_u in the section bc. The force resulting from these stresses would be $T = bc\sigma_u$. Since the element is in equilibrium we have

$$dc = 2T \sin\frac{d\varphi}{2} \approx T d\varphi, \qquad (13\text{-}32a)$$

where $\sin\frac{d\varphi}{2}$ is assumed to be equal to $d\varphi/2$ as the angle considered is very small.

Substituting the values of T and dc in the above equation we have

$$\frac{\gamma}{g} d\varphi bcr^2\omega^2 = bc\sigma_u d\varphi$$

and

$$\sigma_u = \frac{\gamma}{g} r^2\omega^2 = \frac{\gamma}{g} u^2, \qquad (13\text{-}33)$$

where u—circumferential velocity of the rotor.

From equation (13-33) it therefore follows that the hoop stresses are only dependent on the circumferential velocities.

For steels with $\gamma = 0.00785$ kg/cm^3 at u m/sec we have

$$\sigma_u \approx 0.08u^2 \ [\text{kg/cm}^2],$$

from which we may easily calculate the values of σ_u for any given u as is given below:

$u =$ 25	50	75	100	150	200	300 m/sec
$\sigma_u =$ 50	200	450	800	1,250	3,200	7,200 kg/cm^2 .

From what has been shown above we find that even in the absence of external load the stresses induced in hollow rotors are such that only relatively low circumferential speeds can be allowed. For example, thin-shelled carbon steel rotors may have a circumferential speed not higher than 120 m/sec, and nickel-steel rotors not higher than 140 m/sec.

While calculating the stresses in rotors the additional loads imposed by the centrifugal forces of the blades, shrouding and their fastenings must be taken into consideration.

Assuming the additional centrifugal force acting on 1 cm^2 of the element of area to be σ_0 we have the total force acting on this area at the radius of r_0 to be

$$dc_0 = r_0 d\varphi c\sigma_0. \qquad (13\text{-}34)$$

This force induces an additional stress σ_s in the rotor giving the resultant hoop tensions

$$T_0 = cb\sigma_s.$$

Since the element is in equilibrium we have

$$dc_0 = 2T_0 \sin\frac{d\varphi}{2} \approx T_0 d\varphi. \qquad (13\text{-}34a)$$

Substituting the values of dc_0 and T_0 in the above expression we obtain

$$r_0\sigma_0 = b\sigma_s \qquad (13\text{-}35)$$

and

$$\sigma_s = \frac{r_0}{b}\sigma_0.$$

The total of the hoop stresses induced in the rotor will be

$$\sigma=\sigma_u+\sigma_s=\frac{\gamma}{g}u^2+\frac{r_0}{b}\sigma_0. \qquad (13\text{-}36)$$

The stress σ_0 is usually assumed to be uniform along the length of the rotor under consideration and hence its magnitude is determined by equally distributing the centrifugal forces of all the blades on the area $2\pi r_0 c$.

The allowable stresses in rotor drums are usually taken up to 0.4 of the yield point at the given temperature. The yield point of the various steels used in rotor construction depends upon their composition and temperature.

13-7. ROTORS OF IMPULSE TURBINES

The impulse turbine rotors consist of a number of discs mounted on the shaft. Rotors with discs shrunk on the shaft are used only for working temperatures of 400°C and less. Rotors operating under conditions of high temperatures and having relatively small diameters are made of a single-piece solid forging with the discs directly machined on it.

Combined types of rotors are also made in which the first few high-pressure stages are made in one piece with the shaft with the low-pressure stage discs shrunk on it. Examples of such construction are rotors of Kh.T.W. turbines (Fig. 11-6) and rotors of high-pressure turbines made by the L.M.W. (Figs 1-8, 10-16, 11-12, and 11-17).

The dimensions of a turbine with forged steel rotors are comparatively smaller than those of a turbine having a rotor with discs shrunk on the shaft. In the case of separately mounted discs there is always a possibility of their getting loose on the shaft with the consequent stoppage of the set for long periods of maintenance. The disadvantages of a solid forged rotor are that even a single defective disc would necessitate the rejection of the complete rotor, whereas for the rotors with the mounted discs the defective disc may be replaced without the whole rotor being scrapped.

Fig. 13-14 shows the usual methods of fastening the discs on the shaft. In Fig. 13-14,*a* the disc is directly mounted on the shaft. The hole diameter of the disc is made slightly smaller than the shaft diameter and the disc is forced onto the shaft while hot to shrink after cooling. The discs are heated in a boiling water bath or an oil bath for an hour or so after which they are pressed onto the cold shaft with the help of a hydraulic press. The discs are keyed onto the shaft with the help of two longitudinal keys placed at 180° to each other which ensure correct torque transmission. This method of disc mounting is widely used by the L.M.W. and other factories of the U.S.S.R.

Fig. 13-14,*b* shows the method of mounting the discs with the help of special rings. Thin rings *a* are sprung into recesses in the shaft. The rings pass over keys fitted in the grooves provided on the shaft. The discs are heated in boiling water for a period of 2 to 3 hours and are pressed onto the two mounting rings *a* and *b*. A clearance of about 0.1 to 0.2 mm is maintained between two adjacent discs because of the ring *b* projecting out from under the disc. This clearance is necessary for the axial expansion of the discs under operating temperatures. The discs are bored to a diameter slightly larger than the shaft diameter to ensure seating on the rings only, the clearance thus obtained being from 0 3 to 0.5 mm.

Fig. 13-14,*c* shows the modification of this principle, where the discs are mounted on U-shaped chrome-steel rings instead of the ordinary ones. These U-shaped springs press on the shaft from one side and from the other against the recess in the disc. The rigidity of fixture is not affected by the expansion of the mounting ring at the high operating temperatures.

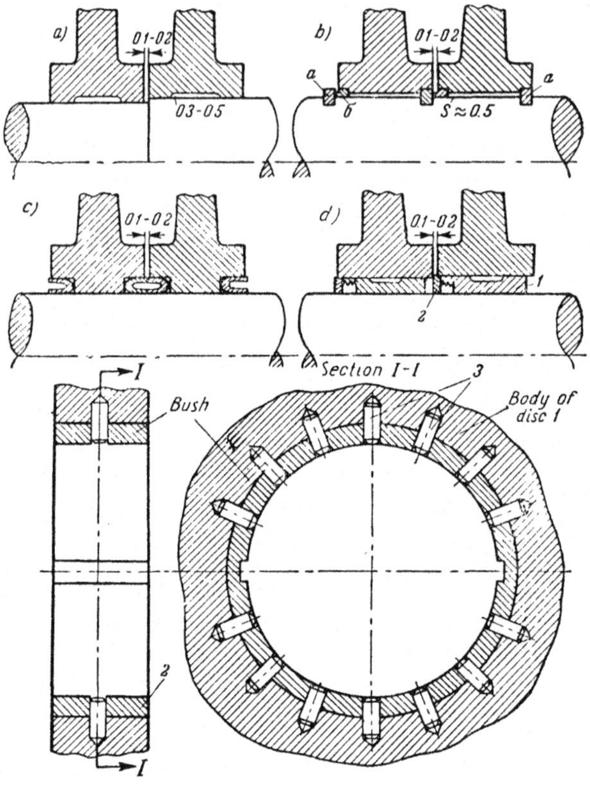

Fig. 13-14. Methods of mounting discs on turbine shaft

Fig. 13-14,*d* shows another method of disc mounting where split conical bushes *1* are pressed hydraulically into the conically bored hub of the discs. The space between the shaft and the conical bushes is packed with hard grease to facilitate easy removal of the disc when necessary. The bushes are slightly shorter than the disc hub in length so that distance rings *2* can be accommodated between two adjacent discs providing a clearance of about 0.1 mm between them.

The removal of the discs from the shaft is carried out as follows. The conical bushes are provided with screw threads at the larger end where the withdrawing gear is attached. By tightening the withdrawing attachment bolts the bushes *1* slide out from the disc hub. This method of disc mounting has been used both by the L.M.W. and the K.T.W. for the construction of turbine AK-3,5.

However, this method of mounting has been found very complicated, the method shown in Fig. 13-14,*a* being widely used at the present time in this country.

Fig. 13-14,*e* shows disc mounting with the help of radial pins. A steel bush *1* is pressed into the disc hub *2*. A number of holes *3* are drilled and radial pins are inserted through them. The bushes are then machined to the required diameter. The required force fit is obtained by mounting the disc when hot. The torque transmission is taken care of by two longitudinal keys. When the discs are in rotation there may be a slight slackening of the disc mounting on the radial pins due to the centrifugal forces present. However, there is a sufficient holding force since the centrifugal forces of the bush are quite small. The radial pins permit only radial expansion preventing any relative displacement between the bush and the disc. This method of disc mounting is used by the Kh.T.W., General Electric Co. and other factories, and has been found quite dependable for rotors operating in high-temperature regions.

Fig. 10-12 shows the method of mounting low-pressure discs on the shaft as used by the L.M.W. for turbines VK-50, VK-100, PVK-200, SKK-300, etc.

In every case of disc mounting a certain amount of initial tension must be provided. This initial tension is theoretically calculated on the assumption of a tight fit between the disc and the shaft at the normal operating speed. The radial displacement of points on the inner surface of the disc hub is obtained from the theory of elastic deformation as

$$\xi = \frac{r_0}{E}(\sigma_{t0} - \nu\sigma_{r0}), \qquad (13\text{-}37)$$

where r_0—shaft dia., mm;

E—modulus of elasticity, kg/cm²;

σ_{t0} and σ_{r0}—tangential and radial stresses on the hub surface, kg/cm²;

ν—coefficient of transverse pressure.

If we assume that at the normal operating speed the initial tension is zero then $\sigma_0 = 0$ and equation (13-37) becomes

$$\xi = \frac{r_0}{E}\sigma_{t0}. \qquad (13\text{-}37a)$$

The stress σ_{t0} is obtained from disc design calculations and r_0 from the calculations of shaft design. Substituting these values in equation (13-37,*a*) we obtain the relative deformation at the radius r_0. The calculated value of ξ is increased 10 to 15% more to obtain some amount of force fit at the operating speed of n r.p.m. It is found from theoretical and practical experience that if 2ξ is assumed to be equal to (0.001 to 0.0015) d_0 (where d_0—shaft diameter) the necessary force fit of the disc on the shaft at the operating speed of n r.p.m. is ensured. In practice the various factories in this country assume a speed n_0 (at which the initial stresses become zero) about 15 to 30% higher than the normal. The speed n_0 must in every case be higher than the speed at which the overspeed trip comes into operation.

Example 13-2. Calculate the initial tension for a disc mounted on a shaft of diameter equal to 300 mm, the tangential stress σ_{t0} at the normal operating speed being 1,600 kg/cm².

From equation (13-37, *a*) the radial deformation is obtained as

$$\xi = \frac{300}{2} \times \frac{1{,}600}{2 \times 100{,}000} = 0.114 \text{ mm}.$$

Adding 10% extra and rounding off the figure we have $\xi = 0.125$ mm on the radius and 0.25 mm on the diameter. Consequently the disc hole diameter must be smaller than the shaft diameter by 0.25 mm, i.e., the hole diameter should be 299.75 mm.

13-8. CONSTRUCTION OF DISCS

When the disc of a turbine is in rotation stresses are induced due to the centrifugal force of the disc and the blades mounted on them. Besides, with load variations and temperature fluctuations the discs in addition experience temperature stresses. The rate of growth of these temperature stresses directly depends upon the rate at which the load or temperature fluctuations occur.

The discs are made in various forms, with constant section along the radius or with a varying thickness thinning down at the periphery.

Some of the theoretical methods of disc design are given below. All these formulas are for discs symmetrical about their vertical axis. The radial stresses in the axial section are assumed to be uniform, and temperature stresses are neglected. Let us consider a small elementary portion of the disc under equilibrium conditions. Fig. 13-15 shows the elementary portion with the directions in which the forces act. For the element to be in equilibrium

$$dR' - dR + dC - 2dT \sin \frac{d\varphi}{2} = 0. \quad (13\text{-}38)$$

The various forces appearing in equation (13-38) may be expressed as follows:

1) force acting on the outer surface

$$dR' = (\sigma_r + d\sigma_r)(y + dy)(r + dr)\, d\varphi; \quad (13\text{-}38a)$$

2) centrifugal force of the element

$$dC = dmr\omega^2 = \frac{\gamma}{g} yr^2 d\varphi dr \omega^2, \quad (13\text{-}38b)$$

where γ—specific gravity of the material of the disc;

g—acceleration due to gravity;

3) force acting on the inner surface of the element

$$dR = \sigma_r yr d\varphi; \quad (13\text{-}38c)$$

4) forces acting on the two side faces of the element

$$dT = \sigma_t y dr. \quad (13\text{-}38d)$$

From equations (13-38a) and (13-38c) we obtain

$$dR' - dR = (\sigma_r r dy + yr d\sigma_r + \sigma_r y dr)\, d\varphi = d(\sigma_r ry)\, d\varphi. \quad (13\text{-}38e)$$

Substituting the values obtained from equations (13-38,*a*, *b*, *c*, *d*, and *e*) in equation (13-38) and cancelling $d\varphi$ from both sides we have

$$d(\sigma_r ry) + \frac{\gamma}{g} yr^2\omega^2 dr - \sigma_t y dr = 0. \quad (13\text{-}39)$$

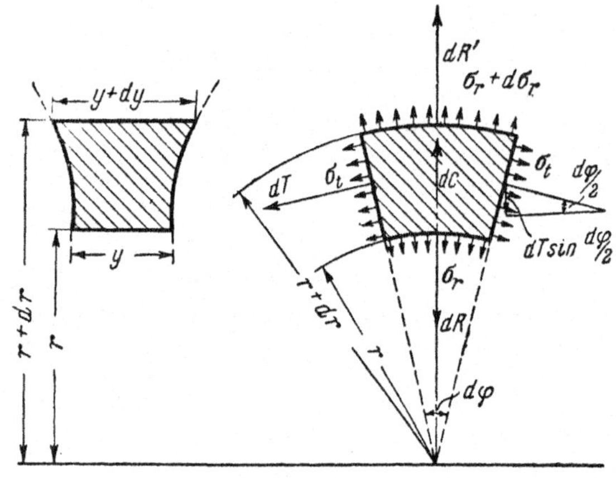

Fig. 13-15. Forces on a disc element

This equation may be further expressed as

$$\frac{d}{dr}(\sigma_r ry) - \sigma_t y + \frac{\gamma}{g} yr^2\omega^2 = 0 \quad (13\text{-}39a)$$

which is the basic differential equation for disc design.

1. Disc of Constant Strength

In a disc of constant strength the stresses acting at any point are the same, i.e.,

$$\sigma = \sigma_r = \sigma_t = \text{const}. \quad (13\text{-}40)$$

For a disc of constant strength equation (13-39a) becomes

$$\sigma \frac{d}{dr}(ry) - \sigma y + \frac{\gamma}{g} yr^2\omega^2 = 0. \quad (13\text{-}41)$$

Differentiating equation (13-41) we have

$$\sigma y + \sigma r \frac{dy}{dr} - \sigma y + \frac{\gamma}{g} yr^2\omega^2 = 0.$$

Simplifying this equation we may write

$$\frac{dy}{y} = -\frac{\gamma\omega^2}{g\sigma} r dr. \quad (13\text{-}42)$$

Integrating this differential equation between the limits of r and r_2 we obtain

$$\log_e y_2 - \log_e y = \frac{\gamma\omega^2}{2g\sigma}(r_2^2 - r^2),$$

or

$$\log_e \frac{y}{y_2} = \frac{\gamma\omega^2}{2g\sigma}(r_2^2 - r^2),$$

and finally

$$y = y_2 e^{\frac{\gamma\omega^2}{2g\sigma}(r_2^2 - r^2)}. \quad (13\text{-}43)$$

Here r_2—disc radius at the periphery;
and y_2—disc thickness at radius r_2.

Discs of constant strength may be designed on the basis of equation (13-43). The thickness y_2 at radius r_2 is determined from the equation

$$y_2 = \frac{C_b + xC_d}{2\pi r_2 \sigma}, \quad (13\text{-}44)$$

where C_b—centrifugal force of the blades and shrouding;

C_d—centrifugal force of the disc at the rim;

x—coefficient accounting for the effect of pressure equalising holes: $x = 0.5$ to 0.7 for solid thickness and 1.0 for the zone of pressure relief hole.

2. Disc of Constant Thickness with Hubs

The use of constant strength discs greatly hampers the construction of multistage turbines. Hence such discs are used only for small single-stage turbines. Multistage turbines are mostly provided with discs of constant thickness for the stages of relatively smaller diameter.

Fig. 13-16 shows the transverse section of such a disc of constant thickness with hub. The disc is in the form of two plates with different thicknesses, the hub portion consisting of the thicker plate. The formula used for the disc is therefore the same for both the disc and the hub. The stresses induced in the disc while in normal operation are calculated as follows:

the tangential and the radial stresses at any point on the disc (from r_1 to r_2) are

$$\sigma_t = k\sigma_u + k_1\sigma_{r2} + k_2\sigma_{t2}; \qquad (13\text{-}45)$$

$$\sigma_r = l\sigma_u + l_1\sigma_{r2} + l_2\sigma_{t2} \qquad (13\text{-}46)$$

where $\sigma_u = \frac{\gamma}{g} u_2^2$ —stress in the thinner section at radius R;

u_2—circumferential velocity at this radius;

σ_{r2} and σ_{t2}—radial and tangential stresses at radius r_2;

k, k_1, k_2, l, l_1, and l_2—coefficients depending upon the ratio r/r_2.

These coefficients are calculated from the formulas:

$$\left.\begin{aligned}
k &= \frac{3.3}{8}\left[0.7875 - 0.575\left(\frac{r}{r_2}\right)^2 - 0.2125\left(\frac{r_2}{r}\right)^2\right];\\
k_1 &= -0.5\left[1 - \left(\frac{r}{r_2}\right)^2\right]\left(\frac{r_2}{r}\right)^2;\\
k_2 &= 0.5\left[1 + \left(\frac{r}{r_2}\right)^2\right]\left(\frac{r_2}{r}\right)^2;\\
l &= \frac{3.3}{8}\left[0.7875 - \left(\frac{r}{r_2}\right)^2 + 0.2125\left(\frac{r_2}{r}\right)^2\right];\\
l_1 &= 0.5\left[1 + \left(\frac{r}{r_2}\right)^2\right]\left(\frac{r_2}{r}\right)^2;\\
l_2 &= -0.5\left[1 - \left(\frac{r}{r_2}\right)^2\right]\left(\frac{r_2}{r}\right)^2.
\end{aligned}\right\} \qquad (13\text{-}47)$$

The tangential and radial stresses in the hub portion (from r_0 to r_{hub}) are

$$\sigma'_{t1} = k\sigma'_u + k_1\sigma_{r\,hub} + k_2\sigma_{t\,hub}; \qquad (13\text{-}48)$$

$$\sigma'_{r1} = l\sigma'_u + l_1\sigma_{r\,hub} + l_2\sigma_{t\,hub}, \qquad (13\text{-}49)$$

where $\sigma'_u = \frac{\gamma}{g} u_1^2$ — stress in the thinner portion of the disc at radius $r_{hub} = r_1$, $u = \pi n r_1/30$;

$\sigma_{r\,hub}$ and $\sigma_{t\,hub}$— radial and tangential stresses at a radius of r_{hub};

k, k_1, k_2, l, l_1, l_2—coefficients depending on the ratio r/r_{hub}.

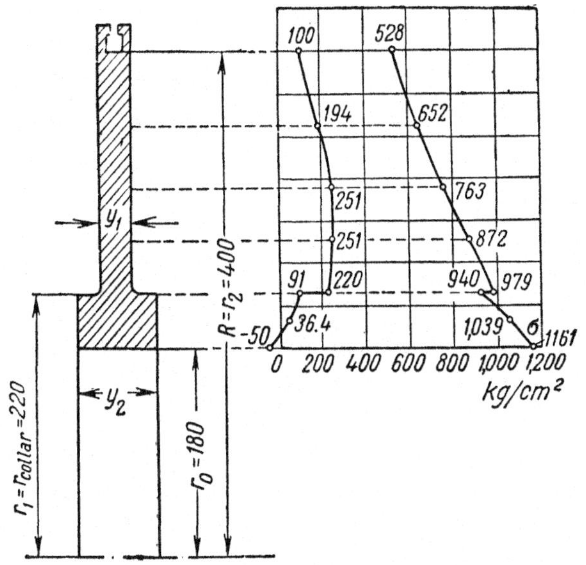

Fig. 13-16. Stresses in a disc with collar

These coefficients are calculated according to equation (13-47) in which the ratios r/r_2 and r_2/r are replaced by r/r_{hub} and r_{hub}/r.

Equations (13-45), (13-46), (13-48) and (13-49) contain the eight unknowns: σ_t, σ_r, σ_{r2}, σ_{t2}, σ'_{t1}, σ'_{r1}, $\sigma_{r\,hub}$ and $\sigma_{t\,hub}$. Consequently additional equations have to be formed for the boundary conditions of the disc.

The radial stress at a radius of r_2 is obtained from the expression

$$\sigma_{r2} = \frac{C_b + xC_d}{2\pi r_2 y_1}, \qquad (13\text{-}50)$$

where y_1—thickness of the disc.

The radial stress at the radius r_0 depends upon the initial tension of the disc and is determined from the equation

$$\sigma_{r0} = l^0\sigma'_u + l_1^0\sigma_{r\,hub} + l_2^0\sigma_{t\,hub}. \qquad (13\text{-}51)$$

The coefficients l^0, l_1^0 and l_2^0 are also calculated from equation (13-47) with the substitution of r_0 and r_{hub} in place of r and r_2.

The stress σ_{r0} is usually assumed to be 50 kg/cm². For heavily loaded discs the value of σ_{r0} may be taken up to 100 or even as much as 150 kg/cm².

The radial deformation of the hub at the collar portion of the disc is obtained from the following expression

$$\xi_{hub} = \frac{r_{hub}}{E}(\sigma_{t\,hub} - \nu\sigma_{r\,hub}). \qquad (13\text{-}52)$$

The radial deformation for the thinner section of the disc at the same place is obtained from the expression

$$\xi_1 = \frac{r_{hub}}{E}(\sigma_{t1} - \nu\sigma_{r1}), \qquad (13\text{-}53)$$

where σ_{t1} and σ_{r1}—tangential and radial stresses in the thinner section of the disc at the radius $r_1 = r_{hub}$. The radial deformations at the hub surface are assumed to be equal, i.e., $\xi_{hub} = \xi_1$.

So that from equations (13-52) and (13-53) we have

$$\sigma_{t\,hub} - \nu\sigma_{r\,hub} = \sigma_{t1} - \nu\sigma_{r1} \text{ from which}$$

$$\sigma_{t1} = \sigma_{t\,hub} + \nu\sigma_{r1} - \nu\sigma_{r\,hub}. \quad (13\text{-}54)$$

Here the coefficient of transverse compression may be assumed as $\nu = 0.3$. If it is assumed that the radial stresses on the cylindrical surfaces at a radius of r_{hub} are uniform then

$$\sigma_{r\,hub} = \frac{y_1}{y_2}\sigma_{r1}, \quad (13\text{-}55)$$

where y_2—thickness of the hub.

Substituting in equations (13-54) and (13-51) the values obtained from equation (13-55) we have

$$\sigma_{r0} = l^0\sigma'_u + l_1^0\frac{y_1}{y_2}\sigma_{r1} + l_2^0\sigma_{t\,hub}; \quad (13\text{-}56)$$

$$\sigma_{t1} = \sigma_{t\,hub} + \left(1 - \frac{y_1}{y_2}\right)\nu\sigma_{r1}. \quad (13\text{-}57)$$

The tangential and the radial stresses in the thinner section of the disc at a radius of $r_1 = r_{hub}$ may be determined by writing equations analogous to equations (13-45) and (13-46) as follows

$$\sigma_{t1} = k\sigma_u + k_1\sigma_{r2} + k_2\sigma_{t2}; \quad (13\text{-}58)$$

$$\sigma_{r1} = l\sigma_u + l_1\sigma_{r2} + l_2\sigma_{t2}. \quad (13\text{-}59)$$

The set of equations (13-56), (13-57), (13-58) and (13-59) forms the basis on which constant thickness discs with hubs are designed.

From these equations the four unknowns σ_{t1}, σ_{r1}, σ_{t2} and $\sigma_{t\,hub}$ are calculated; σ_{r0} is assumed and σ_{r2} is calculated from (13-50). Next from equations (13-45) and (13-46) σ_t and σ_r are calculated for the thinner section of the disc. The stresses in the hub (σ'_{t1} and σ'_{r1} are obtained from equations (13-48) and (13-49).

Example 13-3. Find the stresses in a constant thickness disc with hub from the data given below: $n = 3{,}000$ r.p.m.; centrifugal force of the blades $C_b = 55{,}800$ kg; centrifugal force at the rim of the disc $C_d = 14{,}000$ kg; $y_2 = 60$ mm and $y_1 = 25$ mm.

The various disc dimensions are shown in Fig. 13-16. The radial stresses on the cylindrical surface of the disc at a radius of $r_2 = 400$ mm with $x = 0.5$ are obtained from equation (13-50):

$$\sigma_{r2} = \frac{C_b + xC_d}{2\pi r_2 y_1} = \frac{55{,}800 + 0.5 \times 14{,}000}{2\pi \times 40 \times 2.5} = 100 \text{ kg/cm}^2.$$

The radial stresses on the inner annular surface at a radius of r_0 are assumed to be $\sigma_{r0} = -50$ kg/cm² (depending upon the method of disc mounting).

The stress in the thinner section will be:

a) at a radius of $R = 400$ mm

$$\sigma_u = \frac{\gamma}{g}u_2^2 = 0.08 \times 125^2 = 1{,}250 \text{ kg/cm}^2,$$

where $u_2 = \dfrac{\pi Rn}{30} = \dfrac{\pi \times 0.4 \times 3{,}000}{30} = 125$ m/sec;

b) at a radius of $r_{hub} = 220$ mm

$$\sigma'_u = 0.08u^2_{hub} = 0.08 \times 69^2 = 380 \text{ kg/cm}^2,$$

where $u_{hub} = \dfrac{\pi r_{hub} n}{30} = \dfrac{\pi \times 0.22 \times 3{,}000}{30} = 69$ m/sec.

The stresses at the inner and outer extremes and at the collar are obtained from equations (13-56) and (13-59) as follows

$$\sigma_{r0} = l^0\sigma'_u + l_1^0\frac{y_1}{y_2}\sigma_{r1} + l_2^0\sigma_{t\,hub};$$

$$\sigma_{t1} = \sigma_{t\,hub} + \left(1 - \frac{y_1}{y_2}\right)\nu\sigma_{r1};$$

$$\sigma_{t1} = k\sigma_u + k_1\sigma_{r2} + k_2\sigma_{t2};$$

$$\sigma_{r1} = l\sigma_u + l_1\sigma_{r2} + l_2\sigma_{t2}.$$

The coefficients l^0, l_1^0, l_2^0, l, l_1, l_2, k, k_1 and k_2 are determined from equations (13-47)

$$l^0 = \frac{3.3}{8}\left[0.7875 - \left(\frac{r_0}{r_{hub}}\right)^2 + 0.2125\left(\frac{r_{hub}}{r_0}\right)^2\right] = 0.413\left[0.7875 - \left(\frac{18}{22}\right)^2 + 0.2125\left(\frac{22}{18}\right)^2\right] = 0.18;$$

$$l_1^0 = 0.5\left[1 + \left(\frac{18}{22}\right)^2\right]\left(\frac{22}{18}\right)^2 = 1.243;$$

$$l_2^0 = -0.5\left[1 - \left(\frac{18}{22}\right)^2\right]\left(\frac{22}{18}\right)^2 = -0.247;$$

$$k = 0.413\left[0.7875 - 0.575\left(\frac{22}{40}\right)^2 - 0.2125\left(\frac{40}{22}\right)^2\right] = -0.0367;$$

$$k_1 = -0.5\left[1 - \left(\frac{22}{40}\right)^2\right]\left(\frac{40}{22}\right)^2 = -1.155;$$

$$k_2 = 0.5\left[1 + \left(\frac{22}{40}\right)^2\right]\left(\frac{40}{22}\right)^2 = 2.155;$$

$$l = 0.413\left[0.7875 - \left(\frac{22}{40}\right)^2 + 0.2125\left(\frac{40}{22}\right)^2\right] = 0.491;$$

$$l_1 = 0.5\left[1 + \left(\frac{22}{40}\right)^2\right]\left(\frac{40}{22}\right)^2 = 2.155;$$

$$l_2 = -0.5\left[1 - \left(\frac{22}{40}\right)^2\right]\left(\frac{40}{22}\right)^2 = -1.155.$$

Substituting the values of σ_u and σ'_u and the various coefficients in equations (13-56) and (13-59) we have

$$0.247\sigma_{t\,hub} - 0.518\sigma_{r1} = 118.7; \quad (a)$$

$$\sigma_{t1} - \sigma_{t\,hub} - 0.175\sigma_{r1} = 0; \quad (b)$$

$$\sigma_{t1} - 2.155\sigma_{t2} = -161.4; \quad (c)$$

$$\sigma_{r1} + 1.155\sigma_{t2} = 830. \quad (d)$$

Solving these four equations for the four unknowns we have

$$\sigma_{t2} = 528 \text{ kg/cm}^2;$$

$$\sigma_{r1} = 220 \text{ kg/cm}^2;$$

$$\sigma_{t1} = 979 \text{ kg/cm}^2;$$

$$\sigma_{t\,hub} = 940 \text{ kg/cm}^2.$$

From equation (13-55)

$$\sigma_{r\ \text{hub}} = \frac{y_1}{y_2}\sigma_{r1} = \frac{25}{60} \times 220 = 91 \ \text{kg/cm}^2.$$

The tangential stress at the radius $r_0 = 180$ mm will be, from equation (13-48),

$$\sigma_{t0} = k\sigma'_u + k_1\sigma_{r\ \text{hub}} + k_2\sigma_{t\ \text{hub}}. \qquad \text{(e)}$$

Substituting the ratios $r/r_{\text{hub}} = 18/22$ and $r_{\text{hub}}/r = 22/18$ the coefficients k, k_1, and k_2 are obtained as

$$k = 0.0357;\ k_1 = -0.247 \text{ and } k_2 = 1.243.$$

Substituting the values of σ'_u, $\sigma_{r\ \text{hub}}$, $\sigma_{t\ \text{hub}}$ and the coefficients k, k_1 and k_2 in (e) we have

$$\sigma_{t0} = 0.0357 \times 382 - 0.247 \times 91 + 1.243 \times 940 = \\ = 1{,}161 \ \text{kg/cm}^2.$$

The tangential and radial stresses at any point of the disc are obtained from the expressions:

a) for the thinner section of the disc from equations (13-45) and (13-46)

$$\sigma_t = k\sigma_u + k_1\sigma_{r2} + k_2\sigma_{t2}; \qquad \text{(f)}$$

$$\sigma_r = l\sigma_u + l_1\sigma_{r2} + l_2\sigma_{t2}; \qquad \text{(g)}$$

b) for the hub from equations (13-48) and (13-49)

$$\sigma'_{t1} = k\sigma'_u + k_1\sigma_{r\ \text{hub}} + k_2\sigma_{t\ \text{hub}}; \qquad \text{(h)}$$

$$\sigma'_{r1} = l\sigma'_u + l_1\sigma_{r\ \text{hub}} + l_2\sigma_{t\ \text{hub}}. \qquad \text{(i)}$$

The coefficients in these equations are determined from the set of equations (13-47) with the ratios r/r_2 and r/r_{hub} for the disc and the hub portion respectively. The results of these calculations are given in Tables (13-3) and (13-4).

Table 13-3

Stresses in the Thinner Section of the Disc
$\sigma_u = 1{,}250$ kg/cm², $\sigma_{r2} = 100$ kg/cm², $\sigma_{t2} = 528$ kg/cm²

Tangential stresses

r, cm	22	25	29	34	40
r/r_2	0.55	0.625	0.725	0.85	1
k	—0.0367	0.00806	0.0336	0.0322	0
k_1	—1.155	—0.78	—0.451	—0.193	0
k_2	2.155	1.78	1.45	1.197	1
$k\sigma_u$	—45.9	10.1	42	40.2	0
$k_1\sigma_{r2}$	—115.5	—78	—45	—19.3	0
$k_2\sigma_{t2}$	1,140	941	766	631	528
σ_t, kg/cm²	979	872	763	652	528

Radial stresses

r, cm	22	25	29	34	40
r/r_2	0.55	0.625	0.725	0.85	1
l	0.491	0.388	0.275	0.141	0
l_1	2.155	1.78	1.45	1.197	1
l_2	—1.155	—0.78	—0.451	—0.193	0
$l\sigma_u$	615	485	344	176	0
$l_1\sigma_{r2}$	215.5	178	145	119.7	100
$l_2\sigma_{t2}$	—609	—412	—238	—102	0
σ_r, kg/cm²	220	251	251	194	100

Table 13-4

Stresses in the Hub $\sigma'_u = 382$ kg/cm²,
$\sigma_{r\ \text{hub}} = 91$ kg/cm², and $\sigma_{t\ \text{hub}} = 940$ kg/cm²

Tangential stresses

r, cm	18	20	22
r/r_{hub}	0.819	0.91	1.0
k	0.0357	0.0227	0
k_1	—0.247	—0.1045	0
k_2	1.243	1.105	1
$k\sigma'_u$	13.6	8.7	0
$k_1\sigma_{r\ \text{hub}}$	—22.5	—9.5	0
$k_2\sigma'_{t\ \text{hub}}$	1,179	1,040	940
σ_t, kg/cm²	1,161	1,039	940

Radial stresses

r, cm	18	20	22
r/r_{hub}	0.819	0.91	1
l	0.18	0.09	0
l_1	1.243	1.105	1
l_2	—0.247	—0.1045	0
$l\sigma'_u$	68.7	34.4	0
$l_1\sigma_{r\ \text{hub}}$	113.3	100.5	91
$l_2\sigma_{t\ \text{hub}}$	—232	—98.5	0
σ_r, kg/cm²	—50.0	36.4	91

3. Conical Discs

Discs of constant thickness are not suitable for large diameter stages since the stresses induced at the collar, i.e., where the disc meets the hub, are exceedingly high. Discs of conical form are used for these cases to reduce stress concentration at the collar. A thicker disc section at the hub and a uniformly decreasing thickness towards the rim give a very uniform stress distribution all along the length of the disc. Fig. 13-17 shows a disc of conical section. The stress at any point in the conical portion of the disc may be determined from the equations

$$\sigma_r = \sigma_u p_0 + Ap_1 + Bp_2; \qquad (13\text{-}60)$$

$$\sigma_t = \sigma_u q_0 + Aq_1 + Bq_2, \qquad (13\text{-}61)$$

where A and B—constants of integration obtained from the boundary conditions;
σ_u—stress in the thin ring of radius R (see Fig. 13-17),
p and q—coefficients depending upon the ratio $r/R = x$. The values of these coefficients are shown in Fig. 13-18. To determine the stresses σ_r and σ_t it is first necessary to obtain from boundary conditions the values of the integration constants A and B.

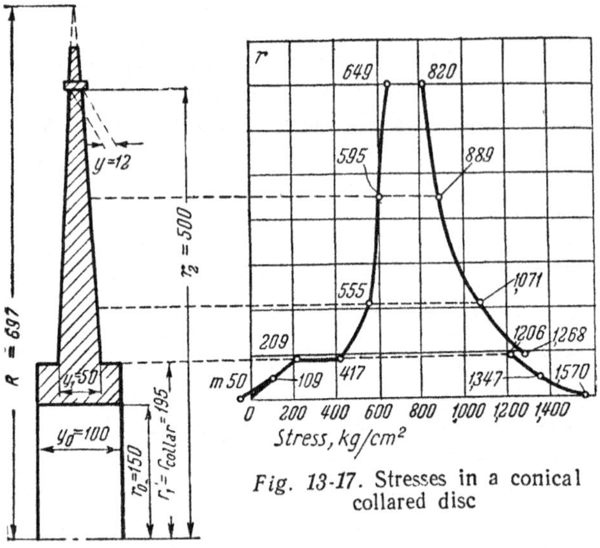

Fig. 13-17. Stresses in a conical collared disc

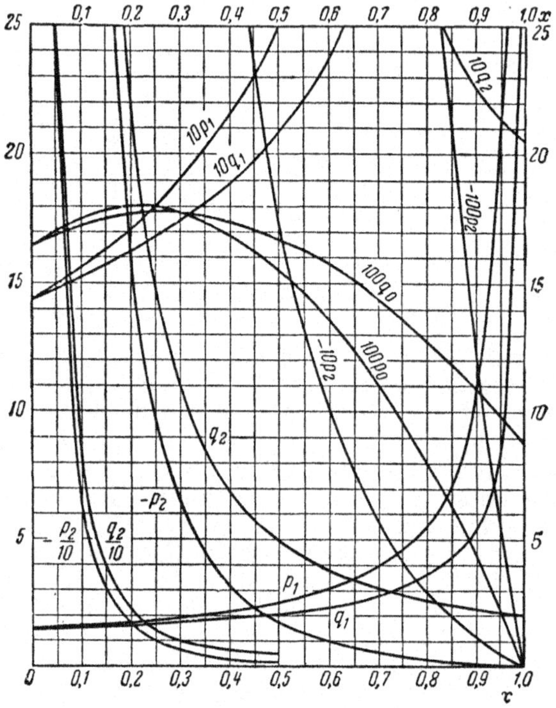

Fig. 13-18. Various coefficients for conical steel discs

The equations for the boundary conditions of the disc are

a) for radial stresses at radius r_2

$$\sigma_{r_2} = \sigma_u p_0 + Ap_1 + Bp_2; \qquad (13\text{-}62)$$

b) for radial and tangential stresses at the collar

$$\sigma_{r_1} = \sigma_u p_0 + Ap_1 + Bp_2; \qquad (13\text{-}63)$$

$$\sigma_{t_1} = \sigma_u q_0 + Aq_1 + Bq_2; \qquad (13\text{-}64)$$

c) for the hub portion the equations (13-55), (13-56) and (13-57) derived earlier will be used.

The radial stress σ_{r_2} is obtained from equation (13-50) as $\sigma_{r_2} = \frac{C_b + xC_d}{2\pi r_2 y}$,

and the stress σ_u from equation (13-33) where $u = \pi Rn/30$. Solving this set of equations (13-55), (13-56), (13-57), (13-62), (13-63) and (13-64) simultaneously we obtain the values of the six unknowns: $\sigma_{r\,\text{hub}}$, $\sigma_{t\,\text{hub}}$, σ_{r_1}, σ_{t_1}, A and B.

Since A and B are known it becomes very easy to calculate the stresses σ_r and σ_t in the conical portion of the disc with the help of equations (13-60) and (13-61) and the stresses in the hub from equations (13-48) and (13-49).

Example 13-4. Design a conical disc from the data given below:

$n = 3{,}300$ r.p.m., $\sigma_{r0} = -50$ kg/cm² (radial stress due to force fit on the shaft). The disc dimensions are shown in Fig. 13-17. The radial stress at radius r_2 due to the centrifugal force of the blades and the disc rim is 650 kg/cm².

First we evaluate several auxiliary quantities:

a) radius of the full cone (R in Fig. 13-17)

$$R = \frac{r_2 y_1 - r_1 y}{y_1 - y} = \frac{50 \times 5 - 19.5 \times 1.2}{5 - 1.2} = 59.7 \text{ cm;}$$

b) stress in the thinner section of the disc at the above radius $R = 59.7$ cm

$$\sigma_u = \frac{\gamma}{g} u^2 = \frac{\gamma}{g} 206.5^2 = 3{,}390 \text{ kg/cm}^2$$

at $\gamma = 0.0078$ kg/cm² and

$$u = \frac{\pi R n}{30} = \frac{0.597 \times 3{,}300}{30} = 206.5 \text{ m/sec;}$$

c) stress in the thinner section of the disc at a radius r_1

$$\sigma_u' = \frac{\gamma}{g} u_1^2 = \frac{\gamma}{g} 67.5^2 = 362 \text{ kg/cm}^2$$

$$\text{at } u_1 = \frac{\pi \times 0.195 \times 3{,}300}{30} = 67.5 \text{ m/sec.}$$

Writing down the various basic equations for the determination of stresses at the important sections we have

a) at radius r_2

$$\sigma_{r_2} = \sigma_u p_0 + Ap_1 + Bp_2; \qquad \text{(a)}$$

b) at radius r_1

$$\sigma_{r_1} = \sigma_u p_0 + Ap_1 + Bp_2; \qquad \text{(b)}$$

$$\sigma_{t_1} = \sigma_u q_0 + Aq_1 + Bq_2. \qquad \text{(c)}$$

For the hub portion:

a) at radius $r_{\text{hub}} = r_1$ $\sigma_{t_1} = \sigma_{t\,\text{hub}} + \left(1 - \frac{y_1}{y}\right)\nu\sigma_{r_1}$; (d)

b) on the annular surface of the disc at radius r_0

$$\sigma r_0 = l^0 \sigma_u' + l_1^0 \frac{y_1}{y_0} \sigma_{r_1} + l_2^0 \sigma_{t\,\text{hub}}. \qquad \text{(e)}$$

The coefficients p_0, p_1, p_2, q_0, q_1 and q_2 are obtained from the curves given in Fig. 13-18.
The coefficients for equation (a) are obtained for
$x = r_2/R = 50/59.7 = 0.838$ (from Fig. 13-18);
$p_0 = 0.0677$; $p_1 = 6.65$; $p_2 = -0.221$.
For equations (b) and (c)

$$x = r_1/R = 19.5/59.7 = 0.327;$$
$$p_0 = 0.1755;\ p_1 = 1.96;\ p_2 = -5.20;$$
$$q_0 = 0.1768;\ q_1 = 1.775;\ q_2 = 9.49.$$

The coefficients l^0, l_1^0 and l_2^0 are determined from equations (13-47) for the ratios

$$r_0/r_{hub} = 15/19.5,\ r_{hub}/r_0 = 19.5/15;$$
$$l^0 = 0.23;\ l_1^0 = 1.35 \text{ and } l_2^0 = -0.344.$$

Substituting these coefficients and the numerical values of y_1, y_0 and y in equations (a), (b), (c), (d) and (e) with the unknowns on the left-hand side we have

$$6.65A - 0.211B = 420; \quad (a')$$
$$1.96A - 5.20B - \sigma_{r1} = -595; \quad (b')$$
$$1.775A + 9.49B - \sigma_{t1} = -598; \quad (c')$$
$$\sigma_{t\,hub} + 0.15\sigma_{r1} - \sigma_{t1} = 0; \quad (d')$$
$$-0.344\sigma_{t\,hub} + 0.675\sigma_{r1} = -133. \quad (e')$$

We shall solve these equations by the successive elimination of the unknowns.

Dividing equation (e') by 0.344 and adding it to (d') we have

$$2.11\sigma_{r1} - \sigma_{t1} = -387.$$

Subtracting the above from equation (c') we have

$$1.775A + 9.49B = 2.11\sigma_{r1} = -211. \quad (f)$$

Dividing the above equation by 2.11 and subtracting it from equation (b') we obtain

$$-1.118A - 9.70B = -495. \quad (g)$$

A and B are now obtained from the equations (a') and (g)

$$A = 65.0 \text{ and } B = 58.6$$

The stresses σ_{r1}, σ_{t1}, $\sigma_{t\,hub}$ and $\sigma_{r\,hub}$ are determined from the equations
$\sigma_{r1} = 1.96A - 5.20B + 595 = 417$ kg/cm² (from equation b');
$\sigma_{t1} = 1.775A + 9.49B + 598 = 1,268$ kg/cm² (from equation c');
$\sigma_{t\,hub} = \sigma_{t1} - 0.15\sigma_{r1} = 1.206$ kg/cm² (from equation d');
$\sigma_{r\,hub} = \frac{y_1}{y_0}\sigma_{r1} = \frac{50}{100} \times 417 = 209$ kg/cm² (from equation (13-55).

Table 13-5

Stresses in the Conical Portion of the Disc

$A = 65$ kg/cm², $B = 58.6$ kg/cm², $\sigma_u = 3,390$ kg/cm²

Radial stresses

r, cm	19.5	25.6	37.8	50.0
$x = r/R$	0.327	0.429	0.634	0.838
p_0	0.1752	0.1662	0.1273	0.0677
p_1	1.969	2.237	3.252	6.65
p_2	−5.15	−2.615	−0.824	−0.221
$\sigma_u p_0$	595	564	432	230
Ap_1	128	145	211	432
Bp_2	−302	−154	−48	−13
σ_r, kg/cm²	417	555	595	649

Tangential stresses

r, cm	19.5	25.6	37.8	50.0
$x = r/R$	0.327	0.429	0.634	0.838
q_0	0.1765	0.1723	0.1536	0.1209
q_1	1.775	1.943	2.496	4.09
q_2	9.49	6.15	3.51	2.45
$\sigma_u q_0$	598	585	521	410
Aq_1	115	126	162	266
Bq_2	555	360	206	144
σ_t, kg/cm²	1,268	1,071	889	820

Table 13-6

Stresses in the Hub

$\sigma'_u = 362$ kg/cm²; $\sigma_{r\,hub} = 209$ kg/cm²;
$\sigma_{t\,hub} = 1,206$ kg/cm²

Radial stresses

r, cm	15	17.25	19.5
r/r_{hub}	0.77	0.885	1.0
l	0.23	0.112	0
l_1	1.35	1.13	1
l_2	−0.344	−0.14	0
$l\sigma'_u$	83	41	0
$l_1\sigma_{r\,hub}$	282	237	209
$l_2\sigma_{t\,hub}$	−415	−169	0
σ_r, kg/cm²	−50	109	209

Tangential stresses

r, cm	15.0	17.25	19.5
r/r_{hub}	0.77	0.885	1
k	0.033	0.029	0
k_1	−0.344	−0.14	0
k_2	1.35	1.13	1
$k\sigma'_u$	12	11	0
$k_1\sigma_{r\,hub}$	−72	−29	0
$k_2\sigma_{t\,hub}$	1,630	1,365	1,206
σ_t, kg/cm²	1,570	1,347	1,206

Table 13 7

Properties of Steels Used for Making Discs (Ref. 2)

Disc category	Ultimate stress, kg/mm²	Yield point, kg/mm²	Relative elongation, %	Relative reduction of area, %	Specific resilience, kg m/cm²	Brinell hardness, kg/mm²
	Not less than					
I	63	32	17	35	4	170 to 207
II	75	40	17	35	4	187 to 223
III	90	75	15	35	3	289 to 321

Stresses in the conical portion of the disc will be determined from equations (13-60) and (13-61)

$$\sigma_r = 3,390p_0 + 65p_1 + 58.6p_2;$$
$$\sigma_t = 3,390q_0 + 65q_1 + 58.6q_2.$$

The results of all the above calculations are shown in Table 13-5.

The stresses in the hub are obtained from equations (13-48) and (13-49)

$$\sigma_r = 362l + 209l_1 + 1,206l_2;$$
$$\sigma_t = 362k + 209k_1 + 1,206k_2.$$

The results of these calculations are also given in Table 13-6.

The types of steels used for the construction of turbine discs depending upon the magnitude of the stresses encountered and the conditions of operation are divided into three categories (see Table 13-7).

Table 13-8 gives the chemical analysis and some of the physical properties of steels commonly used in the manufacture of turbine discs. From the physical properties shown in the table it is seen that the chrome-molybdenum steels are by far the strongest (32XHM).

Discs subjected to heavy loads (the low-pressure stage discs of condensing turbines) are usually made from molybdenum steels of this type. The L. M. W. makes turbine discs with high stresses of 2,700 kg/cm² for the turbines VK-50 and VK-100. According to Kirillov and Kantor the allowable stresses for the molybdenum steel 32XHM at a temperature of 400°C should be taken not more than 1,100 kg/cm². The allowable stresses for each case are decided taking into consideration the physical properties of the steels as well as the temperatures at which the discs are normally supposed to operate. In general the allowable stresses are never assumed more than 0.4 times the yield point stress at the given temperature.

13-9 DESIGN OF TURBINE SHAFTS

The forces acting on a turbine shaft may be divided into four groups:

a) bending forces due to the weight of discs and blades;

b) twisting forces caused by the turning moment of the rotating discs;

c) forces resulting from the non-uniform distribution of pressure on the pads of the thrust bearing;

d) forces induced at short circuiting of alternator. When designing a turbine shaft for mechanical strength the stresses in the weakest section must be taken as the basis for these calculations, i. e., 1) the section at which the bending moment is the largest, 2) the section where the turning moment is the maximum (at the coupling between the turbine and the alternator), 3) the shaft section at the thrust disc. Besides these sections the shaft strength is checked for the stresses appearing in it for the case of alternator short circuiting.

1. Calculations for Bending and Twisting

The design calculations of any given section of the shaft consist of the determination of the bending and the torsional stresses acting at this section. The calculations are carried out for the maximum tangential stress which is obtained from the equation

$$\tau_{max} = \frac{1}{2W}\sqrt{M_b^2 + M_t^2}\ [\text{kg/cm}^2], \qquad (13\text{-}65)$$

where $W = \frac{\pi d^3}{32}$ — moment of resistance of the shaft, cm³;

d—diameter of the shaft, cm;

M_b and M_t—the bending and twisting moments at the section under consideration, kg cm.

The bending moment M_b at any required section may be determined graphically as described in 13-12.

The twisting moment at any required section of the shaft

$$M_t = \frac{\sum P_u d_1}{2} = \frac{102 \sum N_i d_1}{2u}$$

Table 13-8

Chemical Analysis and Some of the Physical Properties of Steels Used for the Construction of Turbine Discs

Disc category	Type of steel	Chemical analysis								Specific gravity, g/cm³	Modulus of elasticity at 20°C, kg/mm²
		C	Si	Mn	Cr	Ni	Mo	S	P		
								not more than			
I	45A	0.42 to 0.47	0.17 to 0.37	0.50 to 0.80	⩽0.20	⩽0.30	—	0.03	0.04	7.81	20,400
	43H	0.40 to 0.45	0.17 to 0.37	0.50 to 0.80	⩽0.2	0.90 to 1.20	—	0.03	0.04	7.84	—
	45X	0.40 to 0.50	0.17 to 0.37	0.50 to 0.80	0.80 to 1.10	⩽0.30	—	0.04	0.025	7.816	21,020
II	34XM	0.30 to 0.38	0.17 to 0.37	0.40 to 0.70	0.90 to 1.20	⩽0.40	0.25 to 0.40	0.03	0.035	—	21,950
	35XHM	0.32 to 0.38	0.17 to 0.37	0.30 to 0.60	0.80 to 1.20	1.40 to 1.80	0.25 to 0.40	0.03	0.035	—	20,400
III	32XHM	0.28 to 0.35	0.17 to 0.37	0.30 to 0.80	0.60 to 0.80	2.75 to 3.00	0.30 to 0.40	0.03	0.035	—	—

or substituting $u = \frac{\pi d_1 n}{60}$,

$$M_t = \frac{102 \times 60 \sum N_i}{2\pi n} = 973 \frac{\sum N_i}{n} \text{ [kg m]} =$$

$$= 97{,}300 \frac{\sum N_i}{n} \text{ [kg cm]} \qquad (13\text{-}66)$$

where ΣN_i—the total internal power developed on the shaft length measured from the front end up to the shaft section under consideration, kW;

ΣP_u —the total force exerted by the steam in these stages;

n—r.p.m.;

d_1—rotor diameter at the mean height of the blades;

u—mean circumferential velocity at the rim of the disc.

For carbon steels τ_{max} may be taken at about 400 kg/cm² and for alloy steels this value may be increased from 600 to 800 kg/cm² or even more.

The bending moment at the coupling between the alternator and the turbine may be approximately taken to be zero in which case equation (13-65) becomes

$$\tau_{max} = \frac{M_t}{2W} \text{ [kg/cm}^2\text{]}.$$

2. Strength of the Shaft at the Thrust Disc

While in operation there may arise non-uniform axial forces on the pads of the thrust bearing of a turbine. In the limiting case it may be

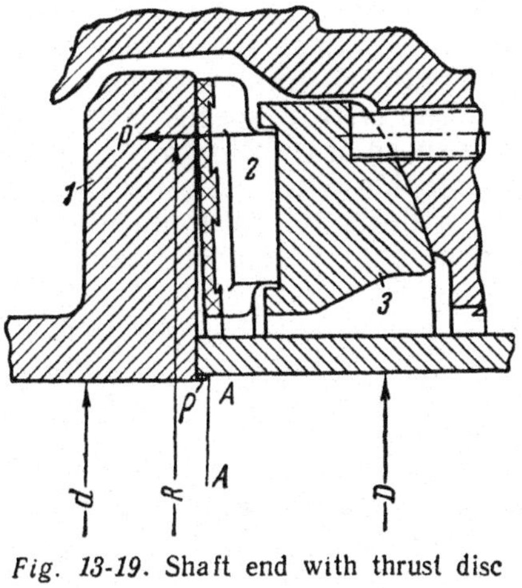

Fig. 13-19. Shaft end with thrust disc

supposed that all the axial thrust is taken up by a single thrust pad giving rise to bending moment at the thrust disc (Fig. 13-19):

$$M_b = PR. \qquad (13\text{-}67)$$

The stress in this section will be

$$\sigma_b = \frac{M_b}{W_1} = \frac{32 M_b}{\pi d^3} \qquad (13\text{-}68)$$

where W_1—moment of resistance of the shaft at this section;

d—shaft diameter;

P—total axial thrust.

If a bearing of the type shown in Fig. 12-37 is assembled carefully and if the spherical pads *2* are sensitive to axial self-alignment the thrust would then be uniformly distributed on all the thrust pads. If, however, the pads become slightly tilted (the stop bolt *5* is tightened too much) or if the spherical pads have poor sensitivity to self-alignment all the axial thrust may, in the limiting case, be transmitted to a single thrust pad *1* giving rise to a couple.

There have been several cases of shaft damage due to unsatisfactory working of the thrust bearings of this type for the Metropolitan Vickers and L.M.W. turbines type AK-25-1. Bending forces appear at the thrust disc because of this couple resulting in shaft breakage (equations [13-67] and [13-68]). As a result of these sad experiences the L.M.W. has now completely stopped the construction of thrust bearings of this type.

The allowable bending stresses for carbon steels may be taken as 1,000 kg/cm² and for alloy steels substantially higher values may be assumed.

3. Shaft Strength at Short Circuiting of Alternator

When the alternator gets short-circuited the turning moment on the alternator shaft instantaneously increases, exceeding by ten times the maximum turning moment at full load operation. Hence the stresses in the shaft at the coupling must be checked for the maximum turning moment $M_{t\,max}$ at alternator short circuit.

The maximum turning moment may be approximately determined from

$$M_{t\,max} \approx 20 M_t \frac{I_t}{I_t + I_g}, \qquad (13\text{-}69)$$

where M_t—twisting moment at nominal load;

I_t and I_g— moments of inertia of the turbine and the alternator rotors.

The maximum tangential stress at short circuit is obtained from equation (13-65) where $M_{t\,max}$ is substituted in place of M_t. This stress in no case should exceed 2/3 of the limiting elastic deformation stress for the material from which the disc is made.

13-10. CRITICAL SPEEDS OF ROTORS

In spite of all the care taken in the construction and balancing of the turbine shaft and the discs, due to some reason or the other, the mass centre of the rotor does not coincide with the geometrical axis of the shaft.

After assembly the rotor always has a certain amount of unbalance. The distance between the geometrical axis of the shaft and the mass centre of the rotor is known as the eccentricity of the rotor. During shaft rotation even a small eccentricity gives rise to a transverse force that increases with shaft r.p.m. and tends to deflect the shaft.

Let us consider a slightly out-of-balance shaft with an eccentricity e (Fig. 13-20).

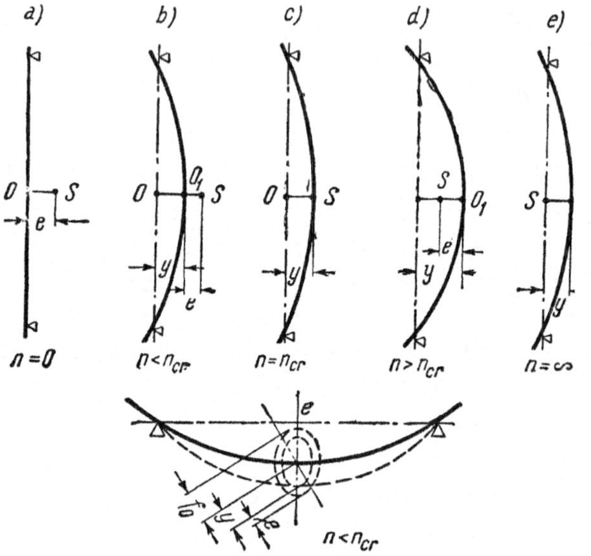

Fig. 13-20. Position of centre of gravity of shaft depending on its r.p.m.

The mass centre of the shaft is at a distance e from the geometrical axis of the shaft. We shall consider the shaft to be supported in the vertical direction to avoid the shaft deflection due to its own weight (Fig. 13-20, *a*). The centrifugal force acting on the shaft while in rotation (Fig. 13-20, *b*) is given by the equation

$$C = m(y+e)\,\omega^2, \qquad (13\text{-}70)$$

where ω—angular velocity of rotation of the shaft 1/sec;

m—mass of the shaft, kg sec²/cm;

y—deflection, cm.

The transverse force appearing as a result of the unbalance is balanced by the elastic force of the shaft which is

$$P = \alpha y, \qquad (13\text{-}71)$$

where α—the force causing the shaft to be deflected through 1 cm. The magnitude of this force depends on the stiffness of the shaft, its length, load distribution and type of end supports of the shaft.

The force balance of P and C is expressed by the relation

$$m(y+e)\,\omega^2 = \alpha y, \qquad (13\text{-}72)$$

from which the deflection of the shaft is obtained as

$$y = \frac{me\omega^2}{\alpha - m\omega^2} = \frac{e}{\dfrac{\alpha}{m\omega^2} - 1}. \qquad (13\text{-}73)$$

From equation (13-73) it follows, that each value of ω conforms to a definite deflection y, e. g., at $\alpha/m\omega^2 = 1$, $y = \infty$. The angular velocity of the shaft at $m\omega^2 = \alpha$ is known as the critical velocity and is determined from the equation

$$\omega_{cr} = \sqrt{\frac{\alpha}{m}} = \frac{981\alpha}{G}, \qquad (13\text{-}74)$$

where $m = \frac{G}{g} = \frac{G}{981}$.

G—weight of the shaft.

Since $\omega_{cr} = \frac{\pi n_{cr}}{30}$, from equation (13-74) the critical speed of the shaft may be determined as

$$n_{cr} = \frac{30}{\pi}\,\omega_{cr} = \frac{30}{\pi}\sqrt{\frac{981\alpha}{G}} \approx$$

$$\approx 300\sqrt{\frac{\alpha}{G}}. \qquad (13\text{-}75)$$

The r.p.m. which numerically coincides with the natural frequency of transverse vibrations of the shaft is known as the critical speed. Theoretically, at the critical speed, the deflection of the shaft tends to infinity. Thus operation of turbines at the critical speeds is to be avoided. The normal speed of rotation of the turbine shaft must not coincide with the critical speed of the shaft, i. e., with its natural frequency. It is found from practice that if the critical speed differs from the normal speed by 15 to 20% safe working of the turbine is completely ensured. However, most of the turbine building factories use a normal speed either higher or lower than the critical by about 30 to 40%.

Turbine shafts having critical speeds less than their normal operating speed are known as flexible shafts and those with critical speeds higher than the normal operating speeds are known as rigid shafts. Rotors of impulse turbines are made of both these types, flexible and rigid. If the critical speed is less than the operating speed, while starting the turbine this speed must be passed over quickly so that there will be hardly any time for the deflection to grow, which, if allowed, could result in a bent shaft and damaged bearings. A characteristic property of the critical speed is that the force vector e due to eccentric loading acts in a direction perpendicular to the plane of the paper at the extreme deflection of the shaft, i. e., the centre of gravity S coincides with the geometrical axis of the bent shaft.

It is found from theoretical and practical considerations that with $n > n_{cr}$ (Fig. 13-20,d) the c. g. S would be between the vertical axis (shown by the dotted line) and the bent shaft. In this case the equilibrium condition for the forces C and P will be expressed as

$$m(y-e)\omega^2 - \alpha y = 0, \qquad (13\text{-}76)$$

from which

$$y = \frac{me\omega^2}{m\omega^2 - \alpha} = \frac{e}{1 - \frac{\alpha}{m\omega^2}}. \qquad (13\text{-}77)$$

From equation (13-74) $\alpha/m = \omega^2_{cr}$, and consequently equation (13-77) can be written as

$$y = \frac{e}{1 - \frac{\omega^2_{cr}}{\omega^2}}. \qquad (13\text{-}77a)$$

From this equation it is seen that with the increase in ω, y has a value less than what it has at the critical speed. This property permits the use of flexible shafts. For example, when $\omega = \omega_{cr}$, $y = \infty$ and when $\omega = \infty$ $y = e$, i. e., at an infinite speed of rotation the centre of gravity of the shaft (point S in Fig. 13-20,e) coincides with the axis of rotation.

It is seen from equations (13-73) and (13-77) that the deflection of the shaft is also a function of eccentricity e and hence while balancing the rotor it is advisable to reduce this quantity to the minimum possible.

If the shaft is in horizontal supports (Fig. 13-20,f) even under static conditions there will be some amount of deflection f_o caused by the weight of the shaft and the discs mounted on it. Thus the shaft will always be slightly bent. Consequently while in rotation there will be an additional amount of deflection y and the shaft begins to vibrate relative to its static geometric axis (Fig. 13-20,f).

The static deflection is given by

$$f_0 = \frac{G_0}{\alpha}, \qquad (13\text{-}78)$$

where G_0—weight of the shaft or rotor.

The deflection f_o depends on the stiffness of the shaft, the distance between the two supports and the load distribution. For simply supported shafts loaded at the centre the deflection will be

$$f_0 = \frac{G_0 l^3}{48EI}$$

and for shafts with fixed ends and loaded at the centre the deflection is given by

$$f_0 = \frac{G_0 l^3}{192EI},$$

where l—distance between the two supports;
E—modulus of elasticity of the material of the shaft;
$I = \pi d^4/64$—moment of inertia of the shaft section;
d—shaft diameter.

We may rewrite the expression for shaft deflection as

$$f_0 = k\frac{G_0 l^3}{EI}, \qquad (13\text{-}79)$$

where k—coefficient depending upon the type of support and the point at which the load is applied (in the cases under consideration $k = 1/48$ and $1/192$).

Having determined the deflection f_0 the critical speed can be easily determined from equations (13-75) and (13-78)

$$n_{cr} \approx \frac{300}{\sqrt{f_0}}. \qquad (13\text{-}80)$$

13-11. DESIGN OF SHAFTS WITH TWO SUPPORTS

Design of shafts from considerations of speed of rotation and mechanical strength was considered separately in 13-9 and 13-10. If the shaft is to operate satisfactorily the above-mentioned designs must be integrated. To begin with the diameter of the shaft must be determined from the critical speed considerations as was shown in 13-10, which should then be further checked for mechanical strength as shown in 13-9.

From equations (13-79) and (13-80) for a shaft supported horizontally we may write

$$f_0 = \left(\frac{300}{n_{cr}}\right)^2 = k\frac{G_0 l^3}{EI}. \qquad (13\text{-}81)$$

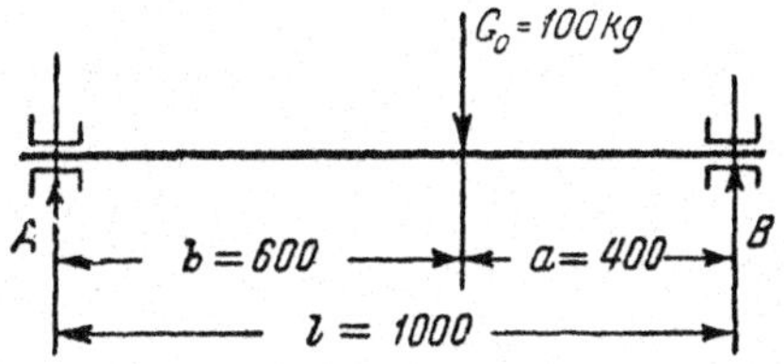

Fig. 13-21. Flexible shaft with a single disc

The moment of inertia I is obtained from this equation as

$$I = \frac{kG_0 l^3}{E}\left(\frac{n_{cr}}{300}\right)^2 \qquad (13\text{-}82)$$

and the diameter

$$d = \sqrt[4]{\frac{64}{\pi} I} = \sqrt[4]{\frac{64kG_0 l^3}{\pi E}\left(\frac{n_{cr}}{300}\right)^2}. \qquad (13\text{-}83)$$

The critical speed is usually assumed for the calculation of shaft diameter.

The coefficient k is obtained from the equations of strength of materials for the various types of loadings.

Example 13-5. Design a flexible shaft for a single-stage impulse turbine from the following data:

$N_e = 300$ kW; $n = 10{,}000$ r. p. m.; $G_0 = 100$ kg; $a = 40$ cm; $b = 60$ cm; $l = a + b = 100$ cm — distance between the two supports.

We shall suppose the shaft to be freely supported at both ends and shall neglect its weight.

The deflection of the shaft under the concentrated load G_0 at the place shown in Fig. 13-21 is

$$f_0 = \frac{a^2 b^2 G_0}{3lEI}.$$

Since in the case under consideration $a = 2/5\,l$, $b = 3/5\,l$,

$$f_0 = \frac{4l^2}{25} \times \frac{9l^2}{25} \times \frac{1G_0}{3lEI} = \frac{12G_0 l^3}{625EI},$$

where $k = \frac{12}{625}$.

We shall assume the critical speed of the rotor to be 1/6 of the operating speed:

$$n_{cr} = \frac{10{,}000}{6} = 1{,}667 \text{ r.p.m.}$$

In the case of single-stage turbines having large speeds of rotation, say $n > 5{,}000$ r. p. m., the possibility of the presence of a second or a third critical speed is not excluded. It is known, theoretically, that for a shaft of constant diameter the ratios of the critical speeds is in the following order

$$\omega_{cr1} : \omega_{cr2} : \omega_{cr3} = 1:4:9,$$

where ω_{cr1}, ω_{cr2}, ω_{cr3} — critical speeds of the shaft at 2/4, 4/4 and 6/4 waves, i. e., without a node for the first speed and with one and two nodal points respectively for the second and third speeds.

From the above relation it follows that the critical speed of the shaft should not be 1,100 or 2,500 r. p. m.

With critical speeds of 1,100 r. p. m. shaft vibration with two nodes will cause the operating speed to coincide with the critical speed. Similarly for the 2,500 r. p. m. the critical and the operating speeds will coincide with single node vibrations of the shaft.

Thus speeds 1/4 and 1/9 of the normal should be avoided

From equation (13-82)

$$I = \frac{12G_0 l^3}{625E}\left(\frac{n_{cr}}{300}\right)^2 = \frac{12 \times 100 \times 100^3}{625 \times 2{,}100{,}000}\left(\frac{1{,}667}{300}\right)^2 =$$
$$= 28.2 \text{ cm}^4.$$

The diameter of the shaft then becomes

$$d = \sqrt[4]{\frac{64}{\pi} I} = \sqrt[4]{\frac{64}{\pi} \times 28.32} = 4.9 \text{ cm} = 49 \text{ mm}.$$

The static deflection of the shaft

$$f_0 = \left(\frac{300}{1{,}667}\right)^2 \approx 0.0325 \text{ cm} = 0.325 \text{ mm}$$

which is permissible for single-stage turbines.

The bending moment caused by the reaction force at the supports

$$M_b = \frac{a}{l} bG_0 = \frac{40}{100} \times 60 \times 100 = 2{,}400 \text{ kg cm}.$$

The twisting moment from equation (13-66) is obtained as

$$M_t = 97{,}300 \times \frac{300}{0.96 \times 0.94 \times 10{,}000} = 3{,}240 \text{ kg/cm},$$

where $\eta_m = 0.96$ and $\eta_g = 0.94$ (assumed).

The maximum tangential stress from equation (13-65)

$$\tau_{max} = \frac{1}{2W}\sqrt{2{,}400^2 + 3{,}240^2} = \frac{1}{2 \times 13.9}\sqrt{1{,}622 \times 10^4} =$$
$$= 145 \text{ kg/cm}^2,$$

where $W = \frac{\pi}{32} \times 4.9^3 = 13.9$ cm^3. The tangential stress obtained here is well within the permissible limits.

13-12. CRITICAL SPEEDS OF SHAFTS WITH SEVERAL DISCS

It may be easily shown that in the simplest case the critical speed of a shaft (rotor) exactly or at least very nearly coincides with the frequency of its natural vibrations. Hence the critical speed may be expressed by the frequency of natural vibrations of the shaft.

Let the shaft be loaded by concentrated loads Q_1, Q_2, Q_3,..., etc. Let the deflection caused by these loads be y_1, y_2, y_3,..., etc. The shaft would then be bent in the form shown in Fig. 13-22 by the line $\alpha\beta\gamma$.

We may assume, with tolerable accuracy, that the whirling line of the shaft while in rotation coincides with the static deflection curve which may be easily determined graphically. From the law of conservation of energy we can assume that the total of the kinetic and potential energies of the shaft (rotor) while in operation would remain unchanged. In other words the change in the kinetic energy T of the vibrating shaft must be equal to the change of the potential energy of deformation of the shaft U, i. e., $T = U$.

When the loads pass through the position of static equilibrium, when their speeds are the maximum, the shaft develops the maximum kinetic energy which is numerically equal to

$$T_{max} = \frac{\lambda^2}{2g}\left(Q_1 y_1^2 + Q_2 y_2^2 + Q_3 y_3^2 + \dots\right),$$

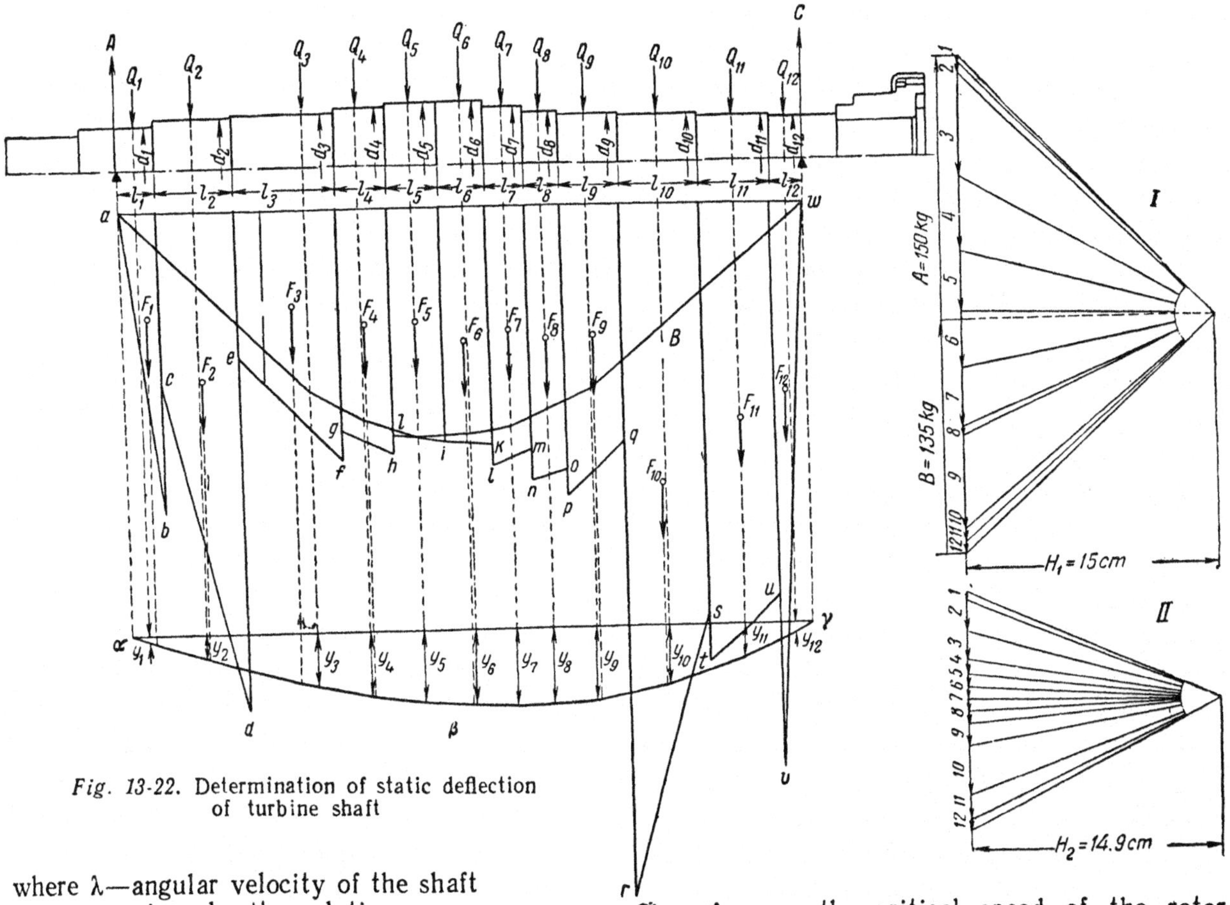

Fig. 13-22. Determination of static deflection of turbine shaft

where λ—angular velocity of the shaft given by the relation

$$\lambda=\frac{2\pi}{\tau};$$

τ—period of vibrations;

g—acceleration due to gravity.

The potential energy of shaft deformation at the position of static equilibrium will be zero, i. e., $U=0$.

When the shaft is at the maximum deflection position from the position of static equilibrium the strain energy (potential energy of shaft deformation) is the greatest.

$$U_{max}=\frac{1}{2}(Q_1y_1+Q_2y_2+Q_3y_3+\ldots).$$

Since the total energy of the system remains constant

$$\frac{\lambda^2}{2g}(Q_1y_1^2+Q_2y_2^2+Q_3y_3^2+\ldots)=$$

$$=\frac{1}{2}(Q_1y_1+Q_2y_2+Q_3y_3\ldots)$$

from which

$$\lambda=\sqrt{\frac{g\sum Q_iy_i}{\sum Q_iy_i^2}}. \qquad (13\text{-}84)$$

Since $\lambda=\omega_{cr}$, the critical speed of the rotor

$$n_{cr}=\frac{60\,\omega_{cr}}{2\pi}\approx 300\sqrt{\frac{\sum Q_iy_i}{\sum Q_iy_i^2}}. \qquad (13\text{-}85)$$

The last equation contains only one unknown quantity, y_i, which may be determined graphically.

While determining the static deflection curve for the shaft the additional stiffening of the shaft due to the disc hubs is usually not taken into account. This may cause the critical speed to be actually higher than what is obtained theoretically.

The critical speeds of solid forged rotors or rotors of the combined type are determined in exactly the same way as is done for the rotors with mounted discs. Increase in the shaft stiffness due to the presence of the discs forged in one piece with the shaft may be considered by increasing the shaft diameter by an amount $b/2$ (where b—thickness of the discs).

Example 13-6. Calculate the critical speed of a shaft loaded as shown in Fig. 13-22 and Table 13-9. The normal operating speed of the rotor is 6,000 r. p. m.

The linear scale shall be assumed to be $\frac{1}{a}=\frac{1}{2}$, i.e., 1 cm on the sketch equals 2 cm length of the shaft.

We shall divide the shaft into 12 parts for the loads acting on it. To the concentrated loads (weights of the discs, glands, etc.) we shall add the weight of the shaft section carrying this load (Fig. 13-22 and Table 13-9 give the gross loads). The effect of the overhanging ends of the shaft shall be neglected.[1] We shall now construct a force polygon I with scale $b=10$ kg/cm, choosing the pole at a distance of 15 cm.

The bending moment diagram B is now drawn. The bending moment at any required point on the shaft may be obtained by multiplying the ordinate at that point in the b. m. diagram measured to scale a by the polar distance H_1 as

$$M=azbH_1.$$

Thus the b. m. diagram B may also be considered as a loading curve where an ordinate of the diagram may be taken as the fictitious load at that point.

The polar distance for the deflection diagram H_2 is usually taken to be proportional to the shaft stiffness EI. For such a condition if the shaft has varying diameters each of the shaft sections of constant diameter d would have its own distance H_2. To simplify the construction of a funicular polygon for the complete shaft with a single polar point the shaft is assumed to be of constant section the moment of inertia of which is usually taken equal to the one at the maximum diameter $d_0=d_5=d_6$.

The bending moment at each of the points is then increased or decreased to keep the deflection at the same value as before. The bending moment is changed in the ratio $I_0/I=d_0^4/d^4$ (where I and d — moment of inertia and the shaft diameter at any point).

The polar distance H_2 is kept constant for the above transformation. If the polar distance H_2 is to be kept constant for all the shaft sections we must take into account the change in the modulus of elasticity the numerical value of which depends on the temperature.

E_0 is selected for any section of the shaft and the ordinates of the b. m. diagram for all the remaining sections are multiplied by the ratio E_0/E (where E — modulus of elasticity of the section under consideration). After obtaining the polar distance of the funicular polygon for the deflection curve in terms of E_0I_0 all the ordinates of the b. m. diagram must be multiplied by $d_0^4/d^4\times E_0/E$ to keep the deflection curve unchanged.

In the present example $d_0=d=12.7$ cm and $E_0=2.1\times10^6$ kg/cm². After multiplying the respective ordinates of the b. m. diagram B by the above quantity we obtain the diagram *abcdefghijklmnopqrstuvw*.

Next we determine the cross-sectional areas of each of the sections and find their c. g. s where these deflecting forces F_1, F_2, F_3 ..., etc., are supposed to act. The second funicular polygon is drawn with these forces to the scale c [cm²/cm] (i. e., 1 cm of the deflection curve conforms to c [cm²] of the b. m. diagram on the drawing). In the example under consideration $c=50$ cm²/cm. Since 1 cm length of the ordinates of the b. m. diagram is equal to abH_1 [kg cm] and each square cm of the diagram equals a^2bH_1 [kg cm²] we have each cm length of the deflection diagram equal to $c\,(a^2bH_1)$ [kg cm²]. If the polar distance E_0I_0 is drawn to the same scale as above, i. e., if we assume

$$H_2'=\frac{E_0I_0}{a^2bcH_1}\ \text{[cm]},$$

the deflections are obtained to linear scale.

Tempera-ture, °C	20	110	200	300	400	500	600
Modulus of elasticity $E\times10^{-3}$	2,070	2,010	1,950	1,880	1,790	1,510	1,340

To obtain greater clarity of the deflection diagram the polar distance is reduced considerably compared to the value of H_2'. In this example we shall have the magnitude of the deflections increased 3,000 times the original ($M=3{,}000$). Since the shaft has been drawn to scale $1/a$ the polar distance will be reduced Ma times, i. e., instead of H_2'

$$H_2=\frac{H_2'}{Ma}=\frac{E_0I_0}{Ma^3bcH_1}\text{[cm]}.$$

In the present case

$$H_2=\frac{2.1\times10^6\times127.7}{3{,}000\times2^3\times10\times50\times15}=14.9\text{ cm}.$$

To simplify the calculations of equation (13-85) we may substitute the shaft deflections as obtained on the drawing in place of the actual deflection values, i. e., $Y_i=My_i$. With this modification equation (13-85) becomes

$$n_{cr}=300\sqrt{\frac{M\sum Q_iY_i}{\sum Q_iY_i^2}},\qquad(13\text{-}85a)$$

$\sum Q_iY_i$ and $\sum Q_iY_i^2$ are obtained by summing the products of each load Q_i and the shaft deflection and the square of the deflection under the loads as obtained on the drawing.

All the above calculations are shown in Table 13-9.

Since the deflection scale has been assumed to be $M=3{,}000$

$$n_{cr}=300\sqrt{\frac{M\sum Q_iY_i}{\sum Q_iY_i^2}}=300\sqrt{\frac{3{,}000\times1{,}088.6}{4{,}316.6}}=$$
$$=8{,}250\text{ r. p. m.}$$

The ratio of the critical speed and the normal speed is

$$n_{cr}/n=8{,}250/6{,}000=1.37,$$

i. e., the shaft is a rigid one and the margin is sufficient.

The maximum static deflection of the shaft is

$$y_{max}=\frac{Y_{max}}{M}=\frac{45}{3{,}000}=0.015\text{ mm}.$$

13-13. COUPLINGS

Turbine couplings are usualy of the following three types: rigid, semi-flexible and flexible. Fig. 13-23 shows a coupling of the first type. It consists of two flanges forged solid with the shafts. The flanges are rigidly held together by

[1] The overhanging ends of the rotor may slightly decrease the critical speed.

Table 13-9

Section No.	Shaft dia., d, cm	Length of section, cm	Load Q, kg	d_0/d	$(d_0/d)^4$	Temp. t, °C	E_0/E	Ordinates of b. m. diagram, z, cm	$z\times(6)\times(8)$	Area of the b. m. diagram for each section, S, cm²	Ordinates of the deflection diagram on the drawing	Y^2, cm	(4)×(12) kg/cm	(4)×(13), kg/cm²
1	2	3	4	5	6	7	8	9	10	11	12	13	14	15
1	7.6	4.45	1.6	1.66	7.70	50	1.0	2.25	17.3/9.7	12.3	0.4	0.16	0.6	0.3
2	8.8	9.20	6.8	1.44	4.33	50	1.0	6.70	29/8.4	8.9	1.43	2.05	9.7	13.9
3	12.0	11.90	59.0	1.06	1.25	50	1.0	11.55	14.5/12.6	67.0	3.1	9.61	182.9	567.0
4	12.4	5.90	42.0	1.02	1.09	75	1.01	12.80	14.6/12.9	39.4	3.7	13.69	155.4	575.0
5	12.7	6.00	36.1	1.0	1.0	127	1.04	13.10	13.5/13.5	39.7	4.2	17.64	151.6	638.8
6	12.7	5.40	30.1	1.0	1.0	165	1.04	12.80	13.6/14.9	37.7	4.4	19.36	132.4	582.7
7	12.4	4.80	34.5	1.02	1.09	193	1.05	12.15	13.9/15.9	34.2	4.5	20.25	155.2	698.6
8	12.0	4.05	5.0	1.06	1.25	282	0.07	11.35	15.2/16.9	31.5	4.45	19.80	22.3	99.0
9	11.7	7.20	56.5	1.09	1.39	282	1.07	9.05	13.5/40.3	54.7	4.3	18.49	243.0	1,045.0
10	8.9	9.30	6.9	1.43	4.15	282	1.07	5.20	23.1/26.5	147.0	3.5	12.25	24.2	84.5
11	8.6	8.40	5.4	1.48	4.75	50	1.0	1.60	7.6/25.7	73.8	2.0	4.00	10.8	21.6
12	6.4	4.00	1.0	2.01	16.05	50	1.0	0.0	0	25.7	0.45	0.20	0.5	0.2

$$\sum Q_iY_i = 1{,}088.6 \text{ kg cm}; \quad \sum Q_iY_i^2 = 4{,}316.6 \text{ kg/cm}^2.$$

From column 8 of Table 13-9 it may be seen that the effect of temperature on the critical speed is very insignificant and could have been neglected. However, in the case of high-pressure high-temperature turbines where the initial temperatures reach 500 to 550°C the variation of modulus of elasticity may substantially reduce the criti cal speed.

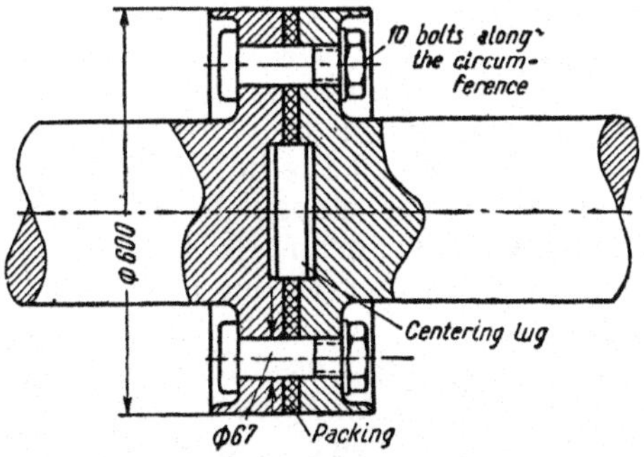

Fig. 13-23. Rigid coupling

bolts. The bolt holes are drilled in position, the flanges being centred by a disc which is removed when the turbine is assembled. The bolts are specially made to exactly fit each of these drilled holes. Since the bolts would not fit in any other hole than the one for which it is made both the bolts and the holes are marked in pairs either by notches or numbers. A packing piece of the required thickness is provided between the two flanges at the time of assembly. Rigid couplings are also made with the flanges keyed on the shaft instead of being in one piece with the shaft as described above. Couplings of this type are mostly used for turbines of smal and medium capacity. With rigid couplings the number of bearings used for a turbo-alternator set may be conveniently reduced to three. These couplings are also used in large turbines for coupling the H. P. and the L. P. rotors so that both of these could be supported by only three journal and one thrust bearings. The main disadvantage of the rigid coupling is that it transmits the vibrations of one rotor to the other and, further, the coupling itself may become a source of vibrations in case of a slight misalignment.

Fig. 13-24 shows a type of semi-flexible coupling used by the L.M.W. for low-pressure turbines. Flanges *3* and *4* are mounted on the turbine and the alternator shafts *1* and *2*. The ends of these shafts are slightly tapered. The flanges

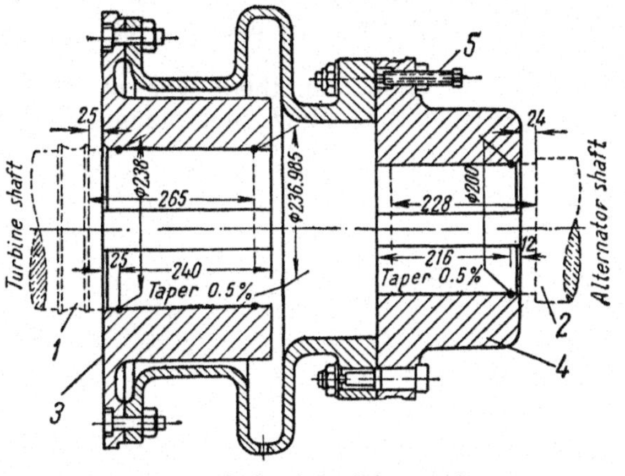

Fig. 13-24. Semi-flexible coupling

1 – turbine shaft; *2* – alternator shaft; *3* and *4* – coupling flanges; *5* – packing bolt

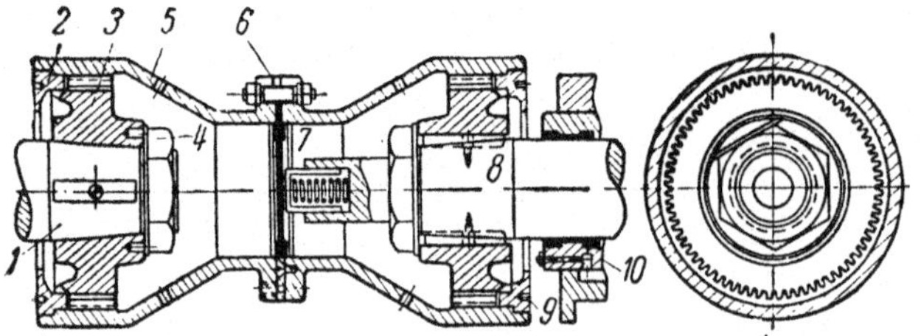

Fig. 13-25. AEG flexible coupling, toothed type
1 — turbine shaft; *2* — thrust washer; *3* — coupling flange; *4* — nut on the shaft; *5* — connecting flange; *6* — centring disc; *7* — spring; *8* — key; *9* — oil hole; *10* — oil supply

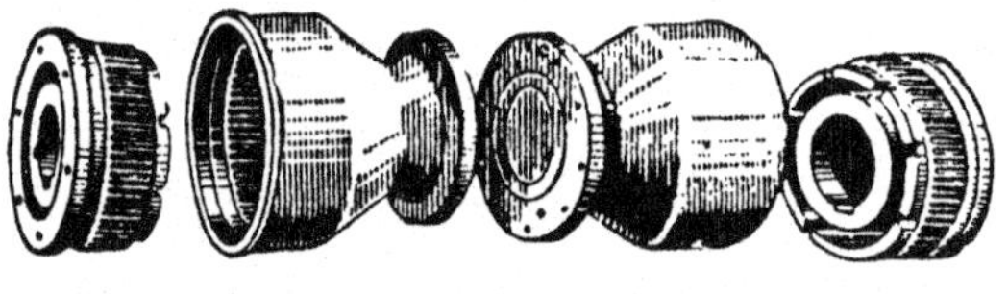

Fig. 13-26. AEG flexible coupling (exploded view)

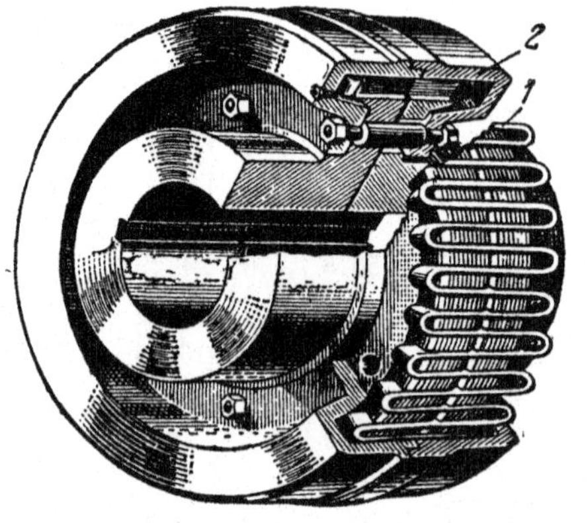

Fig. 13-27. L.M.W. flexible coupling

are keyed on the shafts and are connected together by a forged steel muff which gives rigidity in torsion and a limited degree of flexibility allowing for slight axial misalignment. The use of such a muff for torque transmission considerably reduces the possibility of the transfer of vibrations from one rotor to the other. The muff is bolted to the two flanges and a relief bolt *5* helps in the dismantling of the coupling when required. No lubrication is necessary for both the rigid and the semi-flexible couplings described above.

Fig. 13-25 shows a flexible coupling made by the AEG. These are mostly used for coupling the rotor shafts of two-cylinder turbines and for turbines with reduction gearings.

The tapered portion of the shafts carry toothed flanges *3*. These flanges are fastened on the shafts by two longitudinal keys and a locking nut *4*. The muff *5* consists of two parts rigidly coupled together by bolts. The teeth cut in muff *5* mate with the grooves of the toothed flanges *3*. Some amount of axial displacement is possible in such couplings. The axial displacement, however, is limited by a stop washer *6* and rings *2*. Spring *7* holds the washer *6* in position and helps in aligning the shaft in the required position. Fig. 13-26 shows an exploded view of the coupling. The coupling teeth must be copiously supplied with lubricating oil to reduce wearing away of the teeth.

Fig. 13-27 shows another type of flexible coupling used by the L.M.W. Steel flanges mounted on the turbine and the alternator shafts have long parallel grooves cut in the axial direction. A tempered steel spring *1* is placed in the grooves. For ease of assembly the spring is made up of several sections. The casing *2* is rigidly fixed to one of the flanges of the coupling. Clearance spaces are provided in the bolt hole and between the casing and the flange to allow for axial displacement. The torque is transmitted from one shaft to the other through the steel spring. The coupling is supplied with oil for lubrication. If a flexible coupling is used for joining the shafts of the H. P. and the L. P. cylinders of a turbine, thrust bearings are used on either side. The flanges are made from steels of grades 50, 25H3 and 34XM; bolts and studs from 35 and ЭИ-10. The spring is made from steel 80 with 0.75 to 0.85C, and steel 60C2 with 0.55 to 0.65C. The couplings are made of alloy steels if the loads to be transmitted are very high.

SUBJECT INDEX

CPSIA information can be obtained at www.ICGtesting.com
Printed in the USA
BVOW09s1351140916

462102BV00002B/9/P

9 781410 223487